Dynamics of Machinery

www.contitech.de/aby

ContiTech Air Springs for vibration isolation

Single Convolution Air Springs

Double Convolution Air Springs

Belted Air Springs

Vibration isolation with ContiTech air springs offers unique advantages:

- Constant operating level, adjusted by air pressure
- Load-independent insulating characteristics, proportional to the loading capacity
- User-friendly, automatic control system
- Low structural height, outstanding lateral stability

Contact us. We would be pleased to be of assistance.

ContiTech. Get more with elastic technology.

ContiTech Luftfedersysteme GmbH
Phone +49 (0)511 938 5238
industrial@as.contitech.de

Hans Dresig · Franz Holzweißig

Dynamics of Machinery

Theory and Applications

Translated by Dr. Wolf Grosskopf and
Assoc. Prof. Dr.-Ing. Sven Esche

with CD-ROM

Univ.-Prof. Dr.-Ing. habil. Hans Dresig
Mittelstr.1
D-09244 Lichtenau OT Auerswalde
Germany
hdresig@web.de
Homepage: http://www.dresig.de

Univ.-Prof. Dr.-Ing. habil. Franz Holzweißig
Kaitzer Str. 66
D-01187 Dresden
Germany

Translators

Dr. Wolf Grosskopf (Northridge)
w.grosskopf@sbcglobal.net

Assoc. Prof. Dr.-Ing. Sven Esche
(Stevens Institute of Technology)
sesche@stevens.edu

ISBN 978-3-540-89939-6   e-ISBN 978-3-540-89940-2
DOI 10.1007/978-3-540-89940-2
Springer Heidelberg Dordrecht London New York

Library of Congress Control Number: 2010929624

© Springer-Verlag Berlin Heidelberg 2010

This work is subject to copyright. All rights are reserved, whether the whole or part of the material is concerned, specifically the rights of translation, reprinting, reuse of illustrations, recitation, broadcasting, reproduction on microfilm or in any other way, and storage in data banks. Duplication of this publication or parts thereof is permitted only under the provisions of the German Copyright Law of September 9, 1965, in its current version, and permission for use must always be obtained from Springer. Violations are liable to prosecution under the German Copyright Law.

The use of general descriptive names, registered names, trademarks, etc. in this publication does not imply, even in the absence of a specific statement, that such names are exempt from the relevant protective laws and regulations and therefore free for general use.

Cover design: eStudio Calamar S.L.

Printed on acid-free paper

Springer is part of Springer Science+Business Media (www.springer.com)

Preface to the English Edition

This textbook, the translation of the ninth edition of the German book "Maschinendynamik" (published in its first edition in Germany and Austria in 1979), discusses disciplines that in many countries are presented separately in lectures titled "Theory of Mechanisms and Machines" (TMM) and "Vibrations" or "Theory of Vibrations". The basic idea by my distinguished colleague, Prof. Holzweissig, was to combine these disciplines into the "Dynamics of Machinery". We worked out this concept together in the first three editions, relying on our experience gained in our collaborations with engineers here in Saxony. Since he became a professor emeritus in 1993, I have constantly updated the book since its 4th edition and have added more exercises and sections.

This textbook is based on the four-semester lecture series on "Engineering Mechanics" and confronts students directly with dynamic problems of their field. Dynamics of machinery is viewed both as a typical field for the mathematical modeling of technological processes and as a branch of mechanical engineering that addresses dynamic problems of power machines (reciprocating engines and turbomachines), processing machines (printing machines, textile machines, packaging machines), hoists and conveyors, agricultural machines and vehicles, as well as industrial plants.

The book comprises relatively independent chapters that discuss typical issues of the dynamics of mechinery from the point of view of mechanical engineers. Chapter 6 gives a general overview of linear oscillators and includes methods discussed in Chaps. 3, 4 and 5. We were fully aware and accepted some overlap and repetitions, making many cross-references among sections. Chapter 7 discusses nonlinear and self-excited oscillators, for which calculations are becoming more and more relevant to practical application. Chapter 8 still does not contain any equations, but the rules compiled there are important for engineering practice. We included a new chapter "on interrelations with system dynamics and mechatronics" to enhance the understanding of these adjacent disciplines. This chapter prepares the readers for a more general approach to solving problems of the dynamics of machinery by including sensors and actuators.

The 60 problems with solutions are meant to help in grasping and consolidating the subjects taught. A new feature of this edition is the enclosed CD-ROM, which does not only contain the student version of the SimulationX® software, but also program parts that were used for solving the examples given in the book so that the readers can work with them themselves.

The book was not only written for students, but also for practicing engineers. The examples from many fields of mechanical engineering and the specification of parameter values, as well as references to guidelines and regulations, underline this. We have considered the way in which engineers think by discussing estimation methods, rough calculations and minimal models, and by explaining many instructive dynamic effects (gyroscopic effect, resonance, absorption, self-synchronization, ...) that are relevant for the design of machines that operate under a high dynamic load. The fast development of hardware and software entailed that nowadays there is software available for almost every problem in the dynamics of machinery. While computers always provide some numbers and diagrams, the engineer is still responsible for the results. It remains the engineer's job to provide the respective calculation models, to assess the applicability of a software product, to check the result of a simulation, and to have some idea of the result to be expected before the calculation starts.

Since the current development shows a trend towards refined modeling, we pointed out three aspects: the training of physical understanding, the utilization of modal analysis including sensitivity analysis, and the application of computer-aided methods. We stress the fact that the goal of calculations is not only to arrive at a numerical result, but to achieve a better understanding of the dynamic behavior of real objects and to be able to take design measures that are based in physical theory.

I would like to thank many specialists for their suggestions and tips, in particular my former coworkers at the professorship of Dynamics of Machinery of the Chemnitz University of Technology and my colleagues at ITI GmbH Dresden, in particular Dipl.-Ing. Uwe Schreiber, who helped me work out practical examples. The whole book has been translated by Dr. Wolf Grosskopf (Northridge). I was very pleased that my former student, Prof. Dr.-Ing. Sven Esche (Stevens Institute of Technology), agreed to proof-read the manuscript and to review the terminology. I am in particular grateful for this. I would like to give special praise to the committed work of Dipl.-Ing. Andreas Abel who prepared the final manuscript and thereby took into account all my many extra wishes, so that the manuscript could be submitted to the publisher ready for printing.

Auerswalde, in May 2010   Hans Dresig

Contents

Purpose and Structure of the Dynamics of Machinery 1

1 Model Generation and Parameter Identification 5
   1.1   Classification of Calculation Models 5
        1.1.1   General Principles 5
        1.1.2   Examples.. 10
   1.2   Determination of Mass Parameters 14
        1.2.1   Overview.. 14
        1.2.2   Mass and Position of the Center of Gravity 15
        1.2.3   Moment of Inertia about an Axis 17
        1.2.4   Moment of Inertia Tensor 21
   1.3   Spring Characteristics 25
        1.3.1   General Context 25
        1.3.2   Machine Elements, Sub-Assemblies 29
        1.3.3   Rubber Springs 36
        1.3.4   Problems P1.1 to P1.3 38
        1.3.5   Solutions S1.1 to S1.3 40
   1.4   Damping Characteristics..................................... 42
        1.4.1   General Context 42
        1.4.2   Methods for Determining Characteristic Damping
               Parameters... 47
        1.4.3   Empirical Damping Values 51
   1.5   Characteristic Excitation Parameters........................... 55
        1.5.1   Periodic Excitation 55
        1.5.2   Transient Excitation 56
        1.5.3   Problems P1.4 to P1.6 62
        1.5.4   Solutions S1.4 to S1.6 63

2 Dynamics of Rigid Machines 67
   2.1   Introduction 67
   2.2   Kinematics of a Rigid Body 68

	2.2.1	Coordinate Transformations	68
	2.2.2	Kinematic Parameters	73
	2.2.3	Kinematics of the Gimbal-Mounted Gyroscope	75
	2.2.4	Problems P2.1 and P2.2	76
	2.2.5	Solutions S2.1 and S2.2	78
2.3	Kinetics of the Rigid Body		82
	2.3.1	Kinetic Energy and Moment of Inertia Tensor	82
	2.3.2	Principles of Linear Momentum and of Angular Momentum	87
	2.3.3	Kinetics of Edge Mills	90
	2.3.4	Problems P2.3 and P2.4	93
	2.3.5	Solutions S2.3 and S2.4	95
2.4	Kinetics of Multibody Systems		100
	2.4.1	Mechanisms with Multiple Drives	100
	2.4.2	Planar Mechanisms	112
	2.4.3	States of Motion of a Rigid Machine	122
	2.4.4	Solution of the Equations of Motion	124
	2.4.5	Example: Press Drive	129
	2.4.6	Problems P2.5 to P2.8	133
	2.4.7	Solutions S2.5 to S2.8	136
2.5	Joint Forces and Foundation Loading		141
	2.5.1	General Perspective	141
	2.5.2	Calculating Joint Forces	142
	2.5.3	Calculation of the Forces Acting onto the Frame	145
	2.5.4	Joint Forces in the Linkage of a Processing Machine	148
	2.5.5	Problems P2.9 and P2.10	150
	2.5.6	Solutions S2.9 and S2.10	151
2.6	Methods of Mass Balancing		153
	2.6.1	Objective	153
	2.6.2	Counterbalancing of Rigid Rotors	153
	2.6.3	Mass Balancing of Planar Mechanisms	160
	2.6.4	Problems P2.11 to P2.14	167
	2.6.5	Solutions S2.11 to S2.14	170

3 Foundation and Vibration Isolation 177
3.1 Introductory Remarks 177
3.2 Foundation Loading for Periodic Excitation 181
3.2.1 Minimal Models with One Degree of Freedom 181
3.2.2 Block Foundations 191
3.2.3 Foundations with Two Degrees of Freedom – Vibration Absorption 200
3.2.4 Example: Vibrations of an Engine-Generator System 204
3.2.5 Problems P3.1 to P3.3 206
3.2.6 Solutions to Problems S3.1 to S3.3 208
3.3 Foundations under Impact Loading 211
3.3.1 Modeling Forging Hammers 211

	3.3.2	Calculation Model with Two Degrees of Freedom 213
	3.3.3	Problems P3.4 to P3.6 216
	3.3.4	Solutions S3.4 to S3.6 218

4 Torsional Oscillators and Longitudinal Oscillators 223
   4.1 Introduction 223
   4.2 Free Vibrations of Torsional Oscillators 229
       4.2.1 Models with Two Degrees of Freedom 229
       4.2.2 Oscillator Chains with Multiple Degrees of Freedom 234
       4.2.3 Evaluation of Natural Frequencies and Mode Shapes 238
       4.2.4 Examples 242
       4.2.5 Problems P4.1 to P4.3 251
       4.2.6 Solutions S4.1 to S4.3 253
   4.3 Forced Vibrations of Discrete Torsional Oscillators 259
       4.3.1 Periodic Excitation 259
       4.3.2 Examples 263
       4.3.3 Transient Excitation 272
       4.3.4 Problems P4.4 to P4.6 278
       4.3.5 Solutions S4.4 to S4.6 280
   4.4 Absorbers and Dampers in Drive Systems 283
       4.4.1 Introduction 283
       4.4.2 Design of an Undamped Absorber 284
       4.4.3 Design of a Spring-Constrained Damper 286
       4.4.4 Design of a Springless Damper 289
       4.4.5 Examples 291
   4.5 Parameter-Excited Vibrations by Gear Mechanisms with Varying
       Transmission Ratio 295
       4.5.1 Problem Formulation/Equation of Motion 295
       4.5.2 Solution of the Equation of Motion, Stability Behavior 297
       4.5.3 Examples 299
       4.5.4 Problems P4.7 and P4.8 306
       4.5.5 Solutions S4.7 and S4.8 308

5 Bending Oscillators 311
   5.1 Problem Development 311
   5.2 Fundamentals 312
       5.2.1 Self-Centering in a Symmetrical Rotor 312
       5.2.2 Passing through the Resonance Point 315
       5.2.3 Rotating Shaft with Disk (Gyroscopic Effect) 316
       5.2.4 Bending Oscillators with a Finite Number of Degrees of
              Freedom 326
       5.2.5 Examples 328
       5.2.6 Problems P5.1 to P5.3 335
       5.2.7 Solutions S5.1 to S5.3 336
   5.3 Beam with Distributed Mass 338

	5.3.1	General Perspective 338
	5.3.2	Straight Beam on Two Supports 343
	5.3.3	Estimates by Dunkerley and Neuber 346
5.4	Model Generation for Rotors 347	
	5.4.1	General Considerations 347
	5.4.2	Example: Grinding Spindle 350
5.5	Problems P5.4 to P5.6 351	
5.6	Solutions S5.4 to S5.6 352	

6 Linear Oscillators with Multiple Degrees of Freedom 355
- 6.1 Introduction 355
- 6.2 Equations of Motion 358
  - 6.2.1 Mass, Spring, and Compliance matrix 358
  - 6.2.2 Examples 363
  - 6.2.3 Problems P6.1 to P6.3 372
  - 6.2.4 Solutions S6.1 to S6.3 373
- 6.3 Free Undamped Vibrations 375
  - 6.3.1 Natural Frequencies, Mode Shapes, Eigenforces 375
  - 6.3.2 Orthogonality and Modal Coordinates 378
  - 6.3.3 Initial Conditions, Initial Energy, Estimates 381
  - 6.3.4 Examples 383
  - 6.3.5 Problems P6.4 to P6.6 395
  - 6.3.6 Solutions S6.4 to S6.6 396
- 6.4 Structure and Parameter Changes 399
  - 6.4.1 Rayleigh Quotient 399
  - 6.4.2 Sensitivity of Natural Frequencies and Mode Shapes 400
  - 6.4.3 Reduction of Degrees of Freedom 405
  - 6.4.4 Influence of Constraints on Natural Frequencies and Mode Shapes 407
  - 6.4.5 Examples of the Reduction of Degrees of Freedom 411
  - 6.4.6 Problems P6.7 to P6.9 419
  - 6.4.7 Solutions S6.7 to S6.9 420
- 6.5 Forced Undamped Vibrations 426
  - 6.5.1 General Solution 426
  - 6.5.2 Harmonic Excitation (resonance, absorption) 427
  - 6.5.3 Transient Excitation (Rectangular Impulse) 433
  - 6.5.4 Examples 437
  - 6.5.5 Problems P6.10 to P6.12 440
  - 6.5.6 Solutions S6.10 to S6.12 441
- 6.6 Damped Vibrations 444
  - 6.6.1 Determination of Damping 444
  - 6.6.2 Free Damped Vibrations 445
  - 6.6.3 Harmonic Excitation 447
  - 6.6.4 Periodic Excitation 452
  - 6.6.5 Examples 455

Contents   XI

		6.6.6	Problems P6.13 to P6.16 460
		6.6.7	Solutions S6.13 to S6.16 460

7 Simple Nonlinear and Self-Excited Oscillators 465
7.1 Introduction 465
7.2 Nonlinear Oscillators 468
7.2.1 Undamped Free Nonlinear Oscillators 468
7.2.2 Forced Vibrations with Harmonic Excitation 471
7.2.3 Examples 476
7.2.4 Problems P7.1 to P7.2 488
7.2.5 Solutions S7.1 and S7.2 489
7.3 Self-Excited Oscillators 490
7.3.1 General Perspective 490
7.3.2 Examples 491
7.3.3 Problems P7.3 and P7.4 500
7.3.4 Solutions S7.3 and S7.4 502

8 Rules for Dynamically Favorable Designs 505

9 Relations to System Dynamics and Mechatronics 511
9.1 Introduction 511
9.2 Closed-Loop Controlled Systems 514
9.2.1 General Perspective 514
9.2.2 Example: Influencing Frame Vibrations by a Controller 516

Symbols 525

References 531

Index 533

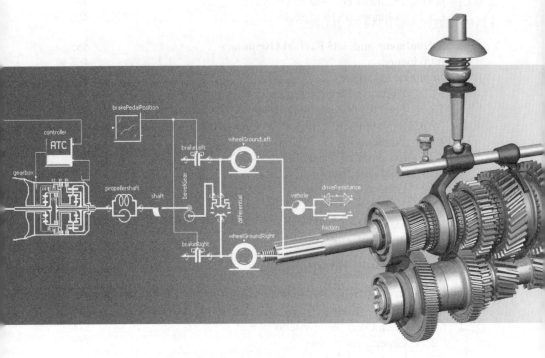

SIMULATION X®
Powered by ITI

The Virtual Experimental Laboratory for the Complete Dynamics of Machines

SimulationX - Open Multi-domain Software Platform for 1D to 3D Machine Dynamics Simulations

Implement your knowledge using ITI simulation solutions for the analysis of dynamics of machines and mechanisms across all industrial applications.

- Modeling of complex vibratory systems
- Simulation of dynamic effects in rigid and elastic machines, vibration chains, foundations and vibration isolation
- Integrated tools e.g. for transient and steady-state analysis of free, forced, self- and parameter-induced vibration, natural frequency computation
- Quick evaluation of physical interactions between mechanical, mechatronic, pneumatic, and hydraulic models
- Automatic formulation and solution of equations of motion
- Superior capabilities for code export (CAx interfaces and integration)
- Extensive application-oriented, ready-to-use model libraries

Test the student edition of the most intuitive, engineer-oriented model assembly - www.simulationx.com

Supporting your visions

ITI Headquarters Webergasse 1 · 01067 Dresden/Germany · www.simulationx.com · sales@simulationx.com · T +49 (351) 26050-0 · F +49 (351) 26050-155
ITI Global For your local representative please visit www.simulationx.com/global

SimulationX is a registered trademark of ITI GmbH Dresden. © ITI GmbH, Dresden, Germany, 2010. All rights reserved.

Purpose and Structure of the Dynamics of Machinery

The purpose of the dynamics of machinery (aka machine dynamics) is to apply the knowledge from the field of dynamics to specific problems in engineering. Its development is closely linked to developments in mechanical engineering.

Dynamic problems first occurred in power and work machines. Torsional vibrations were observed in reciprocating engines, and bending vibrations put turbine components at risk. Explaining such phenomena has long been the only task of the dynamics of machinery, as reflected in the standard publications, e. g. [1].

Knowledge of the dynamics of machinery is also required for designing machines that are based on dynamic principles. This includes hammers, robots, stemmers, oscillating conveyors, screens, vibrators, textile mandrils, centrifuges, etc.

Dynamic problems stepped into the foreground with the ever-increasing operating speed and the enforcement of lightweight design principles in all fields of processing machines, agricultural machines, machine tools, printing machines, and conveyors. To manage their multiplicity, it became necessary to define and answer the basic questions, while abstracting as much as possible from any specific machine. In this way, dynamics of machinery grew into an independent discipline that is part of the tool set of each mechanical engineer.

While 50 years ago people thought that dealing with vibration problems will remain the domain of only a few specialists, today we expect a large number of engineers to have a precise understanding of the dynamic processes in a machine. High-performance machines are to be sized not just based on static, but often primarily based on dynamic considerations. Thus, the use of calculation methods of structural durability depends on the certainty of the load assumptions that result from a machine dynamics calculation. An engineer has to know the physical laws that have an influence on periodic permanent loads, impacts, starting and stopping processes in a machine.

The way in which engineers work has to be taken into account, though. It is governed by the requirement to solve a practical problem in a short time and at economically justifiable expenditure. Frequently, fast decisions have to be made so that waiting for the scientific clarification of detailed questions is not an option. Instead, an engineer has to consider all available information and adjust the solution of the problem to the existing state of the art. An important skill that an engineer should have is to be able to apply an imperfect or incomplete theory as long as there is no better one available. This requires, of course, a wealth of knowledge that will not always include the entire train of thought that leads to an equation or computer program. What is important, however, is to know their scope and make use of ways to check the results obtained using rough estimates.

The diagram below, in Fig. 0.1, outlines the path that is followed when solving a design problem. It is desirable to have someone who is able to go down the external path, that is, someone who glances at the machine (or listens to it) and knows what has to be changed to obtain the desired effect. Chapter 8 provides a few general rules that an experienced designer can rely on. However, they are incomplete and knowing them is not sufficient to solve all problems. Simply put, what matters are decisions to change something in the structure or parameters of a specific object in such a way that the dynamic performance of that object is improved.

Since this selection of decisions is anything but trivial, a step-by-step approach as indicated on the left-hand side of the diagram is preferred. The essential steps include the identification of a mechanical problem in a design (or technological) task, and the formulation of a mathematical problem from it (which is part of modeling). One has to understand the physical behavior before generating a model. Chapter 1 discusses issues of model generation and parameter identification.

In the course of the past two decades, computer-aided analysis using commercial software has become the standard in the dynamics of machinery as well. Today's engineer is spared the many efforts that the solution of the mathematical problems used to involve. In recent years, the interpretation of the mathematical solution and the implementation of mechanical principles in a design solution have become increasingly important engineering tasks. Today, dynamic phenomena can be better calculated in advance and taken advantage of, e. g. dynamic balancing, the effects of gyroscopes, absorbers, dampers, and nonlinear effects (e. g. self-synchronization).

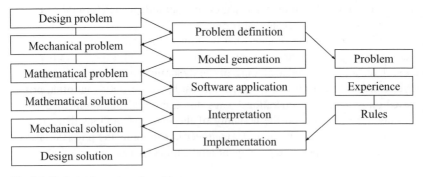

Fig. 0.1 Tasks in dynamics of machinery

Figure 0.2 shows how the field of dynamics of machinery relates to other disciplines. Rotor dynamics [8], or vehicle dynamics [27], could be related to these disciplines in a similar way, and they overlap with the dynamics of machinery. Unlike rotor and vehicle dynamics, machine dynamics deals with many different objects and problems, from machine elements [22] to complex designs where close links to structural dynamics (and not just with regard to foundations) exist [19].

The theoretical foundation of the dynamics of machinery is formed by mathematics and physics [32], including, of course, almost all fields of engineering mechanics [11], [30], and in particular, the theory of vibrations [6], [10], [16], [24]. The findings of machine dynamics studies influence the design and sizing of real machines, and there are close links to design theory, machine elements, structural stability, drive engineering, and acoustics of machines. The most closely related disciplines are drive engineering (for the drives) [4], [9], [21] and structural dynamics (for the frames) [19], [25] if, from a design perspective, one views a machine as a combination of a drive and a support system.

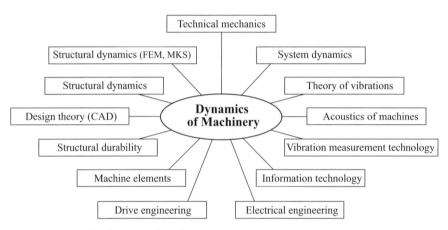

Fig. 0.2 Links of the dynamics of machinery to other disciplines

The dynamics of machinery always involves real objects that are subjected to loads that vary over time, and that are exposed to, or take advantage of, the effects of inertia. Frequently, the purpose is also to avoid malfunctions or damages. The findings of the theory of vibrations provide the basis for understanding many real-world phenomena. The technological development of many machines has influenced the theory of vibrations in that it continuously provided new practical questions that could not be answered using the body of knowledge and theories available.

Problems of modeling in the dynamics of machinery are closely related to system dynamics [2], [3], [14], [18], and vibration measurement technology, but also include methods for evaluating and assessing [15], [17] vibration phenomena. The dynamics of machinery utilizes the findings in fields such as structural dynamics [7] and multibody dynamics [3], [28] that are often found – implemented using numerical mathematics – in commercial software products. Quantitative improvements of the calculation models are not just achieved by mutual perfection of computer and measurement technology. There are crosslinks with electrical engineering when it comes to electric drives and applying findings of control engineering and computer science. Unlike mechatronics, the main task in the dynamics of machinery, however, is the control of inertia forces.

A special field of the dynamics of machinery involves the collection and classification of proven computational models and their associated characteristics that are required as input data for computer programs. Advanced and proven computational models of real machines are often kept confidential by companies. They represent the outcome of a long development process and are very detailed and powerful. If requested (e. g. for calculating specific machine elements), they are often accessible to third parties.

Students and engineers in the field have to understand the computational models commercial software products are based on and know the physical background. Everyone should also have some idea of the scales of masses, spring constants, damping coefficients, forces, frequencies, etc. that they have to deal with. Therefore, this book uses real-world parameter values for most of the problems presented. One should always try to anticipate the results that the computer will provide. With some practice, an engineer will often be able to predict the order of magnitude of real parameter values and results of a calculation.

For clarity reasons, the material discussed in this textbook has been structured from simple to complex concepts. Basic knowledge of engineering mechanics is required. Chapter 2, which deals with the dynamics of rigid bodies, probably also discusses some known phenomena. Understanding the balance of forces in a rigid-body system, however, provides the basis for understanding many other dynamic phenomena. Specific oscillation systems are discussed in Chaps. 3 to 5, but these chapters focus on "'classical'" fields of the dynamics of machinery (foundation design, vibration isolation, torsional and translational vibrations, bending vibrations) and their major applications. Chapter 6 repeats the material presented in Chaps. 3 to 5 at a higher level and includes the theory of linear vibrations, a field that can be applied to many machines. Chapter 7 gives an insight into some nonlinear phenomena and the behavior of self-excited oscillators that are relevant for mechanical engineering. The rules provided in Chap. 8 refer to the qualitative side of the dynamics of machinery.

Chapter 1
Model Generation and Parameter Identification

1.1 Classification of Calculation Models

1.1.1 General Principles

In machine dynamics, calculation models are used mainly for three reasons:

1. They provide time and cost savings in the development of new or improved products by dynamic computer simulations instead of expensive test stations (or measurements at real machines, whose operation would have to be interrupted).
2. They help in clarifying physical causes for interfering phenomena (e.g. resonant oscillations, fracture, noise) or desired effects (e.g. absorption).
3. They enable the determination of optimal parameter values for the given specific criteria (e.g. material expenditure, energy demand, stiffness, safety at work, lifetime and reliability).

In the past years, the possibilities for model-supported analysis have gained importance because the temporal and financial requirements for computer simulations were reduced significantly by the rapidly improving performance of computers and simulation software. In contrast, experiments using test stations still remain time-consuming and expensive.

Therefore, a certain "pool of models", for which he knows the software and its handling, must be available to the engineer.

Note that there is no "calculation of a machine", but rather the first task is the generation of a calculation model. The methods used for that depend on whether the design documents or measurements performed on an actual machine are available as the starting point.

If starting from the design documents, a *structure* must first be defined. The term "structure" refers to the mutual couplings of the elements (topology) and the generation of the calculation model from the following four elements:

Masses: elements for storing kinetic energy
Springs: elements for storing potential energy
Dampers: elements for converting mechanical energy into heat energy
Exciters: elements for supplying energy from an energy source

If the structure has already been defined, then its parameters must be selected. A **parameter** is a geometrical or physical quantity, which is denoted by a letter and appears in calculation models. Defining a structure includes primarily the identification of those parameters that influence a certain phenomenon. Besides the number of degrees of freedom n, the number of parameters K represents the most important characteristic of a calculation model. The selection of the parameters determines how the spatial and temporal boundaries of the model (energy source) are chosen and which internal interactions are considered. In theoretical models, which are derived from design documents only, the parameters are, for instance, lengths, transmission ratios, masses, spring and damping constants, as well as coefficients of kinematic or dynamic excitation. The terms "parameters", "parameter values", "dimensionless parameters" and "characteristic numbers" must be clearly distinguished.

Parameter values (or "characteristics") contain the data of the parameters, i.e. numerical values and measuring units. A specific model is characterized by its parameter values, which can be arranged into a parameter vector and appear, for example, in the elements of the mass and stiffness matrices. Such parameter values are given in the examples and problems in all chapters (see Chaps. 2.4.5, 3.2.4, 4.2.4.1) or for instance P2.5, P3.3, P4.1, S5.5, P6.16 as well as Fig. 4.13. The theoretical or experimental determination of the parameter values is part of the model generation and is covered in Chaps. 1.2 through 1.5.

Especially in the determination of stiffness and damping coefficients, one has to rely largely on experience that results from measurements. If measurements of a machine are available at the start of the model generation, it often becomes clear that only a few natural frequencies and mode shapes play a role. Therefore, it is a fundamental objective to find a model with the lowest number of degrees of freedom, which still allows reliable predictions of the dynamic behavior of the system within a defined range of validity.

The experimental model generation (also denoted as identification) is based on measurements performed on actual machines. It is primarily governed by the state of the art in measurement technology and forms the basis for gaining experience for the analytical model generation. In essential aspects, it relies on procedures of system dynamics and aims to make statements on structures and their associated parameter values.

Besides the numerical values of the "local" parameters associated with a specific machine element, the term "parameter value" can also refer to the "global" parameter values of the overall system, such as the identified natural frequencies and mode shapes, which can be determined, for instance, by an experimental modal analysis. Their numerical values characterize a real system . The quality of a calculation model can be tested by comparing the parameter values determined by different methods.

1.1 Classification of Calculation Models

Dimensionless parameters (similarity numbers) can always be formed by (mostly dimensional) parameters. In machine dynamics, a total of $(K-3)$ dimensionless characteristic numbers (numerical values of parameters) can be formed from K parameters [4]. Calculation models with the same structure and the same dimensionless parameters are physically similar. When constructing enlarged or downsized realistic models, the similarity principles have to be adhered to. Furthermore, these principles can also be used in numerical calculations and in the graphical representation of results. Like in fluid mechanics, also in machine dynamics, results obtained theoretically or experimentally can be generalized with the aid of dimensionless parameters. This applies for example when drawing conclusions about the dynamic behavior of families of machine elements or machines.

As is generally known, the calculation model for the single-mass oscillator with harmonic excitation contains $K = 7$ original physical variables (m, c, b, F, Ω, t, x), of which $K - 3 = 4$ dimensionless parameters ($2D = b/\sqrt{mc}$; $\eta = \Omega\sqrt{m/c}$; $\tau = \Omega t$; $\xi = cx/F$) can be formed for the purpose of presenting the results. Dimensionless parameters are defined, for instance, in (1.140), (4.186), (5.58) and (7.25) as well as in Figs. 4.5, 4.36, 5.17 and 6.25.

In some industrial sectors, sophisticated calculation models are common. These have been developed in decade-long interactions between calculation and measurement. This is true, for example, for turbine manufacturing, shipbuilding, automotive, and aeronautical and aerospace engineering, where hundreds of man-years were invested into the development of appropriate calculation models and their implementation into product-oriented special-purpose programs.

In mechanical engineering in general, adequate calculation models still do not exist for many objects and activities. For many components and assemblies, the process of model generation is currently ongoing.

The calculation models in machine dynamics fall into one of the following *three model levels*:

1. constrained system of rigid bodies ("rigid machine" model)
2. linear vibration system (free vibration or forced excitation)
3. nonlinear or self-excited system.

Vibration systems with forced excitation can be further classified into forced and parameter-excited ones. This classification is based on the following three criteria:

- physical (according to the origin and intensity of the energy supply)
- mathematical (according to the complexity of the resulting equations)
- historical (according to the history of the problem).

A real object (real system), i.e. a machine or one of its components, cannot automatically be associated with one of the calculation models. Instead, one can map it onto all three model levels depending upon which specific loads and motions it is subjected to. The energy source is modeled with increasing precision from the lower to the higher levels.

At level 1 and level 2, the motion and force parameters are given as functions of time, whereby these functions remain unaffected by the reactions of the model. At level 3, the energy source is part of the autonomous system. Then, it is also essential to include the motor in the model (e.g. for electrical assemblies).

The model level 3 represents the autonomous system, from which the models of the other levels are derivable by deduction. It can be shown that under simplifying assumptions each lower level is an approximation for the higher level, i.e. generally, the effects can be described more realistically with a higher-level model than with models of lower levels.

For many objects, the modeling began with a rigid mechanism, see Chap. 2, because the historical development of each machine began with lower velocities. When an engine is started up from zero to its maximum rotational speed, it effectively passes through the different ("historical") model levels, from the simple to the complicated. At lower speeds, the object acts like a rigid-body system, while the nonlinear behavior can be demonstrated no later than upon the destruction of the object. Incidentally, this case must sometimes be analyzed seriously during the reconstruction of an event of damage.

A rigid-body system (model of a "rigid machine") is characterized by its geometry and mass parameters. Chapter 2 of this book addresses objects that correspond to model level 1. The following two simple criteria can be employed to separate the range of applicability of rigid-body models from that of vibration systems:

1. For periodic excitations, the rigid-body system ("model of the rigid machine") is applicable to the steady state of the model if it is excited "slowly". For periodic excitations, this means that the *highest excitation frequency* $f_{\max} = k\Omega/(2\pi)$, that still has an important amplitude in the Fourier spectrum must be significantly smaller than the lowest natural frequency f_1 of the real object. Therefore, the criterion can be written as follows:

$$k\Omega \ll \omega_1 = 2\pi f_1 \tag{1.1}$$

where Ω is the fundamental circular frequency of the excitation and k is the order of the highest relevant harmonic.

2. For transient excitations, , i.e. for typical start-up procedures and brake, acceleration or deceleration processes, the model of rigid-body systems is applicable. This holds true as long as the applied force changes "slowly", i.e. when the longest oscillation period T_1 of the real object is significantly smaller than the *start-up time* t_a of the applied force or motion parameters. This criterion can be written as:

$$\frac{1}{f_1} = T_1 \ll t_a \tag{1.2}$$

Table 1.1 lists the main parameters that are typically associated with the particular model levels.

An essential characteristic of a *vibration system* (model level 2) is its number of degrees of freedom. On the one hand, this number depends on which physical

1.1 Classification of Calculation Models

Table 1.1 Typical parameters of the three model levels

Level	Given parameters	Calculable parameters
1	Geometrical dimensions (lengths, angles, transmission ratios), mass parameters, kinematic sequences of motions and/or input load parameters	reduced moment of inertia, velocity and acceleration of rigid-body motions, joint and bearing forces, foundation load, input or braking torque
2	Translational and torsional spring constants, bending stiffness, translational and torsional damper constants, characteristic material parameters, excitation force curves as functions of time, Fourier coefficients for periodic excitation, parameter change over time	Natural frequencies and mode shapes, time functions of force and kinematic parameters for forced oscillations, higher-order resonance points (critical speeds), locus curves, instability domains of parameter-excited oscillations, absorption
3	Velocity-dependent bearing data (influence of oil film), friction coefficients, characteristics of motor and braking torques, behavior of non-elastic materials (viscous, plastic), non-linear geometrical and material parameters	Non-linear oscillations, self-excited oscillations, combined resonances, limit cycles, interaction of vibration systems and energy source, amplitude-dependent natural frequencies, non-linear interactions

effects have to be considered and how many mode shapes are actually excited in a linear system, but also on the desired spatial resolution for the load and deformation.

One has to ensure that the range of the excitation frequencies lies within the range of the natural frequencies of the model. Hence, the following criterion applies:

The model of a vibration system should have natural frequencies reaching at least up to the highest excitation frequency.

Self-excited oscillators (model level 3) are mostly treated as systems with only a few degrees of freedom. They are always nonlinear systems, but the stability limits can often be determined using only linear systems (see Sect. 7.3).

Each model generation should start with a *minimal model* that

- is spatially and/or temporally limited,
- has only a small number of degrees of freedom,
- includes only a few ("robust") parameters,
- represents the essential physical phenomena qualitatively with sufficient accuracy,
- can be analyzed with relatively little effort (comparison of variants) and
- provides qualitatively (and in their tendency, also quantitatively) correct results.

A minimal model is suitable for verifying a first hypothesis and giving suggestions for further theoretical and experimental steps. Depending upon the requirements, one can either keep the minimal model or, based on the intermediate results, decide after each step on the subsequent progression of the model generation.

Normally, the level of understanding of a phenomenon is inversely proportional to the number of degrees of freedom and model parameters employed. Minimal models are clearly structured and manageable. One should know that the introduction of too many parameters can lead to an apparent but false (incidental) correlation between the solutions from a model calculation and the results of an experiment. Mistakenly, this sometimes triggers the conclusion that the model is accurate. There is also the psychological danger of considering the results of sophisticated computational methods and large computer programs to be especially valuable. At the same time, it should not be forgotten how inaccurate most input data are and how sensitive the corresponding results are. Normally, one does not obtain results from measurements or model calculations that are exact up to three significant digits. If there are more than three significant digits given in the following chapters, then this is mostly for mathematical reasons.

1.1.2 Examples

At first, the problem of modeling will be explained using the example of a tower crane (see Fig. 1.1). Calculation models for demonstrating the resistance of the crane components against breaking and the stability of the entire crane have existed for a long time. Dynamic forces, which obviously occur when accelerating the hoisted load, were historically represented by an additional acceleration acting on the load mass. This approach corresponds to model level 1 (see Fig. 1.1b).

Only since the middle of the 20th century, after cases of damage occurred at cranes that were modeled with these load assumptions, have cranes been treated as vibration systems (see example in Fig. 1.1c). Even with simple vibration models, a qualitatively correct representation of the realistic vibration processes was possible. These simple models contain the upper masses of the tower, which oscillate horizontally, thus causing additional dynamic bending moments in the lower part of the tower. These moments are not proportional to the hoisted load, as assumed by model level 1. Note the different fractions of the dynamic moments at position 1 and position 2 of the crane in Fig. 1.1c. Even with this simple calculation model, it was explained why - from a dynamics point of view - it is more beneficial to attach the counterweight at the bottom (it does not vibrate there) - and there were also fewer incidents of damage to such cranes. Nowadays, very sophisticated calculation models are used for cranes, with which all support parts are modeled very accurately (lightweight construction) and the links between the drives, mechanical engineering and steel construction are taken into consideration.

1.1 Classification of Calculation Models

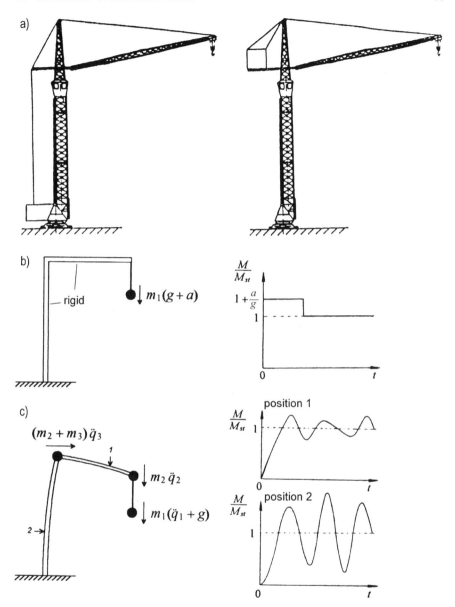

Fig. 1.1 Tower crane; **a)** Schematic of the real system; **b)** Calculation model "rigid-body system"; **c)** Minimal model for a vibration system with typical measured moment curves

The second example explains the problem of calculation models for gear mechanisms. Fig. 1.2 depicts the engineering drawing of a spur gear box. Table 1.2 shows some of the applicable calculation models, which are sorted from the simple to the complicated. Note that the number of degrees of freedom and parameters, which are listed on the right side, increase from level to level. Which model is suitable depends upon the particular objective of the calculation. For each particular model level, the values listed in Table 1.1 can be calculated.

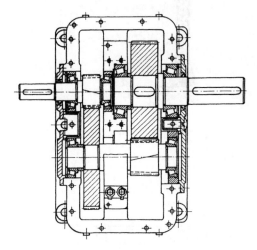

Fig. 1.2 Spur gear box

Model level 1 contains only rigid gears (moments of inertia) and rigid massless shafts. In this model, all internal moments occur at the same time, there are no oscillations, and the time-variant moments are based on the kinetostatic moment distribution (see also Fig. 4.1).

The model of level 2a represents a classical model of a torsional vibration system with forced excitation, which can be used to determine the natural frequencies, mode shapes and dynamic moments (see Sections 4.2 through 4.4). In such models, the determination of the real torsional stiffnesses can be problematic (see Table 1.5). Parameter excitation, which occurs in cases of temporally varying tooth stiffnesses, is considered in model level 2b (see Fig. 1.12 and Sect. 4.5.3.2).

The model shown at the bottom of Table 1.2 includes the coupling of torsional and bending vibrations of the shafts with the elastic bearings. This is still part of the linear theory (see Problem P6.6). In this model, the masses of the gears also play a role. In a calculation model of level 3, the nonlinear characteristics of the driving and output torques, as well as bearing springs, are taken into consideration (see also Sect. 4.5.3.2).

Even more sophisticated calculation models for gear mechanisms have been developed, in which, for instance, the characteristics of the bearings and gears, as well as the vibrations of the housing walls in the acoustic frequency range, are included. A sample model of similar complexity - albeit originating from another area - is shown in Fig. 6.6.

1.1 Classification of Calculation Models

Table 1.2 Calculation model of a gear mechanism

	Model level	model elements, parameter
1	(diagram with J_1, r_1; J_4, r_4; M_{an}; M_{ab}; J_2, r_2; J_3, r_3)	rigid-body models (rigid disks, rigid shafts) Parameter $J_1, J_2, J_3, J_4, r_1, r_2, r_3, r_4$, M_{an}, M_{ab} degree of freedom $f = 1$
2a	(diagram with J_1, r_1; J_4, r_4; M_{an}; c_{T1}; c_{T3}; M_{ab}; J_2, r_2; c_{T2}; J_3, r_3)	Torsional oscillator (rigid disks, torsion-elastic shafts) Parameters $J_1, J_2, J_3, J_4, r_1, r_2, r_3, r_4$, $c_{T1}, c_{T2}, c_{T3}, M_{an}, M_{ab}$ degree of freedom $f = 2$
2b	(diagram with J_1, r_1; J_4, r_4; $M_{an}(t)$; c_{T1}; c_{T3}; $M_{ab}(t)$; $c_{z1}(t)$; $c_{z2}(t)$; c_{T2}; J_2, r_2; J_3, r_3)	Torsional oscillators with parameter excitation (rigid disks, torsion-elastic shafts, variable tooth stiffness) Parameters $J_1, J_2, J_3, J_4, r_1, r_2, r_3, r_4$, $c_{T1}, c_{T2}, c_{T3}, c_{z1}(t), c_{z2}(t)$, $M_{an}(t), M_{ab}(t)$ degree of freedom $f = 4$
3	(diagram with m_1, J_1, r_1; m_4, J_4, r_4; $M_{an}(\varphi)$; c_{T1}; c_{T3}; $M_{ab}(\varphi)$; c_1, c_2, c_5, c_6; $c_{z1}(t)$; $c_{z2}(t)$; EI_1, l_1; EI_2, l_2; EI_3, l_3; c_3; c_{T2}; c_4; m_2, J_2, r_2; m_3, J_3, r_3)	Vibration system with a finite number of degrees of freedom and parameter excitation (rigid disks, torsion-elastic shafts, variable tooth stiffness, bending stiffness of the shafts, bearing stiffness) Parameters $J_1, J_2, J_3, J_4, r_1, r_2, r_3, r_4$, $c_{T1}, c_{T2}, c_{T3}, c_{z1}(\varphi), c_{z2}(\varphi)$, $M_{an}(\varphi), M_{ab}(\varphi)$ degree of freedom $f = 8$ (planar)

1.2 Determination of Mass Parameters

1.2.1 Overview

Knowledge of a rigid body's ten mass parameters is required to describe the dynamic behavior of that rigid body: Mass (m), position of its center of gravity (ξ_S, η_S, ζ_S), position of the principal axes of inertia and principal moments of inertia or the moment of inertia tensor ($J_{\xi\xi}$, $J_{\xi\eta}$, $J_{\xi\zeta}$, $J_{\eta\eta}$, $J_{\eta\zeta}$, $J_{\zeta\zeta}$).

Various methods of determining the parameters are used depending on whether the body is available through design documents, or as a real object (see Table 1.3).

Table 1.3 Methods for determining mass parameters

Parameter	Determination based on the real system	Determination based on drawing
Mass	Weighing, frequency measurements	Volume determination, density, division into elementary bodies
Center of gravity	a) Determination of mass distribution b) Balancing out, suspension c) Dual suspension as a physical pendulum	Determination of the overall center of gravity using individual centers of gravity of elementary bodies
Moment of inertia about a given axis	Rotational body: a) Torsion rod suspension b) Suspension from multiple filaments c) Rolling pendulum	a) Division into ring and disk elements, determination using individual moments of inertia
	arbitrary bodies: d) Dual suspension as a physical pendulum	b) Division into elementary bodies c) Method of cylindrical section
Inertial tensor	Pendulum motion about multiple axes	CAD program

The analytical methods of determining parameters from design documents are always based on breaking the body down into elementary bodies (ring, disk, cuboid, sphere, example: see Fig. 1.6). For example, cylindrical sections with the reference axis of the moment of inertia are used as the cylinder's axis for determining the moments of inertia.

Mass parameters of the rigid bodies are required as input data for computer programs. The accuracy of these parameter values (input data) is of great importance, since the accuracy of the force and motion variables to be calculated depends on it. The mass parameters cannot always be calculated with sufficient accuracy from the data of the structural description in a CAD program (and the specified material density data). Calculating mass parameters is difficult for machine elements that have a complex geometrical shape and/or consist of various materials, the density of which is not known exactly. Sometimes, values for comparison are needed to determine the parameter variation for a mass-produced product. Frequently, the task is to determine the mass parameters for a specific real component experimentally.

1.2.2 Mass and Position of the Center of Gravity

The mass m can mostly, but not always, be easily determined by weighing. The mass of a single-mass oscillator that cannot be separated from the spring can be determined indirectly. The variables needed for this are the natural circular frequency in the original state ($\omega_0^2 = c/m$) and a natural frequency after a defined parameter change.

The natural circular frequency changes if an extra mass Δm is added to the element, or if the element is stiffened using an extra spring with the spring constant Δc:

$$\omega_m^2 = \frac{c}{m + \Delta m} = (2\pi f_m)^2; \qquad \omega_c^2 = \frac{c + \Delta c}{m} = (2\pi f_c)^2. \qquad (1.3)$$

The values of the extra mass or spring constant and the natural frequencies measured can be used to determine the initial mass, since the following equations can be obtained by transformation from (1.3):

$$m = \frac{\omega_m^2 \Delta m}{\omega_0^2 - \omega_m^2} = \frac{f_m^2 \Delta m}{f_0^2 - f_m^2}; \qquad m = \frac{\Delta c}{\omega_c^2 - \omega_m^2} = \frac{\Delta c}{4\pi^2 (f_c^2 - f_0^2)}. \qquad (1.4)$$

The accuracy of the mass m largely depends on the accuracy of the frequency measurement. It is therefore advisable to repeat the measurement several times to be able to determine a mean value. A similar procedure can be applied when determining the moments of inertia with respect to a specific axis, see Fig. 1.7.

Static methods are used to determine the position of the center of gravity. The simplest method is suspension. Since the center of gravity of a freely suspended body is always located below the suspension point, it is located on a vertical axis passing through the suspension point. If there are two suspension points that are not located on the same axis through the center of gravity, the center of gravity will be at the point where these axes intersect. The point of intersection of the vertical axes can be determined using photos (or photogrametrically) after suspending the object on two points.

Balancing can be used for small parts. When putting the part onto a sharp edge, it can be clearly felt when the center of gravity is above the edge. For larger bodies, such as cranes and motor vehicles, the mass distribution is frequently determined by measuring the support forces.

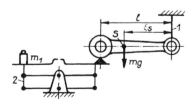

Fig. 1.3 Determination of the center of gravity by determining the mass distribution (*1* Filament; *2* Scales)

Figure 1.3 shows the measurement on a connecting rod. If the support force $m_1 g$ has been determined, the centroidal distance is

$$\xi_S = \frac{m_1 l}{m}. \tag{1.5}$$

For a symmetrical connecting rod, an axis through the center of gravity is provided by the symmetry line. Otherwise, the missing second axis through the center of gravity can be found by tilting the body axis.

This will be demonstrated using the example of a motor vehicle. According to Fig. 1.4, the following are given: $l = 2450$ mm; $h = 500$ mm; $d = 554$ mm.

Measurements were taken

a) in horizontal position $F_1 = 4840$ N
b) in tilted position $F_2 = 5090$ N

The overall weight is $F_G = 9918$ N.

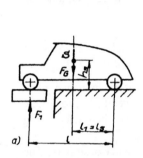
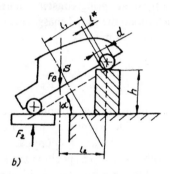

Fig. 1.4 Determining the center of gravity in a motor vehicle; **a)** Horizontal position; **b)** Tilted position

The following results for the horizontal position (a):

$$l_1 = l_S = \frac{F_1 l}{F_G}. \tag{1.6}$$

The following applies to the tilted position (b):

$$l_2 = \frac{F_2 l \cos \alpha}{F_G}; \quad \sin \alpha = \frac{h}{l}. \tag{1.7}$$

The height of the center of gravity h_S is calculated as follows:

$$h_S = \left(\frac{l_2}{\cos \alpha} + l^* - l_1 \right) \frac{1}{\tan \alpha}; \quad l^* = \frac{d}{2} \tan \alpha$$
$$h_S = \frac{d}{2} + \frac{F_2 - F_1}{F_G} \cdot \frac{l^2}{h} \sqrt{1 - \left(\frac{h}{l} \right)^2} \tag{1.8}$$

When we substitute the given numerical values, we obtain $l_S = 1196$ mm; $h_S = 573$ mm. The accuracy of the result is determined by the difference $(F_2 - F_1)$. The height h should therefore be selected as large as possible.

1.2.3 Moment of Inertia about an Axis

In a *simple pendulum test*, the body is suspended as a physical pendulum on an edge in a hole and can perform weakly attenuated oscillations about the contact point. The equation for its small-angle approximation of the natural frequency contains the moment of inertia about the axis of the suspension point and the centroidal distance. Thus, the period of vibration of the pendulum, see Fig. 1.5, for oscillations about A and B, respectively, is:

$$T_A = 2\pi\sqrt{\frac{J_A}{mga}}; \qquad T_B = 2\pi\sqrt{\frac{J_B}{mgb}}. \tag{1.9}$$

The distances a and b extend from the center of gravity to the suspension points A and B.

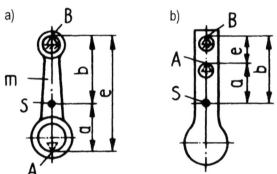

Fig. 1.5 Designations on the physical pendulum; a) Internal center of gravity b) External center of gravity

Furthermore, according to the parallel-axes theorem (*Steiner's theorem:*), the following applies:

$$J_A = J_S + ma^2; \qquad J_B = J_S + mb^2, \tag{1.10}$$

where J_S is the moment of inertia with regard to the center of gravity in S. Taking into account the distance of the pendulum points $e = b \pm a$, it follows that:

$$b = e\frac{\frac{4\pi^2 e}{g} \mp T_A^2}{2\frac{4\pi^2 e}{g} \mp T_A^2 - T_B^2}; \qquad J_S = \frac{T_B^2}{4\pi^2}mgb - mb^2. \tag{1.11}$$

The upper sign (−) applies if S is located between A and B (Fig. 1.5a), the lower sign (+) applies if S falls outside of $\overline{AB}$ (Fig. 1.5b). It should be noted that (1.11) only applies when the center of gravity is located on the connecting line of the two suspension points. If this is not the case, one must first determine the center of gravity using a static method, and then determine the moment of inertia according to (1.9) and (1.10).

An example will be used to compare the result determined by an experiment with a rough calculation. The following has been determined for the connecting rod shown in Fig. 1.6b:

$$e = 156 \text{ mm}; \quad m = 0.225 \text{ kg}; \quad T_A = 0.681 \text{ s}; \quad T_B = 0.709 \text{ s}. \tag{1.12}$$

According to (1.11), these values can be used to determine the centroidal distance $b = 89$ mm and the moment of inertia $J_S = 7.25 \cdot 10^{-4}$ kg $\cdot$ m^2.

For the rough calculation, the body is broken down into elementary bodies, as shown in Fig. 1.6b. It suffices to approximate the shaft as a prismatic rod with the dimensions

$$l_{St} = 129.5 \text{ mm} - \frac{46 + 27}{2} \text{ mm} = 93 \text{ mm}$$
$$f = \frac{18 + 20}{2} \text{ mm} = 19 \text{ mm}; \quad c = 5 \text{ mm} \tag{1.13}$$

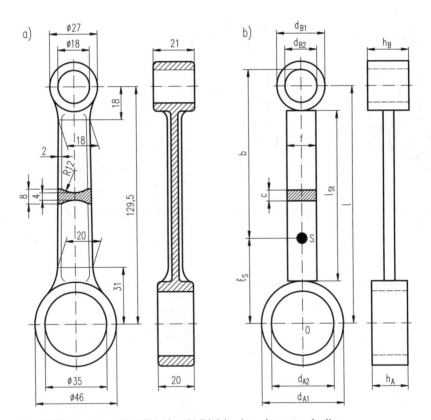

Fig. 1.6 Connecting rod; **a)** Drawing, **b)** Division into elementary bodies

1.2 Determination of Mass Parameters

Starting from axis 0, the centroidal distance ξ_S is:

$$\xi_S = \frac{\frac{l_{St} f c (l_{St} + d_{A1})}{2} + \frac{\pi(d_{B1}^2 - d_{B2}^2) h_B l}{4}}{l_{St} f c + \frac{\pi(d_{B1}^2 - d_{B2}^2) h_B}{4} + \frac{\pi(d_{A1}^2 - d_{A2}^2) h_A}{4}}. \tag{1.14}$$

Substituting the numbers from Fig. 1.6, one finds

$$\xi_S = 50.0 \text{ mm}; \qquad b = l - \xi_S + \frac{d_{B2}}{2}; \qquad b = 88.5 \text{ mm}. \tag{1.15}$$

If a density of $\varrho = 7.85 \text{ g/cm}^3$ is assumed, the moment of inertia about the center of gravity can be calculated from:

$$J_S = \varrho\{\pi h_A (d_{A1}^4 - d_{A2}^4)/32 + \pi(d_{A1}^2 - d_{A2}^2) h_A \xi_S^2/4 + c f^3 l_{St}/12$$
$$+ f c l_{St}^3/12 + c f l_{St}[(l_{St} + d_{A1})/2 - \xi_S]^2 \tag{1.16}$$
$$+ \pi h_B (d_{B1}^4 - d_{B2}^4)/32 + \pi(d_{B1}^2 - d_{B2}^2) h_B (l - \xi_S)^2/4\}$$

The result using the given numbers is $J_S = 7.37 \cdot 10^{-4} \text{ kg} \cdot \text{m}^2$. If the shaft is approximated as a trapezoidal rod, one finds that $\xi_S = 49.85$ mm.

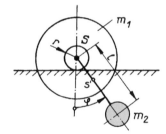

Fig. 1.7 Designations on the rolling pendulum
S Axis through the center of gravity of the cylinder,
S' Axis through the center of gravity of the pendulum system

The method of the *rolling pendulum* can be used to determine the moment of inertia J_S for large cylinders or crankshafts. The ends of the cylinder with a mass of m_1 are placed on two parallel, horizontal edges, and a known point mass m_2 is attached eccentrically from the cylinder axis at a distance of l (Fig. 1.7). The cylinder can then perform a rolling motion. It is required, however, to statically balance the system, i.e. to achieve that the rotor stops at any position. The equation of motion for small angles ($\sin \varphi \approx \varphi$) is:

$$[J_S + m_1 r^2 + m_2 (l - r)^2]\ddot{\varphi} + m_2 g l \varphi = 0. \tag{1.17}$$

The period is

$$T = 2\pi \sqrt{\frac{m_1 r^2 + m_2(l-r)^2 + J_S}{m_2 g l}}. \tag{1.18}$$

It follows from this that the sought after moment of inertia of the cylinder about axis S is:

$$J_S = \frac{T^2}{4\pi^2} m_2 g l - m_1 r^2 - m_2 (l-r)^2. \tag{1.19}$$

A torsional vibration system is frequently used to determine moments of inertia. The restoring moment is achieved by either a torsion rod or a suspension from multiple filaments. It is possible to calculate the spring constant itself (absolute method), however, the sought after moment of inertia is frequently compared to a known one by determining the period using a known extra mass (relative method). In this way, the spring constant is eliminated.

When using a torsion rod suspension (Fig. 1.8a), the following applies to the centroidal axis (which is the axis of rotation):

$$J_S = \frac{T^2}{4\pi^2} c_T; \qquad c_T = \frac{G I_p}{l} \tag{1.20}$$

I_p polar area moment of inertia of the circular cross-section of the rod, see (1.35)
G shear modulus
T period of the torsional vibration without extra mass.

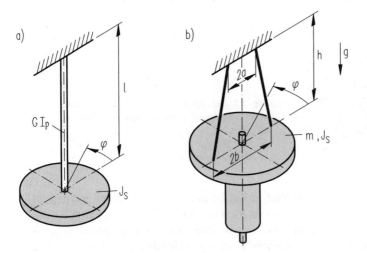

Fig. 1.8 Torsional oscillator; **a)** Torsion rod suspension, **b)** Suspension from multiple filaments

If a known moment of inertia J_Z is attached to the unknown rotating mass J_S, the following applies analogously to (1.20)

$$(J_S + J_Z) = \frac{T_Z^2}{4\pi^2} c_T. \tag{1.21}$$

If one eliminates the torsional spring constant from (1.20) and (1.21), the result is:

1.2 Determination of Mass Parameters

$$J_S = J_Z \frac{T^2}{T_Z^2 - T^2}. \tag{1.22}$$

T_Z is the period with extra mass.

The torsional spring can be replaced by a suspension from multiple (two or three) filaments. Using the terms from Fig. 1.8b, the following results if the vibration amplitudes are small (rotation of the body to be studied about a centroidal axis):

$$J_S = \frac{mg}{4\pi^2} T^2 \frac{ab}{h}. \tag{1.23}$$

Small angles from the vertical axis of the filaments during the oscillations are desirable to stay in the linear range. This can be achieved by using long filaments so that $h \gg a$ and $h \gg b$. The pendulum device is often suspended from the roof structure of the laboratory or from the overhead crane.

An extra mass can also be used in this method. A second oscillation of the pendulum is performed for which a mass m_Z, with a known moment of inertia J_Z, is attached to the mass to be examined. The sought after moment of inertia is then calculated from:

$$J_S = J_Z \frac{mT^2}{(m + m_Z)T_Z^2 - mT^2}. \tag{1.24}$$

The moments of inertia of motors and couplings are given in the catalogs of their manufacturers. The moments of inertia of gear mechanisms are more difficult to find. Therefore, it is often necessary to approximate them. Since the moments of inertia of the individual steps are reduced by the square of the speed ratios when referring them to the fast-running drive shaft, the moments of inertia of the slow shafts are often negligible, see Sect. 2.3. It is, in most cases, sufficient to determine the moment of inertia of the components of the fast running shaft and to multiply it with a factor of 1.1 to 1.2 to estimate the moment of inertia of the entire gear mechanism.

The following empirical correlation (numerical equation) between the moment of inertia of the motor and the driving power was determined for up to 100 kW for slip-ring induction motors (duty rate in percent 60 %, $n = 1000/\text{min}$) that drive cranes:

$$J = 0.0076 P^{1.384}; \quad J \text{ in kg} \cdot \text{m}^2, \quad P \text{ in kW}. \tag{1.25}$$

This equation should be used with caution and viewed as a motivation for establishing empirical formulae for each product group considered.

1.2.4 Moment of Inertia Tensor

The six elements of the moment of inertia tensor can be measured in dynamic tests. There are methods for determining these elements from the inertia forces and inertia moments that are measured during the predefined motion of the actual body. It has been found that the values of the moments of inertia are inaccurate if the moment of inertia tensor is measured for an arbitrary point of reference, because the so-called "Steiner terms" are too large. Therefore, it is useful to calculate it with regard to the center of gravity. Thus, there are three moments of inertia ($J_{\xi\xi}^S$, $J_{\eta\eta}^S$, $J_{\zeta\zeta}^S$), and three products of inertia ($J_{\xi\eta}^S = J_{\eta\xi}^S$, $J_{\eta\zeta}^S = J_{\zeta\eta}^S$, $J_{\zeta\xi}^S = J_{\xi\zeta}^S$) that need to be determined with reference to a body-fixed ξ-η-ζ coordinate system, the origin of which is at the center of gravity. These six mass parameters can also be determined from the principal moments of inertia J_I^S, J_{II}^S and J_{III}^S and the three directional cosines α, β and γ, that indicate the orientation of the principal axes I, II, and III with regard to the body-fixed coordinate system.

The first task to be considered here is the determination of those six variables that are summarized in the moment of inertia tensor with regard to a body-fixed ξ-η-ζ coordinate system with its origin at the center of gravity, see Sect. 2.3:

$$\boldsymbol{J}^S = \begin{bmatrix} J_{\xi\xi}^S & \text{symmetric} & \\ J_{\eta\xi}^S & J_{\eta\eta}^S & \\ J_{\zeta\xi}^S & J_{\zeta\eta}^S & J_{\zeta\zeta}^S \end{bmatrix} \quad (1.26)$$

$$= \begin{bmatrix} \int (\eta^2 + \zeta^2) dm & \text{symmetric} & \\ -\int \eta\xi\, dm & \int (\xi^2 + \zeta^2) dm & \\ -\int \xi\zeta\, dm & -\int \eta\zeta\, dm & \int (\xi^2 + \eta^2) dm \end{bmatrix}.$$

The moments of inertia, with regard to several ($k = 1, 2, \ldots, K$) axes through the center of gravity, can be determined by the test described in 1.2.3, in which the period of a torsional vibration is measured. The moments of inertia J_{kk}^S are obtained for several suspension points.

The position of the kth axis of rotation that passes through the center of gravity can be uniquely described in the body-fixed ξ-η-ζ system using the three angles α_k, β_k and γ_k, see Fig. 1.9. The moment of inertia with regard to the instantaneous axis of rotation that passes through the center of gravity when the body is suspended depends on the six elements of the inertial tensor as follows, see (2.61) in Sect. 2.3.1:

$$\begin{aligned} J_{kk}^S &= \cos^2 \alpha_k\, J_{\xi\xi}^S + \cos^2 \beta_k\, J_{\eta\eta}^S + \cos^2 \gamma_k\, J_{\zeta\zeta}^S \\ &+ 2\cos\alpha_k \cos\beta_k\, J_{\xi\eta}^S + 2\cos\alpha_k \cos\gamma_k\, J_{\xi\zeta}^S + 2\cos\beta_k \cos\gamma_k\, J_{\eta\zeta}^S. \end{aligned} \quad (1.27)$$

Measurement of the angles α_k, β_k, and γ_k that are enclosed by the axis of rotation k–k and the directions of the body-fixed reference system is required. One can evaluate the accuracy of these values by checking the degree to which they meet the Pythagorean theorem, which in this spatial case takes on the following form:

$$\cos^2 \alpha_k + \cos^2 \beta_k + \cos^2 \gamma_k = 1; \quad k = 1, 2, \ldots, K. \quad (1.28)$$

1.2 Determination of Mass Parameters

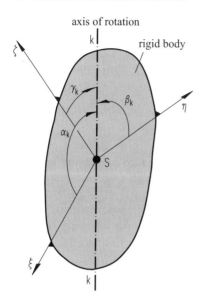

Fig. 1.9 Identification of the position of the kth axis of rotation in the body-fixed ξ-η-ζ system

For an actual body, K vibration tests are performed to subsequently determine the moments of inertia J_{kk}^{S} with regard to the K different axes of rotation that all intersect at the center of gravity S. While $K = 6$ tests should actually suffice, it is useful to perform additional tests to allow for the estimation of the measurement errors.

Furthermore, a vector $\boldsymbol{j}^{\mathrm{S}}$ is introduced that contains the six unknown elements of the moment of inertia tensor:

$$\boldsymbol{j}^{\mathrm{S}} = [J_{\xi\xi}^{\mathrm{S}},\ J_{\eta\eta}^{\mathrm{S}},\ J_{\zeta\zeta}^{\mathrm{S}},\ J_{\xi\eta}^{\mathrm{S}},\ J_{\xi\zeta}^{\mathrm{S}},\ J_{\eta\zeta}^{\mathrm{S}}]^{\mathrm{T}}. \tag{1.29}$$

After evaluating the vibration measurements, the moments of inertia with regard to the kth axis of rotation that are summarized in the vector

$$\boldsymbol{b} = [J_{11}^{\mathrm{S}},\ J_{22}^{\mathrm{S}},\ J_{33}^{\mathrm{S}},\ \ldots,\ J_{KK}^{\mathrm{S}}]^{\mathrm{T}} \tag{1.30}$$

are known. Also known are the directional cosines of all K tests, from which the elements of the following matrix $\boldsymbol{B}$ can be determined.

$$\boldsymbol{B} = \begin{bmatrix} \cos^2\alpha_1 & \cos^2\beta_1 & \cos^2\gamma_1 & 2\cos\alpha_1\cos\beta_1 & 2\cos\alpha_1\cos\gamma_1 & 2\cos\beta_1\cos\gamma_1 \\ \cos^2\alpha_2 & \cos^2\beta_2 & \cos^2\gamma_2 & 2\cos\alpha_2\cos\beta_2 & 2\cos\alpha_2\cos\gamma_2 & 2\cos\beta_2\cos\gamma_2 \\ \vdots & \vdots & \vdots & \vdots & \vdots & \vdots \\ \cos^2\alpha_K & \cos^2\beta_K & \cos^2\gamma_K & 2\cos\alpha_K\cos\beta_K & 2\cos\alpha_K\cos\gamma_K & 2\cos\beta_K\cos\gamma_K \end{bmatrix}$$

$$\tag{1.31}$$

In this way, the following matrix equation arises from (1.27)

$$\boldsymbol{B}\boldsymbol{j}^{\mathrm{S}} = \boldsymbol{b}. \tag{1.32}$$

The rectangular matrix $\boldsymbol{B}$ has K rows and six columns. If $K = 6$ measurements are used, the six unknown variables result from the solution of six linear equations. For $K > 6$, (1.32) represents an overdetermined system of linear equations for the six unknown variables summarized in $\boldsymbol{j}^\mathrm{S}$. According to adjustment calculus [32], the following solution for the sought after elements of the moment of inertia tensor results using the least-square method:

$$\boldsymbol{j}^\mathrm{S} = (\boldsymbol{B}^\mathrm{T} \boldsymbol{B})^{-1} \boldsymbol{B}^\mathrm{T} \boldsymbol{b}. \tag{1.33}$$

For $K > 6$ measurements, it is possible to calculate variance in addition to the mean values [32], i.e. one can use "additional" measurements to gain insight into the number of significant digits of the results obtained.

The principal moments of inertia that are sometimes of interest are calculated from the elements of the moment of inertia tensor by solving the following eigenvalue problem, for which common software can be used, see Sections 2.3.1 and (2.71):

$$\begin{bmatrix} J_{\xi\xi}^\mathrm{S} - J & J_{\xi\eta}^\mathrm{S} & J_{\xi\zeta}^\mathrm{S} \\ J_{\xi\eta}^\mathrm{S} & J_{\eta\eta}^\mathrm{S} - J & J_{\eta\zeta}^\mathrm{S} \\ J_{\xi\zeta}^\mathrm{S} & J_{\eta\zeta}^\mathrm{S} & J_{\zeta\zeta}^\mathrm{S} - J \end{bmatrix} \cdot \begin{bmatrix} \cos\alpha \\ \cos\beta \\ \cos\gamma \end{bmatrix} = \begin{bmatrix} 0 \\ 0 \\ 0. \end{bmatrix} \tag{1.34}$$

The three eigenvalues are the principal moments of inertia: J_I, J_II, and J_III. The components of the three eigenvectors correspond to the three directional cosines that characterize the orthogonal system of the principal axes, which is rotated relative to the body-fixed ξ-η-ζ system. The principal moment of inertia J_I relates to the principal axis I, which is rotated relative to the ξ-η-ζ system, analogously to Fig. 1.9, by angles α_I, β_I, and γ_I that can be calculated from the three directional cosines $\cos\alpha_\mathrm{I}$, $\cos\beta_\mathrm{I}$, and $\cos\gamma_\mathrm{I}$. The orientations of the other two principal axes II and III can be determined analogously.

The calculation becomes simpler when the real body has a plane of symmetry, which in the case on hand is assumed to be the ξ-ζ plane. Each axis that is perpendicular to the plane of symmetry is a principal axis of inertia, and determining the elements of the moment of inertia tensor becomes simpler, see Fig. 1.18.

Figure 1.10 shows a car engine that is suspended for the determination of the principal axes of inertia. The engine is in a frame that can be mounted in various positions to the torsion rod.

It is important to meet the conditions on which the equations are based when performing these tests. These conditions primarily include the linear equation of motion and the neglecting of the damping. Since both the pendulum equations and the equation of motion can only be considered linear for small deflections when an object is suspended using several filaments, the angular amplitude or filament angle, respectively, must not exceed 6°. The damping is sufficiently small if more than 10 vibrations can be counted with ease, see (1.108).

The time should be measured particularly carefully since the period enters the equations (1.19) to (1.24) quadratically. In general, vibrations are so mildly damped that 50 or more vibration periods occur. It is recommended to use the time for a large number of periods for evaluation. Sources for errors can also be found in the

1.3 Spring Characteristics

Fig. 1.10 Suspension of a car engine

analytical determination of the spring constants or the additional moments of inertia, respectively.

If relations in which differences occur are used, see (1.19), (1.22), (1.24), the test parameters should be defined in such a way that differences of variables of almost equal size are avoided.

1.3 Spring Characteristics

1.3.1 General Context

The spring in a vibration system is used to store energy in the form of deformation energy. In multibody structures, the zero-mass springs are located between the masses or between masses and fixed points. Several springs may be arranged in parallel or in series. Table 1.4 provides examples of the overall spring constants.

Spring characteristics of solid components are calculated as part of static deformation calculations, in which the modulus of elasticity E and the shear modulus G are sufficient. Furthermore, they are related to Poisson's ratio ν by $E = 2G(1+\nu)$.

The difficulty is not so much in specifying the dimensions and material parameters but in expressing the effectiveness of the spring attachments in the form of boundary conditions. Questions, such as whether a beam is rigidly clamped, or if a "support spring" acts or whether a coil spring can rotate relative to the contact surfaces, greatly influence the spring performance. These uncertainties often entail that the natural frequencies of models, the parameters of which were determined by calculation only are too high, since the many assumptions made render the springs too rigid.

A torsion shaft is a typical elastic element in mechanical engineering. Most drive shafts have steps where different shaft diameters meet, different transitional radii, various fasteners between shaft and hub, bearing seats, etc. It can be viewed as if a large number of individual springs were arranged in series. The term of reduced length was introduced to be able to better evaluate the influence of these individual springs on the overall spring constant. A shaft segment consisting of two sections (Fig. 1.11), loaded at its ends by the torsional moment M_t, represents an example for this concept. One end is fixed (zero angle of rotation). Now, a shaft segment with constant diameter and the same torsional stiffness is to be found.

Table 1.4 Examples of the coupling of springs

Case	System schematic	Spring coefficients and spring constants
1	(parallel springs, $F = cx$)	$c = c_1 + c_2$
2	(series springs, $F = cx$)	$c = \dfrac{c_1 c_2}{c_1 + c_2}$
3	(springs at angles α_i, $F_x = c_{xx}x + c_{xy}y$, $F_y = c_{yx}x + c_{yy}y$)	$c_{xx} = \sum_{i=1}^{I} c_i \cos^2 \alpha_i; \quad c_{yy} = \sum_{i=1}^{I} c_i \sin^2 \alpha_i$ $c_{xy} = c_{yx} = -\sum_{i=1}^{I} c_i \sin \alpha_i \cos \alpha_i$ **Principal stiffnesses** $c_{\mathrm{I, II}} = \dfrac{c_{xx} + c_{yy}}{2} \left(1 \mp \sqrt{1 - 4\dfrac{c_{xx}c_{yy} - c_{xy}^2}{(c_{xx} + c_{yy})^2}} \right)$ **Principal directions** $\tan \varphi_{\mathrm{I, II}} = \dfrac{c_{\mathrm{I, II}} - c_{xx}}{c_{xy}}$
4	(two springs at angle α, $F_x = c_{xx}x + c_{xy}y$, $F_y = c_{yx}x + c_{yy}y$)	$c_{xx} = (c_1 + c_2) \cos^2 \alpha$ $c_{yy} = (c_1 + c_2) \sin^2 \alpha$ $c_{xy} = c_{yx} = (c_1 - c_2) \sin \alpha \cos \alpha$

1.3 Spring Characteristics

Case 1 and case 4 are special cases of case 3. It should be noted for case 3 that the direction of the components of the force F is indicated by its sign, and that the resulting force can have any direction. The point of application of the force generally does not simply move in the direction of the applied force, but also transversely to this direction, namely when $c_{xy} = c_{yx} \neq 0$.

For the planar case of spring arrangements 3 and 4, there are two principal stiffnesses c_I and c_{II} and two corresponding principal axes, equations for which are given in Table 1.4. Only if a force is acting in the direction of one of the two principal stiffnesses, the displacement occurs in the same direction. The two natural circular frequencies of a mass suspended in a planar spring system are $\omega_1^2 = c_I/m$ and $\omega_2^2 = c_{II}/m$.

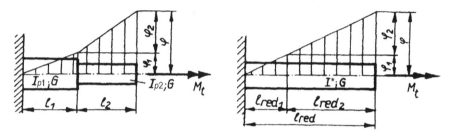

Fig. 1.11 Reduced length for two shaft sections

For a cylindrical shaft, the torsional spring constant can be calculated as follows:

$$c_T = \frac{GI_p}{l} = \frac{\pi G d^4}{32l}; \qquad c_{Ti} = \frac{GI_{pi}}{l_i} = \frac{\pi G d_i^4}{32l_i}. \tag{1.35}$$

G shear modulus, for steel $G = 8 \cdot 10^{10}$ N/m²
I_p polar area moment of inertia of the circular cross-section
l_i lengths of the shaft sections
d_i diameter of the circular cross-section

The total angle of twist is:

$$\varphi = \varphi_1 + \varphi_2 = \frac{M_t}{c_T} = \frac{M_{t1}}{c_{T1}} + \frac{M_{t2}}{c_{T2}}. \tag{1.36}$$

Since the torsional moment is the same in all shaft sections, the following applies:

$$M_t = M_{t1} = M_{t2} \tag{1.37}$$

and therefore

$$\frac{1}{c_T} = \frac{1}{c_{T1}} + \frac{1}{c_{T2}}; \qquad c_T = \frac{c_{T1} c_{T2}}{c_{T1} + c_{T2}}. \tag{1.38}$$

In order to replace the stepped shaft segment by one with constant cross-section, the polar area moment of inertia I^* for the substitute shaft has to be determined first.

Table 1.5 Reduced lengths of various shaft sections

Designation	Labeling	Reduced shaft length
Smooth cylindrical shaft		Solid shaft: $l_{red} = l \dfrac{I_{red}}{I_p} = l \dfrac{D_{red}^4}{D^4}$ Hollow shaft: $l_{red} = l \dfrac{I_{red}}{I_p} = l \dfrac{D_{red}^4}{D^4 - d^4}$
Shaft with keyway		$l_{red} = l_K \dfrac{I_{red}}{I_K} + l \dfrac{I_{red}}{I_p}$
Splined shaft		$I_K = \dfrac{\pi d_K^4}{32} ; I_p = \dfrac{\pi D^4}{32}$
Shaft with taper		$l_{red} = l_K \dfrac{I_{red}}{I_m} + l \dfrac{I_{red}}{I_{p1}}$ $I_m = \dfrac{3 I_{p1}}{\dfrac{D_1}{D_2}\left[\left(\dfrac{D_1}{D_2}\right)^2 + \left(\dfrac{D_1}{D_2}\right) + 1\right]}$ I_m mean polar area moment of inertia of the tapered shaft end
Tapered joint		$l_{red} = \left(l_1 + \dfrac{l_{K1}}{3}\right)\dfrac{I_{red}}{I_{p1}}$ $+ \left(l_2 + \dfrac{l_{K2}}{3}\right)\dfrac{I_{red}}{I_{p2}}$ Section between the assumed power transmission points (x) to be calculated using the above equations
Shaft fillets Press fit joints		$l_{red} = l_{red1} + l_{red2} + \Delta l_{red}$ $l_{red1} = l_1 \dfrac{I_{red}}{I_{p1}}$; $l_{red2} = l_2 \dfrac{I_{red}}{I_{p2}}$ $\Delta l_{red} = \dfrac{\Delta l}{D_1} D_1 \dfrac{I_{red}}{I_{p1}}$ $R_1 = \dfrac{D_1}{2}$

1.3 Spring Characteristics

Thus, the following applies:
$$c_T = \frac{GI^*}{l_{\text{red}}}. \tag{1.39}$$

It follows from (1.38)

$$l_{\text{red}} = l_1 \frac{I^*}{I_{\text{p1}}} + l_2 \frac{I^*}{I_{\text{p2}}} = l_{\text{red1}} + l_{\text{red2}}. \tag{1.40}$$

The reduced length of a shaft consisting of multiple sections is derived from the total of the reduced lengths of all sections. It replaces the overall spring constant and has the advantage that it allows better identification of the effect of each section on the overall system.

The resulting shaft can be drawn using these reduced lengths, see Fig. 4.2. Table 1.5 shows a compilation of various reduced lengths.

1.3.2 Machine Elements, Sub-Assemblies

Machines consist of many sub-assemblies mounted together, which undergo microscopically small relative motions when subjected to a load. All (seemingly immobile) contact points have a lower stiffness than the solid material, and furthermore larger damping occurs there due to microscopic slippage. These lower stiffnesses reduce the natural frequencies, and these influences should be taken into account when generating a calculation model.

In past decades, stiffness and damping values were determined experimentally for many sub-assemblies. These empirical values can be found in company catalogs and special publications [1], [4], [9], [12], [13], [20], [21], [22], [25], [26], [31], [33]). Companies have developed software enabling customers to determine exact data for some sub-assemblies such as crankshafts, clutches, couplings, coil springs, and ball and roller bearings. It is often difficult to obtain such characteristics for novel materials (e. g. fiber-reinforced plastics) or extreme parameter ranges (e. g. very high loading rates, extreme temperatures).

Before we discuss examples, we would like to mention that rather large deviations from such theoretically established values can be found under field conditions. The real (experimentally verified) stiffnesses are typically smaller than the calculated ones. It should be mentioned that this is often due to incorrect modeling. Sometimes a secondary influence becomes significant (see Problem P6.6) or it is incorrect to assume ideal zero backlash and linearity, see Chap. 7. The contact stiffness of the fasteners is frequently underestimated, or static determinateness is assumed that does not exist. Cracks and wear and tear make every machine sub-assembly softer but never stiffer after a longer service life.

Gear mechanisms: For gears, the torsion spring constant is determined using the gear tooth deformation. It is independent of the gear module and can be calculated using numerical equations. If one imagines a gear spring as a longitudinal spring that acts at the pressure line, the following applies to straight-toothed spur gears made of steel with only one tooth engaged, see [22]:

$$c_z = bc'; \qquad c' = 0.8 c'_{\text{th}} \cdot \cos \beta \qquad (1.41)$$
$$c'_{\text{th}} = 1/[0.04723 + 0.15551/z_{n1} + 0.25791/z_{n2} - 0.00635 x_1$$
$$- 0.11654 x_1/z_{n1} - 0.00193 x_2 - 0.24188 x_2/z_{n2} + 0.00529 x_1^2$$
$$+ 0.00182 x_2^2] \, \text{N}/(\mu\text{m} \cdot \text{mm})$$

where $z_{n1,2} = z_{1,2}/(\cos \beta)^3$, b tooth width, z_1, z_2 tooth numbers, x_1, x_2 addendum modification coefficients, β helix angle.

In (1.41), plain disk wheels and the standard reference profile (DIN 867) are assumed for the gearing. In DIN 3990, correction factors are given for using studded wheels and profiles that deviate from the standard profile to take the resulting influences on tooth stiffness into account. A variable number of tooth pairs are engaged due to the contact ratio, which causes the effective tooth stiffness during an engagement period to vary as well.

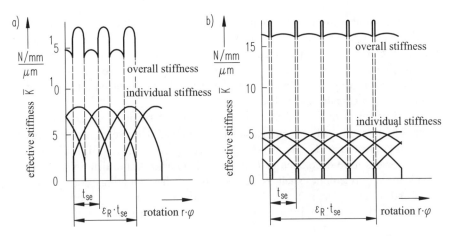

Fig. 1.12 Stiffness curve with engaged teeth; **a)** Contact ratio $\varepsilon = 2.5$, **b)** Contact ratio $\varepsilon = 4.1$

Figure 1.12 shows the stiffness curve for mating gears for two different contact ratios [4]. It can be seen how the resulting stiffness is composed of the stiffnesses of the individual teeth. There is a large jump in the variation of the contact ratio for spur gears. It will be shwon in Sect. 4.5.3.2, how the the position-dependent stiffness influences parametrically excited vibrations in gear mechanisms. In the case of helical gears, there is naturally a larger number of teeth in engagement so that the stiffness jumps are smaller, which has an effect on the excitation of vibrations. Software is available for calculating the gear stiffness $c(t)$ as defined in DIN 3990. The respective data for a specific gearing can be requested from the manufacturers.

1.3 Spring Characteristics

Table 1.6 Spring constants and torsional spring constants

Machine element / sub-assembly	Spring constant c in $\text{N/mm} = \text{kN/m}$		dependent on
Subsoil	$(0.2\ldots1.4) \cdot 10^5 \; (A/\text{m}^2)$		soil type, support area A
		$(0.2\ldots1.4) \cdot 10^7$	e.g. $A = 100\text{ m}^2$
Thread	$(1.5\ldots2) \cdot 10^5 \; (d/\text{mm})$		nominal diameter
		$(1.5\ldots2) \cdot 10^6$	e.g. M10, $d = 10$ mm
Tooth stiffness on steel gear wheels	$(1\ldots2) \cdot 10^4 \; (b/\text{mm})$		tooth width b
		$(3\ldots6) \cdot 10^5$	e.g. $b = 30$ mm
Ball bearings – radial	$(5\ldots10) \cdot 10^3 \; (d/\text{mm})$		inside diameter d
		$(2.5\ldots5) \cdot 10^5$	e.g. $d = 50$ mm
			preload,
			number of rolling elements
Steel tension bar		$1.6 \cdot 10^4$	$c = EA/l$; e.g.:
			$l = 1$ m, $d = 10$ mm
Buffer spring of a craneway		$(0.1\ldots1) \cdot 10^4$	dimensions
Round steel chains		$600\ldots3700$	e.g. 1 m in length
Steel springs for machine foundations	$(30\ldots60) \cdot (F/\text{kN})$		style, static load F
		$600\ldots1200$	e.g. $F = 20$ kN
Car tires		$80\ldots160$	design, air pressure, velocity
Jib head of tower crane (vertical)		$40\ldots400$	overall height, jib length
Suspension strut of a motorbike		$10\ldots20$	spring type
Steel cantilever beam (end point of a cantilever girder)		0.3	$c = 3EI/l^3$; e.g.: $l = 1$ m, $d = 10$ mm
	Torsional spring constant c_T in $\text{Nm/rad} = \text{Nm}$		
Steel block (half space theory) circular clamping point	$3.1 \cdot 10^2 \; (r^3/\text{mm})$		radius r in mm
		$3.1 \cdot 10^5$	e.g. $r = 10$ mm
Steel couplings		$(0.1\ldots2.5) \cdot 10^5$	diameter style,
Rubber spring couplings		$1000\ldots6000$	load
Steel torsion rod		80	$c_T = GI_p/l$; e.g.: $l = 1$ m, $d = 10$ mm

The data provided above should only be used as estimates and help the students to get a feel for these parameters. More accurate values can be taken from manufacturers' catalogs and handbooks or must be calculated by special modeling.

The mean value of the tooth stiffness for one period of engagement (stiffness of the engaging spring) can approximated as follows:

$$c_m = c' \cdot (0.75\varepsilon_\alpha + 0.25); \qquad \varepsilon_\alpha \text{ profile contact ratio.} \tag{1.42}$$

As a rough estimate, the following values can be set for the stiffness of an individual spring and the stiffness of the engaging spring for steel gears:

$$c' = 14 \text{ N}/(\mu\text{m} \cdot \text{mm}); \qquad c_m = 20 \text{ N}/(\mu\text{m} \cdot \text{mm}). \tag{1.43}$$

The spring constants are significantly smaller in the case of gearings that are subjected to low loads ($F_t/b < 100$ N/mm). These can approximated by assuming a linear decline in stiffness:
For $F_t/b < 100$ N/mm:

$$c' = 0.8 c'_{\text{th}} \cdot \cos\beta \cdot F_t/(100b) \tag{1.44}$$

F_t circumferential force acting on the gearing in N
b tooth width in mm

Screws: (according to VDI 2230 [36]) longitudinal spring constant:

$$c = \frac{E}{\sum_i \dfrac{l_i}{A_i}} \tag{1.45}$$

A screw is composed of a set of cylindrical segments with lengths l_i and cross sections A_i. E is the modulus of elasticity of the screw material. The spring constant of an entire threaded connection is much lower due to the contact stiffnesses between the screw, the washer and the sheet metal.

The following applies to the spring constants of a belt:

$$c_{rz} = \frac{E_z A}{L} \tag{1.46}$$

A belt cross-section
L effective free span length

Since the modulus of elasticity E depends on the belt pretensioning force, F_v, E_z and E differ.

Cylindrical coil springs: Quite frequently it is just the longitudinal stiffness that is considered for cylindrical coil springs

$$c = \frac{Gd^4}{8iD^3}, \tag{1.47}$$

see the spring parameters shown in Fig. 1.13:

1.3 Spring Characteristics

d wire diameter
r mean coil diameter
i number of effective coils
G shear modulus (for hardened rolled steel: $G = 7.9 \cdot 10^{10}$ N/m^2)

Coil springs can be designed for any ratio of longitudinal stiffness c to transverse stiffness c_{q}. $y = F/c$ is the deflection due to the longitudinal force F, $x = Q/c_{\mathrm{q}}$ is the spring deflection in the transverse direction due to the transverse force Q.

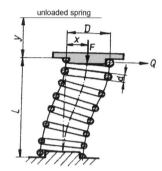

Fig. 1.13 Coil spring, coordinates and dimensions
$l_0 = l + y$ free spring length
l spring length after loading.

The transverse stiffness of a cylindrical coil spring is illustrated in Fig. 1.14, taken from [33].

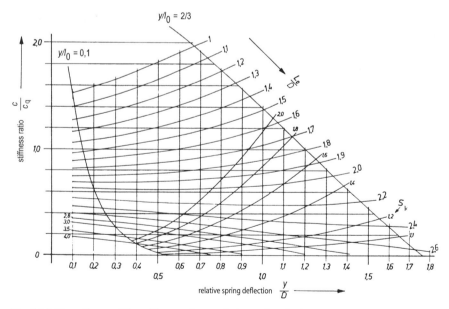

Fig. 1.14 Dependence of the stiffness ratio on relative spring displacement

These curves represent the ratio of the transverse stiffness to the longitudinal stiffness and result from the equation:

$$\frac{c}{c_q} = 1.0613 \frac{D}{y} \sqrt{\frac{l_0}{y} - 0.6142} \cdot \tan\left(0.9422 \frac{y}{D} \sqrt{\frac{l_0}{y} - 0.6142}\right) + 1 - \frac{l_0}{y}. \tag{1.48}$$

Figure 1.14 shows the curves for the range of $l_0/D = 1 \ldots 4$, as calculated by (1.48). Two curves $y/l_0 = 0.1$ and $2/3$ have been outlined. All practically significant loading cases are within the range of this diagram.

A **steel cable** represents another very resilient spring element. The spring constant of a taut steel cable is calculated as follows:

$$c = \frac{E_S A}{l}. \tag{1.49}$$

where A is the metallic cross-section, l the cable length, and E_S the modulus of elasticity. Various values can be found for the latter in the literature, which proves that it is dependent on the make of the cable, the duration of its use, and its preload. The following can be taken as approximate values:

$$E_S = (1 \cdots 1.6) \cdot 10^{11} \text{ N/m}^2 \tag{1.50}$$

The **bearing** of a shaft can have a tremendous effect on its vibration behavior. Bearings are selected based on their load-bearing capacity and service life, so that their spring constant is predefined. A comparison of the shaft stiffness and bearing stiffness should be performed, in particular, for rigid bearing housings. Corresponding options are provided in Table 5.1.

For **radial ball and roller bearings**, the radial bearing stiffness according to WICHE is as follows:

$$c_r = K_L \cdot \frac{F_L}{f_o} \tag{1.51}$$

where K_L is a correction coefficient that can be determined based on the relative bearing clearance, as shown in Fig. 1.15a, F_L is the static bearing load, and f_o is the deflection of the bearing when installed without play. It is calculated from the following numerical equations:

Ball bearings:

$$f_o = \sqrt[3]{\frac{2.08}{d} + \left(\frac{F_L}{i}\right)^2} \; \mu\text{m}; \qquad d \text{ in mm}; F_L \text{ in N}. \tag{1.52}$$

Roller and needle bearings:

$$f_o = \left(\frac{0.252}{L_W - 2r_W}\right)^{0.8} \cdot \left(\frac{F_L}{i}\right)^{0.9} \; \mu\text{m}; \qquad L_W \text{ in mm}; r_W \text{ in mm}. \tag{1.53}$$

1.3 Spring Characteristics

The data regarding the number of rolling elements i, the rolling element diameters d, the rolling element length L_W and their fillet radii r_W can be taken from the data sheets of the bearing manufacturers. The bearing clearance $\Delta r = R_L/2$, which can be positive (play) or negative (preloading), is dependent on the bearing play R_L with which the bearing is delivered. It is specified in DIN 620 as a function of the shaft diameter. However, it is apparent from Fig. 1.15 that the bearing clearance has a great influence, in particular, in the negative range.

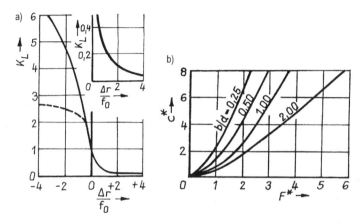

Fig. 1.15 Coefficients to determine bearing stiffness; **a)** K_L for roller bearings, ——— roller bearing, – – – – antifriction bearing; **b)** c^* for journal bearings

The stiffness of **radial journal bearings** is highly dependent on the bearing geometry and the lubricating film. It can be roughly calculated based on the equation:

$$c = c^* \cdot 2\pi \cdot \frac{bd^3 \eta f}{(\Delta r)^3} \qquad (1.54)$$

where

b load-bearing width
d shaft diameter
η dynamic viscosity of the oil at operating temperature
f rotational speed of shaft
Δr radial clearance.

The correction coefficient c^* can be obtained from Fig. 1.15b as a function of the relative bearing force

$$F^* = \frac{2F_L(\Delta r)^2}{\pi b d^3 \eta f} \qquad (1.55)$$

with the width ratio b/d as a parameter. The spring characteristics of ball, roller and journal bearings are highly nonlinear, and the specified relationships, (1.51) and (1.54), are just approximations for a specific bearing load.

The mass of a spring element is generally neglected. It is assumed that the natural frequency of the spring element is high compared with the frequency at which the system vibrates. However, if there are vibrations with a wide frequency range, such as vibrations caused by a superposition of machine and structure-borne sound, the natural frequency of the spring should be estimated to check the spring function. Otherwise it can happen that specific frequencies are transmitted despite operation of the overall system in the subcritical range.

1.3.3 Rubber Springs

Spring elements made of rubber have particular properties as compared to metal springs. For example, their deformation behavior depends on pre-treatment, rubber quality, frequency, number of cycles, geometry, and time (aging of the rubber).

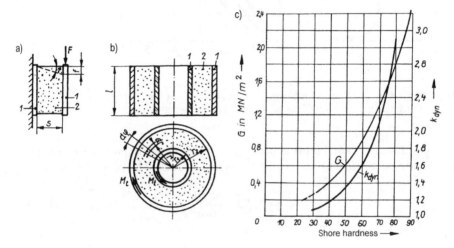

Fig. 1.16 Rubber springs subject to a shear load **a)** Disk-type rubber spring under parallel shear **b)** Sleeve-type rubber spring under torsional load, *1* metal parts; *2* rubber **c)** Shear modulus and factor k_{dyn} as functions of Shore hardness

Rubber springs have a weakly nonlinear force-deflection function that can be considered linear for rough calculations, see P1.3. The spring constant results from the slope of the characteristic curve at the operating point. The primary material characteristic is the shear modulus G that is expressed as a function of Shore hardness, see Fig. 1.16c. The Poisson's ratio of rubber is $\nu \approx 0.5$.

The rubber body must be firmly connected to metal parts, through which the forces enter and exit (these spring elements are called bound "rubber springs"). The boundary conditions have a decisive influence during the large deformation and are dependent on the friction and roughness conditions of the support surfaces for free

1.3 Spring Characteristics

rubber elements. For pure shear loading, the static spring constants can be calculated using the methods of continuum mechanics based on *Hooke*'s law. For example, the following applies to a disk-type rubber spring (Fig. 1.16a):

$$c = \frac{F}{f} = \frac{AG}{s}. \tag{1.56}$$

For a sleeve-type spring under torsional loading, Fig. 1.16b, the static torsional spring constant is:

$$c_\text{T} = \frac{M_\text{t}}{\varphi} = \frac{4\pi l G}{(1/r_1^2) - (1/r_2^2)}. \tag{1.57}$$

If only normal stresses occur under a given load, the conversion stress and strain is performed using the modulus of elasticity E. In the case of rubber elements, however, this is no longer just a pure material parameter but it depends on the shape of the rubber spring. If we introduce a shape factor k_E, it can be approximated from Fig. 1.17b.

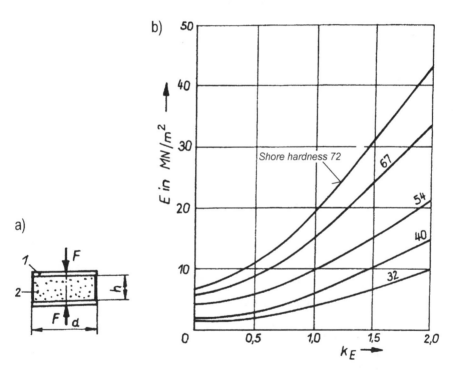

Fig. 1.17 Rubber spring under pressure load; *1* Metal parts, *2* Rubber; **a)** Parameters, **b)** Fictitious modulus of elasticity as a function of the shape factor k_E
(Source: Göbel – Gummifedern)

The following applies to the shape factor:

$$k_E = \frac{\text{a surface under stress}}{\text{entire free surface area}}. \tag{1.58}$$

Thus, the bound cylindrical rubber spring, Fig. 1.17a, has the shape factor

$$k_E = \frac{d^2 \pi}{4 \pi d h} = \frac{d}{4h}. \tag{1.59}$$

Its spring constant is calculated as follows:

$$c_{st} = \frac{AE}{h}. \tag{1.60}$$

It is noted, though, that the modeling of rubber springs under compressive loading is associated with much uncertainty.

It is always advantageous to review the results experimentally. Unlike the spring constants of metal springs, those of rubber springs are frequency-dependent. In models, one takes this remarkable dynamic effect into account by introducing a dynamic spring constant as follows

$$c_{dyn} = k_{dyn} c_{st}. \tag{1.61}$$

The following applies to the typical range of rubber hardness (35 to 95 Shore): $k_{dyn} = 1.1$ to 3.0, see Fig. 1.16. For the operational safety of rubber springs, their strength and heating-up are critical. At frequencies greater than 20 Hz, $k_{dyn} \approx 2.8$ to 3.2.

1.3.4 Problems P1.1 to P1.3

P1.1 Determining the Moment of Inertia

The moment of inertia of a crankshaft with respect to its axis of rotation is to be determined experimentally using a torsion rod suspension system (Fig. 1.8). The following applies to the torsion rod: length $l = 380$ mm; diameter $d = 4$ mm; shear modulus $G = 7.93 \cdot 10^4$ N/mm^2. A time of $T = 41.5$ s was measured for 50 full vibrations.

P1.2 Moment of Inertia Tensor of a Symmetric Body

The moment of inertia tensor of a symmetrical body is to be determined by experiment. Pendulum tests about the three axes shown in Fig. 1.18 ($k = 1, 2$ and 3), which are in the symmetry plane, were performed, from which the three moments of inertia about the axes 1, 2 and 3 could be determined.

Given: Moments of inertia about these axes: J_{11}^S, J_{22}^S, and J_{33}^S

Find: 1. Principal moments of inertia J_I^S, J_{II}^S, J_{III}^S
2. Angles of principal axes $\alpha_1, \gamma_1, \alpha_{II}, \gamma_{II}$

1.3 Spring Characteristics

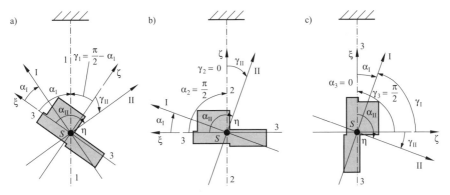

Fig. 1.18 Identification of the position of the three axes through the center of gravity

P1.3 Nonlinear Spring Characteristic

The static characteristic for a rubber compression spring was determined as shown in Table 1.7 and Fig. 1.19. What spring value can be expected in a linear vibration system if the spring load at the static equilibrium position is 9 kN, and the frequency is approximately 20 Hz (Shore rubber hardness larger than 80)?

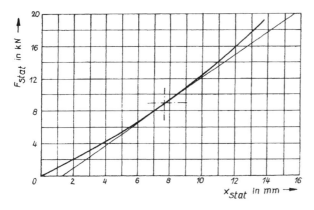

Fig. 1.19 Static spring characteristic of a rubber spring

Table 1.7 Given measured values

k	1	2	3	4	5	6	7
x_k in mm	0	2	4	6	8	10	12
F_k in kN	0	2.00	4.39	6.74	9.26	12.32	16.00

Approximate the characteristic by the polynomial

$$F = c_1 x + c_3 x^3 \tag{1.62}$$

and determine c_1 and c_3 using adjustment calculus.

1.3.5 Solutions S1.1 to S1.3

S1.1 According to (1.20), the following applies:

$$J_S = \frac{T^2}{4\pi^2} c_T; \qquad c_T = \frac{I_p G}{l}. \tag{1.63}$$

One finds

$$T = \frac{41.5}{50}\,\text{s} = 0.83\,\text{s}\,; \qquad I_p = \frac{\pi d^4}{32} = 0.002\,513\,\text{cm}^4; \qquad c_T = 5.244\,\text{N}\cdot\text{m};$$

$$\underline{J_S = 0.0915\,\text{kg}\cdot\text{m}^2} \tag{1.64}$$

S1.2 If the body-fixed coordinate system is placed into the symmetry plane, the following applies to the symmetrical body: $J^S_{\xi\eta} = J^S_{\eta\zeta} = 0$. The system of equations (1.32) is simplified due to $\cos\gamma_1 = \cos(\pi/2 - \alpha_1) = \sin\alpha_1$ to become

$$\begin{bmatrix} \cos^2\alpha_1 & \sin^2\alpha_1 & 2\cos\alpha_1\sin\alpha_1 \\ 0 & 1 & 0 \\ 1 & 0 & 0 \end{bmatrix} \begin{bmatrix} J^S_{\xi\xi} \\ J^S_{\zeta\zeta} \\ J^S_{\xi\zeta} \end{bmatrix} = \begin{bmatrix} J^S_{11} \\ J^S_{22} \\ J^S_{33} \end{bmatrix}. \tag{1.65}$$

It yields the solutions

$$\underline{J^S_{\xi\xi} = J^S_{33}}; \qquad \underline{J^S_{\zeta\zeta} = J^S_{22}}; \qquad \underline{J^S_{\xi\zeta} = \frac{J^S_{11} - J^S_{33}\cos^2\alpha_1 - J^S_{22}\sin^2\alpha_1}{2\cos\alpha_1\sin\alpha_1}}. \tag{1.66}$$

A fourth experiment directly provides the principal moment of inertia from the torsional vibration about the η axis

$$\underline{J^S_{\eta\eta} = J^S_{III}}. \tag{1.67}$$

Four elements of the inertial tensor have been determined by (1.66) and (1.67). In order to determine the two remaining principal moments of inertia, the following eigenvalue problem must be solved, using (1.34):

$$\begin{bmatrix} J^S_{\xi\xi} - J & J^S_{\xi\zeta} \\ J^S_{\xi\zeta} & J^S_{\zeta\zeta} - J \end{bmatrix} \begin{bmatrix} \cos\alpha \\ \cos\gamma \end{bmatrix} = \begin{bmatrix} 0 \\ 0 \end{bmatrix}. \tag{1.68}$$

The two remaining principal moments of inertia can then be derived from a quadratic equation:

$$\underline{J^S_{I,II} = \frac{1}{2}(J^S_{\xi\xi} + J^S_{\zeta\zeta}) \pm \sqrt{\frac{(J^S_{\xi\xi} + J^S_{\zeta\zeta})^2}{4} - J^S_{\xi\xi}J^S_{\zeta\zeta} + (J^S_{\xi\zeta})^2}}. \tag{1.69}$$

The angles of the principal axes α_1 and γ_1 result from (1.68) if the principal moments of inertia J^S_I and J^S_{II} are substituted there:

$$\underline{\tan\alpha_1 = -\frac{J^S_{\xi\zeta}}{J^S_{\zeta\zeta} - J^S_I} = \frac{J^S_I - J^S_{\zeta\zeta}}{J^S_{\xi\zeta}}}; \qquad \underline{\cos\gamma_1 = \sin\alpha_1}$$

$$\underline{\tan\alpha_{II} = -\frac{J^S_{\xi\zeta}}{J^S_{\zeta\zeta} - J^S_{II}} = \frac{J^S_{II} - J^S_{\zeta\zeta}}{J^S_{\xi\zeta}}}; \qquad \underline{\cos\gamma_{II} = \sin\alpha_{II}} \tag{1.70}$$

1.3 Spring Characteristics

S1.3 The cubic polynomial is to approximate the measured values, i.e. meet the conditions

$$c_1 x_k + c_3 x_k^3 = F_k \tag{1.71}$$

for $k = 1$ to 7 as best as possible. These are 7 linear equations for the two unknown quantities c_1 and c_3, which are elements of the parameter vector $\boldsymbol{p}$. If one writes these equations in matrix form

$$\boldsymbol{A}\boldsymbol{p} = \boldsymbol{b}, \tag{1.72}$$

where:

$$\boldsymbol{A}^{\mathrm{T}} = \begin{bmatrix} x_1 & x_2 & \cdots & x_7 \\ x_1^3 & x_2^3 & \cdots & x_7^3 \end{bmatrix}; \quad \boldsymbol{b}^{\mathrm{T}} = \begin{bmatrix} F_1 & F_2 & \cdots & F_7 \end{bmatrix}; \quad \boldsymbol{p} = \begin{bmatrix} c_1 \\ c_3 \end{bmatrix}, \tag{1.73}$$

the condition for the minimum of the mean quadratic deviation results in the following matrix equation [32] (adjustment calculus):

$$\boldsymbol{A}^{\mathrm{T}}\boldsymbol{A}\boldsymbol{p} = \boldsymbol{A}^{\mathrm{T}}\boldsymbol{b}. \tag{1.74}$$

These are the two numeric equations for the two unknown quantities, which, after substituting the measured values from table 1.7, are as follows:

$$\begin{aligned} 0.364\, c_1 + 36.4 \cdot 10^{-6} c_3 m^2 &= 450.92 \text{ kN/m} \\ 36.4\, c_1 + 4.298\,944 \cdot 10^{-3} c_3 m^2 &= 46\,456.16 \text{ kN/m}. \end{aligned} \tag{1.75}$$

Their solutions are:

$$\underline{c_1 = 1032 \text{ kN/m}}; \quad \underline{c_3 = 2.07 \cdot 10^6 \text{ kN/m}^3}. \tag{1.76}$$

The tangent corresponds to the local spring constant, i.e. from (1.62) follows:

$$\frac{\mathrm{d}F}{\mathrm{d}x} = c(x) = c_1 + 3 c_3 x^2. \tag{1.77}$$

For $x_{\mathrm{st}} = 7.66$ mm, therefore, the following applies:

$$\begin{aligned} c(x_{\mathrm{st}}) &= 1032 + 3 \cdot 2.07 \cdot 10^6 \cdot (7.66^2 \cdot 10^{-3})^2 \text{ kN/m} \\ &= (1032 + 364) \text{ kN/m} = 1396 \text{ kN/m}. \end{aligned} \tag{1.78}$$

Since this is a stiff type of rubber, and a frequency of 20 Hz is already considered high for rubber, the factor that lies within the range from $k_{\mathrm{dyn}} = 2.8$ to 3.2 is relevant, see Fig. 1.16c. The dynamic spring constant is approximately

$$\underline{c_{\mathrm{dyn}} \approx 3 \cdot 1400 = 4200 \text{ kN/m}}. \tag{1.79}$$

The assumption of k_{dyn} represents a serious uncertainty in the determination of the dynamic spring constant. It would have been sufficient to use a mean value of a linearized characteristic for the calculation, since the uncertainty of k_{dyn} exceeds that of the nonlinearity.

1.4 Damping Characteristics

1.4.1 General Context

Mechanical energy losses occur with all mechanical movements, i. e., damping is present in all vibrations. However, it has to be decided, when performing a dynamic analysis of a drive system, whether damping is relevant and in what form it is to be included in the calculation model. One can go by the following rules:

1. Damping does not have to be taken into consideration in most cases, when only the following variables are of interest:

   - Low natural frequencies (and resonance ranges) of a drive system,
   - Peak values after impact phenomena,
   - Vibration states outside of the resonance ranges.

2. Damping forces have a noticeable influence and should at least be included via modal or viscous damping if the following variables are of interest:

   - Resonance amplitudes of linear systems under periodic loading,
   - Number of cycles during free-response transients, e. g. after impacts,
   - High natural frequencies and high mode shapes,
   - Predictions of the stability of parameter-excited oscillators.

3. More accurate damping models are recommended if the following variables are of interest

   - Heating-up of the material, e. g. rubber springs,
   - Behavior of intentionally installed damping components, e. g. viscous torsion dampers and damped vibration absorbers,
   - Dynamic behavior of non-metallic materials.

Damping forces emerge on the surface of the moving solid bodies, such as friction in guides and bearings, or inside the solid materials when there are relative movements. The damping forces that act at joints and contact points are calculated for each sub-assembly using experimentally determined damping coefficients.

Calculation models from continuum mechanics should be used for the internal **damping forces** because of material deformations. Their analysis is tedious because systems of partial differential equations must be solved. In most cases, the interaction of mechanical and thermodynamic processes has to be taken into account if the material heats up due to damping work, and the material parameters are changed in the process. It frequently turned out that five to ten parameters are required to accurately capture the material behavior alone.

Therefore, parameter values for the damping have been determined for entire components. It is desirable to those who perform the actual calculations to achieve a good approximation of the damping forces at arbitrary loading patterns with just a few parameters (which should also be easy to determine in experiments).

1.4 Damping Characteristics

Each damping model has to express mathematically that the damping force acts opposite to the direction of the instantaneous velocity, since only then mechanical energy is dissipated from the moving mechanical system. In general, damping forces that act on the various materials depend, in various ways, on forces, deformations and their time derivatives, so that the following functional relationship exists:

$$f(\ldots, \ddot{q}, \dot{q}, q, F, \dot{F}, \ldots) = 0 \tag{1.80}$$

If damping forces depend only on the coordinate and its derivatives, they can be described as follows:

$$F_D = |F(q, \dot{q})|\,\text{sign}(\dot{q}) \tag{1.81}$$

q is the coordinate upon which the damping force F_D acts due to relative movement. The magnitude can be a nonlinear function of the coordinate and/or velocity.

When describing external damping, the spring and damping forces are typically distinguished in the equation of motion. The form obtained for harmonic excitation and linear spring force is:

$$m\ddot{q} + cq + F_D = \hat{F}\sin\Omega t \tag{1.82}$$

The following expressions for the damping force are common:

Coulomb friction:

$$F_D = F_R \frac{\dot{q}}{|\dot{q}|} = F_R \text{sign}(\dot{q}) \tag{1.83}$$

Viscous damping:

$$F_D = b\dot{q} = b|\dot{q}|\text{sign}(\dot{q}) \tag{1.84}$$

Complex damping:

$$F_D = jb^* q \tag{1.85}$$

Frequency-independent damping:

$$F_D = \frac{b^* \dot{q}}{\Omega} = \frac{b^*}{\Omega}|\dot{q}|\text{sign}(\dot{q}) \tag{1.86}$$

Hysteresis damping:

$$F_D = F_R \sqrt{1 - \left(\frac{q}{\hat{q}}\right)^2}\,\text{sign}(\dot{q}). \tag{1.87}$$

Since a linear system in steady-state operation moves at the frequency of excitation, $-q = \hat{q}\sin(\Omega t - \varphi)$; $\dot{q} = \hat{q}\Omega\cos(\Omega t - \varphi)$ – (1.86) is an amplitude-dependent damping force.

Coulomb friction (1.83), however, is neither amplitude- nor frequency-dependent. The term $\dot{q}/|\dot{q}|$ just indicates the velocity-controlled direction of F_D. Unlike the other models, this model does not provide an elliptic hysteresis curve (see Table 1.8) and can be used for any temporal excitation. This model is discussed in more detail together with the viscous damping in Sect. 7.2.2.3.

The direction of the damping force $b|\dot{q}|$ changes with the sign of the velocity $\dot{q}$ for the linear model of viscous damping (1.84). This applies to harmonic motion but not to all motions, which is why this model "will not dampen properly" for more complex force curves, i.e. it provides solutions that deviate from reality. Model (1.84) is used most commonly for applications because it involves the benefits of the mathematical treatment of linear systems (principle of superposition!).

The method of frequency-independent damping (1.86) usually describes the material damping better than model (1.84). Frequency-independent in this context means, that the damping force does not change with Ω. The method only applies to harmonic movements with a given circular frequency of the excitation Ω. Mathematically, it can be treated in a similar way as (1.84). All equations that are derived for viscous damping can be applied if one substitutes b^*/Ω for b, see also Fig. 1.23 and (1.111).

It has been tried to eliminate the restriction of model (1.86) using complex damping (1.85). As the comparison of the two models shows, they use the same constant b^* as the characteristic value, see also the generalization in Sects. 6.6.3 and (6.343) to (6.349). The disadvantage of complex damping is its lack of causality, as S. CRANDALL proved in 1962, i.e., the vibration response can lead (!) the excitation (that is, the effect comes before the cause).

Another frequency-independent damping is described by the nonlinear model (1.87). It models the material damping in a nonlinear way, which means that the displacement is not simply proportional to that of the excitation force. This model also has the disadvantage that it can only easily be applied to harmonic motions. It is discussed in detail in Sect. 7.2.2.3.

Several dimensionless characteristic parameters for describing damping have become internationally accepted, the most common being the damping ratio D (damping factor according to E. LEHR). In the case of harmonic excitation, the following relationships apply between damping parameters and characteristic parameters for the calculation model of the linear oscillator with one degree of freedom for $D \ll 1$(see Table 1.9):

$$D = \frac{\Lambda}{2\pi} = \frac{\psi}{4\pi\eta} = \frac{\varphi}{2\eta} = \frac{\delta}{\omega_0}. \quad (1.88)$$

The four dimensionless characteristic parameters D, Λ, ψ and φ in (1.88) are proportional to the relative loss of mechanical energy per vibration cycle. There is also a connection to the number n of occurring vibration cycles (1.108), (1.109), the coefficient of restitution (1.138) and efficiency, which will not be discussed any further here.

The damping constant

$$b = 2D\sqrt{mc} = \frac{2Dc}{\omega_0} = 2Dm\omega_0 \quad (1.89)$$

has units of $N \cdot s/m$ and can only be determined in conjunction with another dimensional parameter of the oscillator, as (1.89) shows.

1.4 Damping Characteristics

Below, some analytical relationships that are relevant for determining damping parameters are discussed. The equation of motion for free vibrations of a viscously damped single-degree-of-freedom oscillator is as follows:

$$m\ddot{q} + b\dot{q} + cq = 0 \tag{1.90}$$

or with the characteristic parameters

$$\omega_0 = \sqrt{\frac{c}{m}}; \quad 2D = \frac{b}{\sqrt{mc}} \tag{1.91}$$

for the natural circular frequency ω_0 of the undamped oscillator and the damping ratio D:

$$\ddot{q} + 2D\omega_0 \dot{q} + \omega_0^2 q = 0. \tag{1.92}$$

For the initial conditions,

$$t = 0: \quad q(0) = q_0; \quad \dot{q}(0) = v_0 \tag{1.93}$$

it has the following solution (with the natural circular frequency of the damped system $\omega = \omega_0\sqrt{1 - D^2}$ and the decay rate $\delta = D\omega_0$):

$$q(t) = \exp(-\delta t)\left(q_0 \cos \omega t + \frac{v_0 + \delta q_0}{\omega} \sin \omega t\right) \tag{1.94}$$

$$\dot{q}(t) = \exp(-\delta t)\left(v_0 \cos \omega t - \frac{\omega_0^2 q_0 + \delta v_0}{\omega} \sin \omega t\right).$$

A typical time curve is shown in Fig. 1.20.

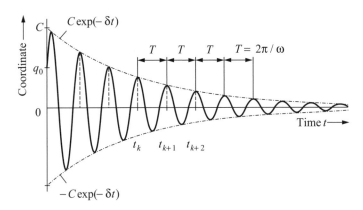

Fig. 1.20 Free response to a free vibration (damping ratio $D = 0.05$)

If one forms the ratio of two local maxima, whose time difference is the n-fold of the period $T = 2\pi/\omega$, the following applies:

$$q(t) = C \exp(-\delta t) \cos(\omega t - \beta) \tag{1.95}$$
$$q(t + nT) = C \exp(-\delta t - \delta nT) \cos(\omega t + n\omega T - \beta). \tag{1.96}$$

Since $\omega T = 2\pi$, the cosine does not change after a full period. Therefore:

$$\frac{q(t)}{q(t+nT)} = \frac{q_k}{q_{k+n}} = \frac{C\exp(-\delta t)}{C\exp(-\delta t - \delta nT)}$$
$$= \exp(\delta nT) = \exp\left(\frac{2\pi n D}{\sqrt{1-D^2}}\right) = \exp(n\Lambda). \tag{1.97}$$

The exact relationship between the damping ratio D and the logarithmic decrement is derived from here, see the approximation in (1.88).

The equation of motion for harmonically excited forced vibrations is as follows:

$$m\ddot{q} + b\dot{q} + cq = \hat{F}\sin(\Omega t). \tag{1.98}$$

The particular solution provides the steady-state curve (permanent state)

$$q(t) = \frac{1}{c} V \hat{F} \sin(\Omega t - \varphi) = \hat{q}\sin(\Omega t - \varphi); \quad \dot{q}(t) = \hat{q}\Omega\cos(\Omega t - \varphi). \tag{1.99}$$

From the frequency ratio $\eta = \Omega/\omega_0$ and the quantities known from (1.91) follows the displacement amplitude $\hat{q} = V\hat{F}/c$ with the nondimensionalized amplitude function

$$V(D, \eta) = \frac{1}{\sqrt{(1-\eta^2)^2 + 4D^2\eta^2}}. \tag{1.100}$$

If the force is plotted as a function of the displacement, the ellipse shown in Table 1.8 is obtained for the case of viscous damping. Note the direction that results from the harmonic time curve of the excitation force.

Due to damping, the displacement curve trails the force curve by the "loss angle" φ given by

$$\sin\varphi = 2D\eta V; \qquad \cos\varphi = (1-\eta^2)V, \tag{1.101}$$

see also Table 1.9. The mechanical work that the excitation force performs during a full cycle ($0 \leq t \leq T$), taking (1.101) into account, is:

$$\Delta W = \oint F dq = \int_0^T F(t)\dot{q}(t)dt = \int_0^T \hat{F}\sin\Omega t \hat{q}\Omega\cos(\Omega t - \varphi)dt$$
$$= \hat{F}\hat{q}\int_0^{2\pi}(\sin\Omega t\cos\Omega t\cos\varphi + \sin^2\Omega t\sin\varphi)d(\Omega t) = \hat{F}\hat{q}\pi\sin\varphi$$
$$= \hat{F}\hat{q}\pi 2D\eta V = 2\pi Dc\eta\hat{q}^2 = \pi b\Omega\hat{q}^2. \tag{1.102}$$

1.4 Damping Characteristics

Table 1.8 Hysteresis curves for various damping approaches

Viscous damping	Friction damping
(hysteresis ellipse with force F vs displacement q, values $c\hat{q}$, $-\hat{q}$, $\hat{q}$)	(hysteresis parallelogram with force F vs displacement q, values $c\hat{q}$, F_R, $-F_R$, $2F_R$, $-\hat{q}$, $\hat{q}$)
(spring c with damper b in parallel, mass q, force F)	(spring c with friction element F_R in parallel, mass q, force F)
$F = cq + b\dot{q}$	$F = cq + F_R \mathrm{sign}(\dot{q})$
$\Delta W = \pi b \Omega \hat{q}^2$	$\Delta W = 4 F_R \hat{q}$

This corresponds to the area inside of the hysteresis curve, see Table 1.8. The relative damping ψ is equal to the ratio of the loss of mechanical energy per period and the mechanical work $W = (\frac{1}{2})c\hat{q}^2$ and, because of (1.102), equals:

$$\psi = \frac{\Delta W}{W} = \frac{2\pi b \Omega}{c} = 4\pi D \eta. \tag{1.103}$$

The hysteresis curves for the damping models (1.84) to (1.86) are also ellipses. A parallelogram results for friction damping, as can be seen in Table 1.8.

1.4.2 Methods for Determining Characteristic Damping Parameters

In addition to theoretical methods that start from rheological characteristic parameters of the material and from the specifics of the workpiece or design, mainly experimental methods are used to determine the characteristic damping parameters. Proven methods include the free-response test and frequency-response tests (forced vibrations with harmonic excitation). An overview of methods for determining damping parameters is given in Table 1.9.

In a *free-response test*, the object to be tested (e. g. a rubber mat with a mass lying on it, or a machine component) is excited by an impact or by the sudden application or removal of a static load to free vibrations. In simple applications, the object is hit by a hammer or a pretensioned cable is suddenly cut. The typical result of a free-

Table 1.9 Elementary methods for determining characteristic damping parameters

	Parameter		Origin, geometrical quantity
Free response	Logarithmic damping decrement $$\Lambda = \frac{1}{n} \ln \left| \frac{q(t_k)}{q(t_k + nT)} \right|$$	(1)	decay curve
Forced harmonic vibration	Relative damping $$\psi = \frac{\Delta W}{W}$$	(2)	Hysteresis curve
Forced harmonic vibration	Loss angle $$\sin \varphi = \frac{b\Omega \hat{q}}{\hat{F}} \approx \varphi$$	(3)	Time curve
Forced harmonic vibration	Damping ratio from half intensity width $$D = \frac{f_2 - f_1}{2 f_0}$$	(4)	Resonance curve
Parameter values	damping ratio $$D = \frac{b}{2\sqrt{cm}}$$	(5)	Calculation model

1.4 Damping Characteristics

response test is the logarithmic decrement Λ. Using a time interval of only a single period between two maxima, it follows for $n = 1$ from (1.97):

$$\Lambda = \ln(q_k/q_{k+1}). \tag{1.104}$$

An evaluation is frequently performed over n full vibrations. One also obtains from (1.97):

$$\Lambda = \frac{1}{n}\ln(q_k/q_{k+n}). \tag{1.105}$$

To evaluate the order of magnitude of the damping ratio from an existing measurement plot from a free-response test, one can count how many full vibrations have occurred until the vibration declined to half of its initial amplitude. It can be derived from (1.97) that a vibration has an amplitude of

$$q(nT) = C \cdot \exp(-\delta nT) = C \cdot \exp(-\omega_0 TDn) = C \cdot \exp(-2\pi Dn) \tag{1.106}$$

after n periods. It also follows from this equation that a vibration only has half of its initial amplitude $q(nT) = \frac{C}{2}$ if

$$\exp(-2\pi Dn) = 0.5; \quad 2\pi nD = \ln 2 \approx 0.6931, \tag{1.107}$$

which for $D \ll 1$ is approximately

$$D \approx \frac{0.11}{n}. \tag{1.108}$$

The damping ratio can be estimated from the number of vibration periods n that occurred until the vibration declined to half of its initial amplitude. The value $n \approx 2$ results, for example, from Fig. 1.20, therefore, from (1.108) the damping ratio $D \approx 0.11/2 = 0.055$. With some practice, one can see the approximate magnitude of the damping ratio from decay curves.

The following estimation provides a mean value if the time until the vibrations have practically decayed is used for counting. The very small amplitudes should be ignored since "yet another theory" applies to those. If the decay can be tracked to about 4% of the initial value, a calculation similar to (1.106) results in the approximation formula

$$D \approx \frac{0.5}{n^*}, \tag{1.109}$$

where n^* is the number of vibrations until complete decay.

It is desirable to determine the spring and damping characteristics at the same frequencies and amplitudes that occur under real operating conditions.
Figure 1.21 shows the schematic of a test station for determining the parameters of a rubber spring and the calculation model.

The energy losses due to internal damping occur because the force/deflection diagram under load does not match the one after load removal. Instead, the two curves form a hysteresis curve for steady-state motion whose area provides a measure of the energy lost during a full vibration cycle since the area in a force/deflection dia-

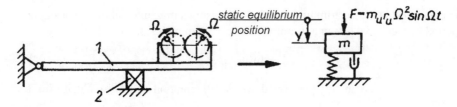

Fig. 1.21 Test stand for determining the characteristic parameters of a rubber spring;
1 Loading and guidance, *2* Rubber spring

gram is proportional to a work. The shape of the area is dependent on the material, the loading, and the shape of the workpiece.

The relative damping ψ in the case of harmonic excitation of a component in the steady state results from the hysteresis curve, see (2) in Table 1.9. The shapes of the respective hysteresis curves result from the specified damping models. Taking into account the equivalent damping capacity, the damping parameters of all models can be converted into each other by equating the relative damping ψ, including into the parameters of a linear oscillator, see (1.103) and Table 1.8.

Measuring the loss angle φ is relatively difficult and mostly yields inaccurate values since this angle is small and cannot be calculated to several significant digits, see (1.101) and (3) in Table 1.9. The evaluation of forced vibrations near the resonance provides relatively accurate values for the damping ratio D. To obtain the most accurate values, it is useful not to measure the height of the resonance peak (which is inversely proportional to $2D$) but the width of the resonance curve at the specified frequencies, see (4) in Table 1.9.

For small damping values, the half width method as derived from (1.100) can be applied. This is how the measurement is performed: Excite the vibration system at resonance and determine the resonance amplitude $\hat{q}_{max}$ and the resonance frequency f_0. Then change the excitation frequency until the vibration amplitude $\hat{q}_{max}/\sqrt{2}$ is reached.

The associated excitation frequencies are f_1 and f_2. The damping ratio D can be determined according to (4) in Table 1.9 from their difference. If the mass m is known, the damping constant can be determined from

$$b = m(\Omega_2 - \Omega_1) = 2\pi m(f_2 - f_1). \tag{1.110}$$

A major advantage of this procedure is that the magnitude of the excitation force does not have to be known.

1.4.3 Empirical Damping Values

The classical way is to measure the parameter values of those components that serve as dampers. This is done for sub-assemblies that were produced as dampers, such as the torsional vibration dampers described in Sect. 4.4 or the commercial dampers, some of which are described in VDI Guideline 3833 [36], see the examples in Fig. 1.22.

Many dampers use the effect that a velocity-dependent force is created when squeezing a viscous medium through a small gap. Proven damping media include oils, bitumen, polybutene and silicone. Some dampers also use friction, see Figs. 4.42 and 4.44. Characteristics and parameter values are available for most commercial dampers. One can request these from the manufacturers.

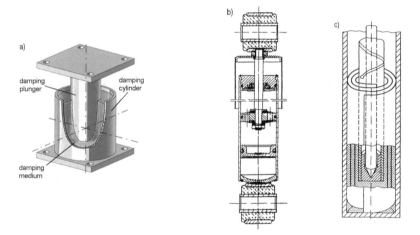

Fig. 1.22 Dampers as sub-assemblies; **a)** VISCO damper (Source: GERB [33]), **b)** Single pipe damper (VDI 3833) (Source: Fichtel & Sachs), **c)** Spiral bearing damper (VDI 3833)

If a free-response test is analyzed, in most cases a dependency of the logarithmic decrement on the amplitude will be found. This means that the assumption of velocity-proportional damping across the entire range is incorrect, see Fig. 1.24.

It is clear that this test is well suited to determine the amplitude dependence of the damping. If it is also desired to determine a frequency dependence, the mass would have to be changed to shift the natural frequency. This can usually only be achieved within narrow limits, though. It should also be noted that the values found apply to vibrations around the static equilibrium position.

Figure 1.23 shows the results of free-response tests of the radial movement of a tractor tire. The result is a strong frequency dependence of the damping constant b. This points to frequency-independent workpiece damping, as it is described by (1.86). If we approximate the test points using the hyperbola drawn into the figure, we find $b^* = b/f = \text{const}$.

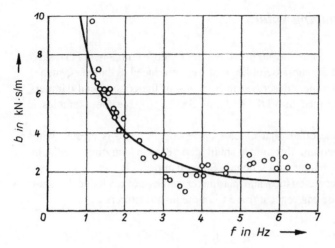

Fig. 1.23 Frequency-dependent "damping constant" of a tractor tire
(Source: Diss. Müller, H., TU Dresden 1977)

There are analytical equations for determining damping constants for some sub-assemblies. Table 1.10 shows some examples. Damping most frequently acts inside the machine sub-assemblies without providing a specific component for the purpose generating damping. In such cases, the damping ratio of the entire system is determined for individual modes and taken into account, for example, as modal damping. An experimental modal analysis provides the modal damping ratios D_i in addition to natural frequencies and mode shapes. Little experience has been gained regarding characteristic damping parameters under non-harmonic excitation.

More precise studies have shown that the damping ratio of a material or component is lower at higher frequencies than at lower frequencies. The models (1.85) to (1.87), which provide an elliptical hysteresis curve of the material, each describe a damping that is independent of frequency. A practical workaround is to specify damping ratios for specific load ranges.

For the natural torsional frequencies of steam turbines, for example, experiments have provided the following numerical equation for the damping ratio in the range from 8 Hz to 150 Hz (HUSTER/ZIEGLER, VDI-Berichte 1749):

$$D = (0.05\ldots 0.08)/f; \qquad \text{natural frequency } f \text{ in Hz.} \qquad (1.111)$$

This results in damping ratios for the same material of $D = 0.0003\ldots 0.01$!

1.4 Damping Characteristics

The following empirical values are known:

Mechanical steel	$D = 0.0008$
High-tensile steel	$D = 0.0003 \ldots 0.0015$
Structural steel	$D = 0.0025$
Cast iron	$D = 0.01 \ldots 0.05$
Drive trains, frames	$D = 0.02 \ldots 0.08$
Concrete, subsoil	$D = 0.01 \ldots 0.1$
Rubber springs	$D = 0.08 \ldots 0.12$

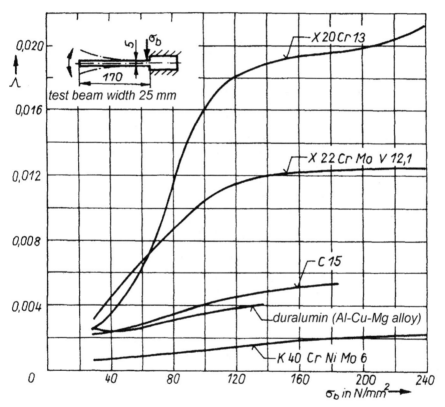

Fig. 1.24 Logarithmic decrement of workpiece damping as a function of tension deflection for various materials

The characteristic damping parameters mentioned in (1.88) are often used for mutual comparison of the damping capacity of materials as shown, for example, in Fig. 1.24. By measuring the logarithmic decrement as a function of the stress amplitude, one can recognize that X20Cr13 steel has considerably better damping properties than K40NiMo6 steel, particularly when the load amplitude is large.

Table 1.10 Empirical values of damping constants

Damping constant	Validity
$b = c_1 B \cdot \eta \left[\dfrac{D}{(D-d)}\right]^3$	Radial journal motion in journal bearings
$b = c_2 \eta \left(\dfrac{B}{h_0}\right)^3$	Lubricated guides when moving perpendicular to the guide direction
c_1, c_2 constant depending on the type of bearing and guide system D bearing diameter d journal diameter B width of bearing and guide trajectory h_0 guide backlash η dynamic oil viscosity	
$b_T = \mu A r^2$ N · cm · s	Crankshaft drives of piston engines during torsional vibrations
A piston area cm^2 r crank radius cm μ damping coefficient N · s · cm^{-3}	Diesel engines: $\mu = 0.04 \cdots 0.05$ N · s · cm^{-3} Car engines: $\mu = 0.015 \cdots 0.02$ N · s · cm^{-3}
$b_T = 19.1 \cdot \dfrac{M_m}{n}$ N · cm · s	centrifugal compressors, fans blowers during torsional vibrations
$b_T = 38.2 \cdot \dfrac{M_m}{n}$ N · cm · s	Marine propellers during torsional vibrations
M_m mean torque absorbed by the rotor N · cm n speed in 1/min	
$b_T = 9.3 \cdot 10^3 \dfrac{EI}{(E-E_s)} n^2$ N · cm · s	Rotors of electric generators during torsional vibrations
E Electromotoric force (EMF) of the generator V E_s EMF of the energy consumer (motor or battery) V, is operated against external resistances, $E_s = 0$ I current A n speed in 1/min	
$b = \dfrac{\pi \eta l d^2}{(D-d)^2}\left[3 + \dfrac{3}{4}\dfrac{d}{(D-d)}\right]$	Piston damper for translational motion
$D-d$ piston backlash d piston diameter l piston length η dynamic oil viscosity	
$b_T = \dfrac{\pi \eta}{\delta}\left[r_a^4 - r_i^4 + 2B(r_a^3 + r_i^3)\right]$	viscous torsional vibration damper according to Fig. 4.45
δ radial backlash r_a outer radius r_i inner radius	B width η dynamic oil viscosity

1.5 Characteristic Excitation Parameters

1.5.1 Periodic Excitation

Excitations can occur as force and motion excitations. Torsional vibrations, for example, are excited by periodic excitation moments, see Chap. 4. The primary excitation of bending vibrations of fast-running rotors are harmonic unbalance forces or periodic movements of bearings. Foundation vibrations can also be excited by periodically moving machine elements (force excitation) or by movement of their installation sites, see Chap. 3.

It is common in dynamics to describe periodic processes in the form of *Fourier series*. If $f(\Omega t)$ is the periodic function, the following periodicity condition applies for a period $T_0 = 2\pi/\Omega$:

$$f(\Omega t) = f[\Omega(t + kT_0)]; \qquad k = 1, 2, \ldots \tag{1.112}$$

The Fourier series is:

$$f(\Omega t) = a_0 + \sum_{k=1}^{\infty} a_k \cos(k\Omega t) + \sum_{k=1}^{\infty} b_k \sin(k\Omega t). \tag{1.113}$$

$$= a_0 + \sum_{k=1}^{\infty} c_k \sin(k\Omega t + \beta_k). \tag{1.114}$$

There are the following relationships:

$$c_0 = a_0; \qquad c_k = \sqrt{a_k^2 + b_k^2}; \qquad \sin \beta_k = \frac{a_k}{c_k}; \qquad \cos \beta_k = \frac{b_k}{c_k}. \tag{1.115}$$

The terms in (1.113) are called *harmonics*. The parameters a_k, b_k, or c_k are the *Fourier coefficients*. It is the purpose of *Fourier analysis* to determine these coefficients.

If the function $f(\Omega t)$ is given in analytical form, the Fourier coefficients can be calculated in closed form. The following applies:

$$a_0 = \frac{1}{2\pi} \int_0^{2\pi} f(\Omega t) \, \mathrm{d}(\Omega t) \qquad \text{(mean value)} \tag{1.116}$$

$$a_k = \frac{1}{\pi} \int_0^{2\pi} f(\Omega t) \cos k\Omega t \, \mathrm{d}(\Omega t) \tag{1.117}$$

$$b_k = \frac{1}{\pi} \int_0^{2\pi} f(\Omega t) \sin k\Omega t \, \mathrm{d}(\Omega t). \tag{1.118}$$

For even functions with $f(\Omega t) = f(-\Omega t)$: $b_k = 0$, for odd funcions with $-f(\Omega t) = f(-\Omega t)$: $a_0 = a_k = 0$.

Table 1.11 lists the Fourier series of some functions, with $\Omega t = \varphi$.

In machine dynamics, $f(\Omega t)$ is often obtained by measurements or numerical calculations in the form of equidistant discrete values. Therefore, the Fourier coefficients are determined numerically, the integrals (1.116) to (1.118) being approximated by sums.

If the function $y(t)$ is to be approximated up to the frequency f_N, the period T_0 has to be scanned in $2N \geq 2f_N T_0$ steps due to:

$$f_N = \frac{N}{T_0} = \frac{N\Omega}{2\pi} \tag{1.119}$$

The number of grid points must therefore be greater than the highest harmonic. The accuracy of the highest harmonic is dependent on the number of grid points. $N = 2m$ grid points and the algorithm of the *Fast-Fourier* transform (FFT) are used in practical calculations.

Figure 1.25 shows the curves for the first seven harmonics into which a periodic function $f(\Omega t)$ was expanded. The higher harmonics ($k \geq 8$) are decreasing with increasing order and are not shown. The function value $f(\Omega t)$ at each point in time results from the sum of the harmonics, see (1.113).

For a slider-crank mechanism, the Fourier coefficients of the input torque M_{an} needed to overcome the inertia forces (the torque $M_T = -M_{\text{an}}$ acts on the crankshaft) can be stated analytically, see Fig. 2.28, Fig. 2.48, and (2.279). The following is found with an oscillating mass m, a crank radius l_2, a connecting rod length l_3, a crank ratio $\lambda = l_2/l_3 < 1$ and an angular velocity Ω of the crankshaft:

$$M_{\text{an}} = -M_T = -ml_2^2\Omega^2(D_1 \sin \Omega t + D_2 \sin 2\Omega t + D_3 \sin 3\Omega t + \cdots) \tag{1.120}$$

with the amplitudes of the harmonics

$$\begin{aligned} D_1 &= \frac{\lambda}{4} + \frac{\lambda^3}{16} + 15\frac{\lambda^5}{512} + \cdots & D_4 &= -\frac{\lambda^2}{4} - \frac{\lambda^4}{8} - \frac{\lambda^6}{16} - \cdots \\ D_2 &= -\frac{1}{2} - \frac{\lambda^4}{32} - \frac{\lambda^6}{32} - \cdots & D_5 &= 5\frac{\lambda^3}{32} + 75\frac{\lambda^5}{512} + \cdots \\ D_3 &= -3\frac{\lambda}{4} - 9\frac{\lambda^3}{32} - 81\frac{\lambda^5}{512} - \cdots & D_6 &= 3\frac{\lambda^4}{32} + 3\frac{\lambda^6}{32} + \cdots \end{aligned} \tag{1.121}$$

1.5.2 Transient Excitation

Non-periodic excitation primarily occurs during start-up, clutch-engaging, and braking processes. The term braking is not limited to the brake component itself but includes the operating resistance that determines the required input torque of a work machine.

1.5 Characteristic Excitation Parameters

Table 1.11 Fourier series of various analytically given functions

Case	Function graph	Function	Fourier coefficient	Fourier series
1	(sawtooth wave)	$f(\varphi) = \frac{h}{\pi}\varphi$; $-\pi < \varphi < \pi$	$a_0 = 0;\ a_k = 0$ $b_k = \pm\frac{2h}{k\pi};\ k = \begin{cases}1;3;5;\ldots \\ 2;4;6;\ldots\end{cases}$	$f(\varphi) = \frac{2h}{\pi}\left(\sin\varphi - \frac{1}{2}\sin 2\varphi + \frac{1}{3}\sin 3\varphi - +\cdots\right)$
2	(triangular wave)	$f(\varphi) = \begin{cases}\frac{2h}{\pi}\varphi; & -\frac{\pi}{2} \leq \varphi \leq \frac{\pi}{2} \\ \frac{2h}{\pi}(\pi - \varphi); & \frac{\pi}{2} \leq \varphi \leq \frac{3\pi}{2}\end{cases}$	$a_0 = 0;\ a_k = 0$ $b_2 = b_4 = b_6 = \cdots = 0$ $b_k = \pm\frac{8h}{k^2\pi^2};\ k = \begin{cases}1;5;9;\ldots \\ 3;7;11;\ldots\end{cases}$	$f(\varphi) = \frac{8h}{\pi^2}\left(\sin\varphi - \frac{1}{3^2}\sin 3\varphi + \frac{1}{5^2}\sin 5\varphi - +\cdots\right)$
3	(rectified sine)	$f(\varphi) = h\lvert\sin\varphi\rvert$	$b_k = 0;\ a_1 = a_3 = a_5 = \cdots = 0$ $a_0 = \frac{2h}{\pi};\ a_k = -\frac{4h}{\pi(k^2-1)}$ $k = 2;4;6;\ldots$	$f(\varphi) = \frac{4h}{\pi}\left(\frac{1}{2} - \frac{1}{1\cdot 3}\cos 2\varphi\right.$ $\left. - \frac{1}{3\cdot 5}\cos 4\varphi - \frac{1}{5\cdot 7}\cos 6\varphi - \cdots\right)$
4	(half-wave rectified)	$f(\varphi) = \begin{cases}h\sin\varphi; & 0 \leq \varphi \leq \pi \\ 0; & \pi \leq \varphi \leq 2\pi\end{cases}$	$a_1 = a_3 = \cdots = 0;\ a_0 = \frac{h}{\pi}$ $b_k = 0;\ k \neq 1;\ b_1 = \frac{h}{2}$ $a_k = -\frac{2h}{\pi(k^2-1)};\ k = 2;4;6$	$f(\varphi) = \frac{h}{\pi} + \frac{h}{2}\sin\varphi$ $- \frac{2h}{\pi}\left(\frac{1}{1\cdot 3}\cos 2\varphi + \frac{1}{3\cdot 5}\cos 4\varphi + \cdots\right)$
5	(rectangular pulse)	$f(\varphi) = \begin{cases}h; & -\frac{\varphi_s}{2} \leq \varphi \leq \frac{\varphi_s}{2} \\ 0; & \frac{\varphi_s}{2} \leq \varphi \leq (2\pi - \frac{\varphi_s}{2})\end{cases}$	$a_0 = \frac{h\varphi_s}{2\pi};\ b_k = 0$ $a_k = \frac{2h}{k\pi}\sin k\frac{\varphi_s}{2}$	$f(\varphi) = \frac{h\varphi_s}{2\pi} + \frac{2h}{\pi}\left(\sin\frac{\varphi_s}{2}\cos\varphi + \frac{1}{2}\sin\varphi_s\cos 2\varphi\right.$ $\left.+ \frac{1}{3}\sin\frac{3\varphi_s}{2}\cos 3\varphi + \cdots\right)$
6	spike pulse	Nr. 5 with $\varphi_s \to 0;\ h \to \infty$ $h\varphi_s \to J$ (finite value)	$a_0 = \frac{J}{2\pi}$ $a_k = \frac{J}{\pi}$	$f(\varphi) = \frac{J}{2\pi} + \frac{J}{\pi}(\cos\varphi + \cos 2\varphi + \cos 3\varphi + \cdots)$

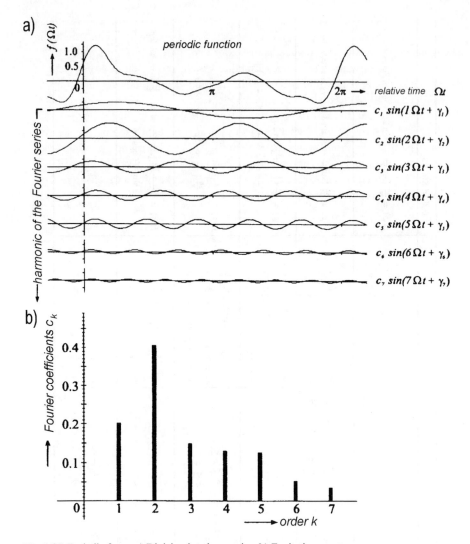

Fig. 1.25 Periodic force; **a)** Division into harmonics; **b)** Excitation spectrum

A torque $M(\varphi; \dot{\varphi})$ mostly occurs in speed-torque characteristics of motors, brakes, fans, pumps, and operating resistances of processing machines.

A special case is the conservative force field in which there is only a dependence on the input coordinate. Figure 1.26a shows some torque characteristics $M(\varphi)$.

When designing drive systems, it is recommended to perform a dynamic analysis with specific programs for simulation, which describes the progressions of forces or torques as functions of the motion variables [4]. Such software is based on mathematical models that describe the characteristics as functions of the respective internal parameters of the drive motors, clutches, or brakes.

1.5 Characteristic Excitation Parameters

A commonly used example is the asynchronous motor. An asynchronous motor is an electromagnetic system that has a characteristic transfer function, which can be calculated from such parameters as the number of pole pairs, stator voltage, armature inductance, stator inductance, armature resistance, synchronous speed and others. This makes it possible to establish differential equations for the coupled electromechanical vibrations.

If one neglects the stator resistance and assumes that the idle slippage is zero, the differential equations of the asynchronous motor can be expressed by the steady-state values of the parameters breakdown slippage s_K and breakdown torque M_K. If there are only small vibrations about a mean torque (and, accordingly, about a mean angular velocity) in a drive, a relationship for the dynamic motor characteristic applies that involves just these two parameters:

$$\ddot{M} + \left(2s_K + \frac{\ddot{\varphi}}{s\Omega^2}\right)\dot{M}\Omega + \left((s_K^2 + s^2)\Omega^2 + \frac{\ddot{\varphi}s_K}{s}\right)M = 2M_K s_K s\Omega^2 \quad (1.122)$$

where:

$$\begin{aligned} &M \quad \text{Motor torque} \\ &\Omega \quad \text{Synchronous angular velocity} \\ &\dot{\varphi} \quad \text{Angular velocity of the motor} \\ &s \quad \text{Slippage } (s = 1 - \dot{\varphi}/\Omega) \\ &s_K \quad \text{Breakdown slippage} \\ &M_K \quad \text{Breakdown torque of the motor.} \end{aligned} \quad (1.123)$$

The breakdown slippage s_K and the breakdown torque M_K are the two parameters that are involved in this differential equation and determine the torque curve. If one linearizes (1.122) with respect to the angular velocity of the motor, one finds:

$$\ddot{M} + 2s_K \dot{M}\Omega + \Omega^2 s_K M = 2M_K s_K s\Omega^2. \quad (1.124)$$

For the "static case" ($\ddot{\varphi} = 0$, $\dot{M} = 0$, $\ddot{M} = 0$), it follows from (1.122):

$$M = 2M_K \frac{s_K s}{s_K^2 + s^2}. \quad (1.125)$$

This equation by KLOSS applies to approximately constant angular velocities $\dot{\varphi}$, that is, under steady-state operating conditions when $\ddot{\varphi} \ll \Omega^2$. (1.125) can be linearized if $s/s_K \ll 1$. This leaves the following equation for the motor characteristic:

$$M = 2M_K \frac{s}{s_K} = M_0 \left(1 - \frac{\dot{\varphi}}{\Omega}\right). \quad (1.126)$$

The dependence of the excitation on the velocity is described by the characteristics of the motors and machines, see Fig. 1.26. When starting the motor according to Fig. 1.27, the characteristic is traversed starting from $n = 0$, passing through the breakdown torque M_K to the nominal torque M_N and to the nominal speed $n_N \approx 970$ rpm. When short-circuit rotors are switched on, currents ranging from four to eight times the nominal current are flowing. The characterisitcs of wound-rotor motors are changed by switching start-up resistors into the rotor circuit.

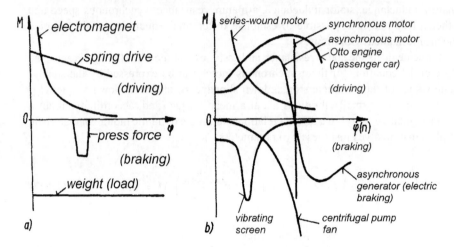

Fig. 1.26 Moment characteristics of various machines; **a)** Machines with a conservative input torque $M(\varphi)$; **b)** Machines with an autonomous drive motor $M(\dot\varphi)$

Figure 1.27 shows the group of characteristics of the motor of a large open-cast mining machine at various rotor resistances. In accordance with the zigzag line in the figure, the motor is started up by switching to another rotor resistance whenever a specific speed is reached. Large dynamic torques that can destroy the drive shaft occur during switching on and switching over [4].

It is difficult to model the curves of the forces and torques that are generated in technological processes such as forming, cutting, pressing as well as in many processes in textile and packaging machines. Usually experimental studies provide the required parameter values.

In many cases, the friction moment of a machine is not known since it depends on many influencing variables in addition to the coefficients of friction in the various joints. It is known that the friction moments (and thus the required input power) are larger when a machine is cold than when it is run in and warm. The absolute magnitude can be determined by measuring the motor output since the motor torque is proportional to the motor current.

It will be shown in Sect. 2.3 that the bearing and joint forces in a machine are proportional to the square of the driving velocity. In addition, there are friction forces that are independent of the motor speed due to static loads. It can therefore be expected that the friction moment that acts on the drive can be described by the following relationship:

$$M_\mathrm{R} = M_1 + M_2 \left(\frac{\dot\varphi}{\omega_0}\right)^2 \tag{1.127}$$

1.5 Characteristic Excitation Parameters

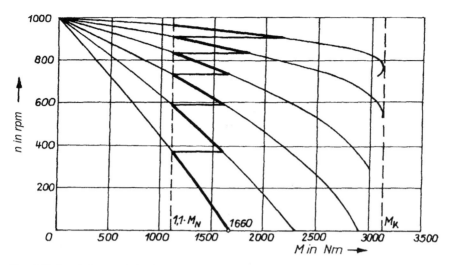

Fig. 1.27 Speed characteristic of an asynchronous motor without slip rings for various rotor resistances with drawn-in start-up characteristic

where ω_0 is the initial angular velocity and M_1 and M_2 are parameter values that can be determined from a coast-down test, see Problem P1.6.

Extreme impact loads can be captured by the **coefficient of restitution** k. This model corresponds to a sudden force that is unlimited in magnitude and acts for a infinitesimally small time, the integral $I = \int F(t)dt$ being finite. Such a model does not apply a force to the point of impact but instead takes into account the initial velocity that was caused by the force. In Sects. 4.3.3.2 and 6.5.3, it is shown that such models are justified if the duration of the impact force is much shorter than the shortest period of an essential mode shape of the impacted vibration system.

The coefficient of restitution k of Newton's collision theory applies to the vertical impact of a mass onto a solid pad and represents the ratio of rebound velocity v_1 (rebounding back to height h_1) to impact velocity v_0 (dropping from initial height h_0), which can be determined by a rebound test. The following equation applies:

$$k = \left|\frac{v_1}{v_0}\right| = \sqrt{\frac{h_1}{h_0}} < 1. \tag{1.128}$$

An impact is perfectly plastic if $k = 0$ and perfectly elastic if $k = 1$. The coefficient of restitution is related to the material parameters of the bodies that come into contact. It equals the square root of the rebound elasticity according to DIN 53 512. One can assume the following approximate values for the following pairings of materials:

Material pairing	Coefficient of restitution k
Wood/wood	0.5
Elastomers	0.4 to 0.8
Forging processes	0.2 to 0.8
Glass/glass	0.9

In some cases, it can be justified to calculate the loss of mechanical energy using the coefficient of restitution (partially plastic deformation). Even if two different perfectly elastic bodies collide, the coefficient of restitution may not be simply related to the loss of mechanical energy.

The coefficient of restitution can be used to calculate the movement of colliding bodies, e. g. the sequences of movements of followers, chisels, hammers, rams, and hydraulic pile drivers. It is difficult to calculate the loads inside of the bodies based on the coefficient of restitution. For specific processes, e. g. forging, empirical values can be analyzed for optimum sizing of hammer foundations, see Sect. 3.3.2.

1.5.3 Problems P1.4 to P1.6

P1.4 Determining Damping from a Resonance Curve

Figure 1.28 shows the measured resonance curve of a vibration system when excited with a constant force amplitude. Use the half width method to determine the damping ratio D.

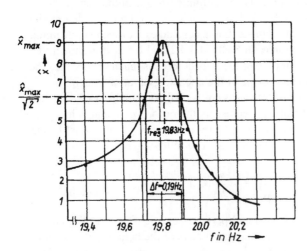

Fig. 1.28 Resonance curve for determining damping

P1.5 Relationship between Coefficient of Restitution and Damping Ratio

According to (1.128), the coefficient of restitution is defined as the ratio of rebound velocity v_1 to impact velocity v_0. Model the impact site using a spring-damper system (c, b), calculate the rebound velocity of a mass m and express the coefficient of restitution k as a function of the damping ratio D.

1.5 Characteristic Excitation Parameters

P1.6 Friction Moment from a Coast-Down Test

Determine the curves of the angular velocity $\dot\varphi(\varphi)$ and $\dot\varphi(t)$ during the coast-down of a machine with the reduced moment of inertia J_m for the speed-dependent moment of friction $M_R = M_1 + M_2 \, (\dot\varphi/\omega_0)^2$. Use this result to calculate the coast-down angle φ_1 at which standstill is reached and the coast-down time t_1 that elapses until standstill. Express the coefficients M_1 and M_2 as functions of J_m and the measured variables φ_1 and t_1. Determine the explicit forms for φ_1 and t_1 for the special case $M_2 = 0$.

1.5.4 Solutions S1.4 to S1.6

S1.4 First, determine the resonance amplitude $\hat{x}_{\max}$. It amounts to about 9 units in Fig. 1.28. Now determine the value $\hat{x}_{\max}/\sqrt{2}$ for the half width and plot it into the resonance curve. The points where it intersects with the resonance curve result in a frequency difference of $\Delta f = 0.19$ Hz. According to (1.110), the following applies to the damping constant:

$$b = m(\Omega_2 - \Omega_1) = 2\pi m \Delta f \tag{1.129}$$

and to the damping ratio according to (1.89):

$$D = b/2m\omega_0. \tag{1.130}$$

ω_0 is the angular frequency of the undampened oscillator, which is equal to the resonance frequency since the damping is that weak. Therefore:

$$D = \frac{2\pi \Delta f}{2m 2\pi f_{res}} = \frac{\Delta f}{2 f_{res}} = 0.0048. \tag{1.131}$$

S1.5 The equation of motion of a dampened oscillator that represents the contact point after the impact of a mass m is expressed according to (1.92):

$$\ddot{q} + 2D\omega_0 \dot{q} + \omega_0^2 q = 0. \tag{1.132}$$

The following initial conditions apply during impact at the contact point:

$$t = 0: \quad q(0) = 0; \quad \dot{q}(0) = v_0. \tag{1.133}$$

The solution is known from (1.94). The following applies with $\omega = \omega_0 \sqrt{1 - D^2}$:

$$q(t) = \frac{v_0}{\omega} \exp(-D\omega_0 t) \sin \omega t \tag{1.134}$$

$$\dot{q}(t) = v_0 \exp(-D\omega_0 t) \left[\cos \omega t - \left(D\frac{\omega_0}{\omega}\right) \sin \omega t\right]. \tag{1.135}$$

As long as $q < 0$ applies, the mass m is in contact with the impact point. We are interested in the velocity at the moment when the mass loses contact. This occurs at the time t_1 when

$$q(t_1) = 0; \quad \sin \omega t_1 = 0, \quad \omega t_1 = \pi. \tag{1.136}$$

At this moment, the velocity is

$$\dot{q}(t_1) = v_0 \exp(-D\omega_0 t_1) \left[\cos \omega t_1 - \left(D\frac{\omega_0}{\omega}\right) \sin \omega t_1\right]$$

$$= v_0 \exp\left(-D\frac{\omega_0 \pi}{\omega}\right)[-1 \; -0] = v_1. \tag{1.137}$$

This is the sought after rebound velocity v_1. The following then applies for the coefficient of restitution, see (1.128):

$$k = \left|\frac{v_1}{v_0}\right| = \exp\left(-D\frac{\omega_0 \pi}{\omega}\right)$$
$$= \exp\left(-\frac{\pi D}{\sqrt{1-D^2}}\right) \approx 1 - \pi D + \frac{1}{2}(\pi D)^2. \quad (1.138)$$

Since the damping ratio is $D \ll 1$, one can expand the exponential function into a Taylor series and use the approximation specified.

Because of (1.138), one could calculate the damping ratio from the coefficient of restitution. It is risky, though, to apply this relationship since both characteristic numbers are really dependent on the velocity, force, and other material parameters of both bodies.

S1.6 For this case, the equation of motion of the machine is:

$$J_m \ddot{\varphi} = \frac{1}{2} J_m \frac{d\dot{\varphi}^2}{d\varphi} = -M_1 - M_2 \left(\frac{\dot{\varphi}}{\omega_0}\right)^2. \quad (1.139)$$

It can be integrated in closed form in both representations with the initial values $\varphi_0 = 0$ and ω_0. When using the dimensionless characteristic parameters,

$$\alpha = \frac{M_2}{M_1} \quad \text{und} \quad \beta = \frac{M_1}{J_m \omega_0^2} \quad (1.140)$$

one finds the solution of the integral derived from the equation above:

$$2\int_0^\varphi d\varphi = \int_{\omega_0}^{\dot\varphi} \frac{J_m d(\dot\varphi^2)}{-M_1 - M_2\left(\frac{\dot\varphi}{\omega_0}\right)^2} = \frac{-1}{\beta}\int_{\omega_0}^{\dot\varphi} \frac{d(\dot\varphi^2)}{\omega_0^2 + \alpha\dot\varphi^2} \quad (1.141)$$

as the inverse function of the position-dependent angular velocity:

$$\dot\varphi(\varphi) = \omega_0 \sqrt{\left(1+\frac{1}{\alpha}\right) e^{-2\alpha\beta\varphi} - \frac{1}{\alpha}}. \quad (1.142)$$

If, however, one starts from $\ddot\varphi = d\dot\varphi/dt$ and the transformation

$$\int_0^t dt = \int_{\omega_0}^{\dot\varphi} \frac{d\dot\varphi}{\ddot\varphi(\dot\varphi^2)} = \int_{\omega_0}^{\dot\varphi} \frac{J_m d\dot\varphi}{-M_1 - M_2\left(\frac{\dot\varphi}{\omega_0}\right)^2} = \frac{-1}{\beta}\int_{\omega_0}^{\dot\varphi} \frac{d\dot\varphi}{\omega_0^2 + \alpha\dot\varphi^2} \quad (1.143)$$

the integration first provides $t = t(\dot\varphi)$ and from there the inverse function with the time-dependent angular velocity

$$\dot\varphi(t) = \omega_0 \sqrt{\frac{1}{\alpha} \tan^2\left(\arctan\sqrt{\alpha} - \beta\sqrt{\alpha}\omega_0 t\right)}. \quad (1.144)$$

The coast-down angle and coast-down time are derived from (1.142) and (1.144) for $\dot\varphi = 0$:

$$\varphi_1 = \frac{\ln(1+\alpha)}{2\alpha\beta}; \quad t_1 = \frac{\arctan\sqrt{\alpha}}{\omega_0 \beta\sqrt{\alpha}}. \quad (1.145)$$

1.5 Characteristic Excitation Parameters

These are two coupled transcendental equations, from which first α and β and then M_1 and M_2 can be calculated from (1.140). If the friction moment M_1 from (1.145) is constant, one finds the following for the special case that $M_2 = 0$ is $\alpha = 0$ and by determining limiting values:

$$\varphi_1 = \frac{1}{2\beta} = \frac{J_m \omega_0^2}{2M_1}; \quad t_1 = \frac{1}{\beta \omega_0} = \frac{J_m \omega_0}{M_1} \tag{1.146}$$

Figure 1.29 shows typical curves for the coast-down movement. The known linear drop over time and the quadratic drop over the angle occur at $M_2 = 0$. The coast-down time shortens with increasing M_2. Under field conditions, the coast-down angle can be determined more accurately than the coast-down time since no such slow increase occurs in the final phase. This behavior can also be seen in Fig. 1.29.

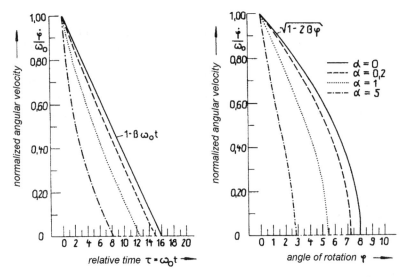

Fig. 1.29 Angular velocity during coast-down at different torque ratios $\alpha = M_2/M_1$ and $\beta = 1/16$

We put theory into practice
Dynamics of Machinery – Structural Dynamics
in

| Mechanical Engineering | Plant Engineering | Automotive Engineering |

- **Analysis and Simulation**
  - Measurement of operational vibrations
  - Experimental Modal Analysis (EMA) and Operational Modal Analysis (OMA)
  - Finite-Element-Analyses, natural frequencies, forced vibrations
  - Simulations of multi-body mechanical systems

- **Vibration Reduction Measures**
  - Vibration isolation
  - Foundation design
  - Development of vibration absorbers
  - Optimisation of machines and plants with regard to vibrations

- **AVR – Active Vibration Reduction**
  - Development of active vibration reduction systems
  - Actuator systems, e. g. piezo-ceramics, voice coils, linear motors, etc.
  - Controller design
  - System integration

AVR: Active control of bending vibrations of a rotor

Wölfel
Beratende Ingenieure

Max-Planck-Str. 15 97204 Höchberg/Germany
Phone +49 931 49 708-600 Fax +49 931 49 708-650
email: wbi@woelfel.de Internet: www.woelfel.de

Chapter 2
Dynamics of Rigid Machines

2.1 Introduction

A "rigid machine" is the simplest calculation model in machine dynamics. It can be defined as a constrained system of rigid bodies, the motion of which is uniquely determined for a given drive motion due to holonomic constraints. This calculation model can be used if the deformations that always exist in reality due to the acting forces are so minor that they have little influence on the motions. It is also assumed that the joints and bearings are ideally backlash-free.

The rigid machine model can be used as the basis for calculating "slow-running" machines, that is, when the lowest natural frequency of the real object under consideration is greater than the largest occurring excitation frequency. The calculation model of the rigid machine can be used for mechanisms with constant transmission ratio such as gearboxes, worm gear systems, belt and chain transmission systems, as well as mechanisms with variable transmission ratio such as linkages, cam drive mechanisms, and geared linkages.

The fundamentals of rigid machine theory go back to works by L. EULER (1707–1783) and J.L. LAGRANGE (1736–1813). When the steam engine was developed, these theories became interesting to mechanical engineers in the second half of the 19th century. Machine designers first used the method of kinetostatics, which considers the inertia forces of moving mechanisms according to d'Alembert's principle as static forces and treats them with known the methods of statics (then primarily graphical methods). The book "Versuch einer grafischen Dynamik" (An Attempt at Graphical Dynamics) by PROELL, which was published in Leipzig in 1874, is an example of the approach taken at that time.

The second volume of "Theoretische Maschinenlehre" (Machine Theory) by F. GRASHOF (1826–1893) that appeared in 1883 also contained fundamentals of machine dynamics. For example, the concept of reduced mass, which was to prove very useful later on, was introduced there. The arising questions about the balancing of masses were first addressed in the book by H. LORENZ (1865–1940) titled "Dynamik der Kurbelgetriebe" (Dynamics of crank mechanisms; Leipzig, 1901).

The works by KARL HEUN (1859–1929), who stressed the mathematical aspects such as the integration of the differential equations), and R. VON MISES (1883–1953) were summarized in 1907 in "Dynamische Probleme der Maschinenlehre" (Dynamic Problems of Machine Theory) so that the rigid machine theory had basically been worked out by the beginning of the 20th century.

For a long time, the authoritative book for mechanical engineers was the one by F. WITTENBAUER (1857–1922) that presented graphical methods suitable for planar mechanisms. The extensions of these methods to spatial mechanisms were provided by K. FEDERHOFER (1885–1960), who published his book "Grafische Kinematik und Kinetostatik des starren räumlichen Systems"(Graphical Kinematics and Kinetostatics of Rigid Spatial Systems) in Vienna in 1928.

These theories have since been included in textbooks on mechanism design, machine dynamics, and mechatronics. The monograph by BIEZENO/GRAMMEL [1] comprehensively discusses the mass balancing of machines. K. MAGNUS [23] wrote a fundamental book on gyroscopes. Rigid machine theory got a fresh impetus when computers emerged and industrial robots raised the question of useful algorithms for calculating rigid-body systems of any given topological structure.

Today, engineers can solve problems in this field using commercial software products, and the mathematical or numerical methods these solutions are based on do not have to be known in detail. However, a user of such programs has to get familiar with the basic ideas of model generation to understand what can be calculated using them and what cannot [3]. The Deutsche Forschungsgemeinschaft (German Research Foundation; DFG) has sponsored a program of key research projects titled "Multibody dynamics", the results of which were summarized by W. SCHIEHLEN in an anthology [28].

2.2 Kinematics of a Rigid Body

2.2.1 Coordinate Transformations

To describe the position and motions of a rigid body in space, it is useful to introduce a body-fixed, that is, a coordinate system $\{\overline{O}; \xi, \eta, \zeta\}$, that moves along with the body in addition to a fixed coordinate system $\{O; x, y, z\}$, see Fig. 2.1. In kinematics and kinetics, there are geometrical and physical quantities that are defined by several components. These are vectors and tensors whose components in the body-fixed system differ from those in the fixed system. They can be converted when switching coordinate systems using specific rules (coordinate transformation).

It is advantageous, in conjunction with other problems of machine dynamics, to use matrix notation for representing the kinematic and dynamic relationships between *vectors* and *tensors*. Vectors are described by bold letters and column matrices and tensors by bold letters and quadratic (3×3) matrices. To apply matrix calculus for the vector product (or cross product), each vector is assigned a skew symmetri-

2.2 Kinematics of a Rigid Body

cal matrix that is labeled by the letter of the vector and a superscript **tilde** (~). For example, the three coordinates of a vector are arranged as follows:

$$\boldsymbol{r} = \begin{bmatrix} x \\ y \\ z \end{bmatrix} \;;\quad \tilde{\boldsymbol{r}} = \begin{bmatrix} 0 & -z & y \\ z & 0 & -x \\ -y & x & 0 \end{bmatrix} \;;$$

$$\boldsymbol{F} = \begin{bmatrix} F_x \\ F_y \\ F_z \end{bmatrix} \;;\quad \tilde{\boldsymbol{F}} = \begin{bmatrix} 0 & -F_z & F_y \\ F_z & 0 & -F_x \\ -F_y & F_x & 0 \end{bmatrix}. \tag{2.1}$$

The **cross product** of the position vector r and the force vector F yields the moment vector and can be expressed as a matrix product as follows:

$$\begin{bmatrix} 0 & -z & y \\ z & 0 & -x \\ -y & x & 0 \end{bmatrix} \cdot \begin{bmatrix} F_x \\ F_y \\ F_z \end{bmatrix} = \begin{bmatrix} -zF_y + yF_z \\ zF_x - xF_z \\ -yF_x + xF_y \end{bmatrix} = \begin{bmatrix} M^O_x \\ M^O_y \\ M^O_z \end{bmatrix} \tag{2.2}$$

The following applies, therefore, in matrix notation (using the tilde operator)

$$\tilde{\boldsymbol{r}} \cdot \boldsymbol{F} = \quad -\tilde{\boldsymbol{F}} \cdot \boldsymbol{r} = \quad \boldsymbol{M}^O \tag{2.3}$$

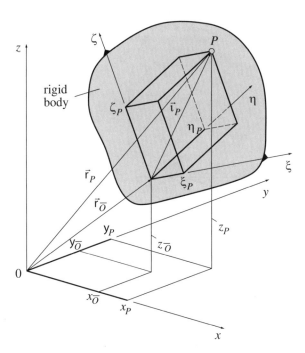

Fig. 2.1 Definition of coordinate systems and position vectors of rigid bodies

2 Dynamics of Rigid Machines

In an inertial system, point O is the origin of a Cartesian coordinate system comprising the fixed coordinate directions x, y, and z, see Fig. 2.1. The position of an arbitrary point P of a rigid body is uniquely characterized by the three coordinates x_P, y_P, and z_P that are summarized in position vector $\boldsymbol{r}_P = (x_P, y_P, z_P)^\mathrm{T}$. A **body-fixed reference point** $\overline{O}$ is selected as the origin of a body-fixed ξ-η-ζ coordinate system. It has the fixed coordinates $\boldsymbol{r}_{\overline{O}} = (x_{\overline{O}}, y_{\overline{O}}, z_{\overline{O}})^\mathrm{T}$. With respect to the directions of the fixed coordinate system, the same point P that is viewed from this reference point $\overline{O}$ has the components

$$\boldsymbol{l}_P = \boldsymbol{r}_P - \boldsymbol{r}_{\overline{O}} = (\Delta x, \Delta y, \Delta z)^\mathrm{T} = (x_P - x_{\overline{O}}, y_P - y_{\overline{O}}, z_P - z_{\overline{O}})^\mathrm{T}. \quad (2.4)$$

In the body-fixed system, the position of that same point P can be given by the following components:

$$\bar{\boldsymbol{l}} = (\xi_P, \eta_P, \zeta_P)^\mathrm{T}. \quad (2.5)$$

The components of $\boldsymbol{l}_P$ and $\bar{\boldsymbol{l}}_P$ differ when the two coordinate systems do not have parallel axes. The index P that characterizes an arbitrary point in the body is omitted in other calculations, i.e. $\bar{\boldsymbol{l}}_P \equiv \bar{\boldsymbol{l}} = (\xi, \eta, \zeta)^\mathrm{T}$. The coordinates x, y, and z as well as ξ, η, and ζ then refer to all points that belong to the rigid body.

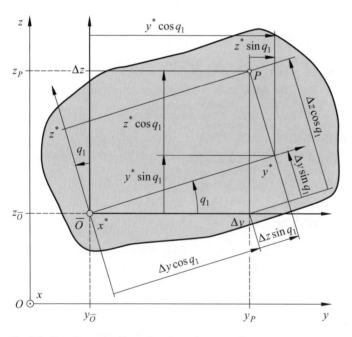

Fig. 2.2 Coordinate transformation for a planar rotation

For motions in three-dimensional space, the rigid body has three rotational degrees of freedom in addition to the three translational degrees of freedom. These rotational ones can be described by three angles. First, the relationships between the

2.2 Kinematics of a Rigid Body

coordinates of a point for a **planar rotation** about the angle q_1 are established. The following relationships can be read for projections of the body-fixed coordinates onto the body-fixed axes (and vice versa) from Fig. 2.2:

$$\Delta x = 1 \cdot x^*, \qquad\qquad x^* = 1 \cdot \Delta x,$$
$$\Delta y = \cos q_1 \cdot y^* - \sin q_1 \cdot z^*, \qquad y^* = \cos q_1 \cdot \Delta y + \sin q_1 \cdot \Delta z, \qquad (2.6)$$
$$\Delta z = \sin q_1 \cdot y^* + \cos q_1 \cdot z^*, \qquad z^* = -\sin q_1 \cdot \Delta y + \cos q_1 \cdot \Delta z.$$

These two times three equations each correspond to a matrix equation if one introduces the vector $\boldsymbol{l}^* = (x^*, y^*, z^*)^\mathrm{T}$ and the rotational transformation matrix $\boldsymbol{A}_1$:

$$\boldsymbol{A}_1 = \begin{bmatrix} 1 & 0 & 0 \\ 0 & \cos q_1 & -\sin q_1 \\ 0 & \sin q_1 & \cos q_1 \end{bmatrix}; \qquad \boldsymbol{A}_1^\mathrm{T} = \begin{bmatrix} 1 & 0 & 0 \\ 0 & \cos q_1 & \sin q_1 \\ 0 & -\sin q_1 & \cos q_1 \end{bmatrix}. \qquad (2.7)$$

The rotational transformation matrix is orthonormal, resulting in the following relationship with the unit matrix $\boldsymbol{E}$

$$\boldsymbol{A}_1(\boldsymbol{A}_1)^\mathrm{T} = \boldsymbol{E}; \qquad (\boldsymbol{A}_1)^\mathrm{T} = (\boldsymbol{A}_1)^{-1}. \qquad (2.8)$$

Thus the relationships (2.6) are as follows:

$$\boldsymbol{l} = \boldsymbol{A}_1 \boldsymbol{l}^*; \qquad \boldsymbol{l}^* = (\boldsymbol{A}_1)^\mathrm{T} \boldsymbol{l}. \qquad (2.9)$$

In a **spatial rotation**, the elements of the rotational transformation matrix $\boldsymbol{A}$ depend on three angles to be defined specifically. The **cardan angles** that are designated q_1, q_2 and q_3 here are used to describe the position of the body, see Fig. 2.3.

The fixed x-y-z system and the body-fixed ξ-η-ζ system coincide in the initial position. When rotating the outer frame about the angle of rotation q_1, the x axis is retained ($x = x^*$), and the plane of the inner frame becomes the new y^*-z^* plane. The angle of rotation q_2 describes the rotation of the inner frame about the positive y^* axis that coincides with the y^{**} axis, so that the x^{**}-z^{**} plane, which is perpendicular to that, takes a new position. The angle of rotation q_3 finally relates to the z^{**} axis that coincides with the ζ axis of the body-fixed coordinate system. The ξ-η plane is perpendicular to $z^{**} = \zeta$. The body-fixed ξ-η-ζ system takes an arbitrary rotated position relative to the fixed x-y-z system upon these three rotations.

Each of the three rotations in itself represents a planar rotation about another axis. According to Fig. 2.3, the following three elementary rotations apply:

$$\boldsymbol{l} = \boldsymbol{A}_1 \boldsymbol{l}^*; \qquad \boldsymbol{l}^* = \boldsymbol{A}_2 \boldsymbol{l}^{**}; \qquad \boldsymbol{l}^{**} = \boldsymbol{A}_3 \bar{\boldsymbol{l}}. \qquad (2.10)$$

The rotational transformation matrices for rotations about the y^* and z^{**} axes can be obtained starting from the projections onto the other planes in analogy to Fig. 2.2. The following matrices implement the rotations about the angles q_2 and q_3 of the respective axes:

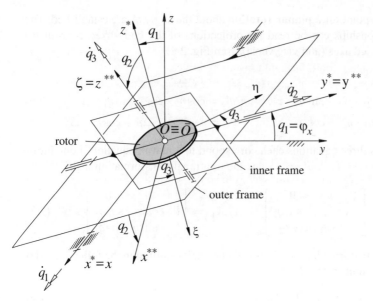

Fig. 2.3 Description of a spatial rotation

$$A_2 = \begin{bmatrix} \cos q_2 & 0 & \sin q_2 \\ 0 & 1 & 0 \\ -\sin q_2 & 0 & \cos q_2 \end{bmatrix}; \quad A_3 = \begin{bmatrix} \cos q_3 & -\sin q_3 & 0 \\ \sin q_3 & \cos q_3 & 0 \\ 0 & 0 & 1 \end{bmatrix}. \quad (2.11)$$

If these relationships are inserted into each other according to (2.10), one obtains:

$$l = A_1 l^* = A_1 A_2 l^{**} = A_1 A_2 A_3 \bar{l} = A\bar{l}; \quad \bar{l} = A^\mathrm{T} l \quad (2.12)$$

and thus the transformation matrix for the spatial rotation

$$A = A_1 A_2 A_3. \quad (2.13)$$

Performing the multiplication using the matrices known from (2.7) and (2.11) according to (2.13), one obtains:

$$A = \begin{bmatrix} \cos q_2 \cos q_3 & -\cos q_2 \sin q_3 & \sin q_2 \\ \sin q_1 \sin q_2 \cos q_3 + \cos q_1 \sin q_3 & -\sin q_1 \sin q_2 \sin q_3 + \cos q_1 \cos q_3 & -\sin q_1 \cos q_2 \\ -\cos q_1 \sin q_2 \cos q_3 + \sin q_1 \sin q_3 & \cos q_1 \sin q_2 \sin q_3 + \sin q_1 \cos q_3 & \cos q_1 \cos q_2 \end{bmatrix}. \quad (2.14)$$

This matrix can not only be used to transform the position vectors, but all vectors, so that the following follows from (2.1) for the force and moment vectors:

2.2 Kinematics of a Rigid Body

$$F = A\overline{F}; \qquad \overline{F} = A^T F; \qquad M^O = A\overline{M}^O; \qquad \overline{M}^O = A^T M^O. \quad (2.15)$$

The components of a vector are typically designated by the same letter as the vector itself but not printed in bold face, and given the indices x, y, z in the fixed system of reference. In the body-fixed system, a bar is added to the bold letter, and its components are given the indices ξ, η and ζ. For example, the same (physical) force vector has the components $\boldsymbol{F} = (F_x, F_y, F_z)^T$ or $\overline{\boldsymbol{F}} = (F_\xi, F_\eta, F_\zeta)^T$ depending on the reference system.

The elements of the rotational transformation matrix $\boldsymbol{A}$ are nonlinear functions of the three angles of rotation q_1, q_2 and q_3, cf. (2.14). The transformation matrices are **orthonormal** for spatial rotations as well, so the following applies in analogy to (2.8):

$$A^T = (A)^{-1}; \qquad A^T A = A A^T = E. \quad (2.16)$$

The coordinates of a body point with respect to fixed directions relative to the origin O can be calculated in matrix notation from those of the reference point $\overline{O}$ and the body-fixed coordinates:

$$\boldsymbol{r} = \boldsymbol{r}_{\overline{O}} + \boldsymbol{l} = \boldsymbol{r}_{\overline{O}} + A\overline{\boldsymbol{l}}. \quad (2.17)$$

2.2.2 Kinematic Parameters

The superordinate term **kinematic parameters** should be understood as to include velocity, acceleration, angular velocity, and angular acceleration. It is used in a similar way as "generalized force" that represents the superordinate term for force and moment/torque.

The three components of absolute **velocity** $\boldsymbol{v} = \dot{\boldsymbol{r}} = (\dot{x}, \dot{y}, \dot{z})^T$ with respect to the fixed directions are derived using the time derivative of $\boldsymbol{r}$ from (2.17). Using (2.12), they are:

$$\boldsymbol{v} = \dot{\boldsymbol{r}} = \dot{\boldsymbol{r}}_{\overline{O}} + \dot{\boldsymbol{l}} = \dot{\boldsymbol{r}}_{\overline{O}} + \dot{A}\overline{\boldsymbol{l}} = \dot{\boldsymbol{r}}_{\overline{O}} + \dot{A}A^T \boldsymbol{l}. \quad (2.18)$$

The time derivative of the relationship given on the right in (2.16) is:

$$\frac{d(AA^T)}{dt} = \dot{A}A^T + A(A^T)\dot{} = \dot{A}A^T + (\dot{A}A^T)^T = \tilde{\omega} + \tilde{\omega}^T = \boldsymbol{o}. \quad (2.19)$$

This equation gives rise to the following conclusion: since the sum of a matrix with its transpose is zero only if the matrix itself is skew symmetrical, the product of $\dot{A}A^T = \tilde{\omega}$ in (2.19) must be a skew symmetrical matrix. One therefore may, according to (2.1), assign a (3 × 3) matrix $\tilde{\omega}$ of the tensor of angular velocity to the vector of angular velocity $\boldsymbol{\omega} = (\omega_x, \omega_y, \omega_z)^T$:

$$\dot{A}A^T = \tilde{\omega} = \begin{bmatrix} 0 & -\omega_z & \omega_y \\ \omega_z & 0 & -\omega_x \\ -\omega_y & \omega_x & 0 \end{bmatrix} = -(\tilde{\omega})^T = -A(A^T)\dot{}. \quad (2.20)$$

The vectors and matrices of the angular velocity are transformed between the fixed and body-fixed coordinate systems in analogy to (2.12):

$$\omega = A\overline{\omega}; \qquad \overline{\omega} = A^T\omega, \qquad (2.21)$$
$$\tilde{\omega} = A\overline{\tilde{\omega}}A^T; \qquad \overline{\tilde{\omega}} = A^T\tilde{\omega}A. \qquad (2.22)$$

It is useful for some applications to determine the body-fixed components of the angular velocity at once. The following applies as an alternative to (2.20)

$$A^T\dot{A} = \overline{\tilde{\omega}} = \begin{bmatrix} 0 & -\omega_\zeta & \omega_\eta \\ \omega_\zeta & 0 & -\omega_\xi \\ -\omega_\eta & \omega_\xi & 0 \end{bmatrix} = \overline{\tilde{\omega}}. \qquad (2.23)$$

The **angular velocity** is the same at all points of the rigid body, it cannot be assigned to any one point of the rigid body and cannot be calculated by a time derivative of an angle for general spatial motion. Without going into more detail, note that each spatial motion of a rigid body can be described as a screw motion about an instantaneous axis, at which the vectors of velocity and angular velocity are proportional to each other ($v = k\omega$).

The magnitude ω of the angular velocity is derived from both (2.20) and (2.23):

$$\omega = \sqrt{\omega^T\omega} = \sqrt{\omega_x^2 + \omega_y^2 + \omega_z^2} = \sqrt{\omega_\xi^2 + \omega_\eta^2 + \omega_\zeta^2} = \sqrt{\overline{\omega}^T\overline{\omega}}. \qquad (2.24)$$

(2.18) can be written in two ways, taking into account (2.20):

$$v = \dot{r} = \dot{r}_{\overline{O}} + \tilde{\omega}l = \dot{r}_{\overline{O}} - \tilde{l}\omega. \qquad (2.25)$$

Therefore, the components of the velocity vector are as follows:

$$\begin{array}{ll} \dot{x} = \dot{x}_{\overline{O}} & -\omega_z\Delta y + \omega_y\Delta z \\ \dot{y} = \dot{y}_{\overline{O}} + \omega_z\Delta x & -\omega_x\Delta z. \\ \dot{z} = \dot{z}_{\overline{O}} - \omega_y\Delta x + \omega_x\Delta y & \end{array} \qquad (2.26)$$

The velocity can also be expressed as a function of the body-fixed components:

$$v = \dot{r} = \dot{r}_{\overline{O}} + \tilde{\omega}A\overline{l} = \dot{r}_{\overline{O}} + A\overline{\tilde{\omega}}\,\overline{l} = \dot{r}_{\overline{O}} - A\overline{\tilde{l}}\,\overline{\omega}. \qquad (2.27)$$

The components of the ω vector can, in actual problems, not only be determined from (2.20) but also by projecting the angular velocity vector onto the directions of the respective system of reference.

Differentiation of the velocity in (2.25) and (2.27) finally provides the absolute **acceleration** of a point in the following form:

$$\ddot{r} = \dot{v} = \frac{\mathrm{d}(\dot{r}_{\overline{O}} + \tilde{\omega}l)}{\mathrm{d}t} = \ddot{r}_{\overline{O}} + \dot{\tilde{\omega}}l + \tilde{\omega}\dot{l}$$

$$\ddot{r} = \dot{v} = \ddot{r}_{\overline{O}} + A(\dot{\overline{\tilde{\omega}}} + \overline{\tilde{\omega}}\,\overline{\tilde{\omega}})\overline{l} = \ddot{r}_{\overline{O}} + \left(\dot{\tilde{\omega}} + \tilde{\omega}\tilde{\omega}\right)l, \qquad (2.28)$$

where the first line uses coordinates in the fixed system and the second line references the rotational transformation matrix and the coordinates in the body-fixed coordinate system.

2.2.3 Kinematics of the Gimbal-Mounted Gyroscope

Figure 2.3 shows a rigid body that can freely rotate (in the sketched massless apparatus) about three axes in space. A rigid body that can only perform three rotations is called a **gyroscope**. The position of the gyroscope can be uniquely described using cardan angles $q = (q_1, q_2, q_3)^\text{T}$, see. (2.14).

If one uses the matrix A and its time derivative $\dot{A}$ to determine the product according to (2.20), one obtains the matrix of the tensor of angular velocity $\tilde{\omega}$, which contains the following components as matrix elements:

$$\begin{aligned} \omega_x &= \dot{q}_1 & &+\dot{q}_3 \sin q_2 \\ \omega_y &= & \dot{q}_2 \cos q_1 &- \dot{q}_3 \sin q_1 \cos q_2 \\ \omega_z &= & \dot{q}_2 \sin q_1 &+ \dot{q}_3 \cos q_1 \cos q_2. \end{aligned} \qquad (2.29)$$

For the body-fixed reference system, the components of the angular velocity are derived according to (2.21) using the $\overline{\omega} = A^\text{T} \omega$ transformation after performing matrix multiplication and some manipulations of the trigonometric functions:

$$\begin{aligned} \omega_\xi &= \dot{q}_1 \cos q_2 \cos q_3 + \dot{q}_2 \sin q_3 \\ \omega_\eta &= -\dot{q}_1 \cos q_2 \sin q_3 + \dot{q}_2 \cos q_3 \\ \omega_\zeta &= \dot{q}_1 \sin q_2 & +\dot{q}_3. \end{aligned} \qquad (2.30)$$

The magnitude ω of the angular velocity ω results from (2.24):

$$\omega = \sqrt{\dot{q}_1^2 + \dot{q}_2^2 + \dot{q}_3^2 + 2\dot{q}_1 \dot{q}_3 \sin q_2}. \qquad (2.31)$$

It follows from here that the magnitude of the angular velocity for gimbal-mounting according to Fig. 2.3, at constant angular velocities is not constant in general, but only if either ($\dot{q}_1$ or $\dot{q}_2$ or $\dot{q}_3$) is zero.

If the condition is met that the fixed x-y-z-system and the body-fixed ξ-η-ζ-system coincide in their initial positions, the angle coordinates

$$\varphi_x \approx q_1; \qquad \varphi_y \approx q_2; \qquad \varphi_z \approx q_3, \qquad (2.32)$$

can be introduced for small angles of rotation

$$|q_1| \ll 1; \qquad |q_2| \ll 1; \qquad |q_3| \ll 1 \qquad (2.33)$$

that describe "small motions". Since $\sin q_k \approx q_k$ and $\cos q_k \approx 1$, it follows from (2.14) and (2.20):

$$A \approx \begin{bmatrix} 1 & -\varphi_z & \varphi_y \\ \varphi_z & 1 & -\varphi_x \\ -\varphi_y & \varphi_x & 1 \end{bmatrix} \quad (2.34)$$

and

$$\dot{A}A^\mathrm{T} = \tilde{\omega} \approx \begin{bmatrix} 0 & -\dot{\varphi}_z & \dot{\varphi}_y \\ \dot{\varphi}_z & 0 & -\dot{\varphi}_x \\ -\dot{\varphi}_y & \dot{\varphi}_x & 0 \end{bmatrix} \quad (2.35)$$

These simple matrices are very popular for some applications. Whoever uses them should be aware of their scope, however.

2.2.4 Problems P2.1 and P2.2

P2.1 Kinematics of a Pivoted Rotor

In many engineering applications, rotating bodies are pivoted about an axis that is perpendicular to their longitudinal axis. Such motions occur during cornering of wheels (bicycle, motorbike, car), when pivoting a carousel, a drilling machine, or a running spin drier. Fig. 2.4 shows a model that describes such motions. A frame that can rotate about the x axis in a fixed x-y-z reference system carries a rotor that can be pivoted therein. Consider the motion of the rotor and one of its points.

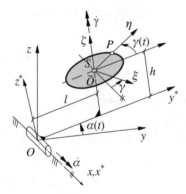

Fig. 2.4 Rotor in a pivotable frame

Given:
    Frame dimensions l and h
    Distance of a point P in the rotor η_P
    Pivoting angle $\alpha(t)$
    Angle of rotation of the rotor $\gamma(t)$

Find:

1. Components of the angular velocity $\bar{\omega}$ and angular acceleration $\dot{\bar{\omega}}$ of the rotor in the co-rotating ξ-η-ζ coordinate system
2. Components of the absolute velocity $\boldsymbol{v}_P$ of point P.

P2.2 Edge Mill

An edge mill is a machine for comminuting, grinding, or mixing, (e. g. ores, coal, clay, corn, etc.) in which rollers are guided along an angular path that compress and comminute the material to be ground.

Figure 2.5 shows the grindstone modeled as a homogeneous cylinder with a center of gravity that is guided at a distance ξ_S along a planar circular path around the fixed vertical z axis. The ξ axis of the grindstone is pivoted horizontally at the angular speed $\dot{\varphi}(t)$. Pure rolling of the center plane of the roller at the grinding level is assumed.

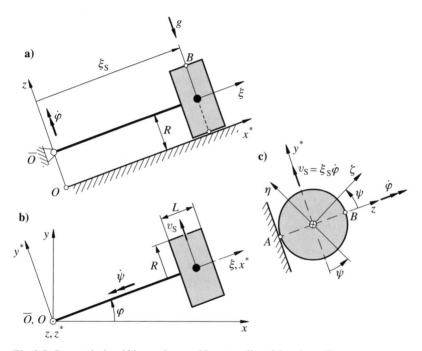

Fig. 2.5 Geometrical and kinematic quantities at a roller of the edge mill

Given:
- Roller radius $\qquad R$
- Distance to the center of gravity ξ_S
- Angular velocity of the axle $\quad \dot{\varphi}(t)$

Find:

1. Rotational transformation matrix $\boldsymbol{A}$
2. Angular velocity vector both in fixed ($\boldsymbol{\omega}$) and in body-fixed ($\overline{\boldsymbol{\omega}}$) coordinate directions
3. Velocity and acceleration distribution along $\overline{AB}$

2.2.5 Solutions S2.1 and S2.2

S2.1 The system in Fig. 2.4 is a special case of the gimbal-mounted gyroscope with regard to the rotational motion, see Fig. 2.3. The body-fixed components of the angular velocity of the body result from (2.30) with $\alpha = q_1$, $\beta = q_2 = 0$ and $\gamma = q_3$ as follows:

$$\bar{\boldsymbol{\omega}} = [\omega_\xi, \omega_\eta, \omega_\zeta]^\mathrm{T} = [\dot\alpha\cos\gamma, -\dot\alpha\sin\gamma, \dot\gamma]^\mathrm{T}. \tag{2.36}$$

The components of the angular acceleration $\dot{\bar{\boldsymbol{\omega}}}$ are the derivatives with respect to time:

$$\begin{aligned}\dot\omega_\xi &= \ddot\alpha\cos\gamma - \dot\alpha\dot\gamma\sin\gamma\\ \dot\omega_\eta &= -\ddot\alpha\sin\gamma - \dot\alpha\dot\gamma\cos\gamma\\ \dot\omega_\zeta &= \ddot\gamma\end{aligned} \tag{2.37}$$

The position of point P is described, according to (2.17), using the body-fixed reference point O, the rotational transformation matrix $\boldsymbol{A}$, and the coordinate in the body-fixed system:

$$\boldsymbol{r}_P = \boldsymbol{r}_{\overline{O}} + \boldsymbol{l}_P = \boldsymbol{r}_{\overline{O}} + \boldsymbol{A}\bar{\boldsymbol{l}}_P. \tag{2.38}$$

The following applies:

$$\boldsymbol{r}_{\overline{O}} = \begin{bmatrix} 0 \\ l\cos\alpha - h\sin\alpha \\ l\sin\alpha + h\cos\alpha \end{bmatrix}; \quad \bar{\boldsymbol{l}}_P = \begin{bmatrix} 0 \\ \eta_P \\ 0 \end{bmatrix}. \tag{2.39}$$

The matrix $\boldsymbol{A}$ is either determined from the product of matrix $\boldsymbol{A}_1$ in (2.7) and matrix $\boldsymbol{A}_3$ in (2.11), or as a special case of (2.14) for $q_2 = 0$:

$$\boldsymbol{A} = \boldsymbol{A}_1\boldsymbol{A}_3 = \begin{bmatrix} \cos\gamma & -\sin\gamma & 0 \\ \cos\alpha\sin\gamma & \cos\alpha\cos\gamma & -\sin\alpha \\ \sin\alpha\sin\gamma & \sin\alpha\cos\gamma & \cos\alpha \end{bmatrix}. \tag{2.40}$$

Using (2.18) or (2.27), the velocity of point P is:

$$\boldsymbol{v}_P = \dot{\boldsymbol{r}}_{\overline{O}} + \dot{\boldsymbol{A}}\bar{\boldsymbol{l}}_P = \dot{\boldsymbol{r}}_{\overline{O}} + \boldsymbol{A}\tilde{\bar{\boldsymbol{\omega}}}\bar{\boldsymbol{l}}_P. \tag{2.41}$$

After inserting (2.39) and a few calculation steps, the result is

$$\begin{bmatrix} \dot x_P \\ \dot y_P \\ \dot z_P \end{bmatrix} = \dot\alpha\begin{bmatrix} 0 \\ -l\sin\alpha - h\cos\alpha \\ l\cos\alpha - h\sin\alpha \end{bmatrix} \\ + \eta_P\begin{bmatrix} -\dot\gamma\cos\gamma \\ -\dot\gamma\cos\alpha\sin\gamma - \dot\alpha\sin\alpha\cos\gamma \\ -\dot\gamma\sin\alpha\sin\gamma + \dot\alpha\cos\alpha\cos\gamma \end{bmatrix}. \tag{2.42}$$

The first expression results from the differentiation of $\boldsymbol{r}_{\overline{O}}$ from (2.39). The second expression is either obtained by differentiation of $\boldsymbol{A}$ and $\bar{\boldsymbol{l}}_P$ from (2.39), or by multiplying $\boldsymbol{A}$ from (2.40) and $\tilde{\bar{\boldsymbol{\omega}}}$ from (2.36) and $\bar{\boldsymbol{l}}_P$.

The acceleration $\boldsymbol{a}_P$ could be determined by another differentiation of $\boldsymbol{v}_P$ with respect to time. This involves terms with the factors $\ddot\alpha$, $\ddot\gamma$, $\dot\alpha^2$, $\dot\gamma^2$, and $\dot\alpha\dot\gamma$.

S2.2 A body-fixed ξ-η-ζ system with its origin in the bearing $\overline{O}$ was introduced in addition to the fixed x-y-z reference system, see Fig. 2.5. In the initial position, the ξ axis is parallel to the x axis, the η axis is parallel to the y axis, and the ζ axis is parallel to the z axis. Pure rolling of a circular cone would be possible if there were a conic support surface. However, a planar

2.2 Kinematics of a Rigid Body

support surface and a circular cylinder are used in the calculations here, assuming that the circle rolls off at the distance ξ_S on the $z = 0$ plane.

The stipulation $\psi(\varphi = 0) = 0$ means for the angles (when pure rolling takes place at the distance ξ_S) that the constraint

$$\xi_S \varphi = R\psi; \qquad \psi = \frac{\xi_S \varphi}{R}. \tag{2.43}$$

applies when taking into account the positive directions of rotation as defined Like in (2.10), the rotation matrix $\boldsymbol{A}$ can be determined using the sequence of the two elementary rotations φ and ψ:

$$\begin{bmatrix} x \\ y \\ z \end{bmatrix} = \underbrace{\begin{bmatrix} \cos\varphi & -\sin\varphi & 0 \\ \sin\varphi & \cos\varphi & 0 \\ 0 & 0 & 1 \end{bmatrix}}_{=\boldsymbol{A}_\varphi} \begin{bmatrix} x^* \\ y^* \\ z^* \end{bmatrix}, \quad \begin{bmatrix} x^* \\ y^* \\ z^* \end{bmatrix} = \underbrace{\begin{bmatrix} 1 & 0 & 0 \\ 0 & \cos\psi & \sin\psi \\ 0 & -\sin\psi & \cos\psi \end{bmatrix}}_{=\boldsymbol{A}_\psi} \begin{bmatrix} \xi \\ \eta \\ \zeta \end{bmatrix} \tag{2.44}$$

If these equations are combined, the rotational transformation relations between the fixed x, y, z"= coordinates and the co-rotating ξ, η, ζ coordinates are obtained:

$$\boldsymbol{A} = \boldsymbol{A}_\varphi \cdot \boldsymbol{A}_\psi = \begin{bmatrix} \cos\varphi & -\sin\varphi\cos\psi & -\sin\varphi\sin\psi \\ \sin\varphi & \cos\varphi\cos\psi & \cos\varphi\sin\psi \\ 0 & -\sin\psi & \cos\psi \end{bmatrix} \tag{2.45}$$

The components of the angular velocity vector with respect to the body-fixed coordinate directions can be read from Fig. 2.5, keeping in mind that the angular velocity $\dot\psi$ opposes the positive ξ-direction, and if the angular velocity $\dot\varphi$ pointing in the z-direction is decomposed into its components in the directions of η and ζ using the angle of rotation ψ:

$$\bar{\boldsymbol{\omega}} = \begin{bmatrix} \omega_\xi \\ \omega_\eta \\ \omega_\zeta \end{bmatrix} = \begin{bmatrix} -\dot\psi \\ -\dot\varphi\sin\psi \\ \dot\varphi\cos\psi \end{bmatrix} = \dot\varphi \begin{bmatrix} -\xi_S/R \\ -\sin\left(\frac{\xi_S\varphi}{R}\right) \\ \cos\left(\frac{\xi_S\varphi}{R}\right) \end{bmatrix} \tag{2.46}$$

The result specified last in (2.46) was obtained by inserting the constraint (2.43) (also in differentiated form). Using the rotation matrix $\boldsymbol{A}$, the following is obtained for the fixed directions:

$$\boldsymbol{\omega} = [\omega_x, \omega_y, \omega_z]^\mathrm{T} = \boldsymbol{A}\bar{\boldsymbol{\omega}} = \dot\varphi \left[-\frac{\xi_S}{R}\cos\varphi, -\frac{\xi_S}{R}\sin\varphi, 1 \right]^\mathrm{T} \tag{2.47}$$

From this follows the skew symmetrical matrix $\tilde{\boldsymbol{\omega}}$, see (2.20), which is needed later:

$$\tilde{\boldsymbol{\omega}} = \dot\varphi \begin{bmatrix} 0 & -1 & -\frac{\xi_S\sin\varphi}{R} \\ 1 & 0 & \frac{\xi_S\cos\varphi}{R} \\ \frac{\xi_S\sin\varphi}{R} & -\frac{\xi_S\cos\varphi}{R} & 0 \end{bmatrix} \tag{2.48}$$

The components of the angular acceleration vector are obtained by differentiating (2.47) with respect to time:

$$\dot{\boldsymbol{\omega}} = \ddot{\varphi} \begin{bmatrix} -\frac{\xi_S}{R}\cos\varphi \\ -\frac{\xi_S}{R}\sin\varphi \\ 1 \end{bmatrix} + \dot{\varphi}^2 \frac{\xi_S}{R} \begin{bmatrix} \sin\varphi \\ -\cos\varphi \\ 0 \end{bmatrix} \qquad (2.49)$$

Differentiation of $\bar{\boldsymbol{\omega}}$ from (2.47), due to the special property $\dot{\bar{\boldsymbol{\omega}}} = \dot{\boldsymbol{\omega}}$, provides the components of the angular acceleration vector, with respect to the body-fixed directions:

$$\dot{\bar{\boldsymbol{\omega}}} = \dot{\boldsymbol{\omega}} = \ddot{\varphi} \begin{bmatrix} -\xi_S/R \\ -\sin\left(\frac{\xi_S\varphi}{R}\right) \\ \cos\left(\frac{\xi_S\varphi}{R}\right) \end{bmatrix} - \dot{\varphi}^2 \frac{\xi_S}{R} \begin{bmatrix} 0 \\ \cos\left(\frac{\xi_S\varphi}{R}\right) \\ \sin\left(\frac{\xi_S\varphi}{R}\right) \end{bmatrix} \qquad (2.50)$$

For the velocity and acceleration distributions, which are most appropriately determined using Euler's kinematic equations, the coordinates of the points located on the line segment $\overline{AB}$ are required, to which

$$\boldsymbol{r} = [x,\ y,\ z]^T = [\xi_S\cos\varphi,\ \xi_S\sin\varphi,\ z]^T \qquad (2.51)$$

applies according to Fig. 2.5. If the center of gravity S of the roller (and not $\overline{O}$!) with

$$\boldsymbol{r}_S = \begin{bmatrix} \xi_S\cos\varphi \\ \xi_S\sin\varphi \\ R \end{bmatrix},\quad \dot{\boldsymbol{r}}_S = \xi_S\dot{\varphi}\begin{bmatrix} -\sin\varphi \\ \cos\varphi \\ 0 \end{bmatrix},$$

$$\ddot{\boldsymbol{r}}_S = \xi_S\ddot{\varphi}\begin{bmatrix} -\sin\varphi \\ \cos\varphi \\ 0 \end{bmatrix} - \xi_S\dot{\varphi}^2\begin{bmatrix} \cos\varphi \\ \sin\varphi \\ 0 \end{bmatrix} \qquad (2.52)$$

is selected as reference point, the following applies to the velocity distribution according to (2.25) with $\tilde{\boldsymbol{\omega}}$ according to (2.48):

$$\dot{\boldsymbol{r}} = \dot{\boldsymbol{r}}_S + \tilde{\boldsymbol{\omega}}(\boldsymbol{r} - \boldsymbol{r}_S)$$

$$= \xi_S\dot{\varphi}\begin{bmatrix} -\sin\varphi \\ \cos\varphi \\ 0 \end{bmatrix} + \xi_S\dot{\varphi}\begin{bmatrix} 0 & -\frac{1}{\xi_S} & -\frac{\sin\varphi}{R} \\ \frac{1}{\xi_S} & 0 & \frac{\cos\varphi}{R} \\ \frac{\sin\varphi}{R} & -\frac{\cos\varphi}{R} & 0 \end{bmatrix}\begin{bmatrix} 0 \\ 0 \\ z-R \end{bmatrix} \qquad (2.53)$$

$$= \xi_S\dot{\varphi}\begin{bmatrix} -\sin\varphi \\ \cos\varphi \\ 0 \end{bmatrix}\frac{z}{R}.$$

Differentiation of (2.51) with respect to time would not have yielded the correct result (which can easily be verified by comparing with (2.53)) since the radius vector $\boldsymbol{r}$ in (2.50) describes the instantaneous position of points that are not body-fixed, that is, the points located on $\overline{AB}$ are always different body points when the cylinder is rolling.

2.2 Kinematics of a Rigid Body

From (2.52) the linear velocity distribution already known from the rolling wheel can be seen: it equals zero at contact point A ($z = 0$) and has its maximum at the upper point B ($z = 2R$).

The following then applies to the acceleration distribution along $\overline{AB}$, see (2.28):

$$\ddot{\boldsymbol{r}} = \ddot{\boldsymbol{r}}_S + (\dot{\tilde{\boldsymbol{\omega}}} + \tilde{\boldsymbol{\omega}}\tilde{\boldsymbol{\omega}})(\boldsymbol{r} - \boldsymbol{r}_S)$$

$$\ddot{\boldsymbol{r}} = \xi_S \ddot{\varphi} \begin{bmatrix} -\sin\varphi \\ \cos\varphi \\ 0 \end{bmatrix} - \xi_S \dot{\varphi}^2 \begin{bmatrix} \cos\varphi \\ \sin\varphi \\ 0 \end{bmatrix}$$

$$+ \left(\ddot{\varphi}\frac{\xi_S}{R} \begin{bmatrix} -\sin\varphi \\ \cos\varphi \\ 0 \end{bmatrix} + \dot{\varphi}^2 \frac{\xi_S}{R}\begin{bmatrix} -2\cos\varphi \\ -2\sin\varphi \\ -\frac{\xi_S}{R} \end{bmatrix} \right)(z - R) \quad (2.54)$$

$$= \xi_S \ddot{\varphi} \frac{z}{R}\begin{bmatrix} -\sin\varphi \\ \cos\varphi \\ 0 \end{bmatrix} - \xi_S \dot{\varphi}^2 \begin{bmatrix} \left(2\frac{z}{R} - 1\right)\cos\varphi \\ \left(2\frac{z}{R} - 1\right)\sin\varphi \\ \frac{\xi_S}{R}\left(\frac{z}{R} - 1\right) \end{bmatrix}$$

The special case $\dot{\varphi} = \Omega = $ const. (i.e. $\ddot{\varphi} \equiv 0$) is shown for the two variants $\xi_S/R = 1$ and $\xi_S/R = 2$ in Fig. 2.6. It can be seen here that the distribution in conjunction with mass causes a moment effect that appears as gyroscopic moment in kinetics, see Sect. 2.3.3.

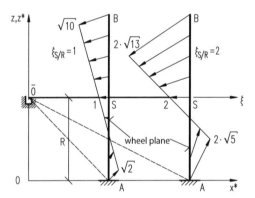

Fig. 2.6 Acceleration distribution $\ddot{\boldsymbol{r}}/(R\Omega^2)$ for two ratios ξ_S/R

In Fig. 2.6, the instantaneous axis of rotation that passes through points $\overline{O}$ and A is shown as a dashed line.

2.3 Kinetics of the Rigid Body

2.3.1 Kinetic Energy and Moment of Inertia Tensor

The kinetic energy of a mass element $\mathrm{d}m$ that moves at a velocity of $\boldsymbol{v} = \dot{\boldsymbol{r}}$ relative to a fixed reference system amounts to

$$\mathrm{d}W_{\mathrm{kin}} = \frac{1}{2}\mathrm{d}mv^2 = \frac{1}{2}\mathrm{d}m(\dot{\boldsymbol{r}})^{\mathrm{T}}\dot{\boldsymbol{r}}. \tag{2.55}$$

The kinetic energy of a rigid body is determined by integrating over the entire body with a velocity distribution according to (2.25), resulting in

$$W_{\mathrm{kin}} = \int \mathrm{d}W_{\mathrm{kin}} = \frac{1}{2}\int v^2 \mathrm{d}m = \frac{1}{2}\int (\dot{\boldsymbol{r}}_{\overline{O}} - \tilde{\boldsymbol{l}}\boldsymbol{\omega})^{\mathrm{T}}(\dot{\boldsymbol{r}}_{\overline{O}} - \tilde{\boldsymbol{l}}\boldsymbol{\omega})\mathrm{d}m. \tag{2.56}$$

Engineering mechanics proves that it is useful to select the center of gravity S as the body-fixed reference point. Starting from (2.56), the kinetic energy of an arbitrarily moving rigid body can then be expressed as follows:

$$\begin{aligned}W_{\mathrm{kin}} &= \frac{1}{2}m\boldsymbol{v}_{\mathrm{S}}^{\mathrm{T}}\boldsymbol{v}_{\mathrm{S}} + \frac{1}{2}\overline{\boldsymbol{\omega}}^{\mathrm{T}}\overline{\boldsymbol{J}}^{\mathrm{S}}\overline{\boldsymbol{\omega}} \\ &= \frac{1}{2}m(\dot{x}_{\mathrm{S}}^2 + \dot{y}_{\mathrm{S}}^2 + \dot{z}_{\mathrm{S}}^2) + \frac{1}{2}(J_{\xi\xi}^{\mathrm{S}}\omega_{\xi}^2 + J_{\eta\eta}^{\mathrm{S}}\omega_{\eta}^2 + J_{\zeta\zeta}^{\mathrm{S}}\omega_{\zeta}^2) \\ &\quad + J_{\xi\eta}^{\mathrm{S}}\omega_{\xi}\omega_{\eta} + J_{\eta\zeta}^{\mathrm{S}}\omega_{\eta}\omega_{\zeta} + J_{\zeta\xi}^{\mathrm{S}}\omega_{\zeta}\omega_{\xi}.\end{aligned} \tag{2.57}$$

The mass of the body is $m = \int \mathrm{d}m$, and $\boldsymbol{v}_{\mathrm{S}} = (\dot{x}_{\mathrm{S}}, \dot{y}_{\mathrm{S}}, \dot{z}_{\mathrm{S}})^{\mathrm{T}}$ is the absolute velocity of the center of gravity S. The translational kinetic energy can be obtained from it, together with the mass m of the body. The vector $\overline{\boldsymbol{\omega}} = (\omega_{\xi}, \omega_{\eta}, \omega_{\zeta})^{\mathrm{T}}$ of the angular velocity is related to the body-fixed ξ-η-ζ coordinate system, see Sects. 2.2.2 and 2.2.3. The rotational energy that corresponds to the other terms in (2.57) can be expressed using the moment of inertia J_{kk}^{S} with respect to the instantaneous axis of rotation labeled with index k. The following applies:

$$J_{kk}^{\mathrm{S}}\omega^2 = J_{\xi\xi}^{\mathrm{S}}\omega_{\xi}^2 + J_{\eta\eta}^{\mathrm{S}}\omega_{\eta}^2 + J_{\zeta\zeta}^{\mathrm{S}}\omega_{\zeta}^2 + 2(J_{\xi\eta}^{\mathrm{S}}\omega_{\xi}\omega_{\eta} + J_{\eta\zeta}^{\mathrm{S}}\omega_{\eta}\omega_{\zeta} + J_{\zeta\xi}^{\mathrm{S}}\omega_{\zeta}\omega_{\xi}). \tag{2.58}$$

Thus the kinetic energy is simply

$$W_{\mathrm{kin}} = \frac{1}{2}mv_{\mathrm{S}}^2 + \frac{1}{2}J_{kk}^{\mathrm{S}}\omega^2. \tag{2.59}$$

The moment of inertia J_{kk}^{S} refers to the direction of the instantaneous axis of rotation, see the application in Sect. 1.2.4. The direction of the instantaneous axis of rotation k can be described with respect to the directions of the body-fixed reference system using the angles α_k, β_k, and γ_k, see Fig. 2.7a. The components of angular velocity with respect to this direction are

2.3 Kinetics of the Rigid Body

$$\omega_\xi = \omega \cos \alpha_k; \qquad \omega_\eta = \omega \cos \beta_k; \qquad \omega_\zeta = \omega \cos \gamma_k. \tag{2.60}$$

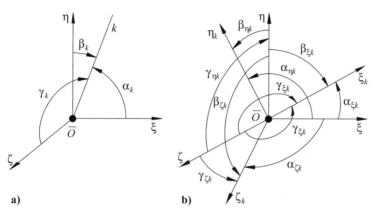

Fig. 2.7 Directional angles within the rigid body; **a)** Identification of a direction k (such as $k =$ I, II, III); **b)** Identification of the position of a ξ_k-η_k-ζ_k system in the ξ-η-ζ system

The dependence of the moment of inertia J_{kk}^S on these angles can be determined from (2.58) and (2.60):

$$\begin{aligned} J_{kk}^S &= J_{\xi\xi}^S \cos^2 \alpha_k + J_{\eta\eta}^S \cos^2 \beta_k + J_{\zeta\zeta}^S \cos^2 \gamma_k \\ &+ 2(J_{\xi\eta}^S \cos \alpha_k \cos \beta_k + J_{\eta\zeta}^S \cos \beta_k \cos \gamma_k + J_{\zeta\xi}^S \cos \gamma_k \cos \alpha_k). \end{aligned} \tag{2.61}$$

The matrix of the moment of inertia tensor with respect to the center of gravity is defined in the body-fixed system by:

$$\overline{\overline{J}}^S = \int (\overline{\overline{i}} - \overline{\overline{i}}_S)^T (\overline{\overline{i}} - \overline{\overline{i}}_S) \mathrm{d}m. \tag{2.62}$$

The integration refers to the entire body volume and is theoretically performed by a triple integral that is hardly solved in closed form in practice because the bodies comprise so many shapes. In most cases, the moment of inertia tensor is calculated by subdividing a body into elementary bodies with small masses (or into bodies with a known moment of inertia tensor) from the CAD programs for any machine parts. When it comes to actually existing parts, it is recommended to determine the moment of inertia tensor from experimental results and to check the theoretical values, see in this context Sect. 1.2.4.

The mass m characterizes the body's inertia during translational motions. Similarly, the moment of inertia tensor captures the respective properties of a rigid body with regard to rotational motions. If the center of gravity is the origin ($S = \overline{O}$), the matrix of the moment of inertia tensor is:

$$\overline{\boldsymbol{J}}^{\mathrm{S}} = \begin{bmatrix} J_{\xi\xi}^{\mathrm{S}} & J_{\xi\eta}^{\mathrm{S}} & J_{\xi\zeta}^{\mathrm{S}} \\ J_{\eta\xi}^{\mathrm{S}} & J_{\eta\eta}^{\mathrm{S}} & J_{\eta\zeta}^{\mathrm{S}} \\ J_{\zeta\xi}^{\mathrm{S}} & J_{\zeta\eta}^{\mathrm{S}} & J_{\zeta\zeta}^{\mathrm{S}} \end{bmatrix}$$

$$= \begin{bmatrix} \int(\eta^2+\zeta^2)\mathrm{d}m & -\int \eta\xi\,\mathrm{d}m & -\int \xi\zeta\,\mathrm{d}m \\ -\int \eta\xi\,\mathrm{d}m & \int(\xi^2+\zeta^2)\mathrm{d}m & -\int \eta\zeta\,\mathrm{d}m \\ -\int \xi\zeta\,\mathrm{d}m & -\int \eta\zeta\,\mathrm{d}m & \int(\xi^2+\eta^2)\mathrm{d}m \end{bmatrix}. \quad (2.63)$$

This matrix is symmetrical. The elements on the principal diagonal are called moments of inertia (in short "rotating masses"), and the elements outside the principal diagonal are called **products of inertia** (also centrifugal moments). Unlike the moments of inertia, the products of inertia can be zero or negative. The moments of inertia are measures of the rotational inertia of a body and the products of inertia are measures of the body's tendency to change its axis of rotation when rotating. They characterize the unsymmetrical mass distribution of the body, see also (2.75).

With respect to the fixed directions, the moment of inertia tensor according to transformation (2.22) results from

$$\boldsymbol{J}^{\mathrm{S}} = \boldsymbol{A}\overline{\boldsymbol{J}}^{\mathrm{S}}\boldsymbol{A}^{\mathrm{T}}. \quad (2.64)$$

In general, it is variable, that is, it depends on the angles of rotation in accordance with the rotational transformation matrix $\boldsymbol{A}$. The matrix

$$\boldsymbol{J}^{\mathrm{S}} = \begin{bmatrix} J_{xx}^{\mathrm{S}} & J_{xy}^{\mathrm{S}} & J_{xz}^{\mathrm{S}} \\ J_{xy}^{\mathrm{S}} & J_{yy}^{\mathrm{S}} & J_{yz}^{\mathrm{S}} \\ J_{xz}^{\mathrm{S}} & J_{yz}^{\mathrm{S}} & J_{zz}^{\mathrm{S}} \end{bmatrix}. \quad (2.65)$$

corresponds to it. When the angles are small, as in (2.33), the linear approximation

$$\boldsymbol{J}^S \approx \overline{\boldsymbol{J}}^{\mathrm{S}} - \overline{\boldsymbol{J}}^{\mathrm{S}}\tilde{\boldsymbol{q}} + \tilde{\boldsymbol{q}}\,\overline{\boldsymbol{J}}^{\mathrm{S}}. \quad (2.66)$$

follows from (2.64) due to $\boldsymbol{A} \approx \boldsymbol{E} + \tilde{\boldsymbol{q}}$ (see (2.34)). The moment of inertia tensor is frequently used in the even more simplified form $\boldsymbol{J}^S \approx \overline{\boldsymbol{J}}^{\mathrm{S}}$ (i.e. $\boldsymbol{A} \approx \boldsymbol{E}$) when calculating linear oscillations, see Sects. 1.2.4, 3.2.2 and 5.2.3.

The static moments and moment of inertia tensor are dependent on the point of reference chosen. The **center of gravity** (or *center of mass*) S is a special body-fixed (reference) point. Its position is defined by the fact that the static moments with respect to it are zero. If it is the origin of the body-fixed ξ-η-ζ coordinate system, the conditions

$$\int \xi\,\mathrm{d}m = \int \eta\,\mathrm{d}m = \int \zeta\,\mathrm{d}m = 0.$$

must be satisfied. When switching from any reference point $\overline{O}$ to the center of gravity S relative to **parallel axes**, the conversion of the matrix elements of the moment

2.3 Kinetics of the Rigid Body

of inertia tensor is governed by the parallel-axis theorem (**Steiner's theorem**):

$$\overline{\overline{J}}^O = \overline{\overline{J}}^S + m(\tilde{\overline{l}}_S)^T \tilde{\overline{l}}_S, \quad \text{with} \tag{2.67}$$

$$\tilde{\overline{l}} = \begin{bmatrix} 0 & -\zeta_S & \eta_S \\ \zeta_S & 0 & -\xi_S \\ -\eta_S & \xi_S & 0 \end{bmatrix}. \tag{2.68}$$

Thus the moments of inertia always have their smallest values with respect to axes of gravity because "Steiner terms" are added for other axes. The components of the moment of inertia tensor also change when switching to **rotated body-fixed axes** ξ_1-η_1-ζ_1. In analogy to (2.22), when it comes to the transformation between fixed and body-fixed directions, a transformation matrix can be used that is designated as A^*. The directional cosines in A^* then refer to the nine angles, $\alpha_{\xi k}$ to $\gamma_{\zeta k}$, that are defined as in Fig. 2.7b between the ξ-η-ζ system and the ξ_k-η_k-ζ_k system that has the same point $\overline{O}$ as its body-fixed origin.

The moment of inertia tensor (here, exemplarily with respect to the center of gravity S – it applies in analogy to each body-fixed point) is transformed when rotating in the body-fixed reference system with the matrix

$$A^* = \begin{bmatrix} \cos \alpha_{\xi k} & \cos \beta_{\xi k} & \cos \gamma_{\xi k} \\ \cos \alpha_{\eta k} & \cos \beta_{\eta k} & \cos \gamma_{\eta k} \\ \cos \alpha_{\zeta k} & \cos \beta_{\zeta k} & \cos \gamma_{\zeta k} \end{bmatrix} \tag{2.69}$$

by the following matrix multiplications:

$$\overline{\overline{J}}^S = A^* J^{*S} A^{*T}; \qquad J^{*S} = A^{*T} \overline{\overline{J}}^S A^*. \tag{2.70}$$

The matrix $\overline{\overline{J}}^S$ contains the components known from (2.63) while the components in J^{*S} relate to the ξ_k-η_k-ζ_k system that is rotated inside the rigid body.

For each reference point $\overline{O}$ there is a special coordinate system with three directions that are perpendicular to one another and for which the moment of inertia tensor becomes a diagonal matrix. These axes are called principal axes. The transformation onto the **central principal axes**, is of particular interest if the center of gravity is selected as the reference point ($O = S$). The principal axes are identified by the Roman numerals I, II, and III. The **principal moments of inertia** J_I^S, J_{II}^S and J_{III}^S are the three eigenvalues of the eigenvalue problem

$$(\overline{\overline{J}}^S - J^S E)a = o, \tag{2.71}$$

which can be solved numerically using known software if parameter values are given. The three eigenvectors associated with the eigenvalues

$$a_k = [\cos \alpha_k, \cos \beta_k, \cos \gamma_k]^T; \qquad k = \text{I, II, III} \tag{2.72}$$

contain, as elements, the directional cosines, which define the orientation of the principal axes with spatial angles α_k, β_k and γ_k relative to the original ξ-η-ζ system, see also Fig. 2.7a. They are normalized in such a way that

$$(a_I)^T a_I = (a_{II})^T a_{II} = (a_{III})^T a_{III} = 1;$$
$$(a_I)^T a_{II} = (a_{II})^T a_{III} = (a_{III})^T a_I = 0 \tag{2.73}$$

and $\det(a_I, a_{II}, a_{III}) = 1$. The transformation matrix

$$A_H^* = [a_I, a_{II}, a_{III}] \tag{2.74}$$

is formed from these three eigenvectors, so that the moment of inertia tensor for the central principal axes can be expressed as follows:

$$\hat{J}^S = A_H^{*T} \overline{J}^S A_H^* = \begin{bmatrix} J_I^S & 0 & 0 \\ 0 & J_{II}^S & 0 \\ 0 & 0 & J_{III}^S \end{bmatrix}. \tag{2.75}$$

The deviation moments with respect to the principal axes are zero. Symmetry axes of a homogeneous rigid body are principal axes. Table 5.2 specifies the moments of inertia with respect to the three principal axes for some bodies of revolution.

The components of the angular velocity with respect to the principal axes are derived from:

$$\omega_H = A_H^{*T} \overline{\omega} = [\omega_I, \omega_{II}, \omega_{III}]^T. \tag{2.76}$$

The expression for the kinetic energy from (2.57) becomes simpler when reference can be made to the principal axes:

$$W_{\text{kin}} = \frac{1}{2} m v_S^T v_S + \frac{1}{2} (\omega_H)^T \hat{J}^S \omega_H$$
$$= \frac{1}{2} m(\dot{x}_S^2 + \dot{y}_S^2 + \dot{z}_S^2) + \frac{1}{2}(J_I^S \omega_I^2 + J_{II}^S \omega_{II}^2 + J_{III}^S \omega_{III}^2). \tag{2.77}$$

If the motion is a rotation about a body point $\overline{O}$ that is fixed in space, the kinetic energy can simply be expressed using the moment of inertia tensor $\overline{J}^{\overline{O}}$ with respect to this point, see (2.67):

$$W_{\text{kin}} = \frac{1}{2}(\omega_H)^T \overline{J}^{\overline{O}} \omega_H = \frac{1}{2}(J_I^{\overline{O}} \omega_I^2 + J_{II}^{\overline{O}} \omega_{II}^2 + J_{III}^{\overline{O}} \omega_{III}^2). \tag{2.78}$$

The principal directions in (2.77) are generally different from those in (2.78), which is why the components of the angular velocity with respect to S and $\overline{O}$ differ as well.

2.3.2 Principles of Linear Momentum and of Angular Momentum

The principle of linear momentum and the principle of conservation of angular momentum are fundamental laws that reveal the interconnection of force quantities and motion quantities of a rigid body.

The principle of linear momentum states that the center of gravity S accelerates ($\ddot{\boldsymbol{r}}_S$) as if the resultant $\boldsymbol{F}$ of the external forces (both the applied forces and the reaction forces) acted on it and as if the mass m was concentrated in S. With respect to a fixed reference system, it is:

$$m\ddot{\boldsymbol{r}}_S = \boldsymbol{F} \qquad (2.79)$$

and for the components in the fixed reference system:

$$m\ddot{x}_S = F_x; \qquad m\ddot{y}_S = F_y; \qquad m\ddot{z}_S = F_z. \qquad (2.80)$$

Newton's second law can also be converted using (2.15) and (2.28) into

$$m\ddot{\boldsymbol{r}}_S = m\left[\ddot{\boldsymbol{r}}_{\overline{O}} + \boldsymbol{A}(\dot{\widetilde{\boldsymbol{\omega}}} + \widetilde{\overline{\boldsymbol{\omega}}}\,\widetilde{\overline{\boldsymbol{\omega}}})\overline{\boldsymbol{l}}_S\right] = m\left[\ddot{\boldsymbol{r}}_{\overline{O}} + (\dot{\widetilde{\boldsymbol{\omega}}} + \widetilde{\boldsymbol{\omega}}\,\widetilde{\boldsymbol{\omega}})\boldsymbol{l}_S\right] = \boldsymbol{F} = \boldsymbol{A}\overline{\boldsymbol{F}} \qquad (2.81)$$

so that it takes the following form for fixed components, see (2.16):

$$m\left[\boldsymbol{A}^{\mathrm{T}}\ddot{\boldsymbol{r}}_{\overline{O}} + (\dot{\widetilde{\overline{\boldsymbol{\omega}}}} + \widetilde{\overline{\boldsymbol{\omega}}}\,\widetilde{\overline{\boldsymbol{\omega}}})\overline{\boldsymbol{l}}_S\right] = \overline{\boldsymbol{F}}. \qquad (2.82)$$

The angular momentum of a mass element $\mathrm{d}m$ with respect to a fixed reference point O is the product of the components of its velocity and their perpendicular distances from the axes that pass through the reference point

$$\mathrm{d}\boldsymbol{L}^O = \mathrm{d}m\,\widetilde{\boldsymbol{r}}\dot{\boldsymbol{r}}. \qquad (2.83)$$

The angular momentum of a rigid body that is arbitrarily moving in space can be found by integration over the entire body:

$$\boldsymbol{L}^O = \int \widetilde{\boldsymbol{r}}\dot{\boldsymbol{r}}\,\mathrm{d}m = \int \widetilde{\boldsymbol{r}}(\boldsymbol{v}_{\overline{O}} - \widetilde{\boldsymbol{l}}\boldsymbol{\omega})\mathrm{d}m. \qquad (2.84)$$

The **principle of conservation of angular momentum**, formulated by L. EULER in 1750 takes the following form for the fixed reference point O and the fixed directions

$$\frac{\mathrm{d}\boldsymbol{L}^O}{\mathrm{d}t} \equiv \frac{\mathrm{d}}{\mathrm{d}t}\left[m\left(\widetilde{\boldsymbol{r}}_{\overline{O}}\dot{\boldsymbol{r}}_S + (\widetilde{\boldsymbol{l}}_S - \widetilde{\boldsymbol{l}}_{\overline{O}})\dot{\boldsymbol{r}}_{\overline{O}}\right) + \boldsymbol{J}^{\overline{O}}\boldsymbol{\omega}\right] = \boldsymbol{M}^O. \qquad (2.85)$$

The vector of the external moments, i.e. the sum of the applied moments $\boldsymbol{M}^{O(e)}$ and the reaction moments $\boldsymbol{M}^{O(z)}$, includes the components $\boldsymbol{M}^O = [M_x^O, M_y^O, M_z^O]^{\mathrm{T}}$ in the fixed reference system. While the principle of linear momentum is mostly used with respect to fixed coordinates, the principle of conservation of angular momentum is frequently applied with respect to body-fixed directions. Therefore, only the cases of most interest will be presented here in the form of **Euler's gyroscope equations**: If the body-fixed point of reference $\overline{O}$ is not accelerated ($\ddot{\vec{r}}_{\overline{O}} \equiv \mathbf{o}$), the principle of conservation of angular momentum is

$$\overline{\bm{M}}^O_{\text{kin}} \equiv \overline{\tilde{\bm{\omega}}}\, \overline{\bm{J}}^O \overline{\bm{\omega}} + \overline{\bm{J}}^O \dot{\overline{\bm{\omega}}} = \overline{\bm{M}}^O. \qquad (2.86)$$

The **kinetic moment** (or moment) due to rotational inertia is on the left-hand side. $\overline{\bm{M}}^O = [M^O_\xi, M^O_\eta, M^O_\zeta]^T$ is the vector of the resultant external moment in the body-fixed reference system with respect to $\overline{O}$. For an arbitrarily moving center of gravity, the principle of conservation of angular momentum is similar to (2.86):

$$\overline{\bm{M}}^S_{\text{kin}} \equiv \overline{\tilde{\bm{\omega}}}\, \overline{\bm{J}}^S \overline{\bm{\omega}} + \overline{\bm{J}}^S \dot{\overline{\bm{\omega}}} = \overline{\bm{M}}^S. \qquad (2.87)$$

$\overline{\bm{M}}^S = [M^S_\xi, M^S_\eta, M^S_\zeta]^T$ is the vector of the resultant external moment in the body-fixed reference system with respect to S. The reader should state each of the three equations described in (2.87) in detail once. It will become evident that the following form results if the central principal axes are selected as the body-fixed reference system:

$$\begin{aligned}
M^S_{\text{kin I}} &\equiv J^S_{\text{I}} \dot{\omega}_{\text{I}} - (J^S_{\text{II}} - J^S_{\text{III}}) \omega_{\text{II}} \omega_{\text{III}} = M^S_{\text{I}} \\
M^S_{\text{kin II}} &\equiv J^S_{\text{II}} \dot{\omega}_{\text{II}} - (J^S_{\text{III}} - J^S_{\text{I}}) \omega_{\text{III}} \omega_{\text{I}} = M^S_{\text{II}} \\
M^S_{\text{kin III}} &\equiv J^S_{\text{III}} \dot{\omega}_{\text{III}} - (J^S_{\text{I}} - J^S_{\text{II}}) \omega_{\text{I}} \omega_{\text{II}} = M^S_{\text{III}}.
\end{aligned} \qquad (2.88)$$

In addition to the term with the angular acceleration, the kinetic moment contains a term that occurs at **constant angular velocities**: the so-called **gyroscopic moment**. The following statement can be noted about the gyroscopic moment, e. g. due to the term $(J_{\text{I}} - J_{\text{II}}) \omega_{\text{I}} \omega_{\text{II}}$: as a result of inertia, the gyroscopic moment will occur about the respective third principal axis that is perpendicular to the two others. The right-hand rule applies to the direction of the gyroscopic moment: If the thumb and index finger of the right hand point in the direction of the vectors of ω_{I} and ω_{II}, the middle finger points in direction III, about which the gyroscopic moment occurs. The body "wants" to turn in direction III. If it is prevented from this rotation, a reaction moment occurs that acts in opposite direction to direction III. If, for example, a wheel (rotation about horizontal component I) rolls around a bend (vertical component II), the gyroscopic moment acts about perpendicular horizontal axis III in such a way that it exerts additional pressure towards the ground. This rule should be noted and checked in all examples, see, for example, problems P2.1, P2.3 and Sect. 2.3.3.

Like in (2.79) to (2.82), where external forces (applied force $\bm{F}^{(e)}$ and reaction force $\bm{F}^{(z)}$) are on the right and the inertia force is on the left side of the equations, the right sides of (2.85) to (2.88) always contain the **external moments**, and the left sides the **kinetic moments** $\bm{M}_{\text{kin}}$ (or "inertia moments" in analogy to "inertia forces"). External moments can be both **applied moments** $\bm{M}^{(e)}$ (e. g. input torques

2.3 Kinetics of the Rigid Body

or moments of friction) and **reaction moments** $M^{(z)}$, such as reaction forces that are absorbed by the bearings.

When solving problems, a free-body diagram for the rigid body is developed and all force quantities that act on it **from outside** are included. Due to inertia, **inertia forces** and **moments** that are also called kinetic forces F_{kin} and kinetic moments M_{kin}, in accordance with the kinetic energy concept, come "from inside". The inertia forces $F_{\text{kin}} \equiv m\ddot{r}_S$ are entered in the free-body diagram **opposite to the positive coordinate direction** of r_S, the kinetic moments $\overline{M}^S_{\text{kin}} \equiv \overline{\omega}\,\overline{J}^S\overline{\omega} + \overline{J}^S\dot{\overline{\omega}}$ opposite to the positive coordinate direction of the body-fixed ξ-η-ζ system.

The formal identity of (2.86) and (2.87) can also be transferred to (2.88) and the special forms (2.90), (2.92), and (2.93) of these equations discussed below. These will not be specified for the case of a fixed body point $\overline{O}$. As usual in engineering calculations, Newton's second law and the principle of conservation of angular momentum can be stated as six conditions of equilibrium using the directions of the generalized forces shown in the free-body diagram:

$$F^{(e)} + F^{(z)} + (-F_{\text{kin}}) = 0; \qquad \overline{M}^{S(e)} + \overline{M}^{S(z)} + (-\overline{M}^S_{\text{kin}}) = 0. \qquad (2.89)$$

Note their application when solving the problems in Sects. 2.3.3 to 2.3.5, see Figs. 2.8, 2.10, and 2.33.

If the body rotates about a single fixed axis only (which is the ζ axis here, such as for a rigid rotor in rigid bearings), the following equations of motion follow from (2.87) for $\omega_\xi \equiv \omega_\eta \equiv 0$:

$$\begin{aligned} M^S_{\text{kin}\,\xi} &\equiv J^S_{\xi\zeta}\dot{\omega}_\zeta - J^S_{\eta\zeta}\omega^2_\zeta = M^S_\xi \\ M^S_{\text{kin}\,\eta} &\equiv J^S_{\eta\zeta}\dot{\omega}_\zeta + J^S_{\xi\zeta}\omega^2_\zeta = M^S_\eta \\ M^S_{\text{kin}\,\zeta} &\equiv J^S_{\zeta\zeta}\dot{\omega}_\zeta \phantom{+ J^S_{\xi\zeta}\omega^2_\zeta} = M^S_\zeta. \end{aligned} \qquad (2.90)$$

This shows that kinetic moments occur about the ξ and η axes (that is, perpendicular to the axis of rotation ζ) if the angular velocity is constant and the products of inertia are not zero. These kinetic moments must be absorbed by bearing forces perpendicular to the axis of rotation in order to force the fixed axis of rotation.

If the fixed x-y-z system and the body-fixed ξ-η-ζ system coincide in their initial positions, small angles of rotation φ_x, φ_y and φ_z can be introduced for the fixed axes so that the following applies because of (2.32) and (2.35):

$$\omega_\xi \approx \dot{\varphi}_x; \qquad \omega_\eta \approx \dot{\varphi}_y; \qquad \omega_\zeta \approx \dot{\varphi}_z. \qquad (2.91)$$

If the products of the angular velocities are neglected with respect to the angular accelerations, because they are small and of second order, the linearized form of the principle of conservation of angular momentum results from (2.87) taking into account (2.91) with a moment of inertia tensor that does change over time. Due to the small angles, the body-fixed and the fixed components approximately coincide if they were congruent in the initial position:

$$M^S_{\text{kin }\xi} \equiv J^S_{\xi\xi}\ddot{\varphi}_x + J^S_{\xi\eta}\ddot{\varphi}_y + J^S_{\xi\zeta}\ddot{\varphi}_z = M^S_\xi \approx M^S_x$$
$$M^S_{\text{kin }\eta} \equiv J^S_{\xi\eta}\ddot{\varphi}_x + J^S_{\eta\eta}\ddot{\varphi}_y + J^S_{\eta\zeta}\ddot{\varphi}_z = M^S_\eta \approx M^S_y \quad (2.92)$$
$$M^S_{\text{kin }\zeta} \equiv J^S_{\xi\zeta}\ddot{\varphi}_x + J^S_{\eta\zeta}\ddot{\varphi}_y + J^S_{\zeta\zeta}\ddot{\varphi}_z = M^S_\zeta \approx M^S_z.$$

If a body rotates at the "large" angular velocity $\omega_\zeta = \Omega = \text{const.}$, another form of the linearized gyroscope equations follows from (2.87) for $|\omega_\xi| \ll \Omega$ and $|\omega_\eta| \ll \Omega$ when neglecting the products of the small components of the angular velocity:

$$M^S_{\text{kin }\xi} \equiv J^S_{\xi\xi}\dot{\omega}_\xi + J^S_{\xi\eta}\dot{\omega}_\eta - [J^S_{\xi\eta}\omega_\xi + (J^S_{\eta\eta} - J^S_{\zeta\zeta})\omega_\eta]\Omega - J^S_{\eta\zeta}\Omega^2 = M^S_\xi$$
$$M^S_{\text{kin }\eta} \equiv J^S_{\eta\xi}\dot{\omega}_\xi + J^S_{\eta\eta}\dot{\omega}_\eta + [J^S_{\xi\eta}\omega_\eta + (J^S_{\xi\xi} - J^S_{\zeta\zeta})\omega_\xi]\Omega + J^S_{\xi\zeta}\Omega^2 = M^S_\eta \quad (2.93)$$
$$M^S_{\text{kin }\zeta} \equiv J^S_{\zeta\xi}\dot{\omega}_\xi + J^S_{\zeta\eta}\dot{\omega}_\eta + (J^S_{\eta\zeta}\omega_\xi - J^S_{\xi\zeta}\omega_\eta)\Omega \qquad = M^S_\zeta.$$

The principle of conservation of angular momentum is often used in the form of (2.92) or (2.93) if a rigid body is part of a vibration system, see also Sects. 3.2.2 and 5.2.3.

For more detailed information on the theory of gyroscopes and its applications, see [23].

2.3.3 Kinetics of Edge Mills

The kinematics of edge mills were discussed in Sect. 2.2.4 in the solution of problem P2.2 so that this discussion refers to the results obtained there.

The rotating body (grindstone) according to Fig. 2.5, which rolls off along a circular path, exerts a force in addition to its own weight on its base that is due to the gyroscopic effect. The problem is to calculate the required input torque, the normal force and the horizontal force on the grindstone for pure rolling motion for a given function of the pivoting angle $\varphi(t)$.

Given:
Gravitational acceleration	g
Roller radius	R
Roller length	L
Distance to the center of gravity	ξ_S
Time function of the pivoting angle	$\varphi(t)$
Mass of the roller (grindstone)	m
Moments of inertia of the roller with respect to S	$J^S_{\zeta\zeta} = J^S_{\eta\eta} = \dfrac{m(3R^2 + L^2)}{12}$, $J^S_{\xi\xi} = \tfrac{1}{2}mR^2$

Since it is assumed here as in S2.2 that a pure rolling motion occurs at $\xi = \xi_S$, a sliding motion will occur along the base at the other contact points between the grindstone and the plane, which may be a desired effect for grinding. The sliding velocity of the contact points in tangential direction between roller and grinding

2.3 Kinetics of the Rigid Body

plane is
$$v_{\rm rel} = (\xi_{\rm S} - \xi)\dot\varphi. \tag{2.94}$$

To simplify, it is assumed in the calculation model that a vertical normal force $F_{\rm N}$ and a horizontal adhesive force $F_{\rm H}$ only act below the center of gravity of the roller. The frictional forces for $\xi \neq \xi_{\rm S}$ are not taken into account.

The components of the angular velocity of the roller with respect to the body-fixed and fixed directions are known from S2.2 (eqs. (2.47), (2.48)), and so is the angular acceleration ((2.49)).

The body-fixed ξ-η-ζ system corresponds to the system of the principal axis of this symmetrical rigid body. Axis I can be assigned to the ξ coordinate, axis II to the η coordinate, and axis III to the ζ coordinate. Euler's gyroscope equations then result from (2.88) with respect to the fixed body point $\overline{O}$:

$$\begin{aligned} M^{\overline{O}}_{\text{kin}\,\xi} &\equiv J^{\overline{O}}_{\xi\xi}\dot\omega_\xi - (J^{\overline{O}}_{\eta\eta} - J^{\overline{O}}_{\zeta\zeta})\omega_\eta\omega_\zeta = M^{\overline{O}}_\xi \\ M^{\overline{O}}_{\text{kin}\,\eta} &\equiv J^{\overline{O}}_{\eta\eta}\dot\omega_\eta - (J^{\overline{O}}_{\zeta\zeta} - J^{\overline{O}}_{\xi\xi})\omega_\zeta\omega_\xi = M^{\overline{O}}_\eta \\ M^{\overline{O}}_{\text{kin}\,\zeta} &\equiv J^{\overline{O}}_{\zeta\zeta}\dot\omega_\zeta - (J^{\overline{O}}_{\xi\xi} - J^{\overline{O}}_{\eta\eta})\omega_\xi\omega_\eta = M^{\overline{O}}_\zeta. \end{aligned} \tag{2.95}$$

These equations state that there is an equilibrium of the kinetic moments from the rotational inertia of the body with the external moments. The moments of inertia given with respect to the center of gravity have to be transformed to the fixed body point $\overline{O}$ using the parallel-axis theorem, see (2.72). They amount to:

$$\begin{aligned} J^{\overline{O}}_{\zeta\zeta} &= J^{\overline{O}}_{\eta\eta} = J^{\rm S}_{\eta\eta} + m\xi_{\rm S}^2 = J_{\rm a} = \frac{m(3R^2 + L^2 + 12\xi_{\rm S}^2)}{12} \\ J^{\overline{O}}_{\xi\xi} &= J^{\rm S}_{\xi\xi} = J_{\rm p} = \frac{1}{2}mR^2. \end{aligned} \tag{2.96}$$

If one takes into account (2.47) and (2.50), the kinetic moments can be calculated first from (2.95):

$$M^{\overline{O}}_{\text{kin}\,\xi} \equiv J^{\overline{O}}_{\xi\xi}\dot\omega_\xi = -J_{\rm p}\ddot\psi \tag{2.97}$$

$$\begin{aligned} M^{\overline{O}}_{\text{kin}\,\eta} &\equiv J^{\overline{O}}_{\eta\eta}\dot\omega_\eta - (J^{\overline{O}}_{\zeta\zeta} - J^{\overline{O}}_{\xi\xi})\omega_\zeta\omega_\xi \\ &= J_{\rm a}(\ddot\varphi\sin\psi + \dot\varphi\dot\psi\cos\psi) - (J_{\rm a} - J_{\rm p})\dot\varphi\dot\psi\cos\psi \\ &= J_{\rm a}\ddot\varphi\sin\psi + J_{\rm p}\dot\varphi\dot\psi\cos\psi \end{aligned} \tag{2.98}$$

$$\begin{aligned} M^{\overline{O}}_{\text{kin}\,\zeta} &\equiv J^{\overline{O}}_{\zeta\zeta}\dot\omega_\zeta - (J^{\overline{O}}_{\xi\xi} - J^{\overline{O}}_{\eta\eta})\omega_\xi\omega_\eta \\ &= J_{\rm a}(\ddot\varphi\cos\psi - \dot\varphi\dot\psi\sin\psi) + (J_{\rm p} - J_{\rm a})\dot\varphi\dot\psi\sin\psi \\ &= J_{\rm a}\ddot\varphi\cos\psi + J_{\rm p}\dot\varphi\dot\psi\sin\psi. \end{aligned} \tag{2.99}$$

The components in (2.95) ($M^{\overline{O}}_\xi$, $M^{\overline{O}}_\eta$ and $M^{\overline{O}}_\zeta$) of the resultant external moment $M^{\overline{O}}$ result from the external forces that act on the body, i.e. the input torque $M_{\rm an}$, the static weight mg and the reaction forces ($F_{\rm N}$ and $F_{\rm H}$) at the contact point. It

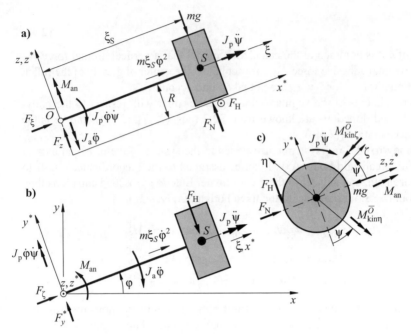

Fig. 2.8 Free-body diagram with forces and moments acting on the roller

is more favorable to write the moment equilibrium about the y^* axis and about the z axis, rather than the moment equilibrium about the η and ζ axes. The following can be derived for the ξ axis both formally from (2.97) to (2.99) and by inspection (Fig. 2.8b and c):

$$\overline{M}^O_{\text{kin}\,\xi} \equiv -J_p\ddot{\psi} = -F_H R \tag{2.100}$$

about the z axis (see Fig. 2.8a and b):

$$\overline{M}^O_{\text{kin}\,\zeta} \cos\psi - \overline{M}^O_{\text{kin}\,\eta} \sin\psi \equiv J_a\ddot{\varphi} = M_{\text{an}} - F_H \xi_S \tag{2.101}$$

and about the y^* axis, see Fig. 2.8a and c:

$$\overline{M}^O_{\text{kin}\,\zeta} \sin\psi + \overline{M}^O_{\text{kin}\,\eta} \cos\psi \equiv J_p\dot{\varphi}\dot{\psi} = (F_N - mg)\xi_S. \tag{2.102}$$

These are the equations for calculating the input torque as well as the reaction forces F_N and F_H. One can express the results using the parameters given in the problem statement – the above form, however, is better suited to recognize the "origin" of each term, such as the gyroscopic effect of the rotor.

The horizontal force that ensures adhesion is

$$F_H = \frac{J_p \ddot{\psi}}{R} = \frac{J_p}{R^2} \xi_S \ddot{\varphi} = \frac{1}{2} m \xi_S \ddot{\varphi}. \tag{2.103}$$

2.3 Kinetics of the Rigid Body

The input torque, that causes the given function $\varphi(t)$ is

$$M_{\text{an}} = \left(J_{\text{a}} + J_{\text{p}}\frac{\xi_{\text{S}}^2}{R^2}\right)\ddot{\varphi} = \frac{m\ddot{\varphi}(3R^2 + L^2 + 18\xi_{\text{S}}^2)}{12} \quad (2.104)$$

and the normal force results from (2.102):

$$F_{\text{N}} = mg + \frac{J_{\text{p}}\dot{\varphi}\dot{\psi}}{\xi_{\text{S}}} = mg\left(1 + \frac{R\dot{\varphi}^2}{2g}\right). \quad (2.105)$$

It is proportional to the radius R and the square of the angular velocity ($\dot{\varphi}^2$), but independent of the length ξ_{S}. It can be considerably larger than the (static) weight.

The influence of the roller radius appears to have been empirically known since ancient times, since one can find such edge mills mostly with great grindstone radii in old mills. The horizontal force F_{H} that results from (2.100) is a reaction force that only occurs with angular accelerations. The individual forces assumed here are the resultants of the actually occurring line loads under the grindstone in both the vertical and horizontal directions. The mechanical behavior of the material to be milled has to be taken into account to calculate their distributions.

2.3.4 Problems P2.3 and P2.4

P2.3 Kinetics of a Pivoting Rotor

The bearing forces of rotating bodies that rotate about their bearing axis and at the same time about an axis perpendicular to this bearing axis are of interest in many engineering applications. Figure 2.4 shows a frame (considered massless) that is pivoted about the x axis in the fixed x-y-z reference system and in which a rotor is pivoted that can rotate about its ζ axis (principal axis III) inside the frame. The center of gravity of the rotor is at the origin of the body-fixed reference system ($\overline{O} = S$). Of interest are general formulae for calculating the moments with respect to the fixed system that occur when rotor and frame are rotated simultaneously.

Given:
    Frame dimensions l and h
    Time functions of angles $\alpha(t)$ and $\gamma(t)$
    Rotor mass m
    Principal moments of inertia of the rotor $J_{\xi\xi}^S = J_{\text{I}}^S = J_{\eta\eta}^S = J_{\text{II}}^S = J_{\text{a}}^S$,
                                                           $J_{\zeta\zeta}^S = J_{\text{III}}^S = J_{\text{p}}^S$

Find:
1. Components of center-of-gravity acceleration
2. Kinetic moments with respect to the center of gravity
3. Moment between rotor and frame (M_{x*}^S, M_{y*}^S, M_{z*}^S)
4. Reaction forces and moments at the origin O (F_y, F_z, M_x^O, M_y^O, M_z^O)
5. Input torques at the rotor (M_{an}^γ) and frame (M_{an}^α)

P2.4 Bearing Forces of a Rotating Body

The bearing forces for the rigid rotor shown in Fig. 2.9 are to be determined. The body has an eccentric center of gravity S with respect to the axis of rotation. A body-fixed ξ-η-ζ coordinate system is used, the origin of which coincides with that of the fixed coordinate system ($O = \overline{O}$) and the ζ axis of which is identical with the fixed z axis.

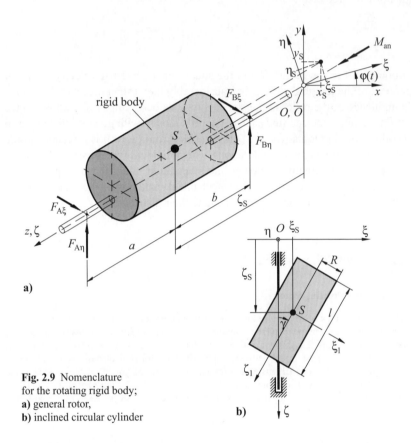

Fig. 2.9 Nomenclature for the rotating rigid body;
a) general rotor,
b) inclined circular cylinder

Note: The topic of "balancing rigid rotors" is discussed in detail in Sect. 2.6.2. The principal purpose of this problem is to illustrate the relationships derived in the previous sections.

2.3 Kinetics of the Rigid Body

Given:
Mass $\quad m$
Moment of inertia $\quad J_{\zeta\zeta}^S$
Products of inertia $\quad J_{\xi\zeta}^S,\ J_{\eta\zeta}^S$
Body-fixed coordinates for the center of gravity $\quad \bar{l}_S = (\xi_S,\ \eta_S,\ \zeta_S)^T$
Angle of rotation $\quad \varphi(t)$
Distances of the bearings from the center of gravity $\quad a,\ b$
Circular cylinder with $\quad$ radius R and length L
Inclination angle of ζ_1 axis relative to the ζ axis $\quad \gamma$
Polar moment of inertia $\quad J_p = \frac{1}{2}mR^2$, see (2.96)
Axial moment of inertia $\quad J_a = m(3R^2 + L^2)/12$

Find:

1. For any function $\varphi(t)$ and a general rotor body
   1.1 Bearing forces $\overline{\boldsymbol{F}}_A$ and $\overline{\boldsymbol{F}}_B$ (body-fixed reference system)
   1.2 Input torque M_{an}
2. Moment of inertia tensor $\boldsymbol{J}^{*S}$ of the circular cylinder that is symmetrically positioned in the ξ_1-η_1-ζ_1 system, inclined in the ξ-ζ plane by the angle γ relative to the axis of rotation, see Fig. 2.9b.

2.3.5 Solutions S2.3 and S2.4

S2.3 The acceleration of the center of gravity can be calculated from the velocities that were determined for a body point in S2.1. For $\eta_P = 0$, $P = S$, and it follows from (2.42):

$$\dot{\boldsymbol{r}}_S = \begin{bmatrix} \dot{x}_S \\ \dot{y}_S \\ \dot{z}_S \end{bmatrix} = \dot{\alpha} \begin{bmatrix} 0 \\ -l\sin\alpha - h\cos\alpha \\ l\cos\alpha - h\sin\alpha \end{bmatrix}. \tag{2.106}$$

The acceleration of the center of gravity, therefore, is

$$\ddot{\boldsymbol{r}}_S = \begin{bmatrix} \ddot{x}_S \\ \ddot{y}_S \\ \ddot{z}_S \end{bmatrix} = \ddot{\alpha} \begin{bmatrix} 0 \\ -l\sin\alpha - h\cos\alpha \\ l\cos\alpha - h\sin\alpha \end{bmatrix} - \dot{\alpha}^2 \begin{bmatrix} 0 \\ l\cos\alpha - h\sin\alpha \\ l\sin\alpha + h\cos\alpha \end{bmatrix}. \tag{2.107}$$

The problem is solved here using a free-body diagram (it could also be solved using the method with α and γ as independent drives as described in Sect. 2.4.1). The inertia forces ($m\ddot{y}_S$, $m\ddot{z}_S$), inertia moments (kinetic moments), the input torque M_{an}^γ that acts on it, the constraint forces (F_y, F_z) and constraint moments (M_{x*}^S, M_{y*}^S, M_{z*}^S) of the frame are included in the free-body diagram of the rotor in Figs. 2.10a and c. The forces in x direction are zero.

The kinetic moments are defined by Euler's gyroscope equations (2.95) and are obtained in conjunction with the angular velocities known from (2.36) and the angular accelerations known from (2.37):

$$M^S_{\text{kin }\xi} \equiv J^S_{\xi\xi}\dot{\omega}_\xi - (J^S_{\eta\eta} - J^S_{\zeta\zeta})\omega_\eta\omega_\zeta$$
$$= J^S_a(\ddot{\alpha}\cos\gamma - \dot{\alpha}\dot{\gamma}\sin\gamma) + (J^S_a - J^S_p)\dot{\alpha}\dot{\gamma}\sin\gamma$$
$$= \underline{\underline{J^S_a\ddot{\alpha}\cos\gamma - J^S_p\dot{\alpha}\dot{\gamma}\sin\gamma}} \quad (2.108)$$
$$M^S_{\text{kin }\eta} \equiv J^S_{\eta\eta}\dot{\omega}_\eta - (J^S_{\zeta\zeta} - J^S_{\xi\xi})\omega_\zeta\omega_\xi$$
$$= J^S_a(-\ddot{\alpha}\sin\gamma - \dot{\alpha}\dot{\gamma}\cos\gamma) - (J^S_p - J^S_a)\dot{\alpha}\dot{\gamma}\cos\gamma$$
$$= \underline{\underline{-J^S_a\ddot{\alpha}\sin\gamma - J^S_p\dot{\alpha}\dot{\gamma}\cos\gamma}} \quad (2.109)$$
$$M^S_{\text{kin }\zeta} \equiv J^S_{\zeta\zeta}\dot{\omega}_\zeta - (J^S_{\xi\xi} - J^S_{\eta\eta})\omega_\xi\omega_\eta$$
$$= \underline{\underline{J^S_p\ddot{\gamma}}}. \quad (2.110)$$

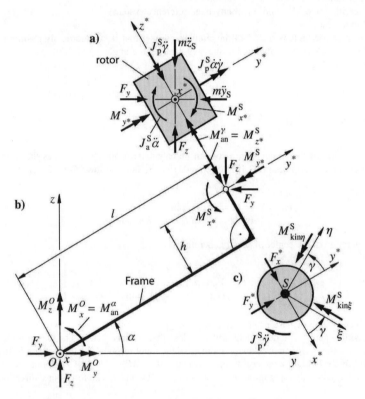

Fig. 2.10 Forces and moments at a pivoting frame with a rotor

The forces and moments are shown in the opposite direction to the body-fixed coordinate directions in Fig. 2.10c. They are transformed into the x^*-y^*-z^* coordinate system and balanced with the applied moments and reaction moments:

2.3 Kinetics of the Rigid Body

$$M^S_{\text{kin}x*} \equiv -M^S_{\text{kin}\eta} \sin\gamma + M^S_{\text{kin}\xi} \cos\gamma \quad = \underline{\underline{J^S_a \ddot{\alpha} = -M^S_{x*}}} \tag{2.111}$$

$$M^S_{\text{kin}y*} \equiv M^S_{\text{kin}\eta} \cos\gamma + M^S_{\text{kin}\xi} \sin\gamma \quad = \underline{\underline{-J^S_p \dot{\alpha}\dot{\gamma} = M^S_{y*}}} \tag{2.112}$$

$$M^S_{\text{kin}z*} \equiv M^S_{\text{kin}\zeta} \quad = \underline{\underline{J^S_p \ddot{\gamma} = M^S_{z*} = M^\gamma_{\text{an}}}} \tag{2.113}$$

See the depiction in Fig. 2.10a. The moment M^γ_{an} causes the angular acceleration $\ddot{\gamma}$ and supports itself against the frame. The components of the kinetic moments of the moving rotor with respect to the x^*-y^*-z^* coordinate system are entered in the opposite direction to the positive coordinate directions. The same sign convention as explained before in Sect. 2.3.2 was also applied to the inertia forces, which can be calculated using (2.107). The following applies at the rotor (and due to the equilibrium of forces at the frame as well) for the reaction forces:

$$\underline{F_y} = m\ddot{y}_S = \underline{\underline{m\left[-\ddot{\alpha}(l\sin\alpha + h\cos\alpha) - \dot{\alpha}^2(l\cos\alpha - h\sin\alpha)\right]}} \tag{2.114}$$

$$\underline{F_z} = m\ddot{z}_S = \underline{\underline{m\left[\ddot{\alpha}(l\cos\alpha - h\sin\alpha) - \dot{\alpha}^2(l\sin\alpha + h\cos\alpha)\right]}}. \tag{2.115}$$

From the equilibrium conditions for the frame, one finds the reaction moments relative to the origin O, see Fig. 2.10a:

$$\underline{M^O_x} = M^\alpha_{\text{an}} = -F_y(l\sin\alpha + h\cos\alpha) + F_z(l\cos\alpha - h\sin\alpha) - M^S_{x*}$$

$$= \underline{\underline{\left[m(l^2 + h^2) + J^S_a\right]\ddot{\alpha}}} \tag{2.116}$$

$$\underline{M^O_y} = M^S_{y*}\cos\alpha - M^S_{z*}\sin\alpha = \underline{\underline{-J^S_p(\dot{\alpha}\dot{\gamma}\cos\alpha + \ddot{\gamma}\sin\alpha)}} \tag{2.117}$$

$$\underline{M^O_z} = M^S_{y*}\sin\alpha + M^S_{z*}\cos\alpha = \underline{\underline{J^S_p(-\dot{\alpha}\dot{\gamma}\sin\alpha + \ddot{\gamma}\cos\alpha)}}. \tag{2.118}$$

It can be seen from (2.116) that the expression in the square brackets represents the moment of inertia about O (parallel-axis theorem). The terms that depend on $\dot{\alpha}^2$ do not exert any influence on the moment about the origin since the resultant centrifugal force acts in radial direction and has no leverage with respect to O. A gyroscopic moment emerges at constant angular velocities that has an effect on the y and z axes, see Section 2.3.

S2.4 The equations of motion for a body that rotates about a fixed ζ axis are given in (2.90) with respect to the center of gravity. Using the component of the angular velocity $\omega_\zeta = \dot{\varphi}$, the result is

$$J^S_{\xi\zeta}\ddot{\varphi} - J^S_{\eta\zeta}\dot{\varphi}^2 = M^S_\xi \tag{2.119}$$

$$J^S_{\eta\zeta}\ddot{\varphi} + J^S_{\xi\zeta}\dot{\varphi}^2 = M^S_\eta \tag{2.120}$$

$$J^S_{\zeta\zeta}\ddot{\varphi} \qquad\qquad = M^S_\zeta. \tag{2.121}$$

The kinetic moments are on the left side of these equations, and the external moments that stem from the bearing forces and the input torque that puts the rotor into this state of motion are on the right side.

The components of the resultant external moment from the bearing forces and input torque are (see Figs. 2.9a and 2.11):

$$M^S_\xi = -F_{A\eta}a + F_{B\eta}b \tag{2.122}$$

$$M^S_\eta = -F_{A\xi}a + F_{B\xi}b \tag{2.123}$$

$$M^S_\zeta = M_{\text{an}} + (F_{A\xi} + F_{B\xi})\eta_S - (F_{A\eta} + F_{B\eta})\xi_S, \tag{2.124}$$

so that one obtains three equations for the unknown components of the bearing forces and the input torque from (2.119) to (2.124):

$$-F_{A\eta}a + F_{B\eta}b = J^S_{\xi\zeta}\ddot{\varphi} - J^S_{\eta\zeta}\dot{\varphi}^2 \tag{2.125}$$

$$F_{A\xi}a - F_{B\xi}b = J^S_{\eta\zeta}\ddot{\varphi} + J^S_{\xi\zeta}\dot{\varphi}^2 \tag{2.126}$$

$$M_{an} + (F_{A\xi} + F_{B\xi})\eta_S - (F_{A\eta} + F_{B\eta})\xi_S = J^S_{\zeta\zeta}\ddot{\varphi} \tag{2.127}$$

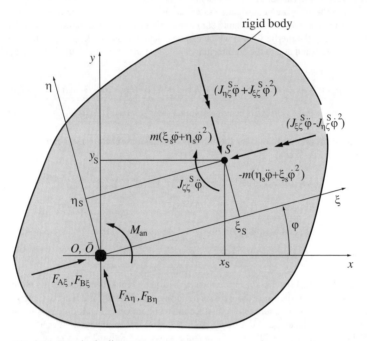

Fig. 2.11 Free-body diagram

Three other equations for the unknown quantities follow from the center-of-gravity theorem. (2.79) applies in the space-fixed reference system. However, it is useful to express it in body-fixed components in accordance with (2.82). The following applies because of $\ddot{\bar{r}}_{\bar{O}} \equiv \mathbf{o}$, see Fig. 2.11:

$$m(\dot{\bar{\boldsymbol{\omega}}} + \bar{\boldsymbol{\omega}}\bar{\boldsymbol{\omega}})\bar{l}_S = \bar{F}. \tag{2.128}$$

Since the external forces are the unknown bearing forces in A and B, one can also write:

$$\bar{F}_A + \bar{F}_B = m(\dot{\bar{\boldsymbol{\omega}}} + \bar{\boldsymbol{\omega}}\bar{\boldsymbol{\omega}})\bar{l}_S. \tag{2.129}$$

In detail, this equation with the vector $\bar{l}_S = (\xi_S, \eta_S, \zeta_S)^T$ and the tensor matrices is as follows:

$$\dot{\bar{\boldsymbol{\omega}}} = \begin{bmatrix} 0 & -\ddot{\varphi} & 0 \\ \ddot{\varphi} & 0 & 0 \\ 0 & 0 & 0 \end{bmatrix}; \quad \bar{\boldsymbol{\omega}} = \begin{bmatrix} 0 & -\dot{\varphi} & 0 \\ \dot{\varphi} & 0 & 0 \\ 0 & 0 & 0 \end{bmatrix} \tag{2.130}$$

and after multiplication of the three matrices, respectively

2.3 Kinetics of the Rigid Body

$$F_{A\xi} + F_{B\xi} = m(-\ddot{\varphi}\eta_S - \dot{\varphi}^2\xi_S) \tag{2.131}$$

$$F_{A\eta} + F_{B\eta} = m(\ddot{\varphi}\xi_S - \dot{\varphi}^2\eta_S). \tag{2.132}$$

The components in the ζ direction are zero. There are two linear equations each for two unknown quantities that can easily be solved: (2.125), (2.126), (2.131), and (2.132). The components of the bearing forces in the body-fixed reference system are derived from

$$
\begin{aligned}
F_{A\xi} &= \frac{J^S_{\eta\zeta}\ddot{\varphi} + J^S_{\xi\zeta}\dot{\varphi}^2 - mb(\ddot{\varphi}\eta_S + \dot{\varphi}^2\xi_S)}{a+b} \\
F_{B\xi} &= \frac{-J^S_{\eta\zeta}\ddot{\varphi} - J^S_{\xi\zeta}\dot{\varphi}^2 - ma(\ddot{\varphi}\eta_S + \dot{\varphi}^2\xi_S)}{a+b} \\
F_{A\eta} &= \frac{-J^S_{\xi\zeta}\ddot{\varphi} + J^S_{\eta\zeta}\dot{\varphi}^2 + mb(\ddot{\varphi}\xi_S - \dot{\varphi}^2\eta_S)}{a+b} \\
F_{B\eta} &= \frac{J^S_{\xi\zeta}\ddot{\varphi} - J^S_{\eta\zeta}\dot{\varphi}^2 + ma(\ddot{\varphi}\xi_S - \dot{\varphi}^2\eta_S)}{a+b}.
\end{aligned}
\tag{2.133}
$$

The respective formulae for the force components with respect to fixed directions can be found in Sect. 2.6.2 (Balancing of rigid rotors) for the special case $\dot{\varphi} = \Omega = \text{const.}$, see (2.327).

The input torque can be found from (2.127) if one inserts the forces from (2.131) and (2.132):

$$M_{\text{an}} = \left[J^S_{\zeta\zeta} + m(\xi_S^2 + \eta_S^2)\right]\ddot{\varphi}. \tag{2.134}$$

The matrix (2.75) with (2.96) corresponds to the moment of inertia tensor of the circular cylinder that is positioned symmetrically in the ξ_1-η_1-ζ_1 system and whose central axis is the ζ_1 axis:

$$\boldsymbol{J}^{*S} = \begin{bmatrix} J_a & 0 & 0 \\ 0 & J_a & 0 \\ 0 & 0 & J_p \end{bmatrix}. \tag{2.135}$$

Transformation into the ξ-η-ζ coordinate system is performed using the matrix $\boldsymbol{A}^*$ from (2.69) that results from the angles

$$
\begin{aligned}
\alpha_{\xi k} &= \gamma; & \beta_{\xi k} &= \frac{\pi}{2}; & \gamma_{\xi k} &= \frac{\pi}{2} - \gamma \\
\alpha_{\eta k} &= \frac{\pi}{2}; & \beta_{\eta k} &= 0; & \gamma_{\eta k} &= \frac{\pi}{2} \\
\alpha_{\zeta k} &= \frac{\pi}{2} + \gamma; & \beta_{\zeta k} &= \frac{\pi}{2}; & \gamma_{\zeta k} &= \gamma
\end{aligned}
\tag{2.136}
$$

and is structured similar to $\boldsymbol{A}_2$ in (2.11):

$$\boldsymbol{A}^* = \begin{bmatrix} \cos\alpha_{\xi k} & \cos\beta_{\xi k} & \cos\gamma_{\xi k} \\ \cos\alpha_{\eta k} & \cos\beta_{\eta k} & \cos\gamma_{\eta k} \\ \cos\alpha_{\zeta k} & \cos\beta_{\zeta k} & \cos\gamma_{\zeta k} \end{bmatrix} = \begin{bmatrix} \cos\gamma & 0 & \sin\gamma \\ 0 & 1 & 0 \\ -\sin\gamma & 0 & \cos\gamma \end{bmatrix}. \tag{2.137}$$

Matrix multiplications according to (2.70) result in the following moment of inertia tensor for the inclined circular cylinder with respect to the directions of the ξ-η-ζ system:

$$\overline{\boldsymbol{J}}^S = \boldsymbol{A}^* \boldsymbol{J}^{*S} \boldsymbol{A}^{*T} = \begin{bmatrix} J_a \cos^2\gamma + J_p \sin^2\gamma & 0 & (J_p - J_a)\sin\gamma\cos\gamma \\ 0 & J_a & 0 \\ (J_p - J_a)\sin\gamma\cos\gamma & 0 & J_a \sin^2\gamma + J_p \cos^2\gamma \end{bmatrix}. \tag{2.138}$$

Thus the elements of the moment of inertia tensor required to calculate the bearing forces in (2.133) and (2.134) are (by comparing coefficients)

$$J^S_{\xi\zeta} = (J_p - J_a)\sin\gamma\cos\gamma; \qquad J^S_{\eta\zeta} = 0; \qquad J^S_{\zeta\zeta} = J_a\sin^2\gamma + J_p\cos^2\gamma. \qquad (2.139)$$

The kinetic moment that acts on the bearings would change its sign if γ became negative, i.e. the tilting were the other way. The product of inertia $J^S_{\xi\zeta}$ is "trying to tilt" the rotor from its axis of rotation. The effect of the products of inertia can vividly be explained by the centrifugal forces that cause a moment about the negative η axis in such a tilted rotor position.

It is also relevant whether the rotor is a flat thin disk ($J_p > J_a$) or a long roller ($J_p < J_a$), see Fig. 5.5 and Table 5.2. For a circular cylinder, the difference of the principal moments of inertia is $J_p - J_a = m(3R^2 - L^2)/12$ in accordance with (2.96) and (2.120), that is, the sign of $J^S_{\xi\zeta}$ and the direction of the "tilting moment" depend on whether the rotor is thick ($L < \sqrt{3}R$) or thin ($L > \sqrt{3}R$).

2.4 Kinetics of Multibody Systems

2.4.1 Mechanisms with Multiple Drives

2.4.1.1 Spatial Rigid-Body Mechanisms

Rigid-body mechanisms are systems of rigid bodies that perform planar or spatial motions, depending on the motions of their input links. So-called **generalized coordinates** q_k are used to describe the motion of such a system, each drive being assigned a coordinate and a **generalized force** Q_k ($k = 1, 2, \ldots, n$). The position of each link of a mechanism with multiple drives depends on these n drive coordinates

$$\boldsymbol{q} = (q_1, q_2, \ldots, q_n)^T. \qquad (2.140)$$

Each q_k is a translation (then Q_k is a force) or an angle (then a moment is assigned to it). The number n of independent drives is termed *mobility* in mechanism theory to distinguish it from the degree of freedom that may, for example, be related to elastic deformations.

A mechanism with rigid links is called a **constrained mechanism**, if – in all positions of the mechanism – the position of any link is unambiguously related to the positions of the other links. The relationship between the number of moving links, the number of kinematic pairs and the degree of freedom of a constrained mechanism is usually called Gruebler's criterion. Mobility is the number of independent coordinates needed to define the configuration of a constrained mechanism. In practice, the conditions of rigid operation are met – and the calculation model of rigid-body mechanisms (*rigid machine*) can be applied to such machines and their sub-assemblies – if the influences of clearances, elastic deformations and oscillations of the links of the mechanism are negligibly small.

2.4 Kinetics of Multibody Systems

The kinematic analysis of planar and spatial mechanisms with multiple drives is typically performed using appropriate software from the field of multibody dynamics. Such programs can be used to analyze the dynamic loads of very complex mechanisms, taking into account arbitrary time functions of the applied forces and moments (such as input forces, spring forces, damper forces, processing forces). In design practice, the engineer has to study the software description thoroughly to be able to utilize these powerful tools efficiently.

The following discussion will touch upon just a few general connections that exist between the drive motions (that is, time functions of the translations or angles) and the generalized forces of the drives. This is relevant for understanding the range of applicability of such programs, for their proper use, and for evaluating the results of calculations. For most practical problems, the driving forces are given by the motor characteristic, so that it becomes necessary to integrate the equations of motion, see Sect. 2.4.3.

A mechanism consists of I links, of which the frame is given the index 1 and the movable bodies are given the indices $i = 2, 3, \ldots, I$, and index I is usually assigned to an output link. Figure 2.12 shows some examples of rigid-body mechanisms with multiple drives. The gyroscope in Fig. 2.14 can also be interpreted in such a way that the position of the rigid body is determined by the three "input coordinates" q_1, q_2, and q_3.

The center-of-gravity coordinates r_{Si} of the ith link of a mechanism show an (often nonlinear) dependence on the so-called kinematic dimensions and the positions of the n input links:

$$r_{Si}(q) = [x_{Si}(q), y_{Si}(q), z_{Si}(q)]^{\mathrm{T}}. \tag{2.141}$$

Their velocities can also be calculated according to the chain rule:

$$\frac{\mathrm{d}(r_{Si})}{\mathrm{d}t} = v_{Si} = \sum_{k=1}^{n} \frac{\partial r_{Si}}{\partial q_k} \dot{q}_k = \sum_{k=1}^{n} r_{Si,k} \dot{q}_k; \qquad i = 2, 3, \ldots, I. \tag{2.142}$$

(2.142) spells out in detail as

$$\dot{x}_{Si} = \sum_{k=1}^{n} x_{Si,k} \dot{q}_k; \qquad \dot{y}_{Si} = \sum_{k=1}^{n} y_{Si,k} \dot{q}_k; \qquad \dot{z}_{Si} = \sum_{k=1}^{n} z_{Si,k} \dot{q}_k. \tag{2.143}$$

The partial derivatives with respect to the coordinates q_k are abbreviated by the letter k after the comma.

The components of the angular velocities of each link in a rigid mechanism show a linear dependence on the velocities of the input coordinates. The following applies to the body-fixed components of the vector $\overline{\omega}_i = (\omega_{\xi i}, \omega_{\eta i}, \omega_{\zeta i})^{\mathrm{T}}$ of the angular velocity of the ith link:

$$\omega_{\xi i} = \sum_{k=1}^{n} u_{\xi i k} \dot{q}_k; \qquad \omega_{\eta i} = \sum_{k=1}^{n} u_{\eta i k} \dot{q}_k; \qquad \omega_{\zeta i} = \sum_{k=1}^{n} u_{\zeta i k} \dot{q}_k. \tag{2.144}$$

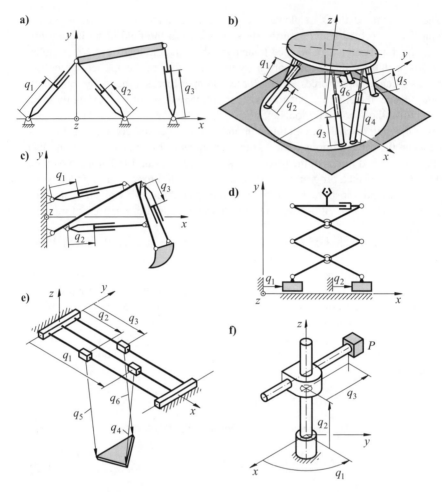

Fig. 2.12 Examples of rigid-body mechanisms with multiple drives; **a)** planar Stewart platform ($n = 3$), **b)** spatial Stewart platform ($n = 6$), **c)** hydraulic drives of a dredging shovel ($n = 3$), **d)** lift truck ($n = 2$), **e)** cable-guided handling system ($n = 6$), **f)** welding robot ($n = 3$)

Compare, for example, (2.163) and (2.164).

The linear relations can be expressed according to (2.142) and (2.144) using a **Jacobian matrix** for each. The following relations apply to the translation of the centers of gravity and to rotation:

$$\dot{\boldsymbol{r}}_{Si} = \boldsymbol{Y}_i(\boldsymbol{q})\dot{\boldsymbol{q}}; \qquad \overline{\boldsymbol{\omega}}_i = \boldsymbol{Z}_i(\boldsymbol{q})\dot{\boldsymbol{q}}; \qquad i = 2, 3, \ldots, I. \qquad (2.145)$$

$\boldsymbol{Y}_i(\boldsymbol{q})$ and $\boldsymbol{Z}_i(\boldsymbol{q})$ are the Jacobian matrices for translation and rotation of the ith rigid body (link). They are rectangular matrices with three rows and n columns. Which elements are contained in these matrices follows from (2.142) and (2.144):

2.4 Kinetics of Multibody Systems

$$\boldsymbol{Y}_i(\boldsymbol{q}) = \begin{bmatrix} x_{Si,1} & x_{Si,2} & \cdots & x_{Si,n} \\ y_{Si,1} & y_{Si,2} & \cdots & y_{Si,n} \\ z_{Si,1} & z_{Si,2} & \cdots & z_{Si,n} \end{bmatrix} \; ; \; \boldsymbol{Z}_i(\boldsymbol{q}) = \begin{bmatrix} u_{\xi i1} & u_{\xi i2} & \cdots & u_{\xi in} \\ u_{\eta i1} & u_{\eta i2} & \cdots & u_{\eta in} \\ u_{\zeta i1} & u_{\zeta i2} & \cdots & u_{\zeta in} \end{bmatrix}$$
(2.146)

$\boldsymbol{Z}_i(\boldsymbol{q})$ is found by differentiating the angular velocities with respect to the input velocities $\dot{q}_k$ or simply by a comparison of coefficients, see for example (2.30) from Sect. 2.2.3.

Position functions and Jacobian matrices can be explicitly stated in analytical form for open linkages, such as in Examples c, d and f in Fig. 2.12. Mechanisms with a loop structure,, such as in cases a, b and e in Fig. 2.12, in which the constraint equations cannot be solved in closed form, the Jacobian matrices can be numerically calculated as a function of position (using a PC and existing software). The elements of the Jacobian matrices typically depend on the position of the input coordinates. They are also called *first-order position functions*.

A mechanism consists of $I - 1$ movable rigid bodies whose dynamic properties are captured by 10 mass parameters, respectively, that are contained in the parameter vector

$$\boldsymbol{p}_i = [m_i, \, \xi_{Si}, \, \eta_{Si}, \, \zeta_{Si}, \, J^S_{\xi\xi i}, \, J^S_{\eta\eta i}, \, J^S_{\zeta\zeta i}, \, J^S_{\xi\eta i}, \, J^S_{\eta\zeta i}, \, J^S_{\xi\zeta i}]^T; \qquad i = 2, 3, \ldots, I$$
(2.147)

For the ith body, these are the mass m_i, the three static moments ($m_i\xi_{Si}$, $m_i\eta_{Si}$, $m_i\zeta_{Si}$) and the six elements of the moment of inertia tensor ($J^S_{\xi\xi i}$, $J^S_{\eta\eta i}$, $J^S_{\zeta\zeta i}$, $J^S_{\xi\eta i}$, $J^S_{\eta\zeta i}$, $J^S_{\xi\zeta i}$), if one refers to the axes of gravity. If one knows the position of the principal axes (through the center of gravity), the moment of inertia tensor includes only the three principal moments of inertia J^S_{Ii}, J^S_{IIi}, and J^S_{IIIi}.

The kinetic energy of the rigid-body system is the sum of the kinetic energies of all its individual bodies consisting of the translational energy and the rotational energy, see (2.57). Using the Jacobian matrices according to (2.146), the kinetic energy with the moment of inertia tensors in analogy with (2.57) becomes

$$W_{\text{kin}} = \tfrac{1}{2}\dot{\boldsymbol{q}}^T \left[\sum_{i=2}^{I}(m_i\boldsymbol{Y}_i^T\boldsymbol{Y}_i + \boldsymbol{Z}_i^T\overline{\boldsymbol{J}}_i^S\boldsymbol{Z}_i)\right]\dot{\boldsymbol{q}} = \tfrac{1}{2}\dot{\boldsymbol{q}}^T\boldsymbol{M}\dot{\boldsymbol{q}} \qquad (2.148)$$

$$= \tfrac{1}{2} \sum_{k=1}^{n}\sum_{l=1}^{n} m_{kl}(\boldsymbol{q})\dot{q}_k\dot{q}_l$$

The symmetrical mass matrix $\boldsymbol{M}$ only depends on $\boldsymbol{q}$ if the Jacobian matrices contain terms that depend on $\boldsymbol{q}$. The matrix $\boldsymbol{M}$ has n^2 elements m_{kl} that are called **generalized masses**:

$$m_{kl}(\boldsymbol{q}) = m_{lk}(\boldsymbol{q}) = \sum_{i=2}^{I} \Big\{ m_i(x_{Si,k}x_{Si,l} + y_{Si,k}y_{Si,l} + z_{Si,k}z_{Si,l})$$

$$+ J^S_{\xi\xi i}u_{\xi ik}u_{\xi il} + J^S_{\eta\eta i}u_{\eta ik}u_{\eta il} + J^S_{\zeta\zeta i}u_{\zeta ik}u_{\zeta il} \quad (2.149)$$

$$+ 2(J^S_{\xi\eta i}u_{\xi ik}u_{\eta ik} + J^S_{\eta\zeta i}u_{\zeta ik}u_{\eta ik} + J^S_{\xi\zeta i}u_{\xi ik}u_{\zeta ik})\Big\}$$

Due to $z_{Si} = \text{const.}$ and $\omega_{\xi i} = \omega_{\eta i} = 0$, $u_{\xi ik} = u_{\eta ik} = 0$ always applies to planar mechanisms, all links of which move parallel to the x-y plane. The angles φ_i are defined in Fig. 2.15. Because of $\omega_{\zeta i} = \dot\varphi_i$, $u_{\zeta ik} = \varphi_{i,k}$ and therefore $J^S_{\zeta\zeta i} = J_{Si}$:

$$m_{kl}(\boldsymbol{q}) = m_{lk}(\boldsymbol{q}) = \sum_{i=2}^{I} [m_i(x_{Si,k}x_{Si,l} + y_{Si,k}y_{Si,l}) + J_{Si}\varphi_{i,k}\varphi_{i,l}] \quad (2.150)$$

The generalized masses are independent of the state of motion, but do depend on the position of the input coordinates in mechanisms with variable transmission ratio.

The partial derivatives are designated by the short notation already introduced in (2.142) (comma and index of the coordinate):

$$\frac{\partial(m_{kl})}{\partial q_p} = m_{kl,p} \quad (2.151)$$

The so-called **Christoffel symbols of the first kind**, that occur when deriving the equations of motion using Lagrange's equations of the second kind are calculated from the partial derivatives of the generalized masses as follows:

$$\Gamma_{klp} = \Gamma_{lkp} = \frac{1}{2}(m_{lp,k} + m_{pk,l} - m_{kl,p}) \quad (2.152)$$

Several applied forces and moments that act on arbitrary points of the ith body are summarized by resultants that act on the ith center of gravity. These resultants of the applied forces $\boldsymbol{F}_i^{(e)}$ and moments $\overline{\boldsymbol{M}}_i^{S(e)}$ that act on all links of the mechanism are then referred to the input coordinates $\boldsymbol{q}$ using the principle of virtual work:

$$\delta W^{(e)} = \delta \boldsymbol{q}^{\mathrm{T}} \sum_i (\boldsymbol{Y}_i^{\mathrm{T}} \boldsymbol{F}_i^{(e)} + \boldsymbol{Z}_i^{\mathrm{T}} \overline{\boldsymbol{M}}_i^{S(e)}) = \delta \boldsymbol{q}^{\mathrm{T}} \boldsymbol{Q} \quad (2.153)$$

The **generalized forces** thus result from:

$$\boldsymbol{Q} = (Q_1, Q_2, \ldots, Q_n)^{\mathrm{T}} = \sum_i (\boldsymbol{Y}_i^{\mathrm{T}} \boldsymbol{F}_i^{(e)} + \boldsymbol{Z}_i^{\mathrm{T}} \overline{\boldsymbol{M}}_i^{S(e)}) \quad (2.154)$$

Each component Q_p of the generalized forces thus follows from the applied forces and the Jacobian matrices that are characteristic for a mechanism. The generalized force quantities are not distinguished based on whether they follow from the derivative of a potential (such as potential energy, deformation energy, magnetic energy) (such as weight, spring forces, electromagnetic forces) or not (such as input forces, input moments, braking torques, friction and damping forces), a distinction that is occasionally found in the technical literature.

The *equations of motion for rigid-body mechanisms* with n drives result from the kinetic energy (2.148) and the generalized forces obtained from (2.154) when using Lagrange's equations of the second kind:

2.4 Kinetics of Multibody Systems

$$Q_{p\,\text{kin}} \equiv \sum_{l=1}^{n} m_{pl}(\boldsymbol{q})\ddot{q}_l + \sum_{k=1}^{n}\sum_{l=1}^{n} \Gamma_{klp}(\boldsymbol{q})\dot{q}_k\dot{q}_l = Q_p; \quad p = 1, 2, \ldots, n \quad (2.155)$$

They each express the equilibrium between the *applied forces* and the kinetic forces (caused by inertia) with respect to the respective coordinates q_p. The forces that result from *inertia of the rigid bodies of the rigid system* are called *kinetic* (or traditionally *"kinetostatic"*) *forces*, in contrast to the general "vibration forces", that arise due to the oscillations of rigid bodies, see the other chapters of this book. The kinetostatic forces can be calculated for given drive motions $\boldsymbol{q}(t)$ – from the left side of (2.155). They depend on the mass parameters, the geometric conditions, and the state of motion, see for example (2.90), (2.92), (2.95), and (2.97) to (2.99).

(2.155) can be looked at from various points of view:

- If the kinematic parameters are given, the generalized forces $\boldsymbol{Q}$ "in the direction of the input coordinates" have to act in such a way that the state of motion described by the kinematic parameters $\dot{\boldsymbol{q}}(t)$ and $\ddot{\boldsymbol{q}}(t)$ is achieved. Or:
- If the generalized forces $\boldsymbol{Q}$ are known, e. g. as functions of the coordinates and velocities, (2.155) represents a system of ordinary nonlinear differential equations that one has to integrate if one wishes to calculate $\boldsymbol{q}(t)$ and its time derivatives.
- If some of the generalized forces are driving forces, the others can be considered reactions and calculated after determining $\boldsymbol{q}(t)$ and its time derivatives, see for example (2.159) and (2.160).

It is evident from (2.155) that the kinetostatic forces do not only depend on the accelerations of the drive motions. This means that inertia forces also act when the drives move at constant velocities. This is not surprising since the masses in rigid-body systems are indeed accelerated and/or decelerated. This phenomenon is known from simple cases: the centrifugal force increases with the square of the angular velocity, and the Coriolis force is proportional to the product of velocity and angular velocity. (2.155) shows that the products of all input velocities can in general occur in combination.

For the special case of a mechanism with two drives ($n = 2$), it follows from (2.155):

$$m_{11}\ddot{q}_1 + m_{12}\ddot{q}_2 + \Gamma_{111}\dot{q}_1^2 + 2\Gamma_{121}\dot{q}_1\dot{q}_2 + \Gamma_{221}\dot{q}_2^2 = Q_1 \quad (2.156)$$

$$m_{21}\ddot{q}_1 + m_{22}\ddot{q}_2 + \Gamma_{112}\dot{q}_1^2 + 2\Gamma_{122}\dot{q}_1\dot{q}_2 + \Gamma_{222}\dot{q}_2^2 = Q_2 \quad (2.157)$$

The following applies

$$\Gamma_{111} = \frac{1}{2}m_{11,1}; \quad \Gamma_{121} = \frac{1}{2}m_{11,2}; \quad \Gamma_{221} = m_{12,2} - \frac{1}{2}m_{22,1}$$

$$\Gamma_{112} = m_{12,1} - \frac{1}{2}m_{11,2}; \quad \Gamma_{122} = \frac{1}{2}m_{22,1}; \quad \Gamma_{222} = \frac{1}{2}m_{22,2} \quad (2.158)$$

If, for example, the input coordinate $q_2 = \text{const.}$, the relationship between the force referred to the drive Q_1 and the kinetostatic forces follows from (2.156)

$$m_{11}\ddot{q}_1 + \Gamma_{111}\dot{q}_1^2 = Q_1 \qquad (2.159)$$

The kinetostatic forces also act in the direction of coordinate q_2:

$$m_{21}\ddot{q}_1 + \Gamma_{112}\dot{q}_1^2 = m_{21}\ddot{q}_1 + \left(m_{12,1} - \frac{1}{2}m_{11,2}\right)\dot{q}_1^2 = Q_2 \qquad (2.160)$$

This lets one draw important conclusions regarding the parameter dependence of inertia forces. Since q_2 can be any coordinate in the direction of which the velocity and acceleration are zero, Q_2 accordingly there is a force or a moment at this "stationary site" of the mechanism. Such a force in the direction of an arbitrary coordinate can be interpreted as a generalized **reaction force inside the mechanism**. So it can be a joint force or an axial force if a coordinate q_2 matches its direction. Equations of the type of (2.160) are rarely used for calculating such internal forces, but they can provide explanations for the following important conclusions:

All reaction forces and moments at all places in any links of mechanisms with a single drive ($n = 1$) **result from two terms**, one of which is proportional to the **acceleration**, the other to the **square of the velocity** of the drive. They all show a **linear dependence on the mass parameters** of each rigid body, see (2.147) and (2.149).

2.4.1.2 Equations of Motion of a Planetary Gear Mechanism

Planetary or epicyclic gear mechanisms are used because of their applicability as speed-transforming, superposition, and switching gear mechanisms in many fields of drive engineering. This type of gear mechanism has proven its worth, in particular, in automotive engineering and shipbuilding, where large outputs have to be transmitted at high speeds. These mechanisms allow extremely high and low gear ratios while requiring little installation space. The distribution of the static and dynamic forces over several gears and the low bearing loads at coaxial positioning of the input and output shafts are advantages. This enables the superposition of the speeds and torques of multiple drives, and they are also used as differential gears, see VDI Guideline 2157 [36].

The equations of motion that describe the relationships between the input moments and angular accelerations shall now be established for a simple planetary gear mechanism as outlined in Fig. 2.13. This superposition gear mechanism (also called gear set, transfer, or differential gear) has two degrees of freedom (mobility $n = 2$). It consists of the sun gear *2*, the ring gear *3*, three planetary gears and the arm that carries the planetary gears *5*, all of which pivot about the z axis. The radii r_2 and r_4, the moments of inertia J_2, J_3 and J_5 with respect to the fixed axis of rotation, the mass m_4 of each gear 4 and the moment of inertia J_4 of one of the gears *4* about its bearing axis are given. The center of gravity S of each planet gear is in its bearing axis. The torques that act on the shafts of the members *2*, *3*, and *5* are to be taken into account.

2.4 Kinetics of Multibody Systems

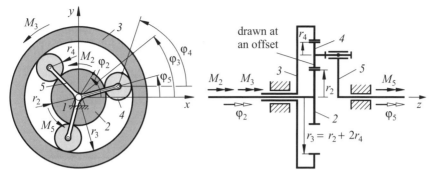

Fig. 2.13 Nomenclature on the spur differential

The constraints form the starting point for the kinematic and dynamic analysis. They are established in general form based on Fig. 2.13 so that they also apply to special cases of spur differentials with one degree of freedom (e. g. $\dot\varphi_3 = 0$ or $\dot\varphi_5 = 0$).

The constraints result from the fact that the relative velocities of the engaged gears are zero at their contact points (the pitch points). Therefore, the following applies:

$$r_2\dot\varphi_2 - [r_2\dot\varphi_5 - r_4(\dot\varphi_4 - \dot\varphi_5)] = 0 \tag{2.161}$$

$$(r_2 + 2r_4)\dot\varphi_3 - [(r_2 + 2r_4)\dot\varphi_5 + r_4(\dot\varphi_4 - \dot\varphi_5)] = 0 \tag{2.162}$$

Four position coordinates were introduced here, but only two independent coordinates exist due to the two constraints. The angles of rotation of the sun gear ($q_1 = \varphi_2$) and ring gear ($q_2 = \varphi_3$) are used as the two independent input coordinates. The dependent angular velocities can be found from (2.161) and (2.162):

$$\dot\varphi_4 = \frac{-r_2}{2r_4}\dot q_1 + \frac{r_2 + 2r_4}{2r_4}\dot q_2 = u_{41}\dot q_1 + u_{42}\dot q_2 \tag{2.163}$$

$$\dot\varphi_5 = \frac{r_2}{2(r_2 + r_4)}\dot q_1 + \frac{r_2 + 2r_4}{2(r_2 + r_4)}\dot q_2 = u_{51}\dot q_1 + u_{52}\dot q_2 \tag{2.164}$$

These equations have the form of (2.144) with $\omega_{\xi i} \equiv 0$, $\omega_{\eta i} \equiv 0$ and $\omega_{\zeta i} = \dot\varphi_i = \sum u_{ik}\dot q_k$. The **gear ratios** u_{ik} express the ratio of angular velocity $\dot\varphi_i$ to angular velocity $\dot q_k$. The kinetic energy is the sum of the rotational energies of the sun gear (2), arm (5), planet gears (4), and ring gear (3), as well as the translational energy of the three planet gears (4):

$$2W_{\text{kin}} = J_2\dot\varphi_2^2 + J_3\dot\varphi_3^2 + 3J_4\dot\varphi_4^2 + J_5\dot\varphi_5^2 + 3m_4(r_2 + r_4)^2\dot\varphi_5^2 \tag{2.165}$$

To specify the kinetic energy as a function of the velocities $\dot{\boldsymbol q} = (\dot q_1, \dot q_2)^{\text{T}}$, (2.163) and (2.164) are used to eliminate the angular velocities $\dot\varphi_4$ and $\dot\varphi_5$. Formally, one could also proceed in accordance with the description in Sect. 2.4.1.1 using the Jacobian matrices according to (2.146), which are independent of the gear position for mechanisms with constant transmission ratio and which upon limiting to $\omega_{\zeta i}$ are reduced to:

$$\boldsymbol{Z}_2 = [1 \quad 0]; \quad \boldsymbol{Z}_3 = [0 \quad 1]; \quad \boldsymbol{Z}_4 = [u_{41} \quad u_{42}]; \quad \boldsymbol{Z}_5 = [u_{51} \quad u_{52}] \quad (2.166)$$

Then, the kinetic energy is, according to (2.148):

$$W_{\text{kin}} = \frac{1}{2}\dot{\boldsymbol{q}}^{\text{T}} \boldsymbol{M} \dot{\boldsymbol{q}} \quad (2.167)$$

with the mass matrix $\boldsymbol{M}$ according to (2.148) that includes the following generalized masses as elements:

$$\begin{aligned}
m_{11} &= J_2 \phantom{{}={}} + 3J_4 u_{41}^2 \phantom{{}={}} + [J_5 + 3m_4(r_2 + r_4)^2]u_{51}^2 \\
m_{12} = m_{21} &= \phantom{J_2 + {}} 3J_4 u_{41} u_{42} + [J_5 + 3m_4(r_2 + r_4)^2]u_{51}u_{52} \\
m_{22} &= \phantom{{}={}} J_3 + 3J_4 u_{42}^2 + [J_5 + 3m_4(r_2 + r_4)^2]u_{52}^2
\end{aligned} \quad (2.168)$$

The generalized masses are constant here, so their partial derivatives and thus all Christoffel symbols equal zero. The generalized forces Q_1 and Q_2 depend on the moments M_2, M_3, and M_5. The virtual work of the input torques must be equal to that of the generalized forces. Therefore,

$$\begin{aligned}
\delta W^{(e)} &= M_2 \delta\varphi_2 + M_3 \delta\varphi_3 + M_5 \delta\varphi_5 \\
&= M_2 \delta q_1 + M_3 \delta q_2 + M_5(u_{51}\delta q_1 + u_{52}\delta q_2) \quad (2.169) \\
&= (M_2 + M_5 u_{51})\delta q_1 + (M_3 + M_5 u_{52})\delta q_2 = Q_1 \delta q_1 + Q_2 \delta q_2
\end{aligned}$$

A comparison of coefficients for δq_1 and δq_2 yields

$$Q_1 = M_2 + M_5 u_{51}; \qquad Q_2 = M_3 + M_5 u_{52} \quad (2.170)$$

Thus the equations of motion are as follows in accordance with (2.156) and (2.157) and taking into consideration (2.168) and (2.170):

$$m_{11}\ddot{q}_1 + m_{12}\ddot{q}_2 = M_2 + M_5 u_{51} \quad (2.171)$$
$$m_{21}\ddot{q}_1 + m_{22}\ddot{q}_2 = M_3 + M_5 u_{52} \quad (2.172)$$

This general relationship leaves it still open which of the three moments or which two angular accelerations are given or sought after. According to (2.163) and (2.164), conditions for the other angles could be taken into account as well. Since this mechanism has a mobility $n = 2$, three out of the five quantities (q_1, q_2, M_2, M_3, M_5) can be prescribed to calculate the remaining two unknown quantities. By integrating the differential equations (2.171) and (2.172), various operating states can be dynamically analyzed, e. g. time functions and dynamic loads during startup, shifting, and braking processes, if the characteristics of the motors or clutches are

2.4 Kinetics of Multibody Systems

given. The driving powers that result from multiplying the moments and angular velocities can also be calculated.

For example, the following operating states can occur:

Operating state a): input at the sun gear *2*, ring gear *3* is fixed, output at the arm *5*

Given are M_2 and M_5 as well as $\dot{q}_2 = \dot{\varphi}_3 = 0$. (2.171) provides the angular acceleration at the input link *2*

$$\ddot{q}_1 = \frac{M_2 + M_5 u_{51}}{m_{11}} \tag{2.173}$$

The time function is obtained, taking into account the initial conditions, by integrating this differential equation, and the moment at the ring gear results after insertion into (2.172):

$$M_3 = -M_5 u_{52} + m_{21} \ddot{q}_1 = -M_5 u_{52} + m_{21} \frac{M_2 + M_5 u_{51}}{m_{11}} \tag{2.174}$$

Operating state b): Input at the ring gear *3* and the sun gear *2*, output at the arm *5*

Given are M_2, M_3 and M_5. The angular accelerations $\ddot{q}_1$ and $\ddot{q}_2$ are obtained from (2.171) and (2.172) by solving the system of linear equations.

The time functions of all angles for given moment functions can then be calculated from q_1 and q_2 and all other angles from (2.163) and (2.164).

The same equations can be used for the operating state "input at the sun gear *2* and the arm *5*, output at the ring gear *3*".

2.4.1.3 Gimbal-Mounted Rotor

The expressions for the kinetic energy and the relationships between the moments ($Q_1 = M_x$, $Q_2 = M_{y*}$, $Q_3 = M_\zeta$) and the three cardan angles q_1, q_2, and q_3 are to be established for the gyroscope that is supported as shown in Fig. 2.14. The center of gravity is assumed to be at the origin of the body-fixed coordinate system ($S = \overline{O}$). Given are all elements of the moment of inertia tensor $\overline{J}^S$ with respect to the center of gravity S and the moments of inertia J_A and J_B of the two frames with respect to their bearing axes.

The kinetic energy of the rotation in (2.57) is used as a starting point and must still be complemented by the kinetic energy of the two frames

$$W_{\text{kin frame}} = \frac{1}{2}(J_A \dot{q}_1^2 + J_B \dot{q}_2^2). \tag{2.175}$$

If one inserts the components of the angular velocity known from (2.30) into (2.57), it is obtained as a function of the generalized coordinates and their time derivatives:

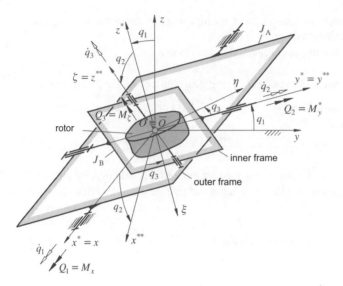

Fig. 2.14 Gimbal-mounted rotor with input torques about three axes

$$W_{\text{kin}} = W_{\text{kin frame}} + W_{\text{kin rotor}} = \frac{1}{2} \dot{\boldsymbol{q}}^{\text{T}} \boldsymbol{M} \dot{\boldsymbol{q}}$$
$$= \frac{1}{2}(m_{11}\dot{q}_1^2 + m_{22}\dot{q}_2^2 + m_{33}\dot{q}_3^2) + m_{12}\dot{q}_1\dot{q}_2 + m_{13}\dot{q}_1\dot{q}_3 + m_{23}\dot{q}_2\dot{q}_3$$
(2.176)

The elements of the mass matrix **M** can be found by comparing coefficients:

$$m_{11} = J_{\text{A}} + (J^{\text{S}}_{\xi\xi} \cos^2 q_3 + J^{\text{S}}_{\eta\eta} \sin^2 q_3) \cos^2 q_2 + J^{\text{S}}_{\zeta\zeta} \sin^2 q_2$$
$$\quad - 2J^{\text{S}}_{\xi\eta} \cos^2 q_2 \cos q_3 \sin q_3 - 2(J^{\text{S}}_{\eta\zeta} \sin q_3 - J^{\text{S}}_{\xi\zeta} \cos q_3) \cos q_2 \sin q_2$$

$$m_{12} = (J^{\text{S}}_{\xi\xi} - J^{\text{S}}_{\eta\eta}) \cos q_2 \sin q_3 \cos q_3$$
$$\quad + J^{\text{S}}_{\xi\eta}(\cos^2 q_3 - \sin^2 q_3) \cos q_2 + (J^{\text{S}}_{\eta\zeta} \cos q_3 + J^{\text{S}}_{\xi\zeta} \sin q_3) \sin q_2$$

$$m_{13} = J^{\text{S}}_{\zeta\zeta} \sin q_2 - (J^{\text{S}}_{\eta\zeta} \sin q_3 - J^{\text{S}}_{\xi\zeta} \cos q_3) \cos q_2$$

$$m_{22} = J_{\text{B}} + J^{\text{S}}_{\xi\xi} \sin^2 q_3 + J^{\text{S}}_{\eta\eta} \cos^2 q_3 + 2J^{\text{S}}_{\xi\eta} \cos q_3 \sin q_3$$

$$m_{23} = J^{\text{S}}_{\eta\zeta} \cos q_3 + J^{\text{S}}_{\xi\zeta} \sin q_3$$

$$m_{33} = J^{\text{S}}_{\zeta\zeta}$$
(2.177)

Now the moments Q_1, Q_2 and Q_3 that act in the direction of the three angular coordinates q_1, q_2 and q_3 are to be calculated from (2.155) for the **special case** $J^{\text{S}}_{\xi\eta} = J^{\text{S}}_{\eta\zeta} = J^{\text{S}}_{\zeta\xi} = 0$ when the gyroscope rotates at the angular speed $\omega_\zeta = \dot{q}_3$ about its ζ axis [principal axis of inertia III ($J^{\text{S}}_{\zeta\zeta} = J^{\text{S}}_{\text{III}}$)] and the angles of the frame

2.4 Kinetics of Multibody Systems

are changed in accordance with $q_1(t)$ and $q_2(t)$. $J_{\xi\xi}^S = J_I^S$ and $J_{\eta\eta}^S = J_{II}^S$ are also principal moments of inertia.

Taking into account the special condition that $m_{23} = m_{32} = 0$ and $m_{33} = J_{\zeta\zeta}^S = $ const., it follows from (2.152) for the Christoffel symbols

$$\Gamma_{111} = \Gamma_{331} = \Gamma_{122} = \Gamma_{222} = \Gamma_{332} = \Gamma_{133} = \Gamma_{233} = \Gamma_{333} = 0. \quad (2.178)$$

It then follows from (2.155)

$$m_{11}\ddot{q}_1 + m_{12}\ddot{q}_2 + m_{13}\ddot{q}_3 + \Gamma_{221}\dot{q}_2^2 + 2\Gamma_{121}\dot{q}_1\dot{q}_2 + 2\Gamma_{131}\dot{q}_1\dot{q}_3 + 2\Gamma_{231}\dot{q}_2\dot{q}_3 = Q_1$$
$$m_{12}\ddot{q}_1 + m_{22}\ddot{q}_2 \qquad\qquad + \Gamma_{112}\dot{q}_1^2 \qquad\qquad + 2\Gamma_{132}\dot{q}_1\dot{q}_3 + 2\Gamma_{232}\dot{q}_2\dot{q}_3 = Q_2$$
$$m_{13}\ddot{q}_1 \qquad\qquad + m_{33}\ddot{q}_3 + \Gamma_{113}\dot{q}_1^2 + 2\Gamma_{123}\dot{q}_1\dot{q}_2 + \Gamma_{223}\dot{q}_2^2 \qquad\qquad = Q_3$$
$$(2.179)$$

The following expressions result from the generalized masses in (2.177) for the Christoffel symbols in this **special case**:

$$\Gamma_{121} = -\Gamma_{112} = \frac{1}{2}m_{11,2} = -(J_{\xi\xi}^S\cos^2 q_3 + J_{\eta\eta}^S\sin^2 q_3 - J_{\zeta\zeta}^S)\sin q_2 \cos q_2$$

$$\Gamma_{221} = m_{12,2} = -(J_{\xi\xi}^S - J_{\eta\eta}^S)\sin q_2 \sin q_3 \cos q_3$$

$$\Gamma_{131} = -\Gamma_{113} = \frac{1}{2}m_{11,3} = -(J_{\xi\xi}^S - J_{\eta\eta}^S)\sin q_3 \cos q_3 \cos^2 q_2$$

$$\Gamma_{231} = \frac{1}{2}m_{13,2} + \frac{1}{2}m_{12,3} = \frac{1}{2}[(J_{\xi\xi}^S - J_{\eta\eta}^S)(\cos^2 q_3 - \sin^2 q_3) + J_{\zeta\zeta}^S]\cos q_2$$

$$\Gamma_{132} = -\Gamma_{123} = \frac{1}{2}m_{12,3} - \frac{1}{2}m_{13,2}$$
$$= \frac{1}{2}[(J_{\xi\xi}^S - J_{\eta\eta}^S)(\cos^2 q_3 - \sin^2 q_3) - J_{\zeta\zeta}^S]\cos q_2$$

$$\Gamma_{232} = -\Gamma_{223} = \frac{1}{2}m_{22,3} = (J_{\xi\xi}^S - J_{\eta\eta}^S)\sin q_3 \cos q_3$$
$$(2.180)$$

The expressions can be simplified further when taking a more specific look at

$$J_{\xi\xi}^S = J_{\eta\eta}^S = J_a^S, \quad \text{so} \quad m_{12} = 0 \quad \text{and} \quad \Gamma_{221} = \Gamma_{131} = \Gamma_{113} = \Gamma_{232} = 0 \quad (2.181)$$

e. g. a rotationally symmetrical rotor. The equations of motion (2.179) then are as follows:

$$(J_A + J_a^S \cos^2 q_2 + J_{\zeta\zeta}^S \sin^2 q_2)\ddot{q}_1 + J_{\zeta\zeta}^S \sin q_2 \ddot{q}_3$$
$$-2(J_a^S - J_{\zeta\zeta}^S)\sin q_2 \cos q_2 \dot{q}_1\dot{q}_2 + J_{\zeta\zeta}^S \cos q_2 \dot{q}_2\dot{q}_3 = Q_1 \quad (2.182)$$
$$(J_B + J_a^S)\ddot{q}_2 + (J_a^S - J_{\zeta\zeta}^S)\sin q_2 \cos q_2 \dot{q}_1^2 - J_{\zeta\zeta}^S \cos q_2 \dot{q}_1\dot{q}_3 = Q_2 \quad (2.183)$$
$$J_{\zeta\zeta}^S \sin q_2 \ddot{q}_1 + J_{\zeta\zeta}^S \ddot{q}_3 + J_{\zeta\zeta}^S \cos q_2 \dot{q}_1\dot{q}_2 = Q_3 \quad (2.184)$$

It is remarkable that varying moments have to act about the three axes to maintain this state of motion even if the angular velocities are constant, that is $\ddot{q}_1 = \ddot{q}_2 = \ddot{q}_3 \equiv 0$.

A rotationally symmetrical body frequently rotates about two axes only. The following special cases result from (2.182) to (2.184):

Case 1: $q_1 = $ const.; $q_2(t)$ and $q_3(t)$ variable.

$$J^S_{\zeta\zeta} \sin q_2 \ddot{q}_3 + J^S_{\zeta\zeta} \cos q_2 \dot{q}_2 \dot{q}_3 = Q_1 \qquad (2.185)$$
$$(J_A + J^S_a)\ddot{q}_2 = Q_2 \qquad (2.186)$$
$$J^S_{\zeta\zeta} \ddot{q}_3 = Q_3 \qquad (2.187)$$

Fall 2: $q_1(t)$ variable; $q_2 = \beta = $ const.; $q_3(t)$ variable

$$(J_A + J^S_a \cos^2 \beta + J^S_{\zeta\zeta} \sin^2 \beta)\ddot{q}_1 + J^S_{\zeta\zeta} \sin \beta \ddot{q}_3 = Q_1 \qquad (2.188)$$
$$(J^S_a - J^S_{\zeta\zeta}) \sin \beta \cos \beta \dot{q}_1^2 - J^S_{\zeta\zeta} \cos \beta \dot{q}_1 \dot{q}_3 = Q_2 \qquad (2.189)$$
$$J^S_{\zeta\zeta} \sin \beta \ddot{q}_1 + J^S_{\zeta\zeta} \ddot{q}_3 = Q_3 \qquad (2.190)$$

All terms of kinetic energy become simpler for the special case defined by (2.181):

$$W_{\text{kin}} = \frac{1}{2}(J_A + J^S_a \cos^2 q_2 + J^S_{\zeta\zeta} \sin^2 q_2)\dot{q}_1^2$$
$$+ \frac{1}{2}(J_B + J^S_a)\dot{q}_2^2 + J^S_{\zeta\zeta} \sin q_2 \dot{q}_1 \dot{q}_3 + \frac{1}{2} J^S_{\zeta\zeta} \dot{q}_3^2 \qquad (2.191)$$

The kinetic energy also is not constant at constant input velocities $\dot{q}_1 = \dot{q}_2 = \dot{q}_3 = \Omega$ in the case of $J^S_{\xi\xi} = J^S_{\eta\eta} = J^S_{\zeta\zeta} = J$ (e.g. rigid body as a homogeneous sphere), but depends on the angle q_2: $W_{\text{kin}} = \frac{1}{2}[J_A + J_B + J(3 + 2 \sin q_2)]\Omega^2$. This is due to the fact that a moment Q_2 according to (2.189) is acting to cause this state of motion, see also (2.31).

2.4.2 Planar Mechanisms

2.4.2.1 General Perspective

Planar mechanisms with their links moving in parallel planes are used more frequently in mechanical engineering than spatial mechanisms, since they are less complicated to build and more conveniently calculated, especially if they have just a single drive. From a mechanical point of view, they are special cases of the mechanisms discussed in Sect. 2.4.1 to which $n = 1$, $q_1 = q$, $\dot{z}_{Si} \equiv 0$, $\omega_{\xi i} = \omega_{\eta i} \equiv 0$ and $\omega_{\zeta i} = \dot{\varphi}_i$ applies. These planar mechanisms are not dismissed as special cases, but discussed in more detail below to spare the reader who is interested in these objects only the reading of Sect. 2.4.1.

The (kinematically) planar mechanism is assumed to consist of a total of I rigid bodies that are numbered in such a way that the frame is number 1. The input link is assigned the number $i = 2$, and the output link the number I. The geometrical

2.4 Kinetics of Multibody Systems

conditions are determined by the structure of the mechanism and the dimensions of its links.

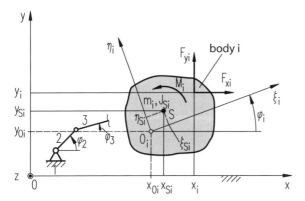

Fig. 2.15 Nomenclature for a rigid body in planar motion (link i)

Given are the characteristic mass parameters for planar motion of all rigid bodies: the positions of the centers of gravity in the body-fixed (ξ_{Si}; η_{Si}) or fixed reference system (x_{Si}; y_{Si}), the masses m_i and the moments of inertia with respect to the body-fixed axes through the center of gravity J_{Si} that were designated as $J_{\zeta\zeta i}^{S}$ in Sect. 2.3.1. External forces and moments such as input and braking torques, friction forces and moments, cutting and pressing forces, etc. can act on each body (link). The forces applied onto the ith body are captured with their components in the fixed coordinate directions and labeled as F_{xi} and F_{yi}. The applied moment on the ith body is M_i.

Figure 2.15 defines the nomenclature of the applied forces and geometrical dimensions at an arbitrary body. Geometrical relations between the position of the input link identified by the generalized coordinate q and those coordinates that specify the position of each rigid body can be formulated based on the structure and dimensions of a machine. For mechanisms with a rotating input link as in Figs. 2.15, 2.18, 2.26, 2.29 and 2.20, $q = \varphi_2$ is often selected, but in principle a translational coordinate can be used as well.

The dependency of the center-of-gravity coordinates and the angles of rotation φ_i on the input coordinate are known in the form of zeroth-order position functions:

$$x_{Si} = x_{Si}(q); \qquad y_{Si} = y_{Si}(q); \qquad \varphi_i = \varphi_i(q); \qquad i = 2, 3, \ldots, I. \quad (2.192)$$

Their calculation will be explained with reference to multiple examples in the following Sections.

Starting from the time dependence of the input coordinate $q = q(t)$, the positions of the links of the mechanism can be determined as time functions, see also (2.142), (2.143) and (2.144):

$$x_{Si}(t) = x_{Si}[q(t)]; \qquad y_{Si}(t) = y_{Si}[q(t)]; \qquad \varphi_i(t) = \varphi_i[q(t)]. \quad (2.193)$$

The velocities result from differentiation with respect to time according to the chain rule

$$\dot{x}_{Si} = \frac{dx_{Si}}{dt} = \frac{dx_{Si}}{dq}\frac{dq}{dt} = x'_{Si}\dot{q}, \qquad \dot{y}_{Si} = y'_{Si}\dot{q}, \qquad \dot{\varphi}_i = \varphi'_i\dot{q} \qquad (2.194)$$

Derivatives with respect to the input coordinate q are denoted by a dash, total derivatives with respect to time by a dot. The accelerations are calculated as follows:

$$\ddot{x}_{Si} = \frac{d^2 x_{Si}}{dt^2} = \frac{d\dot{x}_{Si}}{dt} = \frac{d(x'_{Si}\dot{q})}{dt} = \frac{dx'_{Si}}{dt}\dot{q} + x'_{Si}\frac{d\dot{q}}{dt} = \frac{dx'_{Si}}{dq}\frac{dq}{dt}\dot{q} + x'_{Si}\ddot{q} \quad (2.195)$$

In summary, the following applies:

$$\begin{aligned}\ddot{x}_{Si}(q,t) &= x''_{Si}(q)\dot{q}^2(t) + x'_{Si}(q)\ddot{q}(t)\\ \ddot{y}_{Si}(q,t) &= y''_{Si}(q)\dot{q}^2(t) + y'_{Si}(q)\ddot{q}(t)\\ \ddot{\varphi}_i(q,t) &= \varphi''_i(q)\dot{q}^2(t) + \varphi'_i(q)\ddot{q}(t)\end{aligned} \qquad (2.196)$$

This representation contains a separation of the position functions from zeroth-order position functions $x_{Si}(q)$, $y_{Si}(q)$, $\varphi_i(q)$ to second-order position functions (x''_{Si}, y''_{Si}, φ''_i), which are independent of the state of motion from the time functions $q(t)$, $\dot{q}(t)$, and $\ddot{q}(t)$ of the input link, which characterize the state of motion.

The position functions can be specified analytically in closed form for simple systems, such as gear mechanisms, slider-crank mechanisms, and others; they can be calculated numerically (software) for more complex systems, such as multi-link mechanisms.

The kinetic energy is derived taking into consideration the translational motions of all centers of gravity and the rotations about the axis through the center of gravity of all moving links

$$W_{\text{kin}} = \frac{1}{2}\sum_{i=2}^{I}\left[m_i\left(\dot{x}_{Si}^2 + \dot{y}_{Si}^2\right) + J_{Si}\dot{\varphi}_i^2\right] \qquad (2.197)$$

If the relationships (2.194) are used, the following results from (2.197) as a special case of (2.148)

$$W_{\text{kin}} = \frac{1}{2}\dot{q}^2\sum_{i=2}^{I}\left[m_i\left(x'^2_{Si} + y'^2_{Si}\right) + J_{Si}\varphi'^2_i\right] = \frac{1}{2}J(q)\dot{q}^2, \qquad (2.198)$$

if the generalized mass, that is also called a reduced moment of inertia, is introduced in the form of

$$J_{\text{red}} = J(q) = \sum_{i=2}^{I}\left[m_i\left(x'^2_{Si} + y'^2_{Si}\right) + J_{Si}\varphi'^2_i\right] \qquad (2.199)$$

2.4 Kinetics of Multibody Systems

(special case of m_{11}). If one compares (2.198) with (2.197), it becomes evident that the kinetic energy of the generalized mass is equal to the kinetic energy of all moving links. The generalized mass $J(q)$ has the dimension of a moment of inertia if the generalized coordinate q is an angle, and it has the dimension of a mass if q is a translation. $J(q)$ is always positive. It is worth noting that the first-order position function in (2.199) appear in quadratic form and thus $J(q)$ is independent of the direction of the motion.

Potential energy is often stored in mechanisms in the form of lifting work and/or deformation work of the spring (spring constant c, spring length l, unstretched spring length l_0)

$$W_{\text{pot}} = \sum \left[m_i g y_{Si} + \frac{1}{2} c_i (l_i - l_{0i})^2 \right] \tag{2.200}$$

(the y axis being directed vertically upwards). The total mass of the moving links and the overall center-of-gravity height y_S are

$$m = \sum_{i=2}^{I} m_i \tag{2.201} \qquad y_S = \frac{1}{m} \sum_{i=2}^{I} m_i y_{Si} \tag{2.202}$$

The applied non-potential forces F_{xi}, F_{yi} and moments M_i that act on the links of the mechanism are referred to the generalized coordinate, see (2.153). Their work must be equal to the work of the generalized force Q. Thus

$$dW = Q dq = \sum_{i=2}^{I} (F_{xi} dx_i + F_{yi} dy_i + M_i d\varphi_i) \tag{2.203}$$

It follows for the power of the applied forces:

$$P = Q \frac{dq}{dt} = \sum_{i=2}^{I} \left(F_{xi} \frac{dx_i}{dt} + F_{yi} \frac{dy_i}{dt} + M_i \frac{d\varphi_i}{dt} \right) \tag{2.204}$$

Using (2.194) and after dividing by $\dot{q}$, one finds the sought-after equation for the generalized force

$$Q = \sum_{i=2}^{I} (F_{xi} x'_i + F_{yi} y'_i + M_i \varphi'_i) = Q_{\text{an}} + Q^* \tag{2.205}$$

Q mostly is not constant but depends on the position (q), velocity ($\dot{q}$) and/or time (t). The generalized driving force Q_{an} (input torque M_{an} for a rotary drive and the input force F_{an}) for a linear actuator is not a potential force and contained in Q. It is useful to label and highlight them separately. The other non-potential forces are included in the quantity Q^*.

Lagrange's equation of the second kind for this system with one degree of freedom is

$$\frac{d}{dt}\left(\frac{\partial L}{\partial \dot{q}}\right) - \frac{\partial L}{\partial q} = Q \tag{2.206}$$

with the Lagrangian:
$$L = W_{\text{kin}} - W_{\text{pot}}. \tag{2.207}$$

The differentiations result in the following with (2.198):

$$\frac{\partial L}{\partial \dot{q}} = J(q)\dot{q}$$

$$\frac{\mathrm{d}}{\mathrm{d}t}\left(\frac{\partial L}{\partial \dot{q}}\right) = \frac{\mathrm{d}J(q)}{\mathrm{d}t}\dot{q} + J(q)\ddot{q} = J'(q)\dot{q}^2 + J(q)\ddot{q}$$

$$\frac{\partial L}{\partial q} = \frac{1}{2}J'(q)\dot{q}^2 - W'_{\text{pot}} \tag{2.208}$$

Equations (2.207), (2.208), and (2.206) result in:

$$J(q)\ddot{q} + \frac{1}{2}J'(q)\dot{q}^2 + W'_{\text{pot}} = Q_{\text{an}} + Q^* \tag{2.209}$$

This equation of motion of the rigid machine with one degree of freedom is a special case of (2.155).

2.4.2.2 Hoisting Gear

The hoisting gear of a crane as shown in 2.16 is represents a mechanism with constant transmission ratio, consisting of two gear steps and the translationally moving load. It is assumed that the hoisting cable is massless, has no bending stiffness and is as perfectly rigid (in the axial direction) as all the other components.

The geometrical parameters of the system, i.e. the pitch radii of the gears (r_2, r_{32}, r_{34}, r_4) and the cable length l (hereinafter not relevant). The mass parameters of the system are the moments of inertia of the gears about their axes through the center of gravity that coincide with their axes of rotation (J_2; J_3; J_4) and the mass of the hoisting load m_5. The moments of inertia J_2, J_3, and J_4 include those of the motor, the coupling, the cable drum and all other rotating parts. The external force field consists of the static weight of the load (the force $F_{y5} = -m_5 g$ that acts against the coordinate direction y) and the input torque M_{an}.

The quantities to be found as functions of the coordinate $q = \varphi_2$ are the reduced moment of inertia J_{red} and the equation of motion.

The solution starts with establishing the constraints. The following geometrical relations can be obtained from Figure 2.16:

$$r_2\varphi_2 = -r_{32}\varphi_3, \qquad r_{34}\varphi_3 = -r_4\varphi_4, \qquad y_{S5} = r_4\varphi_4 - l \tag{2.210}$$

After a brief manipulation, this yields the position functions in the form of the equations (2.192):

$$\varphi_3 = -\frac{r_2}{r_{32}}\varphi_2, \qquad \varphi_4 = \frac{r_2 r_{34}}{r_4 r_{32}}\varphi_2, \qquad y_{S5} = \frac{r_{34} r_2}{r_{32}}\varphi_2 - l \tag{2.211}$$

2.4 Kinetics of Multibody Systems

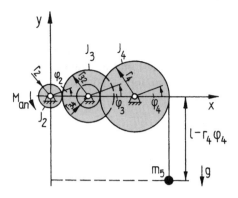

Fig. 2.16 Gear diagram of a hoisting gear

The first-order position functions are:

$$\varphi'_2 = 1, \qquad \varphi'_3 = -\frac{r_2}{r_{32}} = u_{23}, \qquad \varphi'_4 = \frac{r_2 r_{34}}{r_4 r_{32}} = u_{24}, \qquad y'_{S5} = r_2 \frac{r_{34}}{r_{32}} \qquad (2.212)$$

This is where the gear ratios u_{2k} are introduced, see (2.144), (2.163) and (2.164). When changing to a slower motion, the value of a gear ratio is greater than one, and it is smaller than one when changing to a faster motion. The sign indicates the direction of rotation relative to the drive direction. The reduced moment of inertia according to (2.199) is

$$J_{\text{red}} = J_2 \varphi'^2_2 + J_3 \varphi'^2_3 + J_4 \varphi'^2_4 + m_5 y'^2_{S5},$$

i.e., it is

$$J_{\text{red}} = J_2 + J_3 u_{23}^2 + J_4 u_{24}^2 + m_5 \left(\frac{r_2 r_{34}}{r_{32}}\right)^2 = \text{const.} \qquad (2.213)$$

It can be seen that the gear ratios u_{2k} enter quadratically into the calculation of the generalized mass so that their signs (direction of rotation) are not relevant. This squared gear ratio entails that the reduced moment of inertia of many gear mechanisms is primarily determined by the moment of inertia of the fast running gear step and that the total moment of inertia of a gear mechanism can often be estimated by taking the moment of inertia of the first step and multiplying it by a factor (e. g. 1.1 to 1.2).

The change in the center of gravity height of the moving masses that determines the hoisting work is $y'_S = y'_{S5}$, since only the mass m_5 changes its height, see (2.202). The equation of motion thus results from (2.209) with $W'_{\text{pot}} = m_5 g y'_{S5}$ and $Q^* \equiv Q$ to become

$$\left[J_2 + J_3 u_{23}^2 + J_4 u_{24}^2 + m_5 \left(\frac{r_2 r_{34}}{r_{32}}\right)^2\right] \ddot{\varphi}_2 + m_5 g (r_2 r_{34}/r_{32}) = M_{\text{an}} \qquad (2.214)$$

2.4.2.3 Four-Bar Linkage

Four-bar linkages with a rotating input link are used in many machines in the form of crank-rocker mechanisms (output link *4* rocks back and forth) or double-crank mechanisms (output link *4* rotates fully) to generate non-uniform motions.

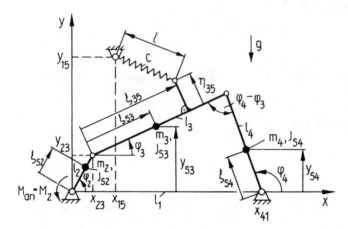

Fig. 2.17 Geometrical and mechanical parameters in a four-bar linkage with a translational spring

The given quantities are the lengths l_i shown in Fig. 2.17, the center-of-gravity coordinates ξ_{Si} in the body-fixed reference system, the masses m_i and moments of inertia J_{Si}, which are causing inertia forces and moments. A spring force, the magnitude of which can be calculated from the spring constant c and the spring deflection, which in turn results from the unstretched spring length l_0 and the instantaneous spring length l, acts in addition to the input torque M_2 and the static weight of the links. The spring length depends on the fixed coordinates x_{15}, y_{15} and the body-fixed coordinates of the pivot point (ξ_{35}, η_{35}).

Sought-after quantities are general formulae for calculating the reduced moment of inertia and the other terms that enter into the equation of motion of the rigid machine for $q = \varphi_2$.

The equations for calculating the angles φ_3 and φ_4 are derived from the constraints. These express the fact that the projections of the coordinates of the joint points onto the two coordinate directions form a solid straight line (loop), see Fig. 2.17:

$$l_3 \cos \varphi_3 = l_4 \cos \varphi_4 + l_1 - l_2 \cos \varphi_2 \tag{2.215}$$

$$l_3 \sin \varphi_3 = l_4 \sin \varphi_4 - l_2 \sin \varphi_2 \tag{2.216}$$

Squaring and adding yields

$$l_3^2 = l_4^2 + l_1^2 + l_2^2 - 2l_1 l_2 \cos \varphi_2 + 2l_4(l_1 - l_2 \cos \varphi_2) \cos \varphi_4 - 2l_4 l_2 \sin \varphi_2 \sin \varphi_4. \tag{2.217}$$

2.4 Kinetics of Multibody Systems

If one solves this equation, the result obtained after some intermediate calculations with the abbreviations

$$a_{34} = \frac{2(l_2 \cos\varphi_2 - l_1)l_4}{N}, \quad b_{34} = \frac{2l_2 l_4 \sin\varphi_2}{N}, \quad w_{34} = a_{34}^2 + b_{34}^2 \quad (2.218)$$

and the denominator

$$N = (l_2 \cos\varphi_2 - l_1)^2 + l_2^2 \sin^2\varphi_2 + l_4^2 - l_3^2 \quad (2.219)$$

is the sine and cosine of φ_4:

$$\sin\varphi_4 = \frac{b_{34} - a_{34}\sqrt{w_{34} - 1}}{w_{34}}, \quad \cos\varphi_4 = \frac{a_{34} + b_{34}\sqrt{w_{34} - 1}}{w_{34}}. \quad (2.220)$$

The other unknown trigonometric functions are most easily derived from (2.215) and (2.216) using (2.220):

$$\cos\varphi_3 = \frac{l_4 \cos\varphi_4 - l_2 \cos\varphi_2 + l_1}{l_3}, \quad \sin\varphi_3 = \frac{l_4 \sin\varphi_4 - l_2 \sin\varphi_2}{l_3}. \quad (2.221)$$

This allows the calculation of all (zeroth-order) position functions of the centers of gravity, see Fig. 2.17:

$$x_{S2} = \xi_{S2} \cos\varphi_2, \quad\quad y_{S2} = \xi_{S2} \sin\varphi_2$$
$$x_{S3} = l_2 \cos\varphi_2 + \xi_{S3} \cos\varphi_3, \quad y_{S3} = l_2 \sin\varphi_2 + \xi_{S3} \sin\varphi_3 \quad (2.222)$$
$$x_{S4} = l_1 + \xi_{S4} \cos\varphi_4, \quad\quad y_{S4} = \xi_{S4} \sin\varphi_4.$$

The first-order position functions are then derived by differentiation with respect to the input coordinate $q = \varphi_2$:

$$x'_{S2} = -\xi_{S2} \sin\varphi_2, \quad\quad y'_{S2} = \xi_{S2} \cos\varphi_2$$
$$x'_{S3} = -l_2 \sin\varphi_2 - \xi_{S3}\varphi'_3 \sin\varphi_3, \quad y'_{S3} = l_2 \cos\varphi_2 + \xi_{S3}\varphi'_3 \cos\varphi_3 \quad (2.223)$$
$$x'_{S4} = -\xi_{S4}\varphi'_4 \sin\varphi_4, \quad\quad y'_{S4} = \xi_{S4}\varphi'_4 \cos\varphi_4.$$

Here, the as yet undetermined first-order position functions of the angles φ_3 and φ_4 appear. They can be calculated from the following system of linear equations that results from differentiating (2.215) and (2.216):

$$-l_3 \sin\varphi_3 \varphi'_3 + l_4 \sin\varphi_4 \varphi'_4 = l_2 \sin\varphi_2 \quad (2.224)$$
$$l_3 \cos\varphi_3 \varphi'_3 + l_4 \sin\varphi_4 \varphi'_4 = -l_2 \cos\varphi_2. \quad (2.225)$$

This results in

$$\varphi'_3 = \frac{l_2 \sin(\varphi_2 - \varphi_4)}{l_3 \sin(\varphi_4 - \varphi_3)}; \quad \varphi'_4 = \frac{l_2 \sin(\varphi_2 - \varphi_3)}{l_4 \sin(\varphi_4 - \varphi_3)}. \quad (2.226)$$

The reduced moment of inertia of the four-bar linkage is derived from (2.199) for $I = 4$

$$J_{\text{red}} = m_2(x'^2_{S2} + y'^2_{S2}) + J_{S2}\varphi'^2_2 + m_3(x'^2_{S3} + y'^2_{S3}) + J_{S3}\varphi'^2_3 \\ + m_4(x'^2_{S4} + y'^2_{S4}) + J_{S4}\varphi'^2_4, \tag{2.227}$$

yielding the following when using (2.223):

$$J_{\text{red}} = m_2\xi^2_{S2} + J_{S2}\varphi'^2_2 + m_3(l^2_2 + 2l_2\xi_{S3}\cos(\varphi_2 - \varphi_3)\varphi'_3 + \xi^2_{S3}\varphi'^2_3) + J_{S3}\varphi'^2_3 \\ + (m_4\xi^2_{S4} + J_{S4})\varphi'^2_4. \tag{2.228}$$

The first-order position functions of the angles that appear herein are known from (2.226). According to (2.209), the input torque necessary to overcome the inertia forces depends on the derivative of the reduced moment of inertia and the square of the angular velocity. The static portions of the moment M_{st} from the weight of the links are derived – without explicitly using the equilibrium conditions – from the position function of the center-of-gravity height:

$$W'_{\text{pot}} = M_{\text{st}} = mgy'_S = (m_2 y'_{S2} + m_3 y'_{S3} + m_4 y'_{S4})g. \tag{2.229}$$

This portion of the moment can be calculated as a function of the crank angle φ_2 using the position functions y'_{Si} known from (2.223).

What remains to be calculated is the portion of the moment that the spring attached to an arbitrary point of the link 3 exerts onto the input link 2. The spring moment is derived from the potential spring energy $W_{\text{pot F}} = c(l - l_0)^2/2$:

$$W'_{\text{pot F}} = M_c = c(l - l_0)l' = cll'\left(1 - \frac{l_0}{\sqrt{l^2}}\right). \tag{2.230}$$

The spring length in the loaded condition is calculated from the coordinates of both spring pivot points using the Pythagorean theorem:

$$l^2 = (x_{35} - x_{15})^2 + (y_{35} - y_{15})^2. \tag{2.231}$$

Implicit differentiation results in:

$$2ll' = 2(x_{35} - x_{15})x'_{35} + 2(y_{35} - y_{15})y'_{35} \tag{2.232}$$

and provides the expression required for (2.230). The position functions of the spring pivot point are required for this. One can gather from Fig. 2.17 that the following geometrical relationships apply:

$$\begin{aligned} x_{35} &= l_2 \cos\varphi_2 + \xi_{35}\cos\varphi_3 - \eta_{35}\sin\varphi_3 \\ y_{35} &= l_2 \sin\varphi_2 + \xi_{35}\sin\varphi_3 + \eta_{35}\cos\varphi_3. \end{aligned} \tag{2.233}$$

Their partial derivatives are

2.4 Kinetics of Multibody Systems

$$x'_{35} = -l_2 \sin \varphi_2 - (\xi_{35} \sin \varphi_3 + \eta_{35} \cos \varphi_3)\varphi'_3$$
$$y'_{35} = l_2 \cos \varphi_2 + (\xi_{35} \cos \varphi_3 - \eta_{35} \sin \varphi_3)\varphi'_3. \quad (2.234)$$

The spring moment according to (2.230) thus becomes:

$$M_c = c\left[(x_{35}-x_{15})\,x'_{35} + (y_{35}-y_{15})\,y'_{35}\right]\left[1 - \frac{l_0}{\sqrt{(x_{35}-x_{15})^2 + (y_{35}-y_{15})^2}}\right]. \quad (2.235)$$

Now all portions of the moment that are included in the equation of motion (2.209) are known for the four-bar linkage shown. Based on the expressions that can be calculated from the given parameters according to equations (2.228), (2.229), and (2.235), the equation of motion for $q = \varphi_2$ is:

$$J(q)\ddot{q} + \frac{1}{2}J'(q)\dot{q}^2 + M_{\text{st}}(q) + M_c(q) = M_{\text{an}}. \quad (2.236)$$

2.4.2.4 Large Press

The 14-link linkage shown as a schematic in Fig. 2.18a is used in a large press. When designing the input elements, the dynamic forces that occur in the operating states of startup, forming, and braking are of particular importance besides the kinematics. First, a software program was used to determine the function of the reduced moment of inertia and its derivative from the given dimensions and mass parameters, Fig. 2.18b. Then the equation of motion (2.209) was numerically integrated and the joint forces were calculated.

Various force fields have to be taken into account depending on the operating states of interest:

1. When braking or engaging the clutch, moments that depend on time occur due to the pneumatically operated friction clutches or brakes: $M(t)$
2. According to the motor characteristic, only a moment that depends on the angular velocity is present under the steady-state operating conditions: $M(\dot{\varphi}_2)$; see Sect. 1.5.2.
3. During pressing, forces occur that are both displacement- and velocity-dependent and that are to be referred to the input shaft: $M(\varphi_2, \dot{\varphi}_2)$.
4. The friction forces and moments depend on the joint forces, the relative velocities, and the friction coefficient and are to be captured in a function $M(\varphi_2, \dot{\varphi}_2)$.

It is often useful for such complex and expensive machines, such as large presses, to develop specific computer programs that capture their design specifics. It is the job of the designer to thoroughly compile the specifications required for such a calculation and to "work" with the software program.

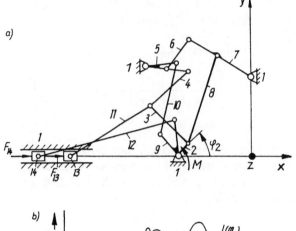

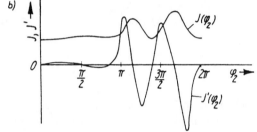

Fig. 2.18 Press drive;
a) schematic; b) Function of the reduced moment of inertia $J(\varphi_2)$ and its derivative $J'(\varphi_2)$

2.4.3 States of Motion of a Rigid Machine

The variation with time of an input motion can generally be obtained by numerical integration of the equation of motion (2.209) if the reduced moment of inertia J_{red} and the moment Q_{an} are given. Its solution in closed form is possible for a conservative force field, see 2.4.4. The result of the integration is the function of the input angle $\varphi(t)$ and its time derivatives $\dot{\varphi}(t)$ and $\ddot{\varphi}(t)$, which are required for calculating all other kinematic and dynamic quantities.

The diagram in Fig. 2.19 represents a typical operating cycle of a mechanism. It consists of start-up process, the steady (or stationary) state, and the coast-down process. Machines that work in an nonstationary operating cycle include cranes, excavators, vehicles, conveyors, presses, actuating and transport systems in which starting and braking processes are frequently repeated.

It is mostly the starting and braking times, the starting and braking distances and angles, and the moment variation with time that are most interesting from a practical point of view. The designer uses these quantities to compare various drive systems or to size motors, brakes, couplings, and clutches. The dynamic forces required to size the links and joints (bolts, bearings, gears, etc.) can also be calculated if the actual sequence of motions is known. These dynamic loads during starting and braking processes will often have to be determined to prove the operating strength.

2.4 Kinetics of Multibody Systems

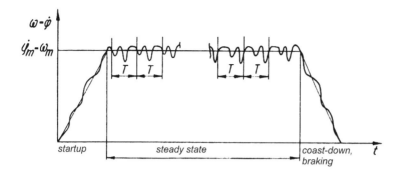

Fig. 2.19 Typical operating cycle of a drive

The friction moment of a machine is hard to calculate in advance since it depends on factors that are only determined during assembly or by the operating state, see (1.127). These include reaction forces for a statically indeterminate support and the operating temperature of a bearing (viscosity of the lubricant). The friction moment and friction loss are often approximated using the efficiency or determined experimentally using coast-down tests, see VDI Guideline 2158 [36].

The mechanical efficiency is defined as the ratio of output power P_m to input power $(P_m + P_v)$, where P_v is the friction loss:

$$\eta = \frac{P_m}{P_m + P_v} < 1. \tag{2.237}$$

The efficiency is specified in the technical literature on machine elements for specific sub-assemblies such as gear mechanisms, block and pulley systems of hoists, etc.

Electric motors are typically selected based on their driving power and heating up, taking into account their duty cycle. The input torque, however, is more meaningful than the driving power when characterizing the mechanical loads on machine elements.

In most machines, multiple drive mechanisms interact in an accurately coordinated sequence of motions. Designers use motion programs that describe the coordinated sequences of motions of all drives of a machine to make major decisions in the blueprint phase that also affect dynamic behavior. Figure 2.20 shows an example of a **motion program**.

Starting from the minimum engineering requirements, a designer has to determine all sequences of motions with consideration to the dynamic aspects to ensure stable a operation even at high operating speeds. Since each mechanism involves a different set of inertia forces, the one that is most dynamically demanding should be designed, for example, such that the unsteady stages of motion are stretched over a longer periods of time.

In the example shown in Fig. 2.20, the motion stages *1*, *3*, *4* and *6* exhibit the greatest accelerations. The reduced moment of inertia changes most in these sec-

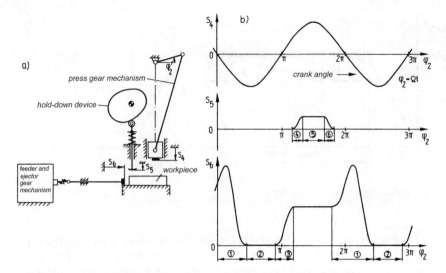

Fig. 2.20 Cutting machine as an example of a machine with multiple mechanisms
a) schematic of mechanism, **b)** motion programs of the three sequences of motion; six stages:
1 eject, *2* take up, *3* feed, *4* hold down, *5* press, *6* release

tions. The designer also has to take into account the influence on the excitation of torsional vibrations, see Sect. 4.3.

2.4.4 Solution of the Equations of Motion

The treatment of starting and braking processes involves the mathematical problem of integrating the differential (2.209) under the initial conditions

$$t = 0: \qquad \varphi(0) = \varphi_0, \qquad \dot{\varphi}(0) = \omega_0 \qquad (2.238)$$

(angle of rotation $\varphi = q$).

In physical terms, this means that the sequence of motions $\varphi(t)$ must be determined if an initial position φ_0 and an initial angular velocity ω_0 are given at a specific time. An analytical solution can be specified for any angular dependence $Q_{an} = M_{an}(\varphi)$, $Q^* = M^*(\varphi)$ that also includes constant values. Since the moments of the inertia forces result from the change in kinetic energy, as can be seen from the following

$$W'_{\text{kin}} = \frac{dW_{\text{kin}}}{d\varphi} = \frac{d\left(J\dot{\varphi}^2/2\right)}{d\varphi} = \frac{1}{2}J'\dot{\varphi}^2 + \frac{1}{2}J\frac{d\dot{\varphi}^2}{d\varphi} = \frac{1}{2}J'\dot{\varphi}^2 + J\ddot{\varphi} \qquad (2.239)$$

(2.209) can be written in this way:

2.4 Kinetics of Multibody Systems

$$W'_{\text{kin}} = M_{\text{an}} + M^* - W'_{\text{pot}}. \tag{2.240}$$

Integration, starting from the initial state according to (2.238) to an arbitrary position φ yields:

$$\int_{W_{\text{kin }0}}^{W_{\text{kin}}} dW_{\text{kin}} = \frac{1}{2}J(\varphi)\dot{\varphi}^2 - \frac{1}{2}J(\varphi_0)\omega_0^2$$

$$= \int_{\varphi_0}^{\varphi} (M_{\text{an}} + M^*)d\overline{\varphi} - W_{\text{pot}}(\varphi) + W_{\text{pot}}(\varphi_0). \tag{2.241}$$

If the **work of the applied force field** and the potential energy are jointly abbreviated by

$$W(\varphi, \varphi_0) = \int_{\varphi_0}^{\varphi} (M_{\text{an}} + M^*)d\overline{\varphi} - W_{\text{pot}}(\varphi) + W_{\text{pot}}(\varphi_0) \tag{2.242}$$

the first result obtained from (2.241) is

$$W_{\text{kin}} = \frac{1}{2}J(\varphi)\dot{\varphi}^2 = \frac{1}{2}J(\varphi_0)\omega_0^2 + W(\varphi, \varphi_0) \tag{2.243}$$

and then the dependence of the angular velocity on the angle of rotation with $W_{\text{kin }0} = J(\varphi_0)\omega_0^2/2$ is determined:

$$\dot{\varphi}(\varphi) = \sqrt{\frac{J(\varphi_0)\omega_0^2 + 2W(\varphi, \varphi_0)}{J(\varphi)}} = \omega_0 \sqrt{\frac{J(\varphi_0)}{J(\varphi)}\left(1 + \frac{W(\varphi, \varphi_0)}{W_{\text{kin }0}}\right)}. \tag{2.244}$$

If one entirely neglects the work W of the applied force field, a special case of (2.244) and its derivative follow:

$$\dot{\varphi}(\varphi) = \omega_0 \sqrt{\frac{J(\varphi_0)}{J(\varphi)}} = \sqrt{\frac{2W_{\text{kin }0}}{J(\varphi)}}, \quad \ddot{\varphi}(\varphi) = \frac{-J'(\varphi)J(\varphi_0)}{2J^2(\varphi)}\omega_0^2 = \frac{-J'W_{\text{kin }0}}{J^2}. \tag{2.245}$$

The state of motion described in (2.245) results when the mechanism is left to its own devices in position φ_0 with the initial energy $W_{\text{kin }0}$. No input torque is required for this **eigenmotion**. Check this by insertion into (2.209)!

The periodic motion that can be caused by the variability of $J(\varphi)$ and/or $M(\varphi)$ is expressed by the coefficient of speed fluctuation δ. Such studies of machines were first conducted in the 19th century in connection with the development of steam engines. The **coefficient of speed fluctuation** expresses the variation of the angular velocity $\omega = \dot{\varphi}$ of the drive during one operating cycle (usually one full revolution) relative to the mean value:

$$\delta = \frac{\omega_{\max} - \omega_{\min}}{\omega_m} \approx \frac{2(\omega_{\max} - \omega_{\min})}{\omega_{\max} + \omega_{\min}}. \qquad (2.246)$$

The coefficient of speed fluctuation is $\delta = 0$ at $\omega_{\max} = \omega_{\min}$ and $\delta = 2$ at $\omega_{\min} = 0$. The smaller its coefficient of speed fluctuation is the smoother a machine operates.

The extreme angular velocities can be specified according to (2.245) for the approximation $W \ll W_{\text{kin 0}}$. If one assumes the mean moment of inertia $J(\varphi_0) = J_m$ and the mean angular velocity $\omega_0 = \omega_m$, one gets:

$$\omega_{\min} = \omega_0 \sqrt{\frac{J_m}{J_{\max}}}; \qquad \omega_{\max} = \omega_0 \sqrt{\frac{J_m}{J_{\min}}}. \qquad (2.247)$$

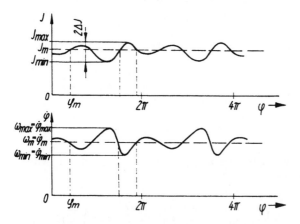

Fig. 2.21 Typical time function of the reduced moment of inertia and the angular velocity of a mechanism without an external force field

These relationships are illustrated in Fig. 2.21. (2.246) and (2.247), after brief manipulation result in

$$\delta = 2\frac{\sqrt{J_{\max}} - \sqrt{J_{\min}}}{\sqrt{J_{\max}} + \sqrt{J_{\min}}} \approx \frac{\Delta J}{J_m}\left[1 + \frac{1}{4}\left(\frac{\Delta J}{J_m}\right)^2 + \cdots\right]. \qquad (2.248)$$

If the external force field ($W \ll W_{\text{kin 0}}$) is negligibly small, one only needs to determine the function $J(\varphi)$ for an operating cycle and use the result to determine the difference ΔJ to be able to specify the coefficient of speed fluctuation. Based on the approximation

$$\Delta J = \frac{J_{\max} - J_{\min}}{2} \ll J_m; \qquad J_m = \frac{1}{2\pi}\int_0^{2\pi} J(\varphi)d\varphi \approx \frac{J_{\max} + J_{\min}}{2} \qquad (2.249)$$

(2.248) can be simplified by series expansion

2.4 Kinetics of Multibody Systems

$$\sqrt{J_{\max/\min}} = \sqrt{J_m \pm \Delta J} = \sqrt{J_m}\left[1 \pm \frac{\Delta J}{2J_m} - \frac{1}{8}\left(\frac{\Delta J}{J_m}\right)^2 + \cdots\right]. \quad (2.250)$$

The approximation for calculating the required mean value for a given coefficient of speed fluctuation results from (2.248):

$$J_m = \frac{\Delta J}{\delta}\left[1 + \frac{\delta^2}{4 + 2\delta^2} + \cdots\right] \approx \frac{\Delta J}{\delta_{\text{zul}}}. \quad (2.251)$$

Consider now the other special case in which the coefficient of speed fluctuation is mainly determined by the work $W(\varphi, \varphi_0)$ and not by $J(\varphi)$. The reduced moment of inertia is captured by its mean value J_m according to (2.249). In the steady state, (2.244) with $J(\varphi_0) = J(\varphi) = J_m$, the mean angular velocity $\omega_0 = \omega_m$ and the mean kinetic energy $J_m \omega_m^2 / 2$ leads to:

$$\omega_{\min} = \omega_m\left(1 + \frac{W_{\min}}{2W_{\text{kin m}}}\right); \quad \omega_{\max} = \omega_m\left(1 + \frac{W_{\max}}{2W_{\text{kin m}}}\right). \quad (2.252)$$

If one inserts the extreme values into (2.246) an alternative for equations (2.248) and (2.251) is obtained:

$$\delta = \frac{\Delta W}{2T_m} = \frac{\Delta W}{J_m \omega_m^2} \quad \text{or} \quad J_m = \frac{\Delta W}{\omega_m^2 \delta}. \quad (2.253)$$

$\Delta W = W_{\max} - W_{\min}$ is the surplus work per period.

It follows from (2.251) and (2.253) that for a given surplus work ΔW, the coefficient of speed fluctuation gets smaller the greater the mean reduced moment of inertia J_m is. To obtain a more uniform motion, the mean moment of inertia must be increased. This can be achieved using a flywheel.

A **flywheel** functions as a storage device for kinetic energy. It compensates the coefficient of speed fluctuation by accumulating kinetic energy in the acceleration phase and releasing it under load conditions. It allows the use of a drive motor with a breakdown moment that is smaller than the reduced static input torque. It is often useful to let the flywheel rotate continuously and to engage the clutch of the mechanism during the working cycles or load phases only. Flywheels are primarily used in machines that work in steady-state mode of operation.

It must be distinguished whether the flywheel is placed between the motor and the mechanism or between the mechanism and the work machine, see Fig. 2.27. Installation between the mechanism and the work machine is useful to protect the mechanism from sudden load increases. The flywheel mass decreases, however, if the flywheel is installed on the fast running shaft between motor and mechanism. A designer has to evaluate the importance of these criteria and decide for the case on hand which arrangement is preferable. A small moment of inertia is preferable for machines with unsteady operation because frequent starting and stopping processes will put larger loads on the motors and brakes (risk of overheating) when the flywheel is bigger.

The kinetic energy of flywheels is not only used to drive toys but also to drive vehicles (gyrobuses in Switzerland since 1945). It has been proven that energy on the order of magnitude of 400 000 N · m/kg can be stored in high-strength fast-rotating flywheels. Values of 1 750 000 N · m/kg are said to be physically possible in theory, which means that superflywheels can store more energy relative to their mass than electrochemical batteries.

In real-world machines, the applied forces and moments depend in a complex manner on the mechanism position, angular velocity, and time. If these functions are known, the equation of motion (2.209) can be solved numerically after solving it for the angular acceleration ($\varphi \widehat{=} q$):

$$\ddot{\varphi} = \ddot{\varphi}(\varphi, \dot{\varphi}, t) = \frac{1}{J(\varphi)} \left[M_{an} + M^* - W'_{pot} - \frac{1}{2} J'(\varphi) \dot{\varphi}^2 \right] \quad (2.254)$$

The sequence of motions can be calculated step-by-step for small time increments Δt under the initial conditions of (2.238).

All simulation programs use numerical integration methods that were worked out by mathematicians. There are many methods of interpretation, and a selection of which that an engineer does not have to know in detail is implemented in the software. An engineer should know, however, that these methods calculate the function values step by step starting from the initial conditions.

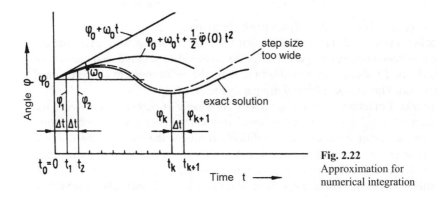

Fig. 2.22 Approximation for numerical integration

Figure 2.22 illustrates the general approach. Starting from the time $t_0 = 0$, the function values with respect to time $t_{k+1} = t_k + \Delta t$ are calculated incrementally from the time values t_k. The accuracy of $\varphi(t_k)$ depends on the correct selection of the increment Δt. It follows from the mathematical analysis that the error changes with $(\Delta t)^5$. This means that the accuracy of the method is improved by a factor of 16 when the increment is cut in half and that it deteriorates by the same factor when the increment is doubled. Since the number of steps is inversely proportional to the increment, this factor is not 32 but 16 for a given interval.

High accuracy is often achieved by selecting small increments but there is a risk that round-off errors add up due to the large number of steps. It can be recommended

2.4.5 Example: Press Drive

A baler consists of a crank-rocker mechanism, the geometrical parameters of which are characterized by $r_4/l_5 \ll 1$ and $r_4/l_4 \ll 1$, see Fig. 2.23a. The dynamic behavior is to be calculated taking into account the motor characteristic, the processing force at the output and the friction moment, focusing particularly on the analysis of the influence exerted by the flywheel. The mass of link 4 is assumed to be distributed across the adjacent links.

Given:
$r_2 = 80$ mm, $r_3 = 320$ mm, $r_4 = 150$ mm,
$l_5 = 1,0$ m, $x_{15} \approx l_4$, $\xi_{S5} = 1,5$ m.
$J_2 = 0,03$ kg · m^2,
$J_3 = 10; 25; 100; 200$ kg · m^2,
$m_5 = 40$ kg,
$J_{S5} = 36$ kg · m^2.
Press force $F_0 = 7,6$ kN, see curve in Fig. 2.23b,
Friction moment $M_R = (7,5 + 0,022\dot{\varphi}_2^2)$ N · m referred to angle φ_2, determined experimentally, see also problem P1.6 ($\dot{\varphi}_2$ in rad/s)
Motor torque $M_{an} = M_0(1 - \dot{\varphi}_2/\Omega)$ with $M_0 = 10\,200$ N · m and $\Omega = 2\pi n$
Synchronous motor speed: $n = 750$ min^{-1}
Forces from static weights are deemed negligible as compared to inertia forces.

Find:
1. Function of reduced moment of inertia $J(\varphi_2)$
2. Functions of angular velocity and input torque in steady-state operation
3. Influence of flywheel size on angular velocity and input torque by varying J_3
4. Effective power, total power and efficiency of this press drive.

The solution starts by establishing the constraints, see Fig. 2.23:

$$\begin{aligned} r_2\varphi_2 &= -r_3\varphi_3, & x_{45} &\approx r_4\cos\varphi_3 + l_4 \\ x_{45} &= x_{15} + l_5\sin\left(\tfrac{\pi}{2} - \varphi_5\right) &\approx x_{15} + l_5\cdot\left(\tfrac{\pi}{2} - \varphi_5\right) \\ x_{S5} &= x_{15} + \xi_{S5}\sin\left(\tfrac{\pi}{2} - \varphi_5\right) &\approx x_{15} + \xi_{S5}\cdot\left(\tfrac{\pi}{2} - \varphi_5\right) \\ y_{S5} &= \xi_{S5}\cos\left(\tfrac{\pi}{2} - \varphi_5\right) &\approx \xi_{S5} \end{aligned} \qquad (2.255)$$

If the motor angle φ_2 is defined as the generalized coordinate, the following applies:

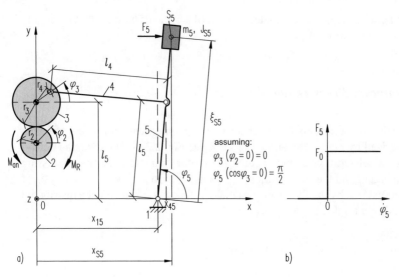

Fig. 2.23 Press drive; **a)** schematic of mechanism, **b)** dependence of press force on angular velocity of rocker

$$\varphi_3 = -\frac{r_2}{r_3}\varphi_2, \qquad\qquad \varphi_3' = -\frac{r_2}{r_3}$$

$$\varphi_5 \approx \frac{\pi}{2} - \frac{r_4}{l_5}\cos\left(\frac{r_2}{r_3}\varphi_2\right), \qquad \varphi_5' \approx \frac{r_4\,r_2}{l_5\,r_3}\sin\left(\frac{r_2}{r_3}\varphi_2\right) \qquad (2.256)$$

$$x_{S5} \approx \xi_{S5}\frac{r_4}{l_5}\cos\left(\frac{r_2}{r_3}\varphi_2\right) + x_{15}, \qquad x_{S5}' \approx -\xi_{S5}\frac{r_4\,r_2}{l_5\,r_3}\sin\left(\frac{r_2}{r_3}\varphi_2\right)$$

$$y_{S5} \approx \xi_{S5}, \qquad\qquad y_{S5}' \approx 0$$

According to (2.199), the reduced moment of inertia is

$$J(\varphi_2) = J_2 + J_3\varphi_3'^2 + m_5(x_{S5}'^2 + y_{S5}'^2) + J_{S5}\varphi_5'^2 \qquad (2.257)$$

After a brief calculation, the above expressions result in

$$J(\varphi_2) = J_2 + J_3\left(\frac{r_2}{r_3}\right)^2 + \left(J_{S5} + m_5\xi_{S5}^2\right)\left(\frac{r_4\,r_2}{l_5\,r_3}\right)^2 \sin^2\left(\frac{r_2\varphi_2}{r_3}\right). \qquad (2.258)$$

When using the given parameter values ($J_3 = 25$ kg $\cdot$ m^2), the following applies

$$\underline{\underline{J(\varphi_2)}} = \left[1,5925 + 0,1772\sin^2\left(\frac{r_2\varphi_2}{r_3}\right)\right] \text{ kg}\cdot\text{m}^2$$

$$= \left(1,6811 - 0,0886\cos\frac{\varphi_2}{2}\right) \text{ kg}\cdot\text{m}^2, \qquad (2.259)$$

because $r_3 = 4r_2$ and based on the identity of $\sin^2\alpha = \frac{1}{2}(1 - \cos 2\alpha)$.

2.4 Kinetics of Multibody Systems

The derivative with respect to φ_2 is

$$J'(\varphi_2) = 0{,}0443 \sin \frac{\varphi_2}{2} \text{ kg} \cdot \text{m}^2. \tag{2.260}$$

The moment of the applied loads referred to the input angle φ_2 is according to (2.205)

$$M(\varphi_2) = M_{\text{an}}(\dot\varphi_2) + F_5(\dot\varphi_5) x'_{\text{S}5} - M_{\text{R}}(\dot\varphi_2). \tag{2.261}$$

Especially the portion that is caused by the processing force $F_5(\dot\varphi_5)$ according to

$$F_5 = \begin{cases} F_0 & \text{for } \dot\varphi_5 \geq 0 \\ 0 & \text{for } \dot\varphi_5 < 0 \end{cases} \tag{2.262}$$

– designated here as processing moment $M_{\text{t}} = F_5 x'_{\text{S}5} \cdot \dot\varphi_2$ can be stated as follows due to $\dot\varphi_2 > 0$ and $\dot\varphi_5 = \varphi'_5 \dot\varphi_2$:

$$M_{\text{t}}(\varphi_2) = F_5 x'_{\text{S}5} = \begin{cases} -F_0 \xi_{\text{S}5} \dfrac{r_4 \, r_2}{l_5 \, r_3} \sin\left(\dfrac{r_2}{r_3}\varphi_2\right) & \text{for } \sin\left(\dfrac{r_2}{r_3}\varphi_2\right) \geq 0 \\ 0 & \text{for } \sin\left(\dfrac{r_2}{r_3}\varphi_2\right) < 0 \end{cases} \tag{2.263}$$

or, using the given parameter values:

$$M_{\text{t}}(\varphi_2) = \begin{cases} -427{,}5 \text{ N} \cdot \text{m} \cdot \sin\left(\dfrac{\varphi_2}{4}\right) & \text{fr } \sin\left(\dfrac{\varphi_2}{4}\right) \geq 0 \\ 0 \text{ N} \cdot \text{m} & \text{fr } \sin\left(\dfrac{\varphi_2}{4}\right) < 0. \end{cases} \tag{2.264}$$

The equation of motion of the rigid machine was integrated numerically after it was converted into the form

$$\ddot\varphi_2 = \frac{1}{J(\varphi_2)} \cdot \left(M_0 \cdot \left(1 - \frac{\dot\varphi_2}{\Omega}\right) + M_{\text{t}}(\varphi_2) - M_{\text{R}}(\dot\varphi_2) - \frac{1}{2} J'(\varphi_2) \dot\varphi_2^2 \right) \tag{2.265}$$

The individual portions of the moment including the input torque were calculated from the functions of $\varphi_2(t)$ and $\dot\varphi_2(t)$ as determined first. Their curves are shown in Fig. 2.24.

As can be seen, a mean angular velocity of $\omega_{2m} = 76{,}4 \text{ s}^{-1}$ is obtained (cycle time $T_0 = 8\pi/\omega_{2m} = 0{,}329$ s) which, as was to be expected, is somewhat lower than the synchronous speed of the motor that amounts to $\Omega = 750\pi/30 = 78{,}5 \text{ s}^{-1}$. The mean input torque $M_{\text{an}} = 272 \text{ N} \cdot \text{m}$, shown in Fig. 2.24c results from an approximate calculation of the sum of the mean friction moment $M_{\text{R}} = 136 \text{ N} \cdot \text{m}$ (Fig. 2.24b) and the mean processing moment $M_{\text{t}} = W_{\text{N}}/8\pi = 136 \text{ N} \cdot \text{m}$. The function of the speed can be interpreted by comparing it to the curve of the input torque and taking the linear motor characteristic into account. The speed rises/falls when the input torque increases/decreases. The sharp drop in speed at a relatively small J_3 is due to the processing moment M_{t}, see Fig. 2.24a.

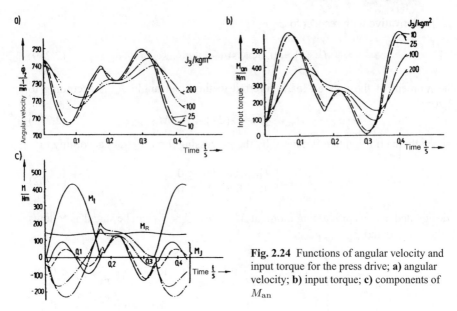

Fig. 2.24 Functions of angular velocity and input torque for the press drive; **a)** angular velocity; **b)** input torque; **c)** components of M_{an}

The input torque M_{an} is composed of the three components that are represented individually in Fig. 2.24c. The significant influence of the flywheel is clearly visible. While M_t and M_R are influenced to a minor extent only by the speed variation (so that this influence is hidden by the thickness of the line in the figure), the kinetic moment is affected considerably by the flywheel size. $M_{kin} = J(\varphi_2)\ddot{\varphi}_2 + 0{,}5 J'(\varphi_2)\dot{\varphi}_2^2$ applies, and both the effects of the angular acceleration and of the angular velocity can be seen from the curve in Fig. 2.24c. When the J_3 values are small, the portion of the moment that originates from the variability of the moment of inertia is dominant. In the curves in Fig. 2.24a, one can notice the dual variation per period that is due to $J'(\varphi_2)$. When the values of J_3 are larger, the influence of the angular acceleration becomes dominant although the angular acceleration itself decreases.

The results show that, as a result of the large flywheel, the peak value of the input torque may be smaller than the one that results from the friction moment and the processing moment. The mean value of the input torque is virtually unaffected by the size of the flywheel. The larger the flywheel is, the smaller are the speed variations in the steady operating state.

A complete operating cycle corresponds to a full crank revolution ($0 < \varphi_3 < 2\pi$), i.e. four revolutions of the motor shaft ($0 < \varphi_2 < 8\pi$). According to (2.242), the effective work during the operating cycle due to the pressing action is

$$W_N = -\int_0^{8\pi} M_t(\varphi_2)d\varphi_2 = F_0 r_4 \frac{\xi_{S5}}{l_5} \frac{r_2}{r_3} \int_0^{4\pi} \sin\frac{\varphi_2}{4} d\varphi_2 + \int_{4\pi}^{8\pi} 0 \, d\varphi_2$$

$$= 8 F_0 r_4 \frac{\xi_{S5}}{l_5} \frac{r_2}{r_3} = 3420 \text{ N} \cdot \text{m}.$$

(2.266)

2.4 Kinetics of Multibody Systems

At the mean angular velocity of $\omega_{2m} = 76,4 \text{ s}^{-1}$, the effective mechanical power is $P_m = W_N/T = 3420 \text{ N} \cdot \text{m}/0,329 \text{ s} = 10,4 \text{ kW}$. The mean input torque $M_{an} = 272 \text{ N} \cdot \text{m}$ that results from all four moment components yields a total power of $P_m + P_v = M_{an}\omega_{2m} = 272 \cdot 76,4 \text{ W} = 20,8 \text{ kW}$. According to (2.237), the efficiency of this press drive is only about $\eta = 0.5$.

2.4.6 Problems P2.5 to P2.8

P2.5 Drive of a Belt-Type Stacker for Strip Mining

The grossly simplified calculation model shown in Fig. 2.25 reflects the drive system of a belt-type stacker used for dumping overburden in strip mining. The slewing gear, which consists of a motor, a coupling, and two gear mechanisms, sets the top section into motion.

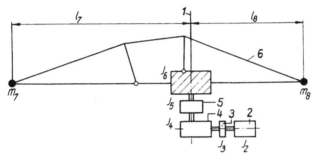

Fig. 2.25 Kinematic schematic of a belt-type stacker for strip mining
1 swivel axis; *2* motor; *3* coupling; *4, 5* gear mechanism; *6* top section

Given: moments of inertia of the
- motor: $J_2 = 2,14 \text{ kg} \cdot \text{m}^2$;
- coupling: $J_3 = 1,12 \text{ kg} \cdot \text{m}^2$;
- gear mechanism 1: $J_4 = 22,6 \text{ kg} \cdot \text{m}^2$; referred to
- gear mechanism 2: $J_5 = 4540 \text{ kg} \cdot \text{m}^2$; transmission output
- engine room: $J_6 = 1,185 \cdot 10^8 \text{ kg} \cdot \text{m}^2$;
- Top section masses: $m_7 = 2,05 \cdot 10^4 \text{ kg}$; $m_8 = 1,85 \cdot 10^5 \text{ kg}$;
- Top section lengths: $l_7 = 110 \text{ m}$; $l_8 = 61 \text{ m}$;
- Gear ratios: $u_{42} = u_{43} = 627$; $u_{54} = u_{64} = 36,2$

Note that the sequence of indices is relevant for the gear ratio ($\varphi'_k = u_{2k} = 1/u_{k2}$) and that it is defined as the ratio of input to output angular velocities, see (2.212).

Find:

1. Input torque of the motor required to move the top section at an angular acceleration of $\ddot{\varphi}_6 = 0.0007 \text{ rad/s}^2$.
2. Input torque M_6 referred to the swivel axis.

P2.6 Slider-Crank Mechanism

Slider-crank mechanisms are used for transforming rotational into translational motions (and vice versa). The moment of inertia referred to the crank angle is required for dynamic calculations.

Given: Dimensions and parameters according to Fig. 2.26a.

Find:

1. Reduced moment of inertia J_red using 2 equivalent masses for the connecting rod
2. Mean value J_m for the reduced moment of inertia
3. Input torque at $\dot\varphi_2 = \Omega$.

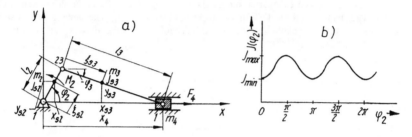

Fig. 2.26 Slider-crank mechanism; **a)** nomenclature, **b)** curve of the reduced moment of inertia

$\lambda = l_2/l_3 \ll 1$ applies to the crank ratio. This fact can be utilized by expanding the appearing root expressions into series and neglecting λ^2 with respect to 1.

P2.7 Flywheel Placement

Two possible placements of the flywheel can be used in a design. Either J_{S1} or J_{S2} are to be used to maintain a specific coefficient of speed fluctuation, see Fig. 2.27.

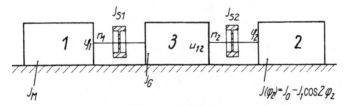

Fig. 2.27 Possible flywheel placements; *1* motor; *2* work machine; *3* gear mechanism, gear ratio u_{12}

Given:
Moments of inertia J_M; J_G; J_0; J_1; $J_1 \ll J_0$; coefficient of speed fluctuation δ_zul; Gear ratio $u_{21} = n_1/n_2 > 1$.

Find:
Formula for calculating the required moments of inertia of the flywheels when the mechanism is idling. Compare the quantities of J_{S1} and J_{S2}.

P2.8 Influence of the Flywheel in a Forming Machine

The forming force only acts in a small range of the operating cycle of presses, cutting machines and other forming machines. The drives of forming machines are therefore equipped with flywheels, which release kinetic energy during the forming process and are "recharged" in the remaining time of each cycle.

The input torque at the motor and the function of the angular velocity of the motor shaft are to be determined for the steady-state operation of a crank press with a basic structure as shown in Fig. 2.28. To simplify the problem, friction can be neglected. It is assumed that the mass of the connecting rod has already been distributed over the adjacent links.

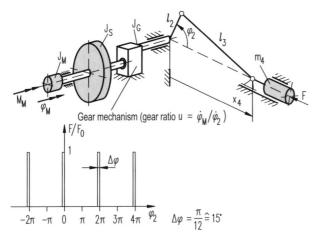

Fig. 2.28 Basic structure of a crank press and function of the forming force

Given:

Link lengths of crank and coupler (connecting rod)	$l_2 = 0.22$ m, $l_3 = 1$ m
Gear ratio	$u = \dot{\varphi}_M / \dot{\varphi}_2 = 70$
Mass of the ram	$m_4 = 8000$ kg
Moment of inertia of the flywheel on motor shaft (2 variants)	$J_S = \begin{cases} 3,5 \text{ kg} \cdot \text{m}^2, \text{ variant A} \\ 39,5 \text{ kg} \cdot \text{m}^2, \text{ variant B} \end{cases}$
Moment of inertia of the motor armature	$J_M = 0.5$ kg $\cdot$ m^2
Moment of inertia of the gear mechanism (referred to the motor shaft)	$J_G = 0.5$ kg $\cdot$ m^2
Breakdown moment	$M_K = 19.5$ N $\cdot$ m
Breakdown slippage, see (1.126)	$s_K = 0.12$
Synchronous speed of the motor	$n_0 = 1500$ rpm ($\Omega_1 = 157.1 \text{ s}^{-1}$)
Angular range of the acting forming force	$\Delta\varphi = \pi/12 \widehat{=} 15°$
Forming force (at $2k\pi - \Delta\varphi \leq \varphi_2 \leq 2k\pi$), with $k = \ldots, -2, -1, 0, 1, 2, \ldots$ (Fig. 2.28)	$F_0 = 3.2$ MN

Find:

1. Using the approximations given in Table 2.1, analytical solutions for
   1.1 the moment of inertia $J(\varphi_2)$ referred to the crank angle
   1.2 the crank moment $M_{\text{St}} = M_{\text{K}}\,(\ddot\varphi_2 = 0)$ required for $\ddot\varphi_2 \approx 0$ and the forming work W to be performed per cycle
   1.3 the mean moment M_{Stm} in the shaft between gear mechanism and crank for $\ddot\varphi_2 \approx 0$
2. The functions (for steady-state operation) of
   2.1 the input torque M_{M} of the motor
   2.2 the angular velocity $\dot\varphi_{\text{M}}$ of the motor shaft for both flywheel variants using the SimulationX® [34] program

2.4.7 Solutions S2.5 to S2.8

S2.5 This is a mechanism with constant transmission ratio ($J' = 0$). Thus the input torque follows from (2.209):

$$M_{\text{an}} = J\ddot\varphi_2 = (J_A + J_O)\ddot\varphi_2 \tag{2.267}$$

According to (2.205), the relationship $M_{\text{an}} = M_6\varphi'_6 = M_6 u_{26} = M_6 u_{24} u_{46}$ applies between the input torque and the moment at the swivel axis. The moment of inertia of the drive referred to the motor shaft is see (2.199) and (2.213):

$$J_A = J_2 + J_3 + J_4 u_{24}^2 + J_5 u_{25}^2 = J_2 + J_3 + \frac{J_4}{u_{42}^2} + \frac{J_5}{u_{42}^2 u_{54}^2} = 3{,}26 \text{ kg} \cdot \text{m}^2. \tag{2.268}$$

The moment of inertia of the top section referred to the motor shaft is

$$J_O = \frac{J_6 + m_7 l_7^2 + m_8 l_8^2}{u_{42}^2 u_{64}^2} = \frac{1055 \cdot 10^6 \text{ kg} \cdot \text{m}^2}{627^2 \cdot 36.2^2} = 2.05 \text{ kg} \cdot \text{m}^2. \tag{2.269}$$

As the above numerical values show, the reduced moment of inertia of the drive system is greater (and has a greater influence on the startup behavior) due to the high gear ratio than that of the entire top section with its huge masses. The input torque of the motor therefore is $\ddot\varphi_2 = \ddot\varphi_6 u_{42} u_{64} = 15.89 \text{ rad/s}^2$ and equals

$$\underline{M_{\text{an}} = J\ddot\varphi_2} = (3.26 + 2.05) \cdot 15.89 \text{ kg} \cdot \text{m}^2/\text{s}^2 = \underline{84.37 \text{ N} \cdot \text{m}}. \tag{2.270}$$

The moment at the swivel axis is

$$\underline{M_6} = u_{42} \cdot u_{64} M_{\text{an}} = 22\,665 \cdot 84{,}37 \text{ N} \cdot \text{m} = \underline{1.915 \text{ MN} \cdot \text{m}}. \tag{2.271}$$

S2.6 A brief calculation produces the exact coordinates given in Table 2.1 for the slider-crank mechanism due to simple geometrical relations.
The second and third columns list values that result for $\lambda = l_2/l_3 < 1$ from a series expansion.
The reduced moment of inertia is obtained according to (2.199):

$$J(\varphi_2) = m_2\left(x'^2_{S2} + y'^2_{S2}\right) + J_{S2}\varphi'^2_2 + m_3\left(x'^2_{S3} + y'^2_{S3}\right) + J_{S3}\varphi'^2_3 + m_4 x'^2_{S4}. \tag{2.272}$$

If one only considers terms up to the second power of λ, the values from the third column of Table 2.1 result in

2.4 Kinetics of Multibody Systems

Table 2.1 Position functions of the slider-crank mechanism, see Fig. 2.26

Zeroth-order position function			first-order position function
exact	approximation $\lambda = l_2/l_3 \ll 1$	for	($\lambda \ll 1$)
$x_{S2} = \xi_{S2} \cos \varphi_2$	$x_{S2} = \xi_{S2} \cos \varphi_2$		$x'_{S2} = -\xi_{S2} \sin \varphi_2$
$y_{S2} = \xi_{S2} \sin \varphi_2$	$y_{S2} = \xi_{S2} \sin \varphi_2$		$y'_{S2} = \xi_{S2} \cos \varphi_2$
$x_{S3} = l_2 \cos \varphi_2$ $+\xi_{S3} \cos \varphi_3$	$x_{S3} = l_2 \cos \varphi_2$ $+\xi_{S3}\left(1 - \frac{\lambda^2}{4}\right)$ $+\xi_{S3}(\frac{\lambda^2}{4}) \cos 2\varphi_2$		$x'_{S3} = -l_2 \sin \varphi_2$ $-\xi_{S3}(\frac{\lambda}{2}) \sin 2\varphi_2$
$y_{S3} = l_2 \sin \varphi_2$ $+\xi_{S3} \sin \varphi_3$	$y_{S3} = (l_2 - \xi_{S3}\lambda) \sin \varphi_2$		$y'_{S3} = (l_2 - \xi_{S3}\lambda) \cos \varphi_2$
$x_4 = x_{S4} = l_2 \cos \varphi_2$ $+l_3 \cos \varphi_3$	$x_{S4} = l_2 \cos \varphi_2$ $+l_3\left(1 - \frac{\lambda^2}{4}\right)$ $+l_2(\frac{\lambda}{4}) \cos 2\varphi_2$		$x'_{S4} = -l_2 \sin \varphi_2$ $-l_2(\frac{\lambda}{2}) \sin 2\varphi_2$
$\varphi_3 = \arcsin(-\lambda \sin \varphi_2)$	$\varphi_3 = -\left(\lambda + \frac{\lambda^3}{8}\right) \sin \varphi_2$ $+(\frac{\lambda^3}{24}) \sin 3\varphi_2$		$\varphi'_3 = -\left(\lambda + \frac{\lambda^3}{8}\right) \cos \varphi_2$ $+(\frac{\lambda^3}{8}) \cos 3\varphi_2$

$$J(\varphi_2) = m_2\xi_{S2}^2 + J_{S2} + m_3 l_2^2 \left\{1 + \left[-2\frac{\xi_{S3}}{l_3} + \left(\frac{\xi_{S3}}{l_3}\right)^2\right] \cos^2 \varphi_2 + \cdots\right\}$$
$$+ J_{S3}\lambda^2 \cos^2 \varphi_2 + m_4 l_2^2 \sin^2 \varphi_2 (1 + \lambda \cos \varphi_2)^2 + \ldots \quad (2.273)$$

One can see from this that the influence of the moment of inertia of the connecting rod J_{S3} is small, since the crank ratio is squared. It is therefore logical to neglect this term just like all the other terms with higher powers of λ. m_3 is distributed over two equivalent masses m_{32} and m_{34} at the crank and piston bolts so that the mass and the center-of-gravity position remain the same (Fig. 2.29). Then:

$$m_{32} + m_{34} = m_3, \qquad m_{34}(l_3 - \xi_{S3}) = m_{32}\xi_{S3}. \quad (2.274)$$

The resulting two equivalent masses are:

$$m_{34} = m_3 \frac{\xi_{S3}}{l_3}, \qquad m_{32} = m_3 \left(1 - \frac{\xi_{S3}}{l_3}\right). \quad (2.275)$$

The reduced moment of inertia can be specified using

$$J_A = J_{S2} + m_2\xi_{S2}^2 \quad (2.276)$$

as follows:

$$\underline{J(\varphi_2) = J_A + m_{32}l_2^2 + (m_4 + m_{34})l_2^2 \sin^2 \varphi_2 (1 + 2\lambda \cos \varphi_2 + \cdots)}. \quad (2.277)$$

Consequently,

$$J'(\varphi_2) = (m_4 + m_{34})l_2^2 \cdot \left[\sin 2\varphi_2 - \frac{\lambda}{2}(\sin \varphi_2 - 3 \sin 3\varphi_2) + 0(\lambda^2)\right]. \quad (2.278)$$

A variable input torque must act on the crankshaft for it to rotate at a constant angular velocity $\dot{\varphi}_2 = \Omega$. If one inserts $J(\varphi_2)$ into (2.209), it follows that:

$$M_{an} = (m_4 + m_{34})l_2^2 \Omega^2 \left[-\frac{1}{4}\lambda \sin \Omega t + \frac{1}{2} \sin 2\Omega t + \frac{3}{4}\lambda \sin 3\Omega t \right] + F_4 x_4'. \quad (2.279)$$

A smaller input torque would occur if one did not force a constant speed $\dot\varphi_2 = \Omega$, but allowed a variation of the speed around its mean value.

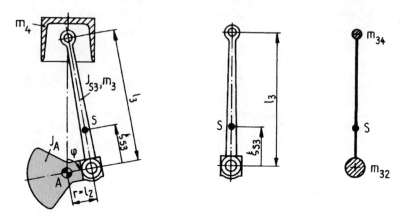

Fig. 2.29 Distribution of the connecting rod mass over two equivalent masses

S2.7 First, the moment of inertia referred to φ_1 is determined according to (2.199):

$$\begin{aligned} J(\varphi_1) &= J_M + J_{S1} + J_G + u_{12}^2 (J_{S2} + J_0 - J_1 \cos 2\varphi_1) \\ J(\varphi_1) &= J_m - u_{12}^2 J_1 \cos 2\varphi_1; \qquad u_{12} = 1/u_{21} \end{aligned} \quad (2.280)$$

Since $J_1 \ll J_0$, $u_{12}^2 J_1 \ll J_m$ applies as well; $\Delta J = u_{12}^2 J_1$, see (2.249). (2.251) can be used as a first approximation for idling:

$$J_m = \frac{\Delta J}{\delta_{zul}} = J_M + J_{S1} + J_G + u_{12}^2(J_{S2} + J_0). \quad (2.281)$$

The following is found for the flywheel on the fast running shaft $J_{S1}(J_{S2} = 0)$:

$$J_{S1} = u_{12}^2 \left(\frac{J_1}{\delta_{zul}} - J_0 \right) - (J_M + J_G) \quad (2.282)$$

and for the flywheel on the slow running shaft $J_{S2}(J_{S1} = 0)$:

$$J_{S2} = \left(\frac{J_1}{\delta_{zul}} - J_0 \right) - \frac{1}{u_{12}^2}(J_M + J_G); \qquad J_{S2} = \frac{J_{S1}}{u_{12}^2} = J_{S1} u_{21}^2 > J_{S1}. \quad (2.283)$$

2.4 Kinetics of Multibody Systems

S2.8 The moment of inertia $J(\varphi_2)$ referred to the crank angle is derived according to (2.199):

$$J(\varphi_2) = (J_\mathrm{M} + J_\mathrm{S})\varphi_\mathrm{M}'^2 + J_\mathrm{G} + m_4 x_4'^2 \tag{2.284}$$
$$\approx (J_\mathrm{M} + J_\mathrm{S})u^2 + J_\mathrm{G} + m_4 l_2^2 \sin^2\varphi_2 (1 + \lambda\cos\varphi_2). \tag{2.285}$$

Based on the formula given in Table 2.1,

$$\frac{\mathrm{d}x_4}{\mathrm{d}\varphi_2} = x_4' \approx -l_2\left(\sin\varphi_2 + \frac{\lambda}{2}\sin 2\varphi_2\right) \tag{2.286}$$

was inserted.

When taking a static view (e. g. for extremely slow operation where $\dot\varphi_2 \approx 0$), the crank moment would have to be large enough to overcome the forming force. According to (2.205), the following applies with k as cycle index ($k = 1, 2, \ldots$)

$$M_\mathrm{St} = M_\mathrm{K}(\dot\varphi_2 \approx 0) \tag{2.287}$$

$$= -F x_4' \approx \begin{cases} 0 & \text{for } 2(k-1)\pi \leq \varphi_2 \leq 2k\pi - \Delta\varphi \\ -F_0 l_2\left(\sin\varphi_2 + \dfrac{\lambda}{2}\sin 2\varphi_2\right) & \text{for } 2k\pi - \Delta\varphi \leq \varphi_2 \leq 2k\pi \end{cases}$$

The maximum value in each cycle occurs at the beginning of the load ($\varphi_2 = 2k\pi - \Delta\varphi$) exerted by the press force:

$$M_{\mathrm{St\,max}} = F_0 l_2\left(\sin\Delta\varphi + \frac{\lambda}{2}\sin 2\Delta\varphi\right) = 0{,}313\,82\,F_0 l_2 = 220{,}9\,\mathrm{kN}\cdot\mathrm{m}. \tag{2.288}$$

The work of the processing force has to be performed by the drive motor. It results from the energy balance for a period of the mean values of the input torque from

$$W = F_0\left[x_4(\varphi_2 = 2\pi - \Delta\varphi) - x_4(\varphi_2 = 2\pi)\right] = 2\pi M_\mathrm{Stm} \tag{2.289}$$

$$= F_0 l_2\left[1 - \cos\Delta\varphi + \frac{\lambda}{4}(1 - \cos 2\Delta\varphi)\right] = 0{,}041\,44\,F_0 l_2 = 29{,}2\,\mathrm{kN}\cdot\mathrm{m}.$$

The mean moment M_Stm in the shaft between the gear mechanism and the crank or the input torque M_anm at the motor shaft can be calculated from this energy requirement for the forming process:

$$M_\mathrm{Stm} = \frac{W}{2\pi} = 4{,}643\,\mathrm{kN}\cdot\mathrm{m}; \qquad M_\mathrm{anm} = \frac{M_\mathrm{Stm}}{u} = 66{,}4\,\mathrm{N}\cdot\mathrm{m}. \tag{2.290}$$

These moments are considerably smaller than the static peak moment, as can be seen from comparing M_Stm with $M_{\mathrm{St\,max}}$, see (2.288). The moment M_anm provides an approximate value for the motor selection because it is around this mean value that the variable moments vary due to the impact-like forming force. Apart from the small inertia forces in the case on hand, only this small moment would be required if the forming work could be applied uniformly over the entire angular range.

In the limiting case of very slow angular velocity, the inertia force of the ram is zero, and the motor would only have to produce the static moment M_Stm.

The steady operating state is described by the equation of motion of the rigid machine that takes the following form when taking into account the motor characteristic:

$$J(\varphi_2)\ddot\varphi_2 + \frac{1}{2}J'(\varphi_2)\dot\varphi_2^2 = M_\mathrm{St} + \frac{2M_\mathrm{K}}{s_\mathrm{K}}\left(1 - \frac{\dot\varphi_2}{\Omega_1}\right) \tag{2.291}$$

The expressions from (2.284) and (2.287) must be inserted. There is no analytical solution. The numerical solution is performed using the SimulationX® program, see the model in Fig. 2.30. The results of the simulation calculation are shown in Fig. 2.31.

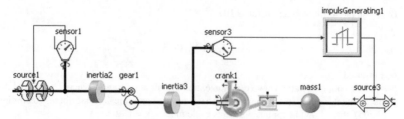

Fig. 2.30 Calculation model of the crank press in SimulationX® [34]

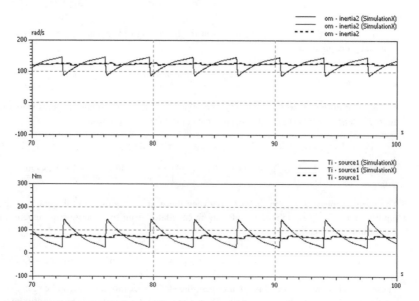

Fig. 2.31 Influence of a flywheel in a forming machine;
solid line: $J_S + J_M = 4.0\,\text{kg}\cdot\text{m}^2$; dashed line: $J_S + J_M = 40\,\text{kg}\cdot\text{m}^2$
a) Angular velocity of the motor shaft; **b)** Input torque of the motor

Figure 2.31 shows the simulation results for the two flywheels of variants A and B in comparison. The motor moment at the input shaft results from the forming force according to (2.287) and the inertia force of the ram. The forming process occurs where the moment rises steeply and the angular velocity drops sharply, when passing through the angle $\Delta\phi$.

The greater the moment of inertia of the flywheel is, the more the peak value of M_{St} is reduced. The smaller the flywheel is in size, the more the angular velocity varies. The mean value of the motor moment matches the amount calculated in (2.290).

The kinetic energy stored in the flywheel is released for the forming work to be performed in the angular range $\Delta\varphi$. The variation of the motor moment in Fig. 2.31 shows how the drive motor re-accelerates the system outside the forming process. The drop in moment that can be observed in this process is due to the linear motor characteristic.

This solution shows that the **mechanical energy requirement** is independent of the size of the flywheel and that the motor can be sized based on the mean moment. The drive motor does not have to generate the large force (or moment, respectively) that the forming process requires for a short time. It can feed mechanical energy for a longer period of time into the flywheel so that it does not heat up excessively in permanent operation. The motor heating-up is proportional to the squared mean value of the moment.

The function of the input torque depends on the speed-torque characteristic of the electric motor. A motor with a "soft" characteristic results in a higher drop in speed than a motor with a "hard" characteristic. The motor will heat up less if a bigger flywheel is used on the fast running motor shaft. It would be uneconomical to size the drive motor based on the peak moment if a small mean input torque is sufficient for permanent operation.

2.5 Joint Forces and Foundation Loading

2.5.1 General Perspective

The inertia forces that develop inside machines during motions often considerably exceed the static forces from the static weights of the components. The accelerations of mechanism links often amount to a multiple of the gravitational acceleration, see Table 2.2. A designer needs methods to determine the bearing and joint forces in any mechanism and machine from given characteristic values, such as mass parameters, geometric parameters, the external force field, and the sequence of motions $q(t)$. This information can be used to configure and size gears, bolts and bearings (surface pressure, deformation, ...), mechanism links (bending, shear, axial force, ...) and foundations (vibrations). The kinetostatic forces and moments that act on the frame are relevant for exciting vibrations of the foundation, see Sect. 3.

All mechanisms that fit into the model of the rigid machine exhibit the same connection between velocity and acceleration of the input link and the joint forces that are caused by inertia forces. This important connection, which was shown in (2.160), applies regardless of the structure of a mechanism to each internal force Q_p for an input motion $q_1(t)$:

$$F_{\text{kin}} = Q_p(t) = m_{p1}(q_1)\ddot{q}_1(t) + \Gamma_{11p}(q_1)\dot{q}_1^2(t). \qquad (2.292)$$

It is important, in this respect, that each joint force is composed of two terms that are associated with the acceleration and the squared velocity of the input link. The factors for these kinematic quantities depend on the mechanism position, the geometrical dimensions, and mass parameters.

For example, it can be concluded from (2.292) that

- for $\dot{q}_1$ = const., all joint forces increase with the square of the input speed
- the joint forces can be influenced using the velocity and acceleration curves, e. g. by controlled drives

- repeated calculation for different states of motion can be simplified since the influence of $\ddot{q}_1$ and $\dot{q}_1^2$ results from (2.292).

Mechanisms with a rotating driving crank perform a periodic motion, during which no harmonic but periodic excitation forces with circular excitation frequencies $\Omega; 2\Omega; 3\Omega; \ldots$ are generated:

$$F_{\text{kin}}(t) = \sum_{k=1}^{\infty} (A_k \cos k\Omega t + B_k \sin k\Omega t) = \sum_{k=1}^{\infty} C_k \sin(k\Omega t + \beta_k). \quad (2.293)$$

The Fourier coefficients A_k and B_k depend on the mass parameters. They are specified for simple examples in VDI Guideline 2149 Part 1 [35]. They can be calculated based on known mass parameters using known software for more complex mechanisms.

To illustrate the influence of speed, Fig. 2.32 shows the function of a joint force for three different speeds at a ratio of 1 : 2 : 3.

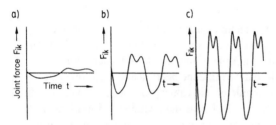

Fig. 2.32 Function of a periodic joint force component at various speeds; **a)** 100/min, **b)** 200/min, **c)** 300/min

Note that the period ($T_0 = 2\pi/\Omega$) **linear** (3 : 2 : 1) shortens, but the maximum force increases **quadratically** (1 : 4 : 9).

2.5.2 Calculating Joint Forces

The dynamic loads in many machines are considerably larger than the static ones, see Table 2.2. Noise caused by vibrations of the mechanism links and the housing, and the risk of interference with the technological flow are the reason why designers have to deal with the occurring dynamic joint forces in greater detail.

Below, a handy method for calculating the joint forces for planar mechanisms that are composed of simple groups of links with revolute joints will be described. The algorithm is based on formulae that apply to a respective dyad, see also VDI Guideline 2729 [36].

The definition of the forces that act onto a mechanism link can be seen from Fig. 2.33. Note that the components of the forces and moments are defined uni-

2.5 Joint Forces and Foundation Loading

Table 2.2 Speeds and maximum relative accelerations for some machine types

Machine type	speed n in 1/min	acceleration ratio a_{max}/g
Cutting machines, presses	30 ... 100	0.3 ... 3
Power looms	200 ... 600	1,0 ... 10
Knitting machines	1500 ... 3500	15 ... 60
Marine diesel engines	400 ... 500	70 ... 80
Domestic sewing machines	1000 ... 2000	50 ... 100
Industrial sewing machines	5000 ... 8000	300 ... 600

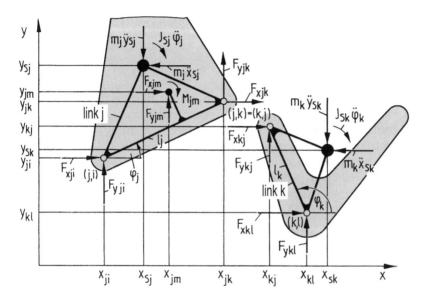

Fig. 2.33 Forces and moments at a dyad

formly in the directions specified for reasons of systematics, which in turn helps meet the requirements of computational processing. The force that is exerted onto link j by link k is designated $\boldsymbol{F}_{jk}$ and its components are defined positive in accordance with the coordinate directions. The equal and opposite counterforce is designated as $\boldsymbol{F}_{kj}$ and is defined in the same way. Therefore,

$$F_{xjk} + F_{xkj} = 0, \qquad F_{yjk} + F_{ykj} = 0 \qquad (2.294)$$

It is assumed that the coordinates of the joint points $(x_{ji}, y_{ji}, x_{jk} \equiv x_{kj}, y_{kj} = y_{jk}, x_{kl}, y_{kl}, x_{jm}, y_{jm})$ and the centers of gravity $(x_{Sj}, y_{Sj}, x_{Sk}, y_{Sk})$ for the outlined dyad are known from a previous kinematic analysis. The joint force (F_{xjm}, F_{yjm}) and the moment M_{jm} act on the joint point (j, m) from the adjacent mechanism link m. The force components are defined positive in the direction of the positive coordinate axes.

The following equations result from the equilibrium of moments about the revolute joint (j, i) and about the revolute joint (k, l):

$$-F_{xjk}(y_{jk}-y_{ji})+F_{yjk}(x_{jk}-x_{ji}) = F_{xjm}(y_{jm}-y_{ji})-F_{yjm}(x_{jm}-x_{ji})$$
$$+M_{jm}+m_j\ddot{y}_{Sj}(x_{Sj}-x_{ji}) \quad (2.295)$$
$$-m_j\ddot{x}_{Sj}(y_{Sj}-y_{ji})+J_{Sj}\ddot{\varphi}_j$$

$$-F_{xkj}(y_{kj}-y_{kl})+F_{ykj}(x_{kj}-x_{kl}) = m_k\ddot{y}_{Sk}(x_{Sk}-x_{kl})$$
$$-m_k\ddot{x}_{Sk}(y_{Sk}-y_{kl})+J_{Sk}\ddot{\varphi}_k \quad (2.296)$$

The four unknown quantities F_{xjk}, F_{xkj}, F_{yjk}, and F_{ykj} can be calculated from (2.294) to (2.296).

If one takes into account the equilibrium of forces in the horizontal and vertical directions, one can determine the remaining joint force components of interest:

$$\begin{aligned} F_{xji} &= m_j\ddot{x}_{Sj}-F_{xjk}-F_{xjm}, & F_{yji} &= m_j\ddot{y}_{Sj}-F_{yjk}-F_{yjm} \\ F_{xkl} &= m_k\ddot{x}_{Sk}-F_{xki}, & F_{ykl} &= m_k\ddot{y}_{Sk}-F_{ykj} \end{aligned} \quad (2.297)$$

Efficient algorithms for calculating the joint forces are based on the decomposition of multilink mechanisms into simple (statically determinate) groups of links.

In mechanisms with varying transmission ratio, the kinetic energy of all moving transmission links varies with the position of the mechanism. There is a permanent exchange of kinetic energy among the mechanism links via the joint forces. The work that is performed at the joint (j, k) by the joint force $\boldsymbol{F}_{jk}$ on link j has the same magnitude as the joint force $\boldsymbol{F}_{kj}$ on joint k, see Fig. 2.33.

The two forces that are applied at the "joint point" in the free-body diagram have the opposite sign ($\boldsymbol{F}_{kj} = -\boldsymbol{F}_{jk}$). In sum, the work of the joint force that acts onto the adjacent links (action = reaction) equals zero. Reaction forces thus do not perform any work in the overall system.

Now this view is generalized for an arbitrary link i ($i = 2, 3, \ldots, I$). The mechanical work that the joint force F_{ik} on a link i (that is viewed independently and cut free) depends on the path along which this joint travels during the motion. The following applies to the work of all joint forces F_{ik}, inertia forces and inertia moments acting on a mechanism link i onto which no applied forces and moments act along differentially small paths and angles according to the conservation of energy principle:

$$\mathrm{d}W_i = \sum_{k^*}(F_{xik}\mathrm{d}x_{ik}+F_{yik}\mathrm{d}y_{ik})-m_j(\ddot{x}_{Si}\mathrm{d}x_{Si}+\ddot{y}_{Si}\mathrm{d}y_{Si})-J_{Si}\ddot{\varphi}_i\mathrm{d}\varphi_i = 0. \quad (2.298)$$

The summation (Index k^*) is performed over all links connected to link i. $\dot{x}_{ik}$ and $\dot{y}_{ik}$ are the velocity components of the joint point (i, k). Furthermore:

$$\mathrm{d}W_i = \left[\sum_{k^*}(F_{xik}\dot{x}_{ik}+F_{yik}\dot{y}_{ik})-m_i(\ddot{x}_{Si}\dot{x}_{Si}+\ddot{y}_{Si}\dot{y}_{Si})-J_{Si}\ddot{\varphi}_i\dot{\varphi}_i\right]\mathrm{d}t = 0. \quad (2.299)$$

The kinetic energy of link i is, see (2.197) and (2.198),

$$W_{\mathrm{kin}\,i} = \frac{1}{2}\left[m_i(\dot{x}_{Si}^2+\dot{y}_{Si}^2)+J_{Si}\dot{\varphi}_i^2\right] = \frac{1}{2}J_{\mathrm{red}\,i}(q)\dot{q}^2. \quad (2.300)$$

2.5 Joint Forces and Foundation Loading

It follows from (2.299) and (2.300) that the time derivative of the kinetic energy, i.e. the kinetic power of the inertia forces and moments is equal to the power that the joint forces of the adjacent links transmit onto the link i:

$$P_{\text{kin } i} = \frac{\mathrm{d}W_{\text{kin } i}}{\mathrm{d}t} = \sum_{k^*}(F_{xik}\dot{x}_{ik} + F_{yik}\dot{y}_{ik}) \qquad (2.301)$$

$$= m_i(\ddot{x}_{Si}\dot{x}_{Si} + \ddot{y}_{Si}\dot{y}_{Si}) + J_{Si}\ddot{\varphi}_i\dot{\varphi}_i = \frac{1}{2}J'_{\text{red } i}(q)\dot{q}^3 + J_{\text{red } i}\ddot{q}\dot{q}.$$

For a constant input speed $\dot{q} = \Omega$, the kinetic power of the ith link is

$$P_{\text{kin } i} = \frac{1}{2}J'_{\text{red } i}(q)\Omega^3. \qquad (2.302)$$

The exchange of kinetic energy that takes place between the links is also of interest for assessing the dynamic behavior of a mechanism. The kinetic power on link i is a measure for the variation of the joint forces F_{ik}. In addition to J_{red}, the function of each portion of each link is of interest, i.e. the summands $J_{\text{red } i}(q)$ and their derivatives, see Fig. 2.36b.

2.5.3 Calculation of the Forces Acting onto the Frame

It is of great practical significance to know the dynamic excitation forces and moments that a machine transmits onto the frame, since these can excite undesirable vibrations in the subsoil or the buildings. The problems of machine foundations and vibration isolation that arise in this respect are discussed in more detail in Sect. 3.

It is not only the maximum value of the periodic forces and moments transmitted by a machine, but the size of each Fourier coefficient that is of interest in conjunction with vibration analyses for foundations, see 3.2.1.3. In multilink mechanisms, even the higher harmonics of the inertia forces are relevant. Often the task is to keep the inertia forces and specific excitation harmonics that are transmitted onto the foundation as small as possible. The respective methods for balancing of mechanisms and balancing of rotors are discussed in 2.6.

Consider a multilink mechanism whose links move in parallel planes that can be offset in the z direction, see, for example, Fig. 2.34. The goal is to determine the resultant forces and moments that are transmitted from the moving machine parts via the frame onto the foundation.

Internal static and kinetostatic forces and moments of the machine, such as spring forces between individual links, processing forces (e. g. cutting and pressing forces in forming machines and polygraphic machines, gas forces in internal combustion engines and compressors), have no influence on the foundation forces since they always occur in pairs and cancel each other out.

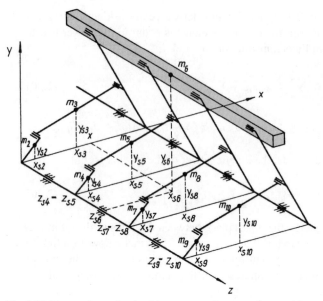

Fig. 2.34 Nomenclature for the drive system of a processing machine with multiple linkages that operate in parallel

In real machines in which the elasticity of the links plays a part, additional inertia forces ("vibration forces") that have an effect on the foundation can occur due to deformation of the links in addition to the kinetostatic forces and moments.

The resultant inertia forces and moments that act from the moving mechanism onto the machine frame are derived from the force and moment balances, see 2.3.2 and the forces and moments in Fig. 2.33. Since the motions are parallel to the x-y plane, $F_z = 0$, and the following forces result:

$$F_x = -\sum_{i=2}^{I} m_i \ddot{x}_{Si} = -m\ddot{x}_S; \qquad F_y = -\sum_{i=2}^{I} m_i \ddot{y}_{Si} = -m\ddot{y}_S. \qquad (2.303)$$

The input moment already known from (2.209) can also be stated as follows, see also (2.199):

$$M_{\text{an}} = \sum_{i=2}^{I} [m_i(\ddot{x}_{Si} x'_{Si} + \ddot{y}_{Si} y'_{Si}) + J_{Si} \ddot{\varphi}_i \varphi'_i] + W'_{\text{pot}} - Q^*. \qquad (2.304)$$

The position of the overall center of gravity of all moving parts of a planar mechanism results from the individual positions of the centers of gravity from the conditions

$$x_S \cdot \sum_{i=2}^{I} m_i = \sum_{i=2}^{I} m_i x_{Si}; \qquad y_S \cdot \sum_{i=2}^{I} m_i = \sum_{i=2}^{I} m_i y_{Si}. \qquad (2.305)$$

2.5 Joint Forces and Foundation Loading

The resultant frame forces can thus be calculated from the acceleration of the overall center of gravity. It follows that these forces only depend on the motion of the overall center of gravity and the overall mass of the links. If the overall center of gravity remains at rest during the motion, the resultant of the frame forces is identical to zero. The individual frame forces have finite values, though, and usually a resulting moment M_z remains, see also Sect. 2.6.3.

The kinetic moments are, see Figs. 2.35 and (2.90):

$$M^O_{\text{kin }x} \equiv \sum_{i=2}^{I} \left[m_i z_{Si} \ddot{y}_{Si} + (J^S_{\eta\zeta i}\ddot{\varphi}_i + J^S_{\xi\zeta i}\dot{\varphi}_i^2)\sin\varphi_i - (J^S_{\xi\zeta i}\ddot{\varphi}_i - J^S_{\eta\zeta i}\dot{\varphi}_i^2)\cos\varphi_i \right]$$

$$M^O_{\text{kin }y} \equiv -\sum_{i=2}^{I} \left[m_i z_{Si} \ddot{x}_{Si} + (J^S_{\eta\zeta i}\ddot{\varphi}_i + J^S_{\xi\zeta i}\dot{\varphi}_i^2)\cos\varphi_i + (J^S_{\xi\zeta i}\ddot{\varphi}_i - J^S_{\eta\zeta i}\dot{\varphi}_i^2)\sin\varphi_i \right]$$

$$M^O_{\text{kin }z} \equiv \sum_{i=2}^{I} \left[m_i(y_{Si}\ddot{x}_{Si} - x_{Si}\ddot{y}_{Si}) - J_{Si}\ddot{\varphi}_i \right] \tag{2.306}$$

Note that they depend on the position of the coordinate system relative to the machine. It is recommended to select the center of gravity of the foundation block on which the mechanism is installed as the origin O of the coordinate system when addressing foundation issues, see Figs. 3.6 and 3.8.

The forces that act on the frame can of course also be calculated from the superposition of all joint forces that act onto the frame. This method is laborious, however, since it requires the calculation of the internal joint forces.

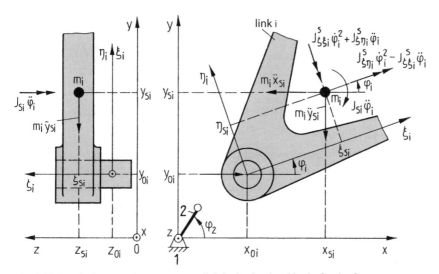

Fig. 2.35 Inertia forces and moments on a link in the fixed and body-fixed reference systems

2.5.4 Joint Forces in the Linkage of a Processing Machine

Thanks to software for dynamic analysis and optimization of mechanisms, designers can obtain an accurate overview of the joint force curves of complicated mechanisms. The compilation of all data from the design documents, e. g. mass parameters of the links, takes the greatest effort in that process.

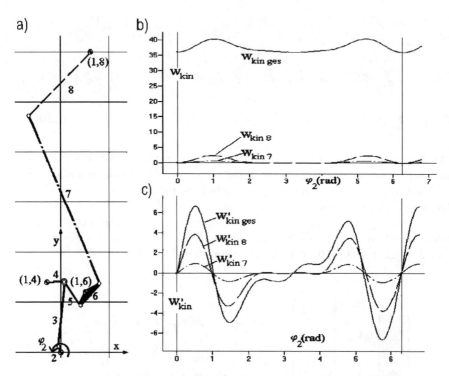

Fig. 2.36 Linkage of a processing machine; **a)** Kinematic schematic; **b)** Kinetic energy and contributions of links 7 and 8, see (2.301); **c)** Kinetic powers of links 7 and 8 according to (2.302)

Figure 2.36b shows the reduced moment of inertia (kinetic energy) and Fig. 2.36c the kinetic power of the overall mechanism as compared to that of two individual links for the eight-bar linkage, shown in Fig. 2.36a. One can see from these curves that the links 7 and 8 cause the variations of the moment of inertia and are essentially responsible for the change in kinetic power. It follows from just this one observation that their joint forces must change dramatically, see Sect. 2.5.2. Figure 2.37 shows the calculated joint forces for three bearings.

Both the representation of the forces with their directions of action (Fig. 2.37a) and the time functions of the resultant force (Fig. 2.37b) are of practical interest. An analysis of these forces provides the basis for drawing conclusions about the dynamic loads in the joint bolts (and thus about friction, lubrication, and wear and

2.5 Joint Forces and Foundation Loading

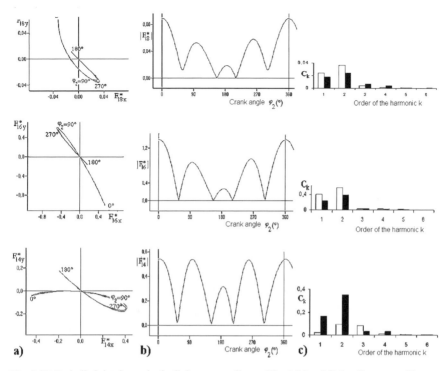

Fig. 2.37 Periodic joint forces in the linkage according to Fig. 2.36a; **a)** Polar diagrams of the joint forces F_{14}, F_{16} and F_{18}; **b)** Magnitude of the joint forces F_{14}, F_{16}, and F_{18} as a function of the crank angle $q = \Omega t$; **c)** Excitation spectrum (Fourier coefficients of the joint forces)

tear of the bearing) and about vibration excitations that will occur. A comparison with values determined experimentally allows decisions on the permissibility of the calculation model used ("rigid machine").

The Fourier coefficients are also of interest for assessing the joint forces (Fig. 2.37), see (2.293). They characterize the periodic excitation and are required for analyzing the forced vibrations, see Sects. 3.2 and 6.6.4. As can be seen from Fig. 2.37c, the higher harmonics are of major significance.

If the measured curves clearly deviate from the calculated ones, which happens frequently in engineering practice, one can deduce the causes for vibrations from the difference between the real and the kinetostatic curves, see VDI Guideline 2149 Part 2 [36].

2.5.5 Problems P2.9 and P2.10

P2.9 Parameter Influences on Joint Forces

It is possible to calculate the joint forces in any rigid-body mechanisms using computer programs. Occasionally the problem arises that the results of such calculations have to be checked. Such checks and plausibility considerations can be performed based on the general relationships.

An arbitrary mechanism is considered assuming that all its links consist of straight rigid rods made of the same material and that the bearing dimensions in the joints have a negligibly small influence on the mass parameters. How do the mass-related joint forces develop when the previous cross-sectional areas A_i of all links are changed by the same factor $\varkappa$ and the speed is changed by the factor $\varkappa_n$?

Given: All geometric and kinematic dimensions

1. Factor $\varkappa$ by which all cross-sectional areas are changed ($A_i^* = \varkappa A_i$)
2. Factor $\varkappa_n$ by which speed is changed ($n^* = \varkappa_n n$)

Find: Influence of the factors $\varkappa$ and $\varkappa_n$ on all joint forces

P2.10 Favorable Distance of the Point of Force Application

When selecting the kinematic dimensions and mass parameters, there are specific freedoms that can be used to reduce the kinetic loads.

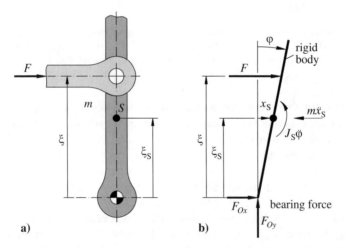

Fig. 2.38 Coupling force at the output link; **a)** Sketch of the link, **b)** Free-body diagram for $|\varphi| \ll 1$

Figure 2.38a shows the output link of a coupler linkage, which is pivoted at the frame. It is driven by a coupling force so that an angular acceleration $\ddot{\varphi}$ occurs. Calculate the horizontal component of the bearing force for small angles $|\varphi| \ll 1$ and state the parameter values for which this bearing force component becomes zero.

2.5 Joint Forces and Foundation Loading

Given:
    Mass m
    Distance to center of gravity ξ_S
    Moment of inertia J_S
    Coupling force F

Find:
    Horizontal force F_{Ox}
    Bearing distance ξ for which the horizontal force becomes zero.

2.5.6 Solutions S2.9 and S2.10

S2.9 The masses and moments of inertia of the slender rod-shaped links with the lengths l_i result from

$$m_i = \varrho A_i l_i; \qquad J_{Si} = \frac{m_i l_i^2}{12}; \qquad i = 2, 3, \ldots, I. \tag{2.307}$$

The modified parameters are denoted by the symbol $*$. If the cross-sections of all links are increased or decreased by the same factor, the following applies to the modified mass parameters because of $A_i^* = \varkappa A_i$:

$$m_i^* = \varrho A_i^* l_i = \varrho \varkappa A_i l_i = \varkappa m_i; \quad J_{Si}^* = \frac{m_i^* l_i^2}{12} = \varkappa J_{Si}; \quad i = 2, 3, \ldots, I. \tag{2.308}$$

The coordinates of the centers of gravity are retained both in the bodies and in the x and y directions. An arbitrary joint force (force F in arbitrary direction q) results for rigid-body mechanisms in the absence of applied forces according to (2.160) or (2.292):

$$m_{21}\ddot{q}(t) + \left(m_{12,\,1} - \frac{1}{2}m_{11,\,2}\right)\dot{q}^2(t) = F(t). \tag{2.309}$$

The generalized masses show a linear dependence on the masses and moments of inertia since the following applies to planar mechanisms according to (2.150)

$$m_{kl}(q) = \sum_{i=2}^{I}\left[m_i(x_{Si,k}x_{Si,l} + y_{Si,k}y_{Si,l}) + J_{Si}\varphi_{i,k}\varphi_{i,l}\right] \tag{2.310}$$

and thus because of (2.308)

$$m_{kl}^*(q) = \sum_{i=2}^{I}\left[m_i^*(x_{Si,k}x_{Si,l} + y_{Si,k}y_{Si,l}) + J_{Si}^*\varphi_{i,k}\varphi_{i,l}\right] = \varkappa m_{kl}(q). \tag{2.311}$$

This results in the time function of a joint force for modified cross-sectional areas:

$$\begin{aligned}F^*(t) &= m_{21}^*\ddot{q}(t) + \left(m_{12,\,1}^* - \frac{1}{2}m_{11,\,2}^*\right)\dot{q}^2(t) = \\ &= \varkappa\left[m_{21}\ddot{q}(t) + \left(m_{12,\,1} - \frac{1}{2}m_{11,\,2}\right)\dot{q}^2(t)\right] = \varkappa F(t).\end{aligned} \tag{2.312}$$

First result:

All joint forces change by the same factor $\varkappa$. If, for example, all widths and heights of the rectangular rod cross sections (or the diameters of circular cross sections) are reduced to $2/3$ of the original values, each cross-sectional area changes by a factor of $\varkappa = (2/3)^2 = 4/9$. It follows also that all joint forces then are reduced to only 44.4 % of their original magnitudes according to (A2.9/6), but their variation over time remains the same, except for factor $\varkappa$.

The speed is proportional to the angular velocity ($\Omega = \pi n/30$). At constant speed, a joint force according to (2.292) or (2.309) amounts to

$$\Gamma_{11\,p}(q)\Omega^2 = F_p \tag{2.313}$$

For a modified speed of ($n^* = \varkappa_n n$), it is $\Omega^* = \varkappa_n \Omega$ and thus

$$F_p^* = \Gamma_{11p}(q)\Omega^{*2} = \Gamma_{11p}(q)\varkappa_n^2\Omega^2 = \varkappa_n^2 F_p \tag{2.314}$$

Second result:

The joint forces change with the square of the speed ratio. When the speed is doubled, for example, all joint forces quadruple if no external forces act and the calculation model of the rigid-body system is still valid.

S2.10 The following results from the equilibrium of moments about point O, see Fig. 2.38b:

$$F\xi \approx (m\xi_S^2 + J_S)\ddot{\varphi}. \tag{2.315}$$

The balance of forces in the horizontal direction yields for $\varphi \ll 1$

$$F + F_{Ox} = m\ddot{x}_S = m\xi_S\ddot{\varphi} \tag{2.316}$$

The horizontal component of the joint force can be calculated from these equations:

$$F_{Ox} = \left[m\xi_S - \frac{m\xi_S^2 + J_S}{\xi}\right]\ddot{\varphi}. \tag{2.317}$$

This horizontal component of the joint force is zero if the expression in the square bracket is zero, i. e. if the coupling force acts at the distance of the so-called center of percussion

$$\xi = \xi_S + \frac{J_S}{m\xi_S}. \tag{2.318}$$

This distance is greater than the distance to the center of gravity.

One can try to place the mass parameters near the relations described by (2.318) by designing the output link accordingly.

The result that may be baffling at first glance becomes physically understandable if one imagines that an individual force moves a free rigid body both translationally (center-of-gravity) and rotationally (moment equilibrium). In this case, an instantaneous center of rotation exists. If the bearing is placed there, it does not have to transmit a force because the body "wants to rotate" about this point. In the design of links, one should first of all take a look at the extreme positions of the output links, since the angular accelerations take on their maximimu values there.

2.6 Methods of Mass Balancing

2.6.1 Objective

It can be achieved by smart distribution of the masses that the resultant inertia forces that a machine transmits onto its foundation become small. All measures aiming at balancing the inertia forces are called mass balancing (or counterbalancing for rotors).

It must be emphasized that mass balancing provides relief for the foundation only. The forces applied to the drive shaft and the dynamic bearing loads on individual joints may even increase due to such balancing and thus limit the capacity of the machine. When using mass balancing methods, the connection of such efforts with other side effects should be considered (such as the influence on the natural frequencies).

In addition to reducing the maximum force

$$|F_{\text{kin}}|_{\max} \stackrel{!}{=} \text{Min}, \qquad (2.319)$$

the task often is to minimize individual kth harmonics:

$$A_k^2 + B_k^2 \stackrel{!}{=} \text{Min}. \qquad (2.320)$$

Problems of mass balancing can be defined using (2.319) and (2.320). Special measures are required when the rotors cannot be assumed to be rigid. Note [29] and the ISO Guideline 11 342 – 1998: Mechanical vibration – Methods and criteria for the mechanical balancing of flexible rotors.

A designer will first attempt to keep the rotating and reciprocating masses small, e. g. by applying the lightweight construction principle or by using light metals or glass fiber-reinforced plastics instead of steel. Additional masses made of heavy metals (sintered tungsten materials) have the smallest dimensions due to their high density ($\varrho = 17 \ldots 19\,\text{g/cm}^3$).

2.6.2 Counterbalancing of Rigid Rotors

Almost all machines include rotors, which is why one should thoroughly study balancing techniques if one has to deal with them [29]. Here, only an introduction to the subject can be provided.

Rotors are rotating bodies, the bearing pins of which are supported by bearings. This term includes many machine elements, such as slender shafts, thin disks, oblong drums, regardless of whether they are rigid or elastic. A rotor is *rigid* if it behaves like an ideal rigid body, i. e. undergoes only negligibly small deformations at its operating speed. For practical purposes, a rotor can often be considered to be rigid as long as its speed is about half of its smallest critical speed, which also depends on the support conditions of the rotor. In an *elastic* rotor, the balancing state changes with its speed due to deformation.

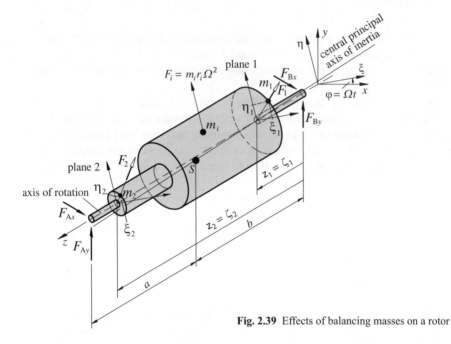

Fig. 2.39 Effects of balancing masses on a rotor

Unbalances often arise as a result of manufacturing inaccuracies and material inhomogeneities. An axisymmetrically designed component does not really have an ideal axisymmetric mass distribution. An **unbalance** is defined as the product of a point mass m_i and its distance r_i from the axis of rotation, see Fig. 2.39:

$$U_i = m_i r_i \tag{2.321}$$

For a single rotating point mass, a centrifugal force

$$F_i = m_i r_i \Omega^2 = U_i \Omega^2 \tag{2.322}$$

occurs at the angular velocity Ω. The unbalances in a rotor are typically distributed unevenly and randomly in space. Since there are always unbalances, dynamic forces develop when rotors rotate. These forces can have an adverse effect on

1. the bearing forces (surface pressure, wear, service life ...)
2. the loads on frame and foundation (excitation of vibrations)
3. loads inside of the rotor.

It is therefore advisable, especially for fast running rotors, to counterbalance any unbalances.

2.6 Methods of Mass Balancing

Balancing is the process in which the mass distribution of a body is checked and adjusted by mass balancing (removal or addition of material) to ensure that the dynamic bearing forces are within their predefined limits at operating speed. Balancing is performed using balancing machines or special measuring instruments in the original bearings.

A rotor is *completely* balanced when its mass is distributed so that it does not transmit any dynamic forces onto the bearings. This ideal state cannot be achieved in practice. Residual unbalances within specific limits are often specified in regulations, see, for example, the standards DIN ISO 1940-1, DIN ISO 11 342, ISO 19 499 or OENORM S9032.

A quantity that is independent of the mass of the balancing body is required to characterize the balancing state. For this purpose, the eccentricity is defined as follows

$$e = \frac{m_u r_u}{m} = \frac{U}{m} \tag{2.323}$$

U is the overall unbalance and m the overall mass of the rotor. e is directly provided by the balancing machine for deflection-measuring balancing machines that are operating below resonance. The evaluation criterion is provided by the product of $e\Omega$. To give an idea of its order of magnitude, some values are listed in Table 2.3.

Table 2.3 Permissible values for the product of eccentricity and angular velocity, see e in (2.323)

$e\Omega$ in mm/s	Rotor or machine
1600	Crank mechanism of rigidly mounted slow-running marine diesel engines
100	Crank mechanisms of rigidly and elastically mounted engines
16	Crank mechanisms of car and truck engines, car tires, rims, wheelsets, cardan shafts
	Centrifuge drums, fans, flywheels,
2,5	Electric motor armatures, machine-tool parts
	Gas and steam turbines, machine-tool drives,
0,4	Capstan drives
	Precision grinding machine armatures, shafts, and disks, gyroscopes

The bearing forces of a rotor for body-fixed coordinates were calculated in Sect. 2.3.5, see (2.133). These can be converted into fixed components using the rotational matrix $A = A_3$ from (2.11). Using

$$A = \begin{bmatrix} \cos\varphi & -\sin\varphi & 0 \\ \sin\varphi & \cos\varphi & 0 \\ 0 & 0 & 1 \end{bmatrix}, \tag{2.324}$$

the components in the fixed reference system result from (2.15) due to the known relation $\boldsymbol{F} = \boldsymbol{A}\overline{\boldsymbol{F}}$:

$$\begin{aligned} F_{Ax} &= F_{A\xi}\cos\varphi - F_{A\eta}\sin\varphi; & F_{Bx} &= F_{B\xi}\cos\varphi - F_{B\eta}\sin\varphi \\ F_{Ay} &= F_{A\xi}\sin\varphi + F_{A\eta}\cos\varphi; & F_{By} &= F_{B\xi}\sin\varphi + F_{B\eta}\cos\varphi. \end{aligned} \tag{2.325}$$

These change harmonically over time with $\varphi(t)$, and

$$F_{Ax}(\varphi - \pi/2) = F_{Ay}; \qquad F_{Bx}(\varphi - \pi/2) = F_{By}. \qquad (2.326)$$

For the special case of a constant angular velocity $\dot{\varphi} = \Omega$, the kinetic forces that are exerted by the rotor onto the bearings, see (2.133) and Fig. 2.40, are:

$$\begin{aligned}
F_{Ax} &= \left[(J^S_{\xi\zeta} - mb\xi_S)\cos\Omega t - (J^S_{\eta\zeta} - mb\eta_S)\sin\Omega t\right]\frac{\Omega^2}{a+b} \\
F_{Ay} &= \left[(J^S_{\xi\zeta} - mb\xi_S)\sin\Omega t + (J^S_{\eta\zeta} - mb\eta_S)\cos\Omega t\right]\frac{\Omega^2}{a+b} \\
F_{Bx} &= \left[-(J^S_{\xi\zeta} + ma\xi_S)\cos\Omega t + (J^S_{\eta\zeta} + ma\eta_S)\sin\Omega t\right]\frac{\Omega^2}{a+b} \\
F_{By} &= \left[-(J^S_{\xi\zeta} + ma\xi_S)\sin\Omega t - (J^S_{\eta\zeta} + ma\eta_S)\cos\Omega t\right]\frac{\Omega^2}{a+b}.
\end{aligned} \qquad (2.327)$$

The factors that determine the dynamic bearing forces of a rigid body can be seen from these equations. Any unbalance distribution in a rigid rotor corresponds to a shift in the center of gravity and an inclined orientation of the principal axes of inertia relative to the axis of rotation. It can be recognized from this that the dynamic forces rotate at the angular frequency in the bearings A and B, but are not in phase when the products of inertia do not equal zero.

Figure 2.40 shows an example of bearing force curves of an unbalanced rigid rotor where different amplitudes and phases occur at the same angular frequency.

A rigid rotor is thus completely balanced if its center of gravity is on the axis of rotation ($\xi_S = \eta_S = 0$) and if its central principal axis of inertia coincides with the axis of rotation ($J^S_{\zeta\xi} = J^S_{\zeta\eta} = 0$). Then $F_A = F_B \equiv 0$.

A *static* unbalance occurs if the center of gravity is not located on the axis of rotation. A *dynamic* unbalance of the rotor occurs when the central principal axis of inertia (that passes through the center of gravity) does not coincide with the axis of rotation. Both phenomena are always superposed in practice. During balancing, the mass distribution of the rigid rotor is adjusted by balancing in *two* planes so that the static and dynamic unbalances are compensated. m_1, m_2, ξ_1, ξ_2, η_1 and η_2 are determined from the measured bearing forces or displacements, see Fig. 2.39.

The following analysis is to show that two balancing masses in two different balancing planes are generally sufficient to completely balance an arbitrary rigid rotor.

According to (2.325), the bearing forces that vary harmonically with the angular frequency Ω depend on four components $F_{A\xi}$, $F_{A\eta}$, $F_{B\xi}$, and $F_{B\eta}$. Balancing is based on the idea that these four components (that have to be determined by experiment for a given real rotor) must be produced in the opposite direction by additional balancing masses and thus to compensate their sum on each bearing. If one defines the position of the balancing planes by the distances ζ_1 and ζ_2, four independent components of the bearing forces can be produced using the four static moments of two balancing masses ($m_1\xi_1$, $m_1\eta_1$, $m_2\xi_2$ and $m_2\eta_2$). The relationship results from

2.6 Methods of Mass Balancing

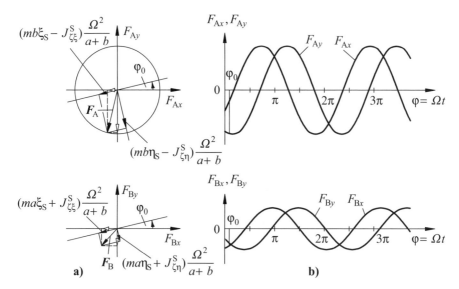

Fig. 2.40 Typical curve of the bearing forces of a rigid rotor for $\dot{\varphi} = \Omega = $ const.; **a)** Components of the bearing forces due to the static and dynamic unbalances; **b)** Curve of the bearing forces

the equilibrium at the rotor:

$$F_{A\xi} = -(m_1\boldsymbol{\xi}_1\zeta_1 + m_2\boldsymbol{\xi}_2\zeta_2)\frac{\Omega^2}{a+b}$$

$$F_{B\xi} = [m_1\boldsymbol{\xi}_1(a+b-\zeta_1) + m_2\boldsymbol{\xi}_2(a+b-\zeta_2)]\frac{\Omega^2}{a+b}$$

$$F_{A\eta} = -(m_1\boldsymbol{\eta}_1\zeta_1 + m_2\boldsymbol{\eta}_2\zeta_2)\frac{\Omega^2}{a+b}$$ (2.328)

$$F_{B\eta} = -[m_1\boldsymbol{\eta}_1(a+b-\zeta_1) + m_2\boldsymbol{\eta}_2(a+b-\zeta_2)]\frac{\Omega^2}{a+b}$$

These are two equations each for two unknown quantities, so one can "generate" the four force components with four static moments (bold print). Balancing machines determine the positions and magnitudes of the balancing masses "automatically", that is, using internal software. Balancing takes a few seconds in the mass production of motors and other small rotors produced in large numbers, while it takes several hours for large turbogenerators.

In view of the arising bending moments within the rotor, unbalances should be compensated, if possible, near the plane in which they occur. Figure 2.41 illustrates the influence of the selected balancing plane on the moment distribution in the rotor for a uniformly distributed unbalance.

Balancing in two arbitrary planes is no longer sufficient for so-called "elastic-shaft rotors" that run close to one of their critical speeds. For these cases, balancing meth-

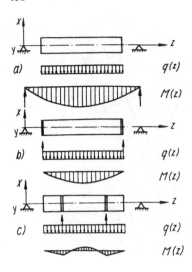

Fig. 2.41 Distribution of the bending moment for various selected balancing planes; **a)** Rotor without balancing; **b)** Balancing planes at the front surfaces of the rotor; **c)** Balancing planes inside of the rotor

ods in three or more planes were developed that require a considerable amount of extra computational and experimental effort, as compared to the balancing of rigid rotors.

The following aspects should be taken into account when selecting the balancing planes:

1. The balancing planes should be as far away from each other as possible.
2. When balancing assembled rotors with balancing planes on different components, it should be ensured by appropriate means that the unique relative positioning is ensured (positive-locking fastening of the parts against each other, e. g. using pins).
3. Balancing should not impair the strength of the component.

Figure 2.42 gives an overview of the adjustment options when balancing: cutting off, grinding or milling off material (Fig. 2.42a, e). Breaking off segments from disks provided for balancing inside (Fig. 2.42d) or outside (Fig. 2.42f) are examples of subtractive adjustment. Dosing the adjustment unbalance by screws of various lengths or diameters (Fig. 2.42c) or by inserting lead wire into rotating grooves (Fig. 2.42b) are some of the additive methods. The technological conditions of mass production should be taken into account when selecting a method. It may also be useful to weld or solder on strips of sheet metal. The designer has to specify the balancing planes from the outset and cannot leave the selection of the balancing method to chance. In practice, alancing is performed using balancing machines with which the position and magnitude of the unbalance is determined from the bearing responses of the rotor. Without going into too much detail, let us just mention that, depending on the size and speed of the rotor, "deflection-measuring" or "force-measuring" balancing machines are common, see Fig. 2.43. The rotor is rigidly supported in "force-measuring" balancing machines, and the bearing forces

2.6 Methods of Mass Balancing

are measured in the subcritical speed range. Their practical operating range is at speeds from 200 to 3000 1/min.

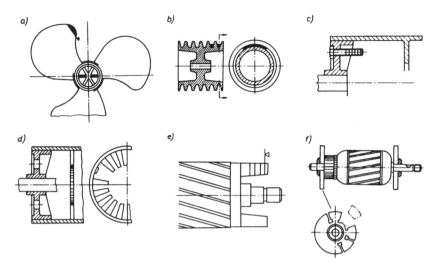

Fig. 2.42 Examples for methods of mass balancing on rotors; **a)** Cutting from fan blade, **b)** Inserting lead wire into groove, **c)** Screwing in bolts of various lengths, **d)** Breaking off segments inside, **e)** Milling off molded-on studs from front end, **f)** Breaking off segments from specially designed outside disks

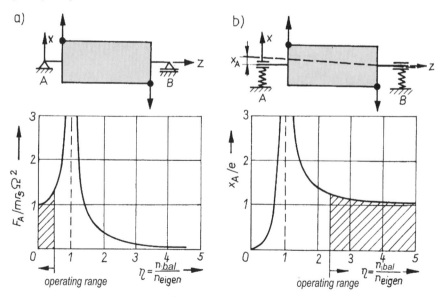

Fig. 2.43 Operating principle of balancing machines; **a)** Force-measuring balancing methods, **b)** Deflection-measuring balancing methods

2.6.3 Mass Balancing of Planar Mechanisms

2.6.3.1 Complete and Harmonic Balancing

The frame forces and moments resulting from inertia forces can be influenced by the mass parameters (m_i, ξ_{Si}, η_{Si}, J_{Si}, $J^S_{\xi\zeta i}$, $J^S_{\eta\zeta i}$), that is, by the mass distribution on the moving links. The goal of mass balancing is to reduce the dynamic forces and moments in such a way that they load the frame within permissible limits only.

These resulting inertia forces and inertia moments can be calculated for planar mechanisms without determining the joint forces inside the mechanism. In the considerations below, reference is made to Fig. 2.15 and to equations (2.192) to (2.196) that will be used again below.

The resultant frame force components F_x, F_y and the component M^0_z of the resulting frame moment known from (2.303) and (2.306) are derived from the principle of conservation of linear momentum and from the principle of conservation of angular momentum

$$F_x = -\frac{dI_x}{dt} = -\frac{d}{dt}\left(\sum_i m_i \dot{x}_{Si}\right)$$
$$= -\sum_i m_i \ddot{x}_{Si} = -m\ddot{x}_S = -m_x(q)\ddot{q} - m'_x(q)\dot{q}^2 \quad (2.329)$$

$$F_y = -\frac{dI_y}{dt} = -\frac{d}{dt}\left(\sum_i m_i \dot{y}_{Si}\right)$$
$$= -\sum_i m_i \ddot{y}_{Si} = -m\ddot{y}_S = -m_y(q)\ddot{q} - m'_y(q)\dot{q}^2 \quad (2.330)$$

$$M^0_z = -\frac{dL^0_z}{dt} = -\frac{d}{dt}\left(\sum_i [m_i(\dot{y}_{Si}x_{Si} - \dot{x}_{Si}y_{Si}) + J_{Si}\dot{\varphi}_i]\right)$$
$$= -m_\varphi(q)\ddot{q} - m'_\varphi(q)\dot{q}^2 \quad (2.331)$$

The generalized masses that depend on the generalized coordinate q (and thus on the position of the mechanism) can be determined as follows:

$$m_x(q) = \sum_i m_i x'_{Si}, \qquad m_y(q) = \sum_i m_i y'_{Si},$$
$$m_\varphi(q) = \sum_i [m_i(y'_{Si}x_{Si} - x'_{Si}y_{Si}) + J_{Si}\varphi'_i] \quad (2.332)$$

Note that in equations (2.329) to (2.331) the coefficients in front of $\dot{q}^2$ represent the derivatives of the ones in front of $\ddot{q}$, with respect to the generalized coordinate q.

The generalized mass $m_\varphi(q)$ referred to the z axis has the dimension of a moment of inertia in the case of an input angle ($q = \varphi$) and a similar form as the reduced moment of inertia known from (2.199), but must not be confused with that!

2.6 Methods of Mass Balancing

While the reduced moment of inertia $J_{\text{red}}(\varphi)$ is linked to the kinetic energy and the input torque, the $m_\varphi(q)$ that results from the angular momentum is required for calculating the frame moment.

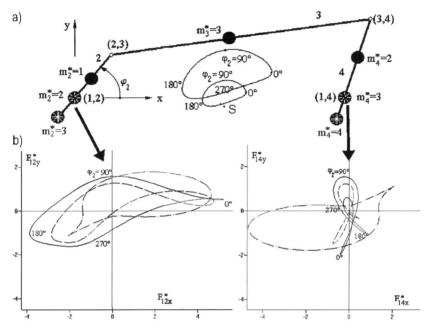

Fig. 2.44 Influence of mass parameters on the frame forces of a crank-rocker mechanism
a) Kinematic schematic and center-of-mass trajectories, including the reduced masses $m_i^* = m_i/m$; **b)** Polar diagrams of joint forces F_{12} and F_{14}
Variant 1: ———, Variant 2: – – – –, Variant 3: – · – · – ·

The overall center of gravity of a mechanism normally moves along a trajectory as shown in Fig. 2.44; for an example, see (2.305). In addition to the center-of-mass trajectories, it shows the polar diagrams of the two frame forces of a crank-rocker mechanism for three variants of mass distribution. The masses m_2 and m_4 and their distances to the center of gravity ξ_{S2} and ξ_{S4} were varied. These curves are obtained at a constant angular velocity of the drive.

The relationship of the center-of-mass trajectories to the polar diagrams is interesting. At a smaller extension of the center-of-mass trajectory, the joint forces for variant 2 also become smaller than for variant 1. In variant 3 where the center-of-mass trajectory contracts into a point, the sum of the joint forces is zero according to (2.303), but the individual joint forces exist.

Any planar mechanism can, in principle, be configured by an appropriate mass distribution of its links such that its center of gravity remains at rest despite arbitrary motion. Complete mass balancing is achieved when the resultant inertia forces and the kinetic inertia moment are zero:

$$F_x \equiv 0; \qquad F_y \equiv 0; \qquad M_z^O \equiv 0 \qquad (2.333)$$

It follows from (2.329) to (2.331) that these conditions are met, if regardless of the state of motion

$$m_x(q) \equiv 0; \qquad m_y(q) \equiv 0; \qquad m_\varphi(q) \equiv 0 \qquad (2.334)$$

Complete inertia **force** balancing is achieved when the center of gravity remains at rest despite arbitrary motion of the mechanism. The condition for this results from (2.329) and (2.330) if the conditional equations for the position of the center of gravity of a multibody system are considered:

$$\ddot{x}_S \equiv 0; \qquad \ddot{y}_S \equiv 0 \qquad (2.335)$$

They can theoretically be satisfied for all planar rigid multilink mechanisms with one drive, see VDI Guideline 2149 Part 1 [35].

A complete balancing of forces and moments is rarely performed in practice because it has the following disadvantages:

- It mostly leads to bulky mechanisms with large dimensions that are hardly realistic from a designer's point of view.
- The mass of the links has often to be changed significantly, which entails a considerable increase of their moments of inertia.
- The individual bearing and joint forces and joint moments may increase.

In cyclically operating mechanisms, the frame forces and moments are periodic, even if the angular velocity of the drive is variable. They often excite forced vibrations in the frame. The prerequisite is that the frame motions only have an insignificant influence on the motions of the mechanism, otherwise parameter-excited oscillations would be generated.

The goal of **harmonic** mass balancing is to reduce the amplitudes of critical excitation harmonics in the spectrum of the dynamic forces. Harmonic mass balancing is effective in steady-state operation and can ensure that no dangerous resonance amplitudes occur in the range of operating speeds. Harmonic balancing requires the use of special software programs because analytical solutions are only possible for simple mechanisms, such as slider-crank mechanisms, see Problem P2.12.

Balancing individual harmonics of the periodic excitation that results from mechanisms with variable transmission ratios in steady-state operation is of the greatest significance for mechanical engineering practice since it minimizes the excitation of vibrations. In cam mechanisms, cams with so-called HS profiles have proven their worth [4]. Mechanisms can be found using computer-aided synthesis, in which specific harmonics of the excitation forces and moments have a minimal magnitude. The solution is not always in balancing the first or second harmonics, often **balancing of higher harmonics** is of practical significance.

In addition to **complete and harmonic mass balancing**, the following practical measures can help to achieve an improvement:

2.6 Methods of Mass Balancing

- Generation of an equivalent countermotion, i. e. compensation by equal and opposite inertia forces of an additional mechanism (Fig. 2.45) or by additional dyad.
- Balancing of specific harmonics using compensatory mechanisms (Fig. 2.45).
- In multicylinder machines by placing counterweights at various crank angles, by offsetting the mechanism planes relative to the axis, and possibly by varying the crank radii and piston masses, see Sect. 2.6.3.3.
- Optimal balancing, taking into account secondary design conditions (requires the use of software).

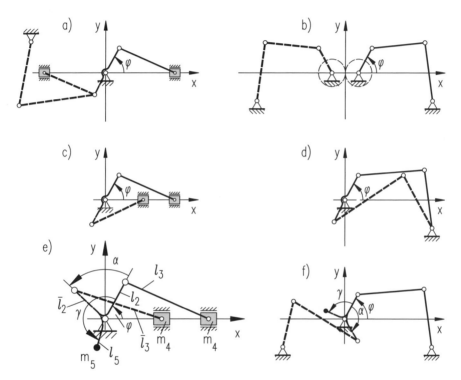

Fig. 2.45 Examples of mass balancing by using a compensatory mechanism moving in the opposite direction (solid line: original mechanism; dashed line: compensatory mechanism)

2.6.3.2 Mass Balancing for a Slider-Crank Mechanism

Slider-crank mechanisms are used in many machines for converting rotating into reciprocating motions (and vice versa) so that mass balancing has met with special interest for quite some time.

Let us first derive the conditions for the complete inertia **force** balancing in a slider-crank mechanism without offset. Complete balancing of the resultant frame moment can be achieved using additional rotational inertia, see VDI Guideline 2149 Part 1 [35].

The position of the center of gravity of the slider-crank mechanism follows from the link positions according to Fig. 2.26:

$$(m_2 + m_3 + m_4)\,r_S = m_2 r_{S2} + m_3 r_{S3} + m_4 r_{S4} \tag{2.336}$$

If the motion plane is identified with the plane of complex numbers for the purpose of compact mathematical treatment, then $r = x + jy = r \cdot e^{j\varphi} = r \cdot (\cos\varphi + j\sin\varphi)$ ($j = \sqrt{-1}$) applies and the position of the centers of gravity of the links can be stated as follows in accordance with Fig. 2.26:

$$r_{S2} = \xi_{S2} e^{j\varphi_2}; \qquad r_{S3} = l_2 e^{j\varphi_2} + \xi_{S3} e^{j\varphi_3}; \qquad r_{S4} = l_2 e^{j\varphi_2} + l_3 e^{j\varphi_3} \tag{2.337}$$

Insertion into (2.336) provides the center-of-mass trajectory:

$$\begin{aligned}(m_2 + m_3 + m_4) r_S &= m_2 \xi_{S2} e^{j\varphi_2} + m_3 (l_2 e^{j\varphi_2} + \xi_{S3} e^{j\varphi_3}) \\ &\quad + m_4 (l_2 e^{j\varphi_2} + l_3 e^{j\varphi_3}) \\ &= e^{j\varphi_2}(m_2 \xi_{S2} + m_3 l_2 + m_4 l_2) + e^{j\varphi_3}(m_3 \xi_{S3} + m_4 l_3)\end{aligned} \tag{2.338}$$

The center of gravity remains at rest ($\ddot{r}_S = o$), and there are no resultant inertia forces that act onto the frame if the following balancing conditions are satisfied (setting the expressions in parentheses to zero):

$$m_2 \xi_{S2} + (m_3 + m_4) l_2 = 0 \tag{2.339}$$

$$m_3 \xi_{S3} + m_4 l_3 = 0 \tag{2.340}$$

This yields the distances to the center of gravity for complete balancing:

$$\xi_{S2} = -\frac{m_3 + m_4}{m_2} l_2; \qquad \xi_{S3} = -\frac{m_4}{m_3} l_3 \tag{2.341}$$

As a result, the common center of gravity of the masses m_3 and m_4 is placed in the joint $(2, 3)$.

If only (2.340) is satisfied, the center of gravity moves along a circular path and causes harmonic excitation forces. If the center of gravity of the masses m_3 and m_4 is at joint $(2, 3)$, it can be shifted by a counterweight m_2^* to point $(1, 2)$ so that it does not change its position. The time functions of the forces F_x, F_y and the moment M_z^0 are composed of several harmonics at a constant input angular velocity $\dot{\varphi}_2 = \Omega$. The first harmonic component is called *first-order inertia force*. The second term in the Fourier expansion varies at twice the frequency and is therefore called *second-order inertia force*. The first-order inertia force, that is, the first harmonic of the force F_x, is balanced when condition (2.339) is satisfied, i. e. if a balancing mass is attached to the crank only. According to (2.341), the balancing mass is then located on the opposite side of the crank. It is, in practice, often designed as a segment of a circle,

2.6 Methods of Mass Balancing

see Fig. 2.29. The first-order inertia force of F_y is balanced when the balancing condition

$$m_2 \xi_{S2} + m_3 l_2 \left(1 - \frac{\xi_{S3}}{l_3}\right) = 0 \qquad (2.342)$$

is satisfied.

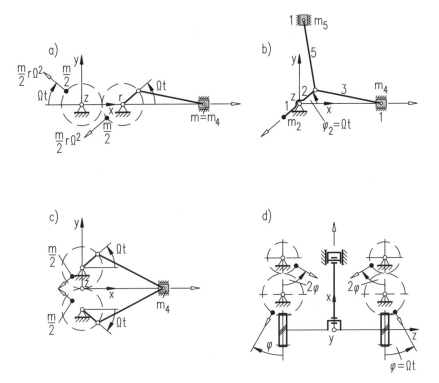

Fig. 2.46 Design options for balancing individual harmonics (compensatory mechanism); **a)** and **c)** forces and moment (1st harmonic), **b)** forces (1st harmonic), **d)** forces (1st and 2nd harmonics)

2.6.3.3 Harmonic Balancing in Multicylinder Machines

Multicylinder machines in which multiple slider-crank mechanisms are connected by a common shaft are frequently used in engines and compressors. Balancing of some harmonics is possible if the relative position of the individual mechanism planes and the relative orientation of the crank angles are favorably selected.

It is assumed for the following derivations that the cylinder axes and the crankshaft axis are in one plane, the y-z plane. This covers the case of an in-line engine with k cylinders. The interesting case of a V-engine or radial engine in which the piston directions are arranged at a specific angle is excluded from these considerations, see [1]. It is also assumed that all rotating masses, that is, the crankshaft with the rotating portions of the connecting rod (see m_{32}, Fig. 2.29), are completely balanced. All cylinders should also be identical (equal masses and geometry) and only have different crank angles, see Fig. 2.47. The angle between the first crank ($j = 1$) and the jth crank is denoted as γ_j.

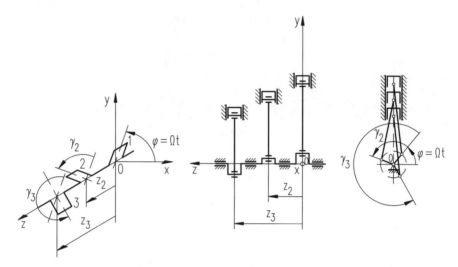

Fig. 2.47 Derivation of the balancing conditions for a multicylinder machine

The inertia forces can be stated in the form of a Fourier series, see (2.293). With ($j = 1, 2, \ldots, J$), the following applies to each mechanism:

$$F_j(t) = \sum_{k=1}^{\infty} [A_k \cos k(\Omega t + \gamma_j) + B_k \sin k(\Omega t + \gamma_j)]. \qquad (2.343)$$

k identifies the order of the harmonic. If it is assumed that the crankshaft revolves at constant angular velocity, the crank angles are $\varphi_j = \Omega t + \gamma_j$. $\gamma_1 = 0$ applies to the first cylinder. The Fourier coefficients (A_k; B_k) of a mechanism are assumed to be known.

The resultant dynamic forces and moments that are transmitted onto the foundation are derived both for the x and for the y components (which is why the index is omitted) for J cylinders:

$$F = \sum_{j=1}^{J} F_j; \qquad M^0 = \sum_{j=1}^{J} F_j z_j \qquad (2.344)$$

z_j is the distance of the respective mechanism plane from the x-y plane of the coordinate system, as shown in Figs. 2.34, 2.35, and 2.47.

2.6 Methods of Mass Balancing

Insertion of F_j from (2.343) into (2.344), using certain trigonometric identities and some conversions, provides

$$F = \sum_j \sum_k [A_k(\cos k\Omega t \cos k\gamma_j - \sin k\Omega t \sin k\gamma_j)$$
$$+ B_k(\sin k\Omega t \cos k\gamma_j + \cos k\Omega t \sin k\gamma_j)]$$

$$F = \sum_k \left[(A_k \cos k\Omega t + B_k \sin k\Omega t) \sum_j \cos(k\gamma_j) \right. \quad (2.345)$$
$$\left. + \sum_k [-A_k \sin k\Omega t + B_k \cos k\Omega t] \sum_j \sin(k\gamma_j) \right]$$

It follows from a comparison of coefficients that the kth-order harmonic of the resultant force of a multicylinder machine is completely balanced when the following two conditions are satisfied:

$$\sum_{j=1}^{J} \cos k\gamma_j = 0; \quad \sum_{j=1}^{J} \sin k\gamma_j = 0 \quad (2.346)$$

Similarly, the kth- order harmonics of the moment M^0 are completely balanced if:

$$\sum_{j=1}^{J} z_j \cos k\gamma_j = 0; \quad \sum_{j=1}^{J} z_j \sin k\gamma_j = 0 \quad (2.347)$$

These are the important balancing conditions of kth-order inertia forces of multicylinder machines. Interestingly, the masses of the links, the speed and geometrical dimensions are not included in these equations. Thus there are four transcendental equations for each order k for calculating the crank angles γ_j and the distances z_j (these are $2J - 1$ unknown quantities for J mechanisms) that are required for complete mass balancing.

2.6.4 Problems P2.11 to P2.14

P2.11 Harmonic Balancing of a Compressor

The inertia forces that occur in mechanisms with a varying transmission ratio can be the cause of frame vibrations. Particularly dangerous are those components of the excitation spectrum, the frequency of which matches the natural frequency of the frame.

The additional balancing mass m_a in the form of an annulus segment of constant thickness, to be attached to the driving crank of the compressor (slider-crank mechanism without offset, see Fig. 2.48), is to be sized in such a way that the first harmonic of the resultant frame force component F_x is completely balanced at constant input angular speed.

Given:

l_2	= 40 mm	crank length
l_3	= 750 mm	coupler length
ξ_{S_2}	= 12 mm	distance of the center of gravity of the crank from O

$J_{S_2} = 6{,}1 \cdot 10^{-3}$ kg·m²   moment of inertia referred to the axis through the center of gravity of the crank

m_2	= 4.8 kg	crank mass
m_4	= 14 kg	piston mass
r	= 20 mm	inner radius of the balancing mass
$R_{\max}$	= 140 mm	maximum outer radius of the balancing mass (installation space!)
b	= 40 mm	thickness of the balancing mass
ϱ_G	= 7250 kg/m³	density of cast iron
ϱ_Z	= 8900 kg/m³	density of tin bronze
ϱ_W	= 9800 kg/m³	density of white metal

Note: The mass parameters of the coupler 3 have been approximately included in the calculation of those of links 2 and 4, see Fig. 2.29.

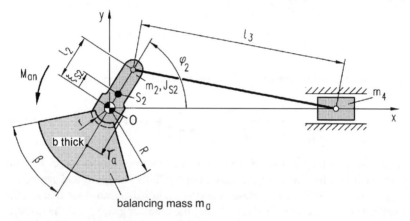

Fig. 2.48 Slider-crank mechanism without offset

Find:

1. Resultant frame force components F_x and F_y in general form for an arbitrary input motion $\varphi_2(t)$, taking into account the balancing mass.
2. Required dimensions (R, β), mass m_a, and moment of inertia J_a^O of the balancing mass so that both the first harmonic of the frame force F_x is balanced and J_a^O becomes as small as possible. The materials that can be selected are cast iron, tin bronze, and white metal.

P2.12 Compensatory Mechanism for Slider-Crank Mechanism

Calculate the x component of the bearing force (F_{x12}) and the input torque M_{an} for a slider-crank mechanism for which the inertia forces are to be balanced using an balancing mass and a compensatory mechanism arranged as shown in Fig. 2.45e. State the balancing conditions for the first and second harmonics of F_{x12} and M_{an} in general form. What values should be selected for the angles α and γ and the balancing masses $\overline{m}_4$ and m_5 so that the first two harmonics of these forces are compensated? Assume $\lambda = l_2/l_3 = \bar{l}_2/\bar{l}_3 \ll 1$ and $\varphi = \Omega t$.

2.6 Methods of Mass Balancing

P2.13 Crankshaft of a Four-Cylinder Machine

The following quantities are up for discussion for balancing individual harmonics in a four-cylinder machine, see Fig. 2.51:
Variante a) : $\gamma_1 = 0°$; $\gamma_2 = 90°$; $\gamma_3 = 270°$; $\gamma_4 = 180°$
Variante b) : $\gamma_1 = 0°$; $\gamma_2 = 180°$; $\gamma_3 = 180°$; $\gamma_4 = 0°$
Find out which orders of the forces and moments are balanced by these variants if the cylinder distances and cylinders are the same.

P2.14 Mass Balancing of Crank Shears

Increasing the speed of rolling stock also requires higher speeds of the shears that cut a rolled bar "on the fly" to a specified length. The inertia forces of the crank shears excite undesirable vibrations in the machine frame and cause inadmissibly large loads to the anchoring of the machine, so that measures for mass balancing are required.

Perform mass balancing on the crank shears as shown in Fig. 2.49 in such a way that complete balancing of the forces is achieved by adding balancing masses to the crank and to the rocker.

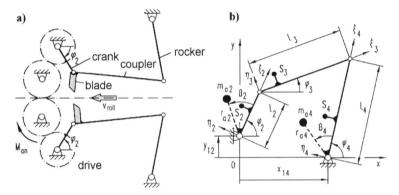

Fig. 2.49 Crank shears; a) Kinematic schematic, b) Calculation model with mass parameters

The crank shears consist of two rigidly coupled crank-rocker mechanisms that are symmetrically arranged relative to the rolled bar and move in opposite directions. Since the two crank-rocker mechanisms are almost the same with regard to their dimensions and mass parameters and are arranged symmetrically, it is sufficient to consider just one mechanism, see the calculation model in Fig. 2.49.

Given:

Parameter values:

i	Name	l_i in m	m_i in kg	ξ_{Si} in m	η_{Si} in m
2	Crank	0,100	41,5	0,021	0
3	Coupler (blade holder)	0,205	53,2	0,074	0,018
4	Rocker	0,147	17,7	0,065	0

Find:

Magnitudes ($U_2 = m_{a2} \cdot r_{a2}$; $U_4 = m_{a4} \cdot r_{a4}$) and angular positions (β_2, β_4) of the balancing masses to be attached to the crank and to the rocker to achieve complete balancing of the resultant frame forces.

2.6.5 Solutions S2.11 to S2.14

S2.11 According to (2.303), and taking into account the functions $x_{Si} = x_{Si}(\varphi_2(t))$ and $y_{Si} = y_{Si}(\varphi_2(t))$, one can write:

$$F_x = -\ddot{\varphi}_2 \sum_{i=2}^{I} m_i x'_{Si} - \dot{\varphi}_2^2 \sum_{i=2}^{I} m_i x''_{Si}; \quad F_y = -\ddot{\varphi}_2 \sum_{i=2}^{I} m_i y'_{Si} - \dot{\varphi}_2^2 \sum_{i=2}^{I} m_i y''_{Si} \quad (2.348)$$

Since the nomenclature used here mostly coincides with that in Fig. 2.26a, the first-order position functions can be taken directly from Table 2.1; the prerequisite stated there regarding the crank ratio λ is satisfied.

$\lambda = l_2/l_3 = 0{,}04\text{ m}/0{,}75\text{ m} = 0{,}053\bar{3} \ll 1$ is obtained for the given parameter values. Using the functions from Table 2.1, and after another differentiation with respect to φ_2, one obtains:

$$\begin{aligned} x''_{S_2} &= -\xi_{S_2} \cdot \cos\varphi_2; & y''_{S_2} &= -\xi_{S_2} \cdot \sin\varphi_2 \\ x''_{S_4} &= -l_2 \cdot (\cos\varphi_2 + \lambda \cos 2\varphi_2) \end{aligned} \quad (2.349)$$

The balancing mass has the following center-of-gravity coordinates in the fixed system, see Fig. 2.48:

$$\begin{aligned} x_{Sa} &= r_a \cdot \cos(\varphi_2 + \pi) = -r_a \cdot \cos\varphi_2 \\ y_{Sa} &= r_a \cdot \sin(\varphi_2 + \pi) = -r_a \cdot \sin\varphi_2 \end{aligned} \quad (2.350)$$

The derivatives with respect to the crank angle are

$$\begin{aligned} x'_{Sa} &= r_a \cdot \sin\varphi_2; & x''_{Sa} &= r_a \cdot \cos\varphi_2 \\ y'_{Sa} &= -r_a \cdot \cos\varphi_2; & y''_{Sa} &= r_a \cdot \sin\varphi_2 \end{aligned} \quad (2.351)$$

If one inserts the expressions from (2.349) and (2.351) into the relationship (2.348) for the forces, the following is obtained:

$$\begin{aligned} F_x &= \left[(m_2 \xi_{S_2} - m_a r_a + m_4 l_2) \cdot \sin\varphi_2 + m_4 l_2 \frac{\lambda}{2} \cdot \sin 2\varphi_2 \right] \cdot \ddot{\varphi}_2 \\ &\quad + \left[(m_2 \xi_{S_2} - m_a r_a + m_4 l_2) \cdot \cos\varphi_2 + m_4 l_2 \lambda \cdot \cos 2\varphi_2 \right] \cdot \dot{\varphi}_2^2 \quad (2.352) \\ F_y &= -(m_2 \xi_{S_2} - m_a r_a) \cdot (\ddot{\varphi}_2 \cos\varphi_2 - \dot{\varphi}_2^2 \sin\varphi_2) \end{aligned}$$

The equation for the force F_x is an approximation because due to $\lambda \ll 1$, the terms that contain higher powers of λ have already been neglected in the equations for x'_{S_4}, x''_{S_4}.

The first harmonic of F_x becomes identical to zero if the expression in parentheses vanishes. Thus the balancing condition is:

$$m_2 \xi_{S_2} - m_a r_a + m_4 l_2 \stackrel{!}{=} 0 \quad (2.353)$$

The unbalance of the balancing mass is derived as:

$$U_a = m_a r_a = m_2 \xi_{S_2} + m_4 l_2 = 0{,}6176 \text{ kg} \cdot \text{m} \quad (2.354)$$

The general equations for mass, position of the center of gravity, and moment of inertia for the given form of an annulus segment are required for sizing the balancing mass. One can find the following in handbooks:

2.6 Methods of Mass Balancing

$$m_a = \varrho b \cdot (R^2 - r^2) \cdot \beta; \qquad r_a = \frac{2}{3} \cdot \frac{R^3 - r^3}{R^2 - r^2} \cdot \frac{\sin \beta}{\beta} \qquad (2.355)$$

$$J_a^O = \frac{\varrho b}{2} \cdot (R^4 - r^4) \cdot \beta \qquad (2.356)$$

It follows from (2.355) for the unbalance, the required magnitude of which is known from (2.354):

$$U_a = m_a r_a = \frac{2}{3} \varrho b \cdot (R^3 - r^3) \cdot \sin \beta \qquad (2.357)$$

n It still contains two variables, namely R and β. If one solves (2.357) for $\sin \beta$, the following relation is obtained:

$$\sin \beta = \frac{3 \cdot U_a}{2 \varrho b \cdot (R^3 - r^3)} \leqq 1 \qquad (2.358)$$

This inequation results in the following for the outer radius R in conjunction with the limitation set in the problem statement:

$$R_{\min} = \sqrt[3]{r^3 + \frac{3 \cdot U_a}{2 \varrho b}} \leq R \leq R_{\max} \qquad (2.359)$$

Taking the given parameter values and the unbalance according to (2.354), the minimum radii are:

cast iron $R_{\min} = 147.4$ mm
tin bronze $R_{\min} = 137.7$ mm
white metal $R_{\min} = 133.4$ mm.

A comparison of these minimum radii with the permissible maximum value $R_{\max} = 140$ mm shows that cast iron is not an option here. White metal ($\varrho b = 392$ kg/m^2) is selected as the material and $R = 135$ mm as the radius. These selections let one determine the angle β from (2.358). First, $\sin \beta = 0{,}963\,66$ is obtained. That angle of the two possible solutions

$$\beta_1 = 1{,}3004 \text{ rad} \quad (\widehat{=}74{,}5°) \quad \text{and} \quad \beta_2 = 1{,}8412 \text{ rad} \quad (\widehat{=}105{,}5°) \qquad (2.360)$$

is used for which the mass and the moment of inertia are the smallest. As both have a linear dependency on the angle β, only the smaller of the two angles qualifies.

Now one can calculate the mass and moment of inertia of the balancing mass from (2.355) and (2.356):

$$m_a = 9{,}09 \text{ kg}; \qquad J_a^O = 0{,}0846 \text{ kg} \cdot \text{m}^2 \qquad (2.361)$$

Only the **first harmonic** of a force component can be balanced in a mechanism by a balancing mass that rotates with the drive. More complex balancing mechanisms are required to balance multiple force components and harmonics.

S2.12 The bearing force of a slider-crank mechanism with an unbalance mass m_5 has (see Table 2.1) the x component

$$\begin{aligned}F_{x12} &= -m_5 \ddot{x}_5 - m_4 \ddot{x}_4 \\ &= \Omega^2 \left[m_5 l_5 \cos(\varphi + \gamma) + m_4 l_2 (\cos\varphi + \lambda \cos 2\varphi + \cdots) \right]\end{aligned} \qquad (2.362)$$

Only the first two harmonics of the Fourier series were given since the higher ones are smaller than the λ^2 order.

Therefore, the following results for the compensatory mechanism with a crank offset by an angle α (since the angle $\varphi + \alpha$ is written here instead of φ):

$$\overline{F}_{x12} = \overline{m}_4 \overline{l}_2 \Omega^2 \left[\cos(\varphi + \alpha) + \lambda \cos 2(\varphi + \alpha) + \cdots\right] \qquad (2.363)$$

The sum $F_x = F_{x12} + \overline{F}_{x12}$ is then, after using some trigonometric identities, sorted by the order of the harmonics:

$$F_x = \Omega^2[\cos\varphi \left(m_4 l_2 + \overline{m}_4 \overline{l}_2 \cos\alpha + m_5 l_5 \cos\gamma\right) \\ - \sin\varphi \left(\overline{m}_4 \overline{l}_2 \sin\alpha + m_5 l_5 \sin\gamma\right) \\ + \lambda \cos 2\varphi \left(m_4 l_2 + \overline{m}_4 \overline{l}_2 \cos 2\alpha\right) + \lambda \sin 2\varphi \left(\overline{m}_4 \overline{l}_2 \sin 2\alpha\right) + \cdots]. \quad (2.364)$$

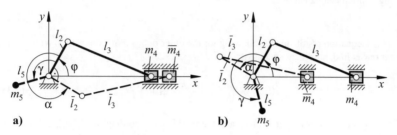

Fig. 2.50 Mechanisms for balancing the 1st and 2nd harmonics of F_x and the 2nd harmonic of M_{an} of the slider-crank mechanism; a) $\alpha = \frac{3\pi}{2}, \gamma = \frac{3\pi}{4}$; b) $\alpha = \frac{\pi}{2}, \gamma = \frac{5\pi}{4}$

The input torque of the slider-crank mechanism coupled with the compensatory mechanism is obtained (at $\varphi = \Omega t$ independent of m_5) using (2.304):

$$M_{\mathrm{an}} = m_4 \ddot{x}_4 x'_4 + \overline{m}_4 \ddot{\overline{x}}_4 \overline{x}'_4 \\ = -\frac{\Omega^2}{4}\left[\lambda \cos\varphi (\overline{m}_4 \overline{l}_2^2 \sin\alpha) \right. \\ \left. + \lambda \sin\varphi (m_4 l_2^2 + \overline{m}_4 \overline{l}_2^2 \cos\alpha) - 2\cos 2\varphi (\overline{m}_4 \overline{l}_2^2 \sin 2\alpha) \right. \\ \left. - 2\sin 2\varphi (m_4 l_2^2 + \overline{m}_4 \overline{l}_2^2 \cos 2\alpha) + \cdots\right]. \quad (2.365)$$

Those harmonics of F_x and M_{an} for which the expressions in parentheses are set to zero vanish. One can select those balancing conditions that are relevant for the respective application and take design measures accordingly.

The first harmonic of F_x can be balanced without a compensatory mechanism using an unbalance mass m_5. The balancing conditions are derived from the first two parentheses of F_x with $\overline{m}_4 = 0$

$$m_4 l_2 + m_5 l_5 \cos\gamma = 0, \qquad m_5 l_5 \sin\gamma = 0. \quad (2.366)$$

Both are satisfied for the values $\gamma = \pi$ and $m_5 l_5 = m_4 l_2$, see Fig. 2.29.

Combined mass and power balancing is achieved when setting the first harmonic of F_x and the dominant second harmonic of M_{an} to zero. The solution follows from the corresponding four balancing conditions:

$$m_4 l_2 + \overline{m}_4 \overline{l}_2 \cos\alpha + m_5 l_5 \cos\gamma = 0 \\ \overline{m}_4 \overline{l}_2 \sin\alpha + m_5 l_5 \sin\gamma = 0 \\ \overline{m}_4 \overline{l}_2^2 \sin 2\alpha = 0 \\ m_4 l_2^2 + \overline{m}_4 \overline{l}_2^2 \cos 2\alpha = 0 \quad (2.367)$$

2.6 Methods of Mass Balancing

as $\bar{m}_4 = m_4$, $\bar{l}_2 = l_2$, $m_5 l_5 = \sqrt{2} m_4 l_2$ and either $\alpha = 3\pi/2$, $\gamma = 3\pi/4$ (Fig. 2.50a) or $\alpha = \pi/2$, $\gamma = 5\pi/4$ (Fig. 2.50b). This balances the second harmonic of F_x as well. Check by insertion whether the balancing conditions are satisfied.

S2.13 If the coordinate system is placed in the first cylinder (Fig. 2.51), $z_1 = 0$ applies. The general balancing conditions according to (2.346) and (2.347) for a four-cylinder machine are then:

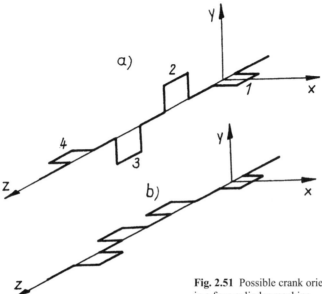

Fig. 2.51 Possible crank orientations in a four-cylinder machine

kth-order forces

$$\left. \begin{array}{l} 1 + \cos k\gamma_2 + \cos k\gamma_3 + \cos k\gamma_4 = 0 \\ 0 + \sin k\gamma_2 + \sin k\gamma_3 + \sin k\gamma_4 = 0 \end{array} \right\}, \quad (2.368)$$

kth-order moments

$$\left. \begin{array}{l} 0 + z_2 \cos k\gamma_2 + z_3 \cos k\gamma_3 + z_4 \cos k\gamma_4 = 0 \\ 0 + z_2 \sin k\gamma_2 + z_3 \sin k\gamma_3 + z_4 \sin k\gamma_4 = 0 \end{array} \right\}. \quad (2.369)$$

The following applies to variant a) with the specified angles for

$k = 1$:

$1 + 0 + 0 - 1 = 0$
$0 + 1 - 1 + 0 = 0$
$0 + 0 + 0 - z_4 \neq 0$
$0 + z_2 - z_3 + 0 \neq 0$

$k = 2$:

$1 - 1 - 1 + 1 = 0$
$0 + 0 + 0 + 0 = 0$ (2.370)
$0 - z_2 - z_3 + z_4 = 0$
$0 + 0 + 0 + 0 = 0$

Thus the first-order and second-order forces and second-order moments can be balanced. The first-order moments are not balanced because the conditions $z_4 = 0$ and $z_2 = z_3$ cannot be satisfied.

The following results for variant b) with $\gamma_2 = \gamma_3 = \pi$, $\gamma_4 = 0$:

$$\begin{array}{ll}
k=1: & k=2: \\
1-1-1+1=0 & 1+1+1+1\neq 0 \\
0+0+0+0=0 & 0+0+0+0=0 \\
0-z_2-z_3+z_4=0 & 0+z_2+z_3+z_4\neq 0 \\
0+0+0+0=0 & 0+0+0+0=0
\end{array} \qquad (2.371)$$

The first-order forces and moments are balanced in this variant, while the second-order forces and moments cannot be balanced.

S2.14 Analogous to (2.338) for the slider-crank mechanism, a complex equation can be formulated for the center-of-mass trajectory of the four-bar linkage shown in Fig. 2.51b. With the unbalances U_2 and U_4 defined in the problem statement, see VDI Guideline 2149 Part 1 [35], it takes the following form:

$$\begin{aligned}
(m_2 + m_{a2} + m_3 + m_4 + m_{a4})\,r_S \\
= m_2\left[jy_{12} + (\xi_{S2} + j\eta_{S2})\,\mathrm{e}^{\mathrm{j}\varphi_2}\right] + U_2\left(\mathrm{j}\frac{y_{12}}{r_{a2}} + \mathrm{e}^{\mathrm{j}(\varphi_2+\beta_2)}\right) \\
+ m_3\left[jy_{12} + l_2\mathrm{e}^{\mathrm{j}\varphi_2} + (\xi_{S3} + j\eta_{S3})\,\mathrm{e}^{\mathrm{j}\varphi_3}\right] \\
+ m_4\left[x_{14} + (\xi_{S4} + j\eta_{S4})\,\mathrm{e}^{\mathrm{j}\varphi_4}\right] + U_4\left(\frac{x_{14}}{r_{a4}} + \mathrm{e}^{\mathrm{j}(\varphi_4+\beta_4)}\right)
\end{aligned} \qquad (2.372)$$

The constraint in complex form resulting from Fig. 2.51b must be considered:

$$jy_{12} + l_2\mathrm{e}^{\mathrm{j}\varphi_2} + l_3\mathrm{e}^{\mathrm{j}\varphi_3} = x_{14} + l_4\mathrm{e}^{\mathrm{j}\varphi_4} \qquad (2.373)$$

If $\mathrm{e}^{\mathrm{j}\varphi_4}$ is obtained from this and inserted into (2.372), the center-of-mass trajectory results as a function of the two angles φ_2 and φ_3. The center of gravity remains stationary during any motion if the factors in front of the variable terms of $\mathrm{e}^{\mathrm{j}\varphi_2}$ and $\mathrm{e}^{\mathrm{j}\varphi_3}$ are set to zero. Four real balancing conditions result from separating the real and imaginary parts:

$$m_2\xi_{S2} + U_2\cdot\cos\beta_2 + m_3l_2 + (m_4\xi_{S4} + U_4\cdot\cos\beta_4)\cdot\frac{l_2}{l_4} = 0 \qquad (2.374)$$

$$m_2\eta_{S2} + U_2\cdot\sin\beta_2 + (m_4\eta_{S4} + U_4\cdot\sin\beta_4)\cdot\frac{l_2}{l_4} = 0 \qquad (2.375)$$

$$m_3\xi_{S3} + (m_4\xi_{S4} + U_4\cdot\cos\beta_4)\cdot\frac{l_3}{l_4} = 0 \qquad (2.376)$$

$$m_3\eta_{S3} + (m_4\eta_{S4} + U_4\cdot\sin\beta_4)\cdot\frac{l_3}{l_4} = 0. \qquad (2.377)$$

Their solution for the unbalance components and insertion of terms from (2.376) and (2.377) into (2.374) and (2.375) provides:

$$U_{2\xi} = U_2\cdot\cos\beta_2 = -m_2\xi_{S2} - m_3l_2\cdot\left(1 - \frac{\xi_{S3}}{l_3}\right) \qquad (2.378)$$

$$U_{2\eta} = U_2\cdot\sin\beta_2 = -m_2\eta_{S2} + m_3\eta_{S3}\cdot\frac{l_2}{l_3} \qquad (2.379)$$

$$U_{4\xi} = U_4\cdot\cos\beta_4 = -m_4\xi_{S4} - m_3\xi_{S3}\cdot\frac{l_4}{l_3} \qquad (2.380)$$

$$U_{4\eta} = U_4\cdot\sin\beta_4 = -m_4\eta_{S4} - m_3\eta_{S3}\cdot\frac{l_4}{l_3}. \qquad (2.381)$$

Now the unbalance values and their angular positions can be calculated from the unbalance components. From (2.378) to (2.381), it follows with $k = 2$ and 4: that

$$U_k = \sqrt{U_{k\xi}^2 + U_{k\eta}^2};\qquad \cos\beta_k = \frac{U_{k\xi}}{U_k};\qquad \sin\beta_k = \frac{U_{k\eta}}{U_k}. \qquad (2.382)$$

2.6 Methods of Mass Balancing

Using the parameter values given in the problem statement, the result is:
$$U_{2\xi} = -4,2711 \text{ kg} \cdot \text{m}; \qquad U_{2\eta} = 0,467\,12 \text{ kg} \cdot \text{m}. \tag{2.383}$$

The magnitude of the unbalance at the crank, therefore, is
$$\underline{U_2 = m_{a2} \cdot r_{a2} = 4,2966 \text{ kg} \cdot \text{m}}. \tag{2.384}$$

and its angular position is defined as:
$$\cos \beta_2 = -0,994\,07; \qquad \sin \beta_2 = 0,108\,72 \quad \Rightarrow \quad \underline{\beta_2 = 173,76°}. \tag{2.385}$$

The following results for the corresponding quantities at the rocker:
$$U_{4\xi} = -3,9735 \text{ kg} \cdot \text{m}; \qquad U_{4\eta} = -0,686\,67 \text{ kg} \cdot \text{m}. \tag{2.386}$$

This provides the basis for calculating the magnitude of the unbalance at the rocker:
$$\underline{U_4 = m_{a4} \cdot r_{a4} = 4,0324 \text{ kg} \cdot \text{m}} \tag{2.387}$$

at an angular position:
$$\cos \beta_4 = -0,985\,39; \qquad \sin \beta_4 = -0,170\,29 \quad \Rightarrow \quad \underline{\beta_4 = 189,8°}. \tag{2.388}$$

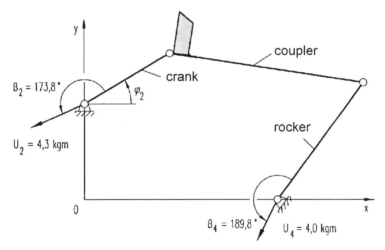

Fig. 2.52 Mechanism with additional unbalances

The mechanism is shown true to scale with additional unbalances in Fig. 2.52.

The design of the balancing masses depends on the specific conditions, such as the available installation space. Balancing masses are often designed so that their moments of inertia are minimized.

In the four-bar linkage, complete mass balancing can be achieved by adding one additional unbalance each to the crank and to the rocker. Their influence on the input torque as well as the individual bearing and joint forces should be checked by proper calculations.

Chapter 3
Foundation and Vibration Isolation

3.1 Introductory Remarks

Unfortunately, machines often cause vibrations at their mounting site and, consequently, interruptions, damages and disturbances to people. Such damages that are caused or suspected to be caused by vibrations frequently spawn negotiations on compensation or legal disputes. If the complete system were modeled and its vibration potential was assessed in due time, and the various project participants from various companies (or divisions) coordinated their efforts, many of such disputes could be avoided. The compensation amounts involved are often substantial since foundations do not just weigh a few dozen but hundreds (forging hammers, printing machines) or thousands of metric tons (groups of buildings) [25], [33].

This chapter discusses only some issues from the vast field of machine foundations [15], [19], [25], [31]: foundation blocks. It should be noted that many dynamic problems arise when configuring machine platforms as they are used, for example, for turbogenerators and printing machines, see the example shown in Fig. 3.1.

Such large foundation designs require comprehensive calculations in advance using models with many degrees of freedom, see Chap. 6. The steel structure that carries the actual machine sub-assemblies is connected with a concrete foundation, which in turn rests on the subsoil, the characteristic values of which depend on the geological conditions at the mounting site. Sometimes a huge effort must be made to install a large machine, the foundation of which weighs several hundred tons, in such a way that it is operationally safe and isolated against vibrations. In areas prone to earthquakes, design solutions have to be found to prevent any disruption of, or damage to, machinery in the case of an earthquake.

In addition to the adverse dynamic loads on the mounting site, vibrations often put stress on the human body (e. g. vibrations of the floor or the driver's seat) or on the hand-arm-system when guiding vibrating equipment (hammers, drills, tampers). From a dynamics point of view, issues of foundation design and vibration isolation are related. Measuring principles and evaluation methods have been formulated with respect to the impact on people (industrial safety requirements), buildings, and machines in standards and guidelines that a designer has to comply with. Table 3.1 lists some of these regulations, but is not a complete list. The empirical data reflected in standards or guidelines can be viewed as reference values, even though they may not apply exactly to the case on hand or may sometimes contradict each other.

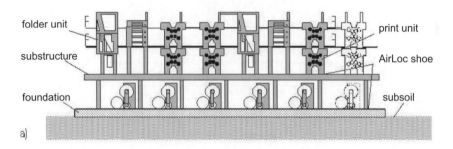

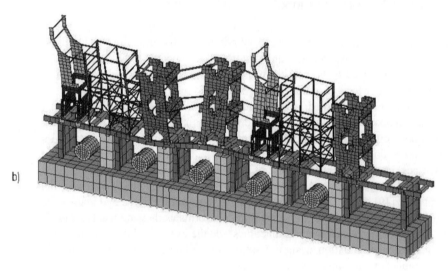

Fig. 3.1 Rotary offset press with foundation; **a)** Schematic, **b)** FEM model and calculated third mode shape at $f_3 = 5.01$ Hz (Source: Doctoral thesis *Xingliang Gao*, TU Chemnitz, 2001)

Foundation vibrations are often excited periodically or intermittently. Periodic excitations occur due to unbalanced inertia forces. Impact excitations occur, for example, on presses, punch presses, shears, and hammers.

The set-up of machines is of interest to machine manufacturers, designers, and operators. Each installation site, whether it is directly onto the subsoil, a ceiling of a structure, or a supporting structure, is elastic. Any machine setup therefore represents a vibration system. The first task in foundation design is to prove that the machine setup is permissible without special isolation measures. The static and dynamic forces transmitted onto the ground and the movement of the machine play a decisive role in this respect.

3.1 Introductory Remarks

Table 3.1 Standards and guidelines on vibration isolation (selection)

BS 7385	(British Standard): Measurement and Evaluation of Vibration in Buildings
DIN ISO 10 816-1	Evaluation of machine vibrations by measurements on non-rotating parts
DIN 4150 Part 2	Structural vibrations, human exposure to vibrations in buildings
DIN 30 786	Transportation loads, Part 1: Fundamentals, Part 2: Vibrations and impact loads during road transport
E DIN ISO 8002	Mechanical vibrations: Land vehicles
DIN EN ISO 5349-1	Measurement and evaluation of the effect of vibrations on the the human hand-arm system
DIN EN 28 662-1	Hand-held portable power tools. Measurement of vibrations at the handle. – General human hand-arm system
SN 640 323	(Swiss standard) Vibration impact on structures
VDI 2057 1-4 [36]	Human exposure to mechanical vibrations
VDI 2059 1-5 [36]	Shaft vibrations of turbosets
VDI 2062-1 [36]	Vibration isolation; Terminology and methods
VDI 2062-2 [36]	Vibration isolation; Isolation elements
VDI 3831 [36]	Protective measures against the effects of mechanical vibrations on humans

The protection against vibrations starts with methods in which the oscillating motions are balanced by influencing the excitation mechanisms or by controlled countermotions. This includes measures for mass balancing, see Chap. 2.

The second step includes all methods by which the proper set-up of the object influences the transmission of forces or vibrations. **Vibration isolation** is aimed at protecting the installation site from dynamic forces from the machine or at keeping vibrations of the installation site away from the machine (or from a person or measuring instrument). In the first case, the calculation model is based on force excitation, in the second, on motion excitation.

In active vibration compensation, a time signal of the vibration is measured by a sensor, "processed" and used to control an actuator. Actuators can be piezoelectric, electromagnetic, hydraulic, or pneumatic control elements. The signal is usually "processed" by a control circuit that evaluates the difference between setpoint value and actual value [11], [18], [33].

Engineers often face the task of evaluating the effects of vibrations on humans, structures, or machines. For this purpose, evaluation criteria can be used if no exact proof can be provided. These are provided in the form of standards, guidelines, and recommendations, both with respect to the quantities to be evaluated and the way in which these quantities are measured. The best agreement internationally has been achieved when it comes to the impact of vibrations on humans. Data for structures often refer to traditional building styles and cannot always be easily applied to modern industrial buildings. For the effect of vibrations on machines, there are a large number of special company standards or data for machine groups, but they have the status of reference values only.

The vibration intensity scale by RISCH and ZELLER that goes back to oscillatory power is the most common measure for evaluating the effect of vibrations on structures and the subsoil:

$$S = 10 \lg \frac{x}{x_0} \quad \text{in vibrar;} \qquad x = \frac{a^2}{f}; \qquad x_0 = 10^{-5} \, \text{m}^2/\text{s}^3 \tag{3.1}$$

a Amplitude of acceleration in m/s^2; f Frequency in Hz

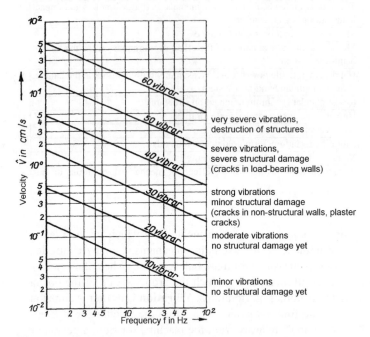

Fig. 3.2 Vibration intensity scale for evaluating structure endangerment

Since the international standard ISO 2631-1 (1997) was issued, new frequency evaluation curves apply to whole-body vibrations. In Germany, these are also part of national regulations for evaluating the effect of mechanical vibrations on humans. The K values that were common in Germany only have been replaced by effective values (r.m.s. values) of the frequency-weighted vibrational acceleration a_w.

DIN 45 671, DIN 45 675 and the VDI Guidelines 2057 and 3831 [36] apply to human exposure to loads from mechanical vibrations. The stresses caused by human exposure to mechanical vibrations depend on various parameters such as amplitude, frequency (spectrum), and direction of the acting vibrations, the duration of the exposure and the site on the human body that is exposed to them. The relationship (3.1) is plotted in the velocity-frequency diagram shown in Fig. 3.2. A permissible maximum velocity (effective value in mm/s) is given depending on a density index I_D for evaluating the effect of vibrations on the subsoil. The following applies:

$$\tilde{v}_{\text{Gr}} = 1.5 \cdot e^{2{,}85 I_D} \text{ mm/s}. \tag{3.2}$$

I_D is in the range from 0.1 to 0.7. Depending on the soil condition and type, $\tilde{v}_{\text{Gr}}$ is thus in the range

$$\tilde{v}_{\text{Gr}} = (2 \ldots 11.0) \text{ mm/s}. \tag{3.3}$$

3.2 Foundation Loading for Periodic Excitation

3.2.1 Minimal Models with One Degree of Freedom

3.2.1.1 Model Description

A calculation model with six degrees of freedom would result if assuming a rigid machine on an elastic foundation. However, a model with just one degree of freedom is sufficient for answering basic questions of vibration isolation with respect to periodic excitation (force or motion excitation), while impact excitation mostly requires a minimal model of two masses with one degree of freedom each. In these models, the parameters mass m, spring constant c, and damping constant b have different associations with the system. Table 3.3 lists these associations for periodic vibrations.

Each foundation design aims at limiting the dynamic forces that are transmitted onto the installation site and at keeping the motions of the machine within specific limits. This is achieved by tuning the natural frequencies of the foundation with the excitation frequencies set by the machine. System damping plays a relatively minor role here.

The principles specified in Table 3.2 depend on the machine speed, which typically corresponds to the lowest excitation speed. This is due to various implementation options since low tuning requires a substantial design effort if the excitation frequency is small. If one starts from a minimal model of an elastically mounted machine (Table 3.3), the natural circular frequency when neglecting damping is:

$$(2\pi f)^2 = \omega_0^2 = \frac{c}{m} = \frac{g}{\frac{mg}{c}} = \frac{g}{x_{\text{st}}}. \tag{3.4}$$

Spring stiffness for a linear spring can also be shown by the static deflection x_{st}. It follows for the natural frequency

$$f = \frac{1}{2\pi}\sqrt{\frac{g}{x_{\text{st}}}}. \tag{3.5}$$

If the gravitational acceleration is approximated by $g \approx 100\pi^2$ cm/s^2, with x_{st} in cm, the resulting approximate formulae for the fundamental frequency f and for the respective speed are:

$$f \approx \frac{5}{\sqrt{\frac{x_{\text{st}}}{\text{cm}}}} \text{ in Hz}; \qquad n \approx \frac{300}{\sqrt{\frac{x_{\text{st}}}{\text{cm}}}} \text{ in rpm.} \tag{3.6}$$

Figure 3.3 shows this relationship, which also points to optional designs using resilient intermediate layers.

Table 3.2 Overview of principles of vibration isolation for various setup types

Machine speed in rpm	Setup directly on subsoil		Setup on building ceilings or support structures	
	Small excitation forces (well balanced, mass balancing)	Large excitation forces (not balanced, no mass balancing)	Small excitation forces (well balanced, mass balancing)	Large excitation forces (not balanced, no mass balancing)
0 to 500	Foundation slab; static calculation in absence of resonances	High tuning; small foundation block; subsoil with large base area	Anchoring; static calculation in absence of resonances	Low tuning; large foundation block; steel spring
300 to 1000	High or low tuning or mixed tuning; small foundation block; subsoil spring; ensure absence of resonances	High tuning or mixed tuning, small foundation blocks; subsoil spring or low tuning; large foundation block; steel or rubber springs	Low tuning; small or no foundation block; steel or rubber springs or high tuning; anchoring	Low tuning; large foundation mass; steel or rubber springs
Above 1000	Low tuning; small foundation block; subsoil cushioned with a small base area; resilient intermediate layers or individual springs	Low tuning; large foundation block; subsoil cushioning; resilient intermediate layers or individual springs	Low tuning; small or no foundation block; steel or rubber springs; resilient intermediate layers; support structures	Low tuning; large foundation block; steel or rubber springs

Definition of terms:

Low tuning: The highest natural frequency of foundation vibration is smaller than the lowest excitation frequency.
High tuning: The natural frequencies of foundation vibration are above the excitation frequency spectrum.
Mixed tuning: The spectra of natural and excitation frequencies show partial overlay but no resonance occurs.

3.2 Foundation Loading for Periodic Excitation

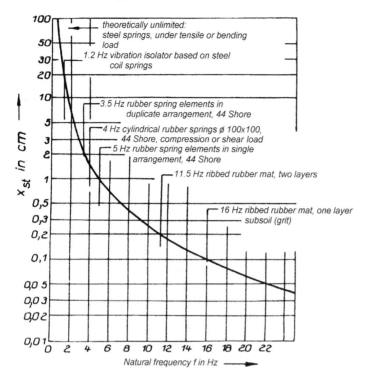

Fig. 3.3 Natural frequencies of machine setup as a function of the static spring deflection due to its own weight

Required are the motion parameters for evaluating the subsoil loads or the loads on a building ceiling for direct mounting. The vibration amplitude is decisive for the building loads, the vibration velocity is a measure of the machine loads, and the vibrational acceleration is a measure of the subsoil loads. If a resilient intermediate layer with a small stiffness, as compared to that of the installation site, is placed between the installation site and machine, the force that acts onto the ground is of interest and used for evaluating the loads. Resilient intermediate layers can be implemented by individual springs (coil springs, rubber springs, air springs) or isolating mats.

3.2.1.2 Harmonic Excitation

The following equation of motion applies to the calculation model with motion or force excitation shown in Table 3.3

$$m\ddot{x} + b\dot{x} + cx = F(t) + cs(t) + b\dot{s}(t) = Q(t) \tag{3.7}$$

or for a harmonic excitation force (period of oscillation $T_0 = 2\pi/\Omega$)

Table 3.3 Parameter associations for the minimal model for harmonic excitation

Model		Set-up on subsoil	Set-up directly on ceiling of building	Set-up using an elastic intermediate layer
	m	Mass of machine and foundation	Mass of machine and portion of ceiling	Mass of machine and foundation
	c	Subsoil stiffness	Ceiling resilience	Resilience of the elastic intermediate
	b	Subsoil damping	Ceiling damping. Influence of machine fastening system	Damping of the elastic intermediate layer
Quantities of interest		Vibration amplitude, vibration velocity, vibration acceleration		dynamic force acting on the ground

$$\ddot{x} + 2D\omega_0\dot{x} + \omega_0^2 x = \frac{\hat{Q}}{m}\sin\Omega t. \qquad (3.8)$$

The abbreviations

$$D = \frac{\delta}{\omega_0} = \frac{b}{2m\omega_0}; \qquad \eta = \frac{\Omega}{\omega_0}; \qquad \omega_0^2 = \frac{c}{m} \qquad (3.9)$$

provide the second representation (3.8) from (3.7). The steady-state solution is:

$$x = \frac{\hat{Q}}{c}\frac{1}{\sqrt{(1-\eta^2)^2 + 4D^2\eta^2}}\sin(\Omega t - \varphi); \qquad \tan\varphi = \frac{2D\eta}{1-\eta^2}. \qquad (3.10)$$

The dynamic force that acts onto the ground for $s(t) \equiv 0$ and $\hat{Q} = \hat{F}$ is

$$F_B = cx + b\dot{x} = \hat{F}\sqrt{\frac{1 + 4D^2\eta^2}{(1-\eta^2)^2 + 4D^2\eta^2}}\sin(\Omega t + \gamma - \varphi). \qquad (3.11)$$

γ is the phase shift that occurs due to the combination of the spring force cx and the damping force $b\dot{x}$, while the phase shift φ occurs between the excitation force and the motion. The motion with harmonic excitation is at excitation frequency.

The following applies to the amplitudes:

$$\hat{x} = \frac{\hat{F}}{c}\frac{1}{\sqrt{(1-\eta^2)^2 + 4D^2\eta^2}} = \frac{\hat{F}}{c}V_1 \qquad (3.12)$$

3.2 Foundation Loading for Periodic Excitation

$$\hat{F}_B = \hat{F}\sqrt{\frac{1+4\eta^2 D^2}{(1-\eta^2)^2 + 4D^2\eta^2}} = \hat{F}V_2. \tag{3.13}$$

For unbalance excitation $\hat{Q} = m_u r_u \Omega^2$ follows:

$$\hat{x} = \frac{m_u r_u}{m} \frac{\eta^2}{\sqrt{(1-\eta^2)^2 + 4D^2\eta^2}} = \frac{m_u r_u}{m} V_3 \tag{3.14}$$

$$\hat{F}_B = \frac{m_u r_u}{m} c\eta^2 \sqrt{\frac{1+4D^2\eta^2}{(1-\eta^2)^2 + 4D^2\eta^2}} = \frac{m_u r_u}{m} cV_4. \tag{3.15}$$

The machine vibration and the force acting onto the ground are determined by the *nondimensionalized amplitude functions* V_1, V_2, V_3, V_4. They are shown in Fig. 3.4.

The equation of motion (3.7) for $F \equiv 0$ is valid for the calculation model shown in Table 3.3 involving pure motion excitation $s = \hat{s}\sin\Omega t$. Thus:

$$\begin{aligned}Q(t) &= m(2\delta\hat{s}\Omega\cos\Omega t + \omega_0^2 \hat{s}\sin\Omega t) \\ &= m\left[\hat{s}\sqrt{\omega_0^4 + (2\delta\Omega)^2}\sin(\Omega t + \gamma)\right].\end{aligned} \tag{3.16}$$

When comparing (3.16) and (3.8), it becomes evident that the solution (3.10) can be used if:

$$\hat{Q} = m\hat{s}\sqrt{\omega_0^4 + (2\delta\Omega)^2} = m\hat{s}\omega_0^2\sqrt{1+4D^2\eta^2}. \tag{3.17}$$

Using the abbreviations (3.9), the following results for the amplitude of the absolute motion, which is a criterion for the vibration isolation of the mass:

$$\hat{x} = \hat{s}\sqrt{\frac{1+4\eta^2 D^2}{(1-\eta^2)^2 + 4\eta^2 D^2}} = \hat{s}V_2. \tag{3.18}$$

(3.12) is relevant for direct setup on the subsoil or a building ceiling. In most cases, one has to assume that damping is low and cannot be increased by suitable components. One can see in Fig. 3.4 that the amplitudes become very large in the range of $\eta = 1$, that is, near resonance. Only high tuning or low tuning are practical options.

For $\eta \ll 1$ (high tuning) $V_1 \to 1$. This is achieved by a stiff spring and by doing without a foundation mass. A large supporting surface is required when installing directly onto the subsoil, and for ceilings this is achieved by appropriate reinforcement of the ceiling. $\hat{x} = \hat{F}/c$ in the most favorable case, which corresponds to the static deflection under the force. Low tuning ($\eta \gg 1$) is another option, based on the nondimensionalized amplitude function. $\eta \to \infty$ is assumed to evaluate the maximum achievable effect.

The limit is obtained from (3.12):

$$\lim_{\eta \to \infty} \hat{x} = \frac{\hat{F}}{c\eta^2} = \frac{\hat{F}}{m\Omega^2}. \tag{3.19}$$

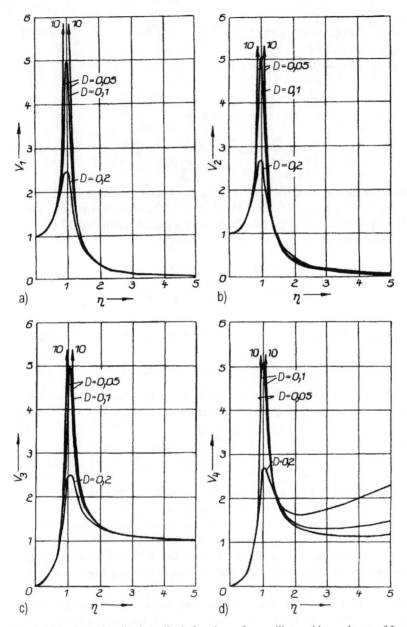

Fig. 3.4 Nondimensionalized amplitude functions of an oscillator with one degree of freedom according to (3.12) to (3.15)

Small vibration amplitudes in the low tuning range are achieved with a large mass. The machine is therefore placed on foundation blocks of multiple times larger mass to increase the machine mass. In most cases, however, this exceeds the static load ca-

3.2 Foundation Loading for Periodic Excitation

pacity of the installation site (such as the ceiling of a structure). This creates the necessity for separating the load-bearing function from the cushioning function, which is achieved by resilient intermediate layers (individual springs or isolating mats) between the foundation and the installation site. If it is assumed that their stiffness is small, as compared to the stiffness of the installation site, a second question about the force that acts onto the ground arises. According to (3.13), it is directly proportional to the nondimensionalized amplitude function V_2. It can be seen from Fig. 3.4b that the smallest dynamic ground force occurs if damping is small. Therefore, low-damping cushioning has to be achieved, most of all for small η values in the low tuning range ($\eta \approx 3$). Individual springs, particularly coil springs or spring packets, will be preferred over isolating mats.

As long as the unbalancing mass moves along a circular path, a revolving excitation force $\hat{Q} = m_u r_u \Omega^2$ acts and the nondimensionalized amplitude function V_3 applies. This circular path is changed as a result of the motion of the foundation mass, and the excitation force is noticeably influenced by large resonance motions, see the Sommerfeld effect described in [4].

The nondimensionalized amplitude function V_3 reaches the value of $V_3 = 1$ for a low-tuned setup. It stays below this value for high tuning for $\eta < \sqrt{1/2} \approx 0.7$. Larger values lie in between, which increase considerably near resonance ($\eta = 1$). Only the ranges of $0 < \eta < \sqrt{1/2}$ or $\eta > 3$ are suitable for foundations. If $V_3 = 1$, the following results from the definition of the center of gravity for the motion amplitude: $\hat{x}m = m_u r_u$.

One usually first tries to operate the machine in the high tuned range. This is difficult, however, if the excitation force has several harmonics, such as a piston engine, since the highest harmonic has to be used for the high tuning, see Fig. 3.5. This cannot be achieved in a vehicle, so that once again, resilient intermediate layers in the form of individual springs are selected. However, the dynamic force that acts onto the ground (3.15) is decisive for this. It depends on the nondimensionalized amplitude function V_4. One can see the strong influence of damping for low tuned setups from Fig. 3.4d. This is a reason why the noise level of the engine portion in a motor vehicle increases with increasing speed.

(3.18) applies to motion excitation. The best isolating effect occurs if the setup is just weakly damped at low tuning. Note, however, that (3.12) to (3.15) and (3.18) only apply to the steady state. Damping virtually always ensures that start-up problems wear off. The importance of low tuning became clear for all cases of vibration isolation discussed. Note that free movement of the foundation block has to be ensured for any practical design. Pipes and other connections should either be very flexible or be taken into account when calculating the spring constants. Note, at any rate, that high loads are put on stiff connections when passing through resonance.

The peak values for these nondimensionalized amplitude functions are in the vicinity of $\eta = 1$ for

$$V_{\max} = \frac{1}{2D}. \tag{3.20}$$

3.2.1.3 Periodic Excitation/Fourier Series

Cyclically operating machines such as power machines and work machines (e. g. pumps, compressors), processing machines (e. g. textile machines, packaging machines, polygraphic machines), robots, and transportation equipment are mainly loaded by periodic forces and moments. The focus of this discussion is not the start-up and braking process, but steady-state operation, which is characterized by a specific (kinematic) cycle time $T_0 = 2\pi/\Omega$. A periodic excitation force can be expanded into a Fourier series according to (1.112):

$$F(t) = F(t+T_0) = \sum_{k=0}^{\infty} \left(F_{ak} \cos \frac{2\pi k t}{T_0} + F_{bk} \sin \frac{2\pi k t}{T_0} \right). \tag{3.21}$$

The summands are called kth-order harmonics of the excitation force.

The mean angular velocity $\Omega = 2\pi/T_0$ of the drive can be used for calculation; it may be variable and characterized by a coefficient of speed fluctuation. The Fourier series then takes one of the following forms:

$$F(t) = \sum_{k=1}^{\infty} \hat{F}_k \sin(k\Omega t + \beta_k) = \sum_{k=1}^{\infty} (F_{ak} \cos k\Omega t + F_{bk} \sin k\Omega t). \tag{3.22}$$

The Fourier coefficients $\hat{F}_k$ and the phase angles β_k are related to the coefficients F_{ak} and F_{bk} according to (1.115).

The equation of motion of the damped oscillator with one degree of freedom that is excited periodically is, see (3.7):

$$m\ddot{x} + b\dot{x} + cx = F(t) = \sum_{k=1}^{\infty} \hat{F}_k \sin(k\Omega t + \beta_k). \tag{3.23}$$

The steady-state solution of this equation of motion is determined in analogy to (3.10)

$$x(t) = \frac{1}{c} \sum_{k=1}^{\infty} V_k \hat{F}_k \sin(k\Omega t + \beta_k - \varphi_k) = \sum_{k=1}^{\infty} \hat{x}_k \sin(k\Omega t + \beta_k - \varphi_k). \tag{3.24}$$

It is useful to apply the frequency ratio η and the damping ratio D defined in (3.9). The Fourier coefficients of the deflection that result in analogy to (3.12)

$$\hat{x}_k = \frac{V(k, D, \eta)\hat{F}_k}{c} \tag{3.25}$$

are "distorted", as compared to those of excitation ($\hat{F}_k$) by the *nondimensionalized amplitude functions*

$$V(k, D, \eta) = \frac{1}{\sqrt{(1-k^2\eta^2)^2 + 4D^2k^2\eta^2}}; \quad k = 1, 2, \ldots, \tag{3.26}$$

3.2 Foundation Loading for Periodic Excitation

i. e. they can be enlarged or diminished in comparison to the "static" values ($\hat{F}_k/c$). The periodic time function of the solution x(t) generally does not show any geometric similarity to that of the excitation, see the example in Fig. 3.5. As a result of damping, the deflection function trails each harmonic of the force function. The phase angles result from

$$\sin \varphi_k = \frac{2Dk\eta}{\sqrt{(1-k^2\eta^2)^2 + 4D^2k^2\eta^2}}$$
$$\cos \varphi_k = \frac{1-k^2\eta^2}{\sqrt{(1-k^2\eta^2)^2 + 4D^2k^2\eta^2}}.$$
(3.27)

Figure 3.5a shows the function of the periodic excitation force according to (3.22), as compared to the functions of the deflections according to (3.24) in Fig. 3.5b. The time functions are shown on the left-hand side of the figure, and the corresponding Fourier coefficients $\hat{F}_k$ and $\hat{x}_k$, to scale, on the right-hand side. The figure illustrates the significant influence that the frequency ratio has with reference to two examples, see Fig. 3.5b.

The frequency ratio $\eta = 0.18$ is between the 5th-order ($\eta = 1/5$) and 6th-order ($\eta = 1/6$) resonance points. That is why the amplitudes of the fifth and sixth harmonic increase particularly strongly, as is shown in Fig. 3.5b. The number of maxima (or minima) per period matches the order of the harmonic. The original function of the force is so strongly dominated by the oscillations with the excited natural frequency that it can hardly be detected. The situation is similar for the frequency ratio $\eta = 0.36$. The proximity to the third-order resonance is clearly visible. It is important to note that, when interpreting measurement results, other time functions for the deflections (and accelerations, which are even more "roughened") may result depending on the speed range. It is sometimes difficult to recognize the excitation force curve at all in a measured acceleration curve.

The resonance curve is shown in Fig. 3.5c. It shows the maximum deflection as a function of the frequency ratio. Deflection maxima occur in the immediate vicinity of the resonance points of the kth-order harmonic, and their height and the magnitude of the kth Fourier coefficient of the force are approximately proportional, as a comparison with Fig. 3.5a shows.

For continuous operation, machines should be running in a resonance-free range! The operating speed often lies between two resonance areas, and the operators are unaware of this. Thus, it may be that the dynamic behavior of a machine at first deteriorates with a speed increase, but improves at even higher speeds. This is the case, for example, when a machine running at $\eta = 0.15$, at $\eta = 0.165$ (10% speed increase) vibrates heavily but runs more smoothly at $\eta = 0.18$ (that is at a 20% speed increase). One should not be astonished that unexpectedly strong vibrations occur after a speed increase which later vanish at even higher speeds, see the resonance curve in Fig. 3.5c.

It should be checked in which "valley" of the resonance curve a machine is operating. No higher harmonics need to be detectable as excitations at first glance. Periodic forces and several resonance points can occur even if the motion appears

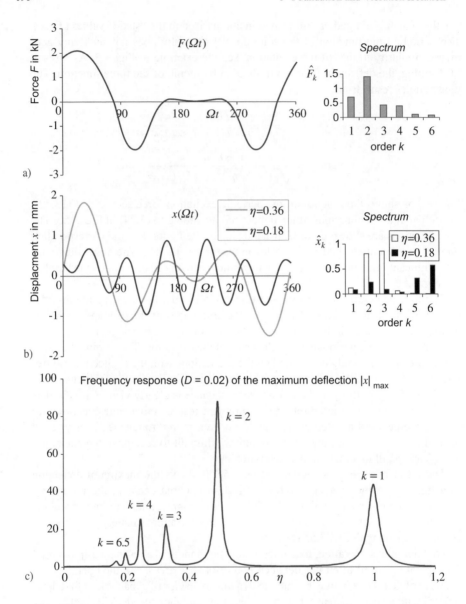

Fig. 3.5 Periodic excitation (time and frequency domain); **a)** Excitation force, **b)** Deflection (frequency ratio $\eta = 0.18$ and $\eta = 0.36$), **c)** resonance curve

to be harmonic, e. g. if the machine speed varies as is often the case, see Sect. 2.3. One has to conclude that higher order resonance points should always be watched. The lower speed ranges to which higher order resonance areas correspond have to be passed through during machine start-up, before the operating speed range is reached.

The reduction of the amplitudes of the higher harmonics of the excitation force is an important vibration control measure, see also the discussion in Sect. 4.3.2.3.

3.2.1.4 Shock Isolation

Highly elastic supports also reduce the transmission intermittent forces. The spring-supported mass responds to impact excitations in its natural frequency. Impacts are mainly characterized by their peak force F_{SO}, the impact duration Δt and the shape of the impact pulse. Figure 6.25c illustrates the dynamic responses for an impact with rectangular pulse shape (Fig. 6.25a) where the peak force is reached immediately and an impact with half-sine shape $F = F_{SO}\sin(\pi t/\Delta t)$ where the force raises gradually from zero to its maximum value.

Figure 6.25c shows the ratios between the maximum dynamic force F_{Bmax} acting on the ground and the peak of the impact force F_{SO} for the two pulse shapes. The undamped spring-mass system has the natural period $T = 1/f$. The isolation effect significantly depends on the ratio between the impact duration Δt and the natural period T. The transmitted force F_B will be less than the maximum impact force F_{SO} if $\Delta t/T < 0.167$ for rectangular pulses or $\Delta t/T < 0.267$ for pulses with half-sine shape, respectively. In order to ensure that the ground force becomes smaller than the impact force, the natural period of the isolating system must be at least 6 times the pulse duration for a rectangular pulse or 3.75 times the pulse duration for a half-sine pulse. The force propagated to the ground decreases, if the natural frequency of the spring-mass system is lowered, i.e., if the mass m is increased or the spring rate c is decreased.

3.2.2 Block Foundations

3.2.2.1 Natural Frequencies and Mode Shapes

Every elastically mounted rigid machine has six degrees of freedom. A calculation model with six degrees of freedom has six natural frequencies. If low tuning is required, it must be ensured that the highest natural frequency is smaller than the lowest excitation frequency. For mixed tuning, all natural frequencies have to be far enough away from the excitation frequencies.

The calculation model is shown in Fig. 3.6. The following assumptions are made:

1. The coordinate origin is at the center of gravity of the rigid system, composed of rigid machine and foundation mass (in short: foundation block), which is in its static equilibrium position.
2. In the equilibrium position, the body-fixed coordinate system ξ-η-ζ is identical with the fixed x-y-z system in which the displacements x, y, z and the small angles φ_x, φ_y, φ_z about these axes are measured.

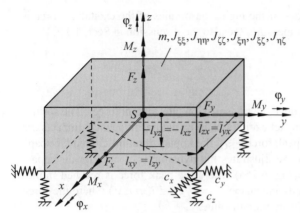

Fig. 3.6 Calculation model of a foundation block

3. All external forces and moments are referred to this coordinate system.
4. All springs, which are assumed to have no mass, have spring constants in the three coordinate directions. The determination of the transverse spring constants is discussed in Sect. 1.3.2.
5. The point of application of the spring forces on the foundation block is at the elastic center, which is located in the middle of the statically compressed springs in the case of cylindrical coil springs. It is determined for each spring by three distances l_{mn}. m is the coordinate direction of the spring force and n the coordinate direction of the distance. A specific spring with the spring constant c_x has the (signed) distances l_{xy}, l_{xz}. The principal elastic axes are parallel to the axes of the system, see Table 1.4.
6. In addition to the position of the axis through the center of gravity, the given mass parameters are the mass m, the moments of inertia $J_{\xi\xi}$, $J_{\eta\eta}$, $J_{\zeta\zeta}$ and the products of inertia $J_{\xi\eta}$, $J_{\xi\zeta}$, $J_{\eta\zeta}$ ($\overline{\boldsymbol{J}}^S \approx \boldsymbol{J}^S$, see Sect. 2.3.1).
7. Damping is neglected.
8. Small amplitudes are assumed (linear equations of motion).
9. The influence of the gravitational force on the vibrational motion is not taken into account.

The *Lagrange's* equations are used to establish the six equations of motion. They are summarized in the matrix equation

$$\boldsymbol{M}\ddot{\boldsymbol{q}} + \boldsymbol{C}\boldsymbol{q} = \boldsymbol{f} \tag{3.28}$$

. The following applies for the coordinate vector: $\boldsymbol{q} = [x\ y\ z\ \varphi_x\ \varphi_y\ \varphi_z]^{\mathrm{T}} = [q_1\ q_2\ q_3\ q_4\ q_5\ q_6]^{\mathrm{T}}$ and the excitation force vector $\boldsymbol{f} = [F_x\ F_y\ F_z\ M_x\ M_y\ M_z]^{\mathrm{T}}$.

The elements m_{kl} of the mass matrix $\boldsymbol{M}$ and the elements c_{kl} of the stiffness matrix $\boldsymbol{C}$ can be determined from the kinetic and potential energies, see (6.10) and (6.15):

$$m_{kl} = m_{lk} = \frac{\partial^2 W_{\mathrm{kin}}}{\partial \dot{q}_k \partial \dot{q}_l}; \qquad c_{kl} = c_{lk} = \frac{\partial^2 W_{\mathrm{pot}}}{\partial q_k \partial q_l}. \tag{3.29}$$

3.2 Foundation Loading for Periodic Excitation

The kinetic energy is known from Sect. 2.3.1:

$$2W_{\text{kin}} = m(\dot{x}^2 + \dot{y}^2 + \dot{z}^2) + J_{\xi\xi}\dot{\varphi}_x^2 + J_{\eta\eta}\dot{\varphi}_y^2 + J_{\zeta\zeta}\dot{\varphi}_z^2 \\ + 2J_{\xi\eta}\dot{\varphi}_x\dot{\varphi}_y + 2J_{\xi\zeta}\dot{\varphi}_x\dot{\varphi}_z + 2J_{\eta\zeta}\dot{\varphi}_y\dot{\varphi}_z. \quad (3.30)$$

The potential energy results from (omitting the indices of the spring coefficients)

$$2W_{\text{pot}} = \sum c_x(x + \varphi_y l_{xz} - \varphi_z l_{xy})^2 + \sum c_y(y - \varphi_x l_{yz} + \varphi_z l_{yx})^2 \\ + \sum c_z(z + \varphi_x l_{zy} - \varphi_y l_{zx})^2. \quad (3.31)$$

(3.29) yields the mass matrix and the stiffness matrix

$$M = \begin{bmatrix} m & 0 & 0 & 0 & 0 & 0 \\ 0 & m & 0 & 0 & 0 & 0 \\ 0 & 0 & m & 0 & 0 & 0 \\ 0 & 0 & 0 & J_{\xi\xi} & J_{\xi\eta} & J_{\xi\zeta} \\ 0 & 0 & 0 & J_{\eta\xi} & J_{\eta\eta} & J_{\eta\zeta} \\ 0 & 0 & 0 & J_{\zeta\xi} & J_{\zeta\eta} & J_{\zeta\zeta} \end{bmatrix}; \quad C = \begin{bmatrix} c_{11} & 0 & 0 & 0 & c_{15} & c_{16} \\ 0 & c_{22} & 0 & c_{24} & 0 & c_{26} \\ 0 & 0 & c_{33} & c_{34} & c_{35} & 0 \\ 0 & c_{42} & c_{43} & c_{44} & c_{45} & c_{46} \\ c_{51} & 0 & c_{53} & c_{54} & c_{55} & c_{56} \\ c_{61} & c_{62} & 0 & c_{64} & c_{65} & c_{66} \end{bmatrix}. \quad (3.32)$$

The following applies to the elements of the spring matrix C:

$$c_{11} = \sum c_x; \quad c_{15} = c_{51} = \sum c_x l_{xz}; \quad c_{16} = c_{61} = -\sum c_x l_{xy}$$

$$c_{22} = \sum c_y; \quad c_{24} = c_{42} = \sum c_y l_{yz}; \quad c_{26} = c_{62} = -\sum c_y l_{yx}$$

$$c_{33} = \sum c_z; \quad c_{34} = c_{43} = \sum c_z l_{zy}; \quad c_{35} = c_{53} = -\sum c_z l_{zx} \quad (3.33)$$

$$c_{44} = \sum c_y l_{yz}^2 + \sum c_z l_{zy}^2; \quad c_{45} = c_{54} = -\sum c_z l_{zx} l_{zy}$$

$$c_{46} = c_{64} = -\sum c_y l_{yx} l_{yz}; \quad c_{55} = \sum c_x l_{xz}^2 + \sum c_z l_{zx}^2$$

$$c_{56} = c_{65} = -\sum c_x l_{xy} l_{xz}; \quad c_{66} = \sum c_x l_{xy}^2 + \sum c_y l_{yx}^2$$

The eigenvalue problem for the free vibrations ($f = o$) is obtained by assuming the solution to be of the form $q = v \sin \omega t$ from (3.28):

$$(C - \omega^2 M)v = o. \quad (3.34)$$

The vector $v = [v_1 \; v_2 \; v_3 \; v_4 \; v_5 \; v_6]^T$ contains the normalized amplitudes of the mode shapes. The solution of (3.34) can be obtained using commercial software, and it provides the natural frequencies and mode shapes, see 3.2.2.2 and Chap. 6. The position of the natural frequencies is interesting in conjunction with the speed range in which the machine that is mounted on the foundation operates. However, the mode shapes are important to evaluate the risk potential of resonances. It has to be pointed out once again here that the "correct" natural frequencies do not only have to be calculated for the mass of the foundation block, but for the entire system (together with the machine mounted onto it).

Unsymmetrical foundation blocks have complicated mode shapes since typically more than two translational and rotational motions are combined, i. e. free helical motions occur about axes that are askew in space. The free motions of damped systems become even more complicated, see Sect. 6.5.

3.2.2.2 Model Decomposition for a Symmetrical Arrangement

It is beneficial for several reasons to decouple the six equations of motion for a foundation block. This is feasible if the ξ", η and ζ axes are principal axes of inertia and the springs are arranged symmetrically. The products of inertia are then $J_{\xi\eta} = J_{\eta\zeta} = J_{\zeta\eta} = 0$ (see Sect. 2.5) and the mass matrix becomes a diagonal matrix. Certain spring coefficients become zero in a symmetrical spring arrangement because the spring distances enter the calculation with their respective signs. For $c_{16} = c_{34} = c_{36} = c_{45} = c_{56} = 0$, two independent systems are derived from (3.28):

$$\begin{bmatrix} m & 0 & 0 \\ 0 & m & 0 \\ 0 & 0 & J_{\eta\eta} \end{bmatrix} \begin{bmatrix} \ddot{x} \\ \ddot{z} \\ \ddot{\varphi}_y \end{bmatrix} + \begin{bmatrix} c_{11} & 0 & c_{15} \\ 0 & c_{33} & c_{35} \\ c_{15} & c_{35} & c_{55} \end{bmatrix} \begin{bmatrix} x \\ z \\ \varphi_y \end{bmatrix} = \begin{bmatrix} 0 \\ 0 \\ 0 \end{bmatrix} \quad (3.35)$$

$$\begin{bmatrix} m & 0 & 0 \\ 0 & J_{\xi\xi} & 0 \\ 0 & 0 & J_{\zeta\zeta} \end{bmatrix} \begin{bmatrix} \ddot{y} \\ \ddot{\varphi}_x \\ \ddot{\varphi}_z \end{bmatrix} + \begin{bmatrix} c_{22} & c_{24} & c_{26} \\ c_{24} & c_{44} & c_{46} \\ c_{26} & c_{46} & c_{66} \end{bmatrix} \begin{bmatrix} y \\ \varphi_x \\ \varphi_z \end{bmatrix} = \begin{bmatrix} 0 \\ 0 \\ 0 \end{bmatrix} \quad (3.36)$$

that is, the determinant of the eigenvalue problem (3.34) $\det(\boldsymbol{C} - \omega^2 \boldsymbol{M}) = 0$ is divided into the two determinants

$$\begin{vmatrix} c_{11} - m\omega^2 & 0 & c_{15} \\ 0 & c_{33} - m\omega^2 & c_{35} \\ c_{15} & c_{35} & c_{55} - J_{\eta\eta}\omega^2 \end{vmatrix} = 0$$

$$\begin{vmatrix} c_{22} - m\omega^2 & c_{24} & c_{26} \\ c_{24} & c_{44} - J_{\xi\xi}\omega^2 & c_{46} \\ c_{26} & c_{46} & c_{66} - J_{\zeta\zeta}\omega^2 \end{vmatrix} = 0,$$

(3.37)

from each of which two cubic equations result for the squares of the natural circular frequencies. If one, in addition, requires symmetry with respect to the x-z plane and the y-z plane, the spring coefficients become $c_{35} = c_{26} = c_{46} = 0$, and the following relations result from the two determinants:

$$c_{33} - m\omega^2 = 0; \quad c_{66} - J_{\zeta\zeta}\omega^2 = 0$$
$$(c_{11} - m\omega^2)(c_{55} - J_{\eta\eta}\omega^2) - c_{15}^2 = 0 \quad (3.38)$$
$$(c_{22} - m\omega^2)(c_{44} - J_{\xi\xi}\omega^2) - c_{24}^2 = 0.$$

3.2 Foundation Loading for Periodic Excitation

Thus two linear and two quadratic equations for ω^2 are found. The solutions are identified using the coordinate index k:

$$\omega_3^2 = \frac{c_{33}}{m}; \qquad \omega_6^2 = \frac{c_{66}}{J_{\zeta\zeta}}$$

$$\omega_{1,5}^2 = \frac{1}{2}\left(\frac{c_{55}}{J_{\eta\eta}} + \frac{c_{11}}{m}\right) \pm \sqrt{\frac{1}{4}\left(\frac{c_{55}}{J_{\eta\eta}} + \frac{c_{11}}{m}\right)^2 + \frac{c_{15}^2 - c_{11}c_{55}}{J_{\eta\eta}m}} \qquad (3.39)$$

$$\omega_{2,4}^2 = \frac{1}{2}\left(\frac{c_{44}}{J_{\xi\xi}} + \frac{c_{22}}{m}\right) \pm \sqrt{\frac{1}{4}\left(\frac{c_{44}}{J_{\xi\xi}} + \frac{c_{22}}{m}\right)^2 + \frac{c_{24}^2 - c_{22}c_{44}}{J_{\xi\xi}m}}.$$

ω_3 is the natural circular frequency of the vibration in z direction and ω_6 that of a rotation about the z axis. The motions are in the x and φ_y directions for ω_1 and ω_5 and in the y and φ_x directions for ω_2 and ω_4.

Figure 3.7 shows such a result, but the natural frequencies have different indices there (by size, index i). A pure up-and-down motion resulted at the natural frequency of 4.75 Hz and a pure torsional vibration at 7.78 Hz. Note that torsional and translational motions are coupled at other frequencies.

The thick and thin arrows in Fig. 3.7 indicate the synchronous motion of the respective coordinates. One can conclude from the directions of these arrows that the **representation of two mode shapes is incorrect**. This should serve as a warning that one should always take a hard look at (doubt) all results and check them critically. Is it possible that the inversely phased translational and torsional motions at 4,42 Hz occurs at lower natural frequencies than the in-phase vibration at the natural frequency of 8.47 Hz? Figure 3.7 contains an error: The direction of one of the arrows has to be reversed in both of these two cases! Compare, in this respect, the mode shapes at 4.02 Hz and 7.79 Hz.

If one assumes that only one spring plane exists, which is parallel to the x, y plane, when mounting on individual vibration isolators ($l_{xz} = l_{yz} = l_z$) and only n equal spring elements $c_x = c_y = c_H$; $c_z = c$ are used, the following applies:

$$c_{11} = c_{22} = nc_H; \qquad c_{33} = nc; \qquad c_{15} = c_{51} = nc_H l_z; \qquad c_{24} = c_{42} = -nc_H l_z$$

$$c_{44} = c\sum_{i=1}^{n} l_{yi}^2 + nc_H l_z^2; \qquad c_{55} = c\sum_{i=1}^{n} l_{xi}^2 + nc_H l_z^2; \qquad c_{66} = c_H \sum_{i=1}^{n}(l_{xi}^2 + l_{yi}^2)$$

(3.40)

Complete decoupling can be achieved on a purely formal level by symmetry with respect to the x, y plane. However, apart from special cases, this could only be achieved by two spring planes that would have to be above and below the x, y plane, which does not make much sense in terms of design. Therefore, such approaches are typically limited to symmetry of the setup with respect to two planes, which makes it easier to influence the natural frequencies. Foundations are often structured symmetrically so that their principal axes are easily determined. One should also try to arrange the springs symmetrically with respect to the principal axes.

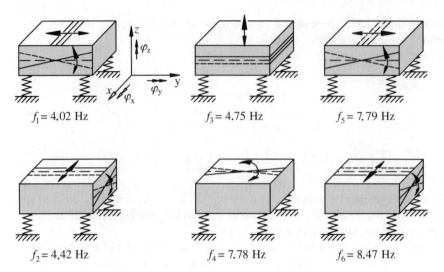

Fig. 3.7 Natural frequencies and mode shapes (attention: error) of a foundation block

A major simplification occurs for planar foundation blocks. This model can be used for approximate calculations if the excitation forces are in a symmetry plane with respect to the center of gravity.

There are software products available for modeling foundations that allow both the sizing of foundation blocks and the determination of the excitation forces for measured motions.

3.2.2.3 Types of Foundation Blocks

There are three major types of foundation blocks . The first group is formed by foundations that are placed directly onto the subsoil. The foundation mass must be small for high tuning (see Table 3.2). However, a large base area is required to obtain a sufficiently stiff "soil spring". This foundation form is primarily determined by the required base area and the required stiffness of the block. The second group includes those machines that are mounted without a specific foundation using spring elements. These must satisfy various conditions.

The excitation amplitude may only be of such a magnitude that the machine mass on the low tuned mounting pad does not exceed the permitted vibration amplitude. The excitation frequency must be large enough to allow a stable low tuned setup. Furthermore, the machine frame should be stiff enough to prevent malfunctioning due to the mounting on individual springs.

Examples are the mounting of vehicle engines in the frame by means of three or four individual spring elements, the mounting of fast-running machine tools, fans, etc. In most cases, these are high-speed machines with small excitation forces. The third group requires a foundation mass in addition to the machine mass. This design

3.2 Foundation Loading for Periodic Excitation

is also suited for low excitation frequencies and large excitation forces. Figure 3.8 shows the mounting of a marine diesel engine with generator and exciter in a peak-load power plant (operating speed $n = 300$ rpm).

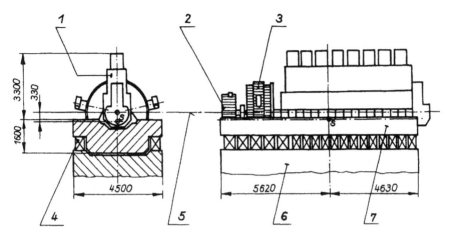

Fig. 3.8 Example of a foundation block with a peak-load diesel engine; *1* Diesel engine; *2* Exciter; *3* Generator; *4* Spring packets; *5* Crankshaft; *6* Bottom foundation; *7* Top foundation

Machine builders often have stringent requirements with regard to the permissible vibration amplitude of the foundation block. These require checking the prerequisite that the foundation block can be viewed as a rigid body. A rough calculation has to be performed to determine its natural frequency.

Massless springs may only be assumed as long as the frequencies of the foundation vibration are low as compared to the natural frequencies of the pre-loaded spring. Especially in the case of structure-borne noise, this condition is often violated. Then, vibrations at resonance may occur at the spring. Vibration isolation has no effect at these frequencies ranging from $f = 100$ to $1\,000$ Hz for steel springs and from $f = 200$ to $4\,000$ Hz for rubber springs. It should also be noted that compression springs such as they are frequently used for foundation blocks also have to be checked for safety against buckling.

The size and design of the foundation block depend on the mounting type. If it is placed directly onto the subsoil, the permissible surface pressure is decisive. It is on the order of magnitude of $\sigma = 20$ N/cm^2 (sand); $\sigma = 30$ N/cm^2 (clay); $\sigma > 60$ N/cm^2 (rock). Soil stability issues are addressed, for example, in DIN 4019. In addition, the stiffness of the soil and the achievable natural frequency depend on the foundation surface. Furthermore, a foundation block has to be designed in such a way that the overall center of gravity (machine and foundation) is above the center of gravity of the foundation surface. The following can be applied as a guideline for the size of the foundation mass m_F as a function of the machine speed n:

for $n < 300$ rpm: $m_F = (5\ldots 10)m_M$
for $n > 1000$ rpm: $m_F = (10\ldots 20)m_M$

(m_M is the mass of the machine).

The overall mass of low tuned foundations is determined by the permissible vibration amplitude of the foundation. The previous derivations were performed based on the assumption of an elastically mounted rigid machine. This means that the lowest natural frequency of machine and foundation block has to be large as compared to the highest excitation frequency. For a foundation block, the calculation is based on a frequency ratio

$$\eta = \frac{f_{\text{err max}}}{f_1} \leq 0.2 \ldots 0.33. \tag{3.41}$$

An estimate may sometimes suffice for determining the lowest natural frequency. While the natural frequencies are typically high enough for foundation blocks where height, width, and length are of the same order of magnitude, tuning problems do occur in slender and particularly long foundations. The foundation is assumed to be a beam with a constant cross-section and a homogeneous mass distribution for the approximate calculation. Since the foundation block is low tuned, the model of a free-free beam applies. The oscillating beam is discussed in detail in 5.3.

A calculation for the foundation shown in Fig. 3.8 is to be performed. For the given dimensions $h = 1.6$ m, $l = 10.25$ m, the density $\varrho = 1.8$ t/m^3, the modulus of elasticity $E = 3 \cdot 10^{10}$ N/m^2 (concrete grade B 25 according to DIN 1045) the following is obtained for a rectangular cross-section:

$$f_1 = 1.029 \frac{h}{l^2} \sqrt{\frac{E}{\varrho}} = 1.029 \frac{1.6}{10.25^2} \sqrt{\frac{3 \cdot 10^{10}}{1.8 \cdot 10^3}} \text{ Hz} \approx 64 \text{Hz} \tag{3.42}$$

Potential excitations are the inertia forces of an eight-cylinder engine. Measurements show that the first and second excitation orders occur even with theoretically completely balanced engines due to manufacturing-related deviations. If the second order is used as the highest excitation, the following applies for a rotational frequency of $f = n/60 = 300/60 = 5$ Hz: $f_{\text{err max}} = 10$ Hz $\ll$ 64 Hz. Thus, the requirement (3.41) for a "rigid" foundation is met.

As can be seen from Fig. 3.3, the lowest achievable natural frequency of a foundation is related to the *ductility* of the spring. The lowest natural frequencies are achieved using *coil springs*; they amount to around 1 Hz. Their calculation was discussed in 1.3.2. The damping ratio is of the order of magnitude of $D = 0.001$ to 0.01.

Rubber springs or rubber-metal combinations are also used as discrete spring elements. Their ductility is lower due to their low maximum permissible load, and only natural frequencies of foundations over 5 Hz can be reached. The damping ratio of plants equipped like that is $D = 0.01$ to 0.1.

1.3.3 discusses the calculation of rubber springs. It should just be pointed out that, for large values of $k_{\text{dyn}} = c_{\text{dyn}}/c_{\text{stat}}$, the static deflection is not suitable as a measure of the frequency to be isolated. Figure 3.9a shows a rubber spring that is commonly used for foundation designs.

3.2 Foundation Loading for Periodic Excitation

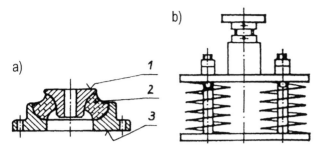

Fig. 3.9 Spring designs; **a)** Basic structure of a rubber spring; *1* Foundation or machine, *2* Rubber spring, *3* Mounting site; **b)** Steel spring arrangement

The dynamic stiffness of the *subsoil* or of *isolating plates* is approximated by foundation moduli. Calculation rules only account for the contact between foundation and subsoil at the base area of the foundation.

The foundation moduli are spring constants per unit area and depend on the type of soil, density index, base area, and the static base pressure. The reference value is the foundation modulus C_z that acts in the vertical direction. It amounts to

$$C_z = (2 \cdots 15) \cdot 10^4 \text{ kN/m}^3 \tag{3.43}$$

and is specified in the subsoil investigation report. These data relate to base areas $A \geq 10 \text{ m}^2$ and apply for a static base pressure of $p_\text{stat} \leq 5 \text{ N/cm}^2$. For $A < 10 \text{ m}^2$, multiply C_z by the factor $a_1 = \sqrt{10/A}$ (A in m^2). A factor $a_2 = \sqrt{p_\text{stat}/5 \text{ N/cm}^2}$ applies for $p_\text{stat} > 5 \text{ N/cm}^2$.

Derived from this value are the following:

$$C_x = C_y = 0.7 C_z; \qquad C_{\varphi x} = C_{\varphi y} = 2 C_z \tag{3.44}$$

(Rotation of the foundation about an axis through the center of gravity of the base area),

$$C_{\varphi z} = 1.05 C_z \tag{3.45}$$

(Rotation of the foundation about the vertical axis).

If the center of gravity (coordinate origin) is above the center of gravity of the supporting area, decoupling by symmetry into two planes applies, which leads to (3.39). Note that the spring distances $l_{xz} = l_{yz} = l_z$, which are measured up to the base area, are the same for each surface element and are thus constant, while l_{zx}, l_{zy} depend on the position of the surface elements. The following applies for the elements of the stiffness matrix in (3.32):

$$c_{11} = c_{22} = \int 0.7 C_z \mathrm{d}A = 0.7 C_z A; \qquad c_{33} = \int C_z \mathrm{d}A = C_z A \qquad (3.46)$$

$$c_{44} = \int 0.7 C_z l_z^2 \mathrm{d}A + \int C_{\varphi x} l_{zy}^2 \mathrm{d}A = 0.7 C_z l_z^2 A + C_{\varphi x} I_x = C_z (2 I_x + 0.7 A l_z^2)$$
$$(3.47)$$

$$c_{55} = C_z (2 I_y + 0.7 A l_z^2); \qquad c_{66} = C_{\varphi z} I_z = 1.05 C_z I_z \qquad (3.48)$$

$$c_{15} = c_{51} = -c_{24} = -c_{42} = -C_x A l_z = -0.7 C_z A l_z \qquad (3.49)$$

I_x and I_y are the area moments of inertia of the base area with respect to the x and y axes, $I_z = I_x + I_y$ is the polar area moment of inertia. It was taken into account for the value for c_{44} (and similarly c_{55}) that not C_z but $C_{\varphi x}$ becomes effective in the case of tilting. The six natural circular frequencies are obtained if one inserts these relationships into (3.39).

3.2.3 Foundations with Two Degrees of Freedom – Vibration Absorption

Figure 3.10a shows a longitudinal oscillator with two degrees of freedom. It can be used as the calculation model for an elastically supported machine (on a foundation block embedded in resilient subsoil or with an additional vibration absorber). The external spring-mass system then corresponds to the absorber. Initially, the damping is neglected to make some major relations clearer. The equations of motion have the same form as those of a torsional oscillator, which is discussed in greater detail in Sect. 4.4.3 taking into account the effect of damping.

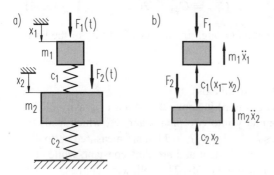

Fig. 3.10 Oscillator with two degrees of freedom
a) Calculation model,
b) Free-body diagram

The equations of motion for forced vibrations result from the dynamic equilibrium of forces, see Fig. 3.10b:

$$\begin{aligned} m_1 \ddot{x}_1 + c_1 (x_1 - x_2) &= F_1(t) \\ m_2 \ddot{x}_2 - c_1 (x_1 - x_2) + c_2 x_2 &= F_2(t). \end{aligned} \qquad (3.50)$$

3.2 Foundation Loading for Periodic Excitation

In the simplest case of harmonic excitation forces

$$F_1(t) = \hat{F}_1 \sin \Omega t; \qquad F_2(t) = \hat{F}_2 \sin \Omega t \qquad (3.51)$$

the vibration with the circular frequency of the excitation Ω prevails in the steady state so that an in-phase steady-state solution of the form is assumed:

$$x_1(t) = \hat{x}_1 \sin \Omega t; \qquad x_2(t) = \hat{x}_2 \sin \Omega t. \qquad (3.52)$$

In the steady state, the system oscillates at the excitation frequency. If one inserts the excitation forces (3.51) and the solution (3.52) into (3.50), one obtains a system of linear equations for calculating the two amplitudes:

$$\begin{aligned}(c_1 - m_1\Omega^2)\hat{x}_1 - c_1\hat{x}_2 &= \hat{F}_1 \\ -c_1\hat{x}_1 + (c_1 + c_2 - m_2\Omega^2)\hat{x}_2 &= \hat{F}_2.\end{aligned} \qquad (3.53)$$

It has the following solutions:

$$\hat{x}_1 = \frac{(c_1 + c_2 - m_2\Omega^2)\hat{F}_1 + c_1\hat{F}_2}{\Delta}; \qquad \hat{x}_2 = \frac{c_1\hat{F}_1 + (c_1 - m_1\Omega^2)\hat{F}_2}{\Delta}. \qquad (3.54)$$

The denominator

$$\begin{aligned}\Delta &= (c_1 - m_1\Omega^2)(c_1 + c_2 - m_2\Omega^2) - c_1^2 \\ &= m_1 m_2 (\Omega^2 - \omega_1^2)(\Omega^2 - \omega_2^2)\end{aligned} \qquad (3.55)$$

represents the principal determinant of the equation system (3.53). The roots of the denominator are the two natural circular frequencies of this vibration system, see also (3.39). The sign of the amplitudes $\hat{x}_1$ and $\hat{x}_2$ describes the phasing (since there is no damping).

The dependence of the amplitudes on the excitation frequency, that is, the *amplitude response* can be calculated from (3.54).

The important correlation between the deflection and force amplitudes

$$\hat{x}_1 = \frac{c_1\hat{F}_2}{\Delta} = x_{\text{st}} V_1(\Omega); \qquad \hat{x}_2 = \frac{(c_1 - m_1\Omega^2)\hat{F}_2}{\Delta} = x_{\text{st}} V_2(\Omega) \qquad (3.56)$$

is depicted in Fig. 3.11 for the special case that only the force F_2 is applied. The static deflection relationship $x_{\text{st}} = \hat{F}_2/c_2$ results from (3.56) so that the amplitude responses $\hat{x}_1(\Omega)$ and $\hat{x}_2(\Omega)$ can be viewed as **nondimensionalized amplitude function**. This representation generalizes the statements obtained for oscillators with one degree of freedom, see Fig. 3.4 and Fig. 3.5. Note the difference, though: The oscillator with one degree of freedom, when excited *periodically* with a fraction of the fundamental excitation frequency (at $\Omega = \omega/k$), oscillates at its own (and only) natural frequency at the resonance points in Fig. 3.5c, while in a two-mass system two resonance points occur even at *harmonic* excitation when the excitation frequency coincides with one of the two natural frequencies.

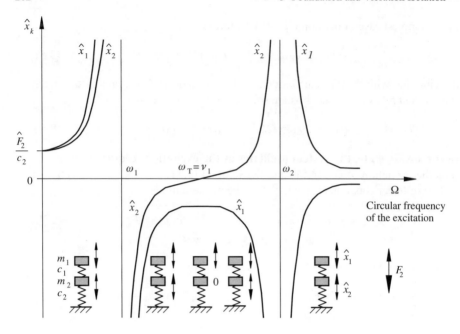

Fig. 3.11 Amplitude response of an oscillator with two degrees of freedom at $F_1 \equiv 0$ (thin and thick arrows characterize the same phase positions, respectively)

At very small excitation frequencies ($f_{\text{err}} = \Omega/(2\pi)$), the two masses move in the same direction as in the static case, and in the limiting case $\Omega = 0$, the deflection of the two masses is even equal, see Fig. 3.11. When the excitation frequency increases, the inertia forces first increase the amplitudes, and $\hat{x}_1 > \hat{x}_2$ applies for the lower frequency range. The directions of both deflections only coincide with the excitation force F_2 below the first natural circular frequency (in the range of $0 < \Omega < \omega_1$) at every point in time. The vibrations of the two masses are in phase for the undamped case in the range of $0 < \Omega < \omega_T$, but above the first resonance point they are in the opposite direction to the excitation force.

An interesting phenomenon occurs for the circular frequency of absorption

$$\Omega = \omega_T = \sqrt{\frac{c_1}{m_1}}. \tag{3.57}$$

So-called *vibration absorption* occurs here, i. e., the point of application of the force remains at rest, the spring c_2 is not loaded, and only the mass m_1 continues to vibrate in opposite phase to the excitation force. The inertia force that is introduced into the lower part of the oscillator by the vibrating mass m_1 via the spring c_1 is of exactly the same size as the excitation force (but pointing in the opposite direction) – in accordance with the assumptions for this ideal undamped system. The damping

3.2 Foundation Loading for Periodic Excitation

ensures that this ideal absorption does not occur, but instead it expands the range of small amplitudes, see Sect. 4.4.

The absorbers always have to be adapted to the design for structures with vibration excitation. There are vibration absorbers with absorber masses from 40 kg to 4500 kg that are effective at frequencies from 16 Hz to about 0.3 Hz. In most cases, these are vertical absorbers with helical compression springs and viscous damping, see the example in Fig. 3.12. Absorbers with leaf springs or pendulum suspension are often used against horizontal vibrations of television and church towers. Practical absorber designs are contained in VDI Guideline 3833 [36], Sheet E.

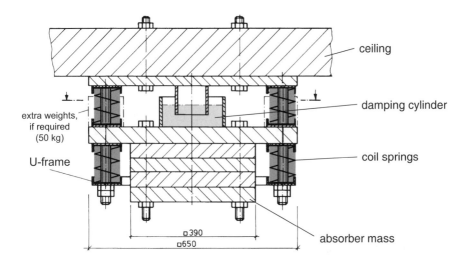

Fig. 3.12 Ceiling-mounted absorber (Source: GERB, Berlin)

The two masses vibrate in opposite directions in the range of $\omega_T < \Omega$, which is indicated by the thin and thick arrows for the phasing in the bottom part of Fig. 3.11. When the excitation frequency comes close to the second natural frequency, very large amplitudes occur again, and the phase shifts, once again, above ω_2. The amplitudes are getting smaller and smaller at very high excitation frequencies because the inert masses oppose the excitation force with larger and larger inertia forces (increasing with Ω^2). The force acting onto the mounting site is $F(t) = c_2 x_2(t)$, that is, proportional to the magnitude of the amplitude $\hat{x}_2$. Figure 3.11 thus also characterizes the dynamic foundation load.

The *absorption effect* in conjunction with dampers is discussed in more detail in Sect. 4.4. The physical relations described there for torsional oscillators also apply to longitudinal oscillators. Note that vibration absorption measures are not limited to harmonic excitations. Absorbers can also be tuned to higher harmonics of the excitation or, in conjunction with damping, to any other excitations for oscillators with multiple degrees of freedom, see Sect. 6.5.2.

3.2.4 Example: Vibrations of an Engine-Generator System

Figure 3.13 shows a foundation block with symmetrically arranged foundation springs.

The following parameter values are given:

Foundation mass	$m = 20$ t
Principal moments of inertia with respect to the axes through the center of gravity S	$J_{xx} = 48 \cdot 10^3$ kg $\cdot$ m^2; $J_{yy} = 14 \cdot 10^3$ kg $\cdot$ m^2; $J_{zz} = 52 \cdot 10^3$ kg $\cdot$ m^2
Horizontal spring constant of a vibration isolator	$c_x = c_y = c_H = 1.5 \cdot 10^5$ N/m
Spring constant of a vibration isolator in z direction	$c_z = 3.0 \cdot 10^5$ N/m
Distances of the spring attachment points	$a = h = 1$ m; $l = 0.5$ m; $l_0 = 0.4l$; $l_1 = 1.6l$
Number of vibration isolators	$n = 16$.

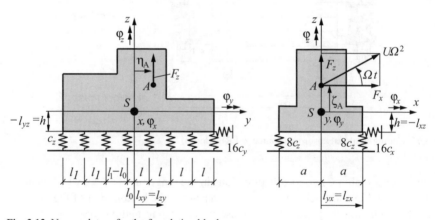

Fig. 3.13 Nomenclature for the foundation block

The spring coefficients result from the given data from (3.33):

$$c_{11} = c_{22} = 16c_H; \quad c_{24} = -c_{15} = 16hc_H; \quad c_{33} = 16c_z$$
$$c_{16} = c_{26} = c_{34} = c_{35} = c_{45} = c_{46} = c_{56} = 0$$
$$c_{44} = 65.86l^2 c_z + 16c_H h^2; \quad c_{55} = 16(c_H h^2 + c_z a^2)$$
$$c_{66} = c_H(65.86l^2 + 16a^2). \tag{3.58}$$

First, the natural frequencies and mode shapes are calculated. The equations (3.28) are divided into (3.35) and (3.36). The corresponding eigenvalue equations from (3.34) are further simplified to

3.2 Foundation Loading for Periodic Excitation

$$\begin{bmatrix} c_{11} - m\omega^2 & c_{15} \\ c_{15} & c_{55} - J_{\eta\eta}\omega^2 \end{bmatrix} \begin{bmatrix} v_1 \\ v_5 \end{bmatrix} = \begin{bmatrix} 0 \\ 0 \end{bmatrix}$$

(3.59)

$$\begin{bmatrix} c_{22} - m\omega^2 & c_{24} \\ c_{24} & c_{44} - J_{\xi\xi}\omega^2 \end{bmatrix} \begin{bmatrix} v_2 \\ v_4 \end{bmatrix} = \begin{bmatrix} 0 \\ 0 \end{bmatrix}.$$

The amplitude ratios of the associated mode shapes follow from (3.59)

$$(\hat{x}/\hat{\varphi}_y)_i = v_{1i}/v_{5i} = (c_{55} - J_{\eta\eta}\omega_i^2)/c_{15}; \quad \text{for } i = 1 \text{ and } 5$$
$$(\hat{y}/\hat{\varphi}_z)_i = v_{2i}/v_{4i} = (c_{44} - J_{\xi\xi}\omega_i^2)/c_{24}; \quad \text{for } i = 2 \text{ and } 4.$$

(3.60)

The natural frequencies result from (3.39) after a brief calculation with the indices provided there (Exception: They are normally sorted by size using the index i!) with $f_i = \omega_i/(2\pi)$:

Vertical vibration: $f_3 = 2.466$ Hz
Torsional vibration about the z axis: $f_6 = 1.892$ Hz
Pitching vibration in the x-z plane: $f_1 = 3.769$ Hz; $f_5 = 1.365$ Hz;
     $v_{51} = 0.3888$; $v_{15} = -0.2722$
Pitching vibration in the y-z plane: $f_2 = 2.716$ Hz; $f_4 = 1.467$ Hz;
     $v_{42} = -0.2919$; $v_{24} = 0.7006$

Note them in Fig. 3.14. The amplitude ratios of the mode shapes mentioned are obtained when normalizing $v_{ii} = 1$ from (3.60). The reader is encouraged to sketch these shapes. Knowledge of the natural frequencies and mode shapes is not sufficient to evaluate the vibration behavior under operating conditions. Calculation of the forced vibrations, while taking damping into account, is required.

It is assumed that the system has an unbalance U that rotates in the x-z plane about an axis A, see the body-fixed distances η_A and ζ_A in Fig. 3.13. The following components of the centrifugal force resulting from this unbalance act onto the foundation:

$$F_x = U\Omega^2 \sin \Omega t; \qquad M_x = F_z\eta = \eta_A U\Omega^2 \sin \Omega t$$
$$F_y = 0; \qquad M_y = \zeta F_x = \zeta_A U\Omega^2 \sin \Omega t \qquad (3.61)$$
$$F_z = U\Omega^2 \cos \Omega t; \qquad M_z = -F_x\eta = -\eta_A U\Omega^2 \sin \Omega t.$$

The steady-state amplitudes of the forced vibrations were calculated using the method described in Sect. 6.5 for specific distances ($\eta_A = 0.5$ m; $\zeta_A = 0.3$ m), an unbalance $U = 0.3$ kg · m and a damping constant for each spring ($b = 3000$ N · s/m). Figure 3.14 shows the results in the form of the amplitude response curves, see also (3.14) and V_3 in Fig. 3.4. The reference value $e = U/m = 0.015$ mm represents the displacement with which the unrestricted block would move due to the rotating unbalance (center-of-gravity theorem).

One can see that all natural frequencies are below the operating speed of $300 rpm$ ($f_{\text{err}} = 5$ Hz) and that the foundation is **low tuned**. The resonance points have

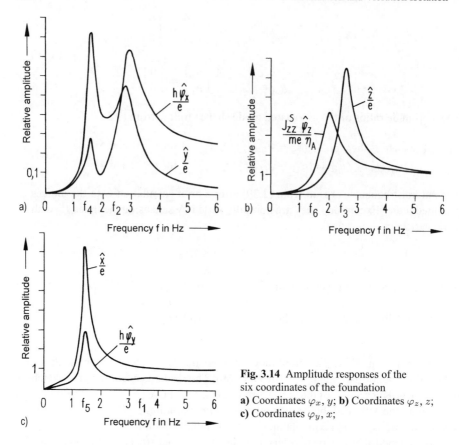

Fig. 3.14 Amplitude responses of the six coordinates of the foundation
a) Coordinates φ_x, y; **b)** Coordinates φ_z, z;
c) Coordinates φ_y, x;

to be passed, but the vibration amplitudes that occur during operation are insignificant. The resonance points are at the calculated natural frequencies. Check if the amplitude of the free unbalance-excited system occurs at very high excitation frequencies (when the spring forces are small, as compared to the inertia forces), i. e. if $\hat{x}/e = \hat{z}/e = J_{zz}\hat{\varphi}_z/(U\eta_A) = 1$ and $h\hat{\varphi}_y/e = \hat{y}/e = 0$ for $f \to \infty$.

3.2.5 Problems P3.1 to P3.3

P3.1 Mounting a Reciprocating Saw

A reciprocating saw is to be mounted at with low tuning. The reciprocating gate causes the force $F = \hat{F}\sin\Omega t$. The reciprocating saw runs at an angular speed n. How large must the mass of the foundation and reciprocating saw be and which overall spring constant has to be used to ensure that only 5 % of the excitation force enters the subsoil and that the amplitude does not exceed $\hat{x}$?

3.2 Foundation Loading for Periodic Excitation

Given:

$\hat{F} = 16 \cdot 10^4$ N; $n = 600$ rpm; $\hat{x} = 1$ mm

Find:

1. Frequency ratio η
2. Spring constant c
3. Static deflection of the spring due to its own weight

P3.2 Minimum Dynamic Housing Load

A vertical shaft rotates about the ζ axis in the compressor of a refrigerator. One end of the shaft (at distance ζ_1) drives the piston, which in turn causes the horizontal inertia force $F_1(t)$. For design reasons, the desired mass balancing cannot be performed in the plane at the distance ζ_1. A balancing mass can only be attached at the distance ζ_2, but the installation space is so narrow that a mass of the required size cannot be fitted and that it can only muster a portion of p percent of the inertia force (phase shifted by 180 degrees) that would be required for complete balancing. The compressor block (including the motor shaft) has a mass m and a moment of inertia $J_S = J_{yy}$, see Fig. 3.15.

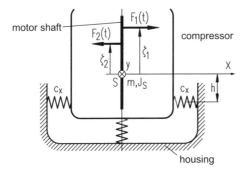

Fig. 3.15
Compressor block with excitation forces

The compressor is to be elastically supported in such a way that, if possible, it does not transmit any horizontal dynamic forces onto the housing of the refrigerator. The structure-borne sound would result in undesirable noise. The compressor is supported in the housing on a central vertical spring (which absorbs its static weight) and horizontally acting springs at a distance h from the center of gravity.

Given:

Horizontal excitation force at a distance ζ_1	$F_1(t)$
Balancing force at a distance ζ_2	$F_2(t) = pF_1(t)$
Distances of the planes of action of the forces	ζ_1 and ζ_2
Mass of the motor block	m
Moment of inertia of the motor block	J_S

Find:

1. Derive the conditions for minimum loading of the housing (after deriving the equations of motion)
2. Optimum distance h for placing the horizontal springs
3. Distance h for the parameter values $J_S = ml^2$; $\zeta_1 = l$; $\zeta_2 = 0.8l$; $p = 0.8$

For simplification, consider only the equilibrium in the x-z plane without taking the vertical force components into account.

P3.3 Vibration-Isolating Suspension of a Motorbike engine

A motorbike engine is to be elastically suspended, see Fig. 3.16. The vibration amplitude $\hat{x}_P$ of the crankshaft center point P, caused by the piston motion, is not to exceed a predetermined value $\hat{x}_{zul}$. Small vibrations may be assumed.

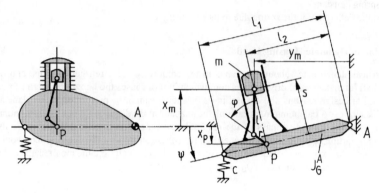

Fig. 3.16 Suspension of a motorbike engine

Given:

Piston motion	$s = l + r \cos \Omega t$
Piston mass including reduced connecting rod mass (center of gravity is at the connecting-rod bearing of the piston)	$m = 0.6$ kg
Moment of inertia of the housing with built-in parts	$J_G^A = 0.523$ kg·m^2
Lengths	$l_1 = 360$ mm;
	$l_2 = 150$ mm;
	$l = 150$ mm;
	$r = 32$ mm
Permissible amplitude of point P	$\hat{x}_{zul} = 1$ mm
Speed range	$n \geq 1000$ rpm

Find:

1. Equation of motion for $x_P(t)$
2. Required spring stiffness c for the condition $|\hat{x}_P| \leq \hat{x}_{zul}$ over the entire speed range

3.2.6 Solutions to Problems S3.1 to S3.3

S3.1 Damping-free mounting is assumed since damping has no noticeable influence outside of a resonance, see the nondimensionalized amplitude functions V_1 and V_2 in Fig. 3.4.

The calculation is based on the model with force excitation (Table 3.3). Away from the resonance condition, the damping is neglected. Therefore, $D = 0$, and according to (3.12), (3.13) the amplitudes of the deflection and the ground force are:

$$\hat{x} = \frac{\hat{F}}{c} \frac{1}{|1 - \eta^2|}; \qquad \hat{F}_B = \hat{F} \frac{1}{|1 - \eta^2|}. \tag{3.62}$$

3.2 Foundation Loading for Periodic Excitation

Since $\eta > 1$ for low tuning, the negative sign of the ground force is taken into account. (3.62) yields

$$\eta^2 = 1 + \frac{\hat{F}}{\hat{F}_B} = 1 + \frac{1}{0,05} = 21; \qquad \underline{\eta = 4.58}. \tag{3.63}$$

Now the spring constant at a given amplitude $\hat{x}$ can be determined.

$$c = \frac{\hat{F}}{\hat{x}} \frac{1}{\eta^2 - 1} = \frac{\hat{F}}{\hat{x}} 0.05 = \frac{16 \cdot 10^4 \text{ N}}{1 \cdot 10^{-3} \text{ m}} \cdot 0.05 = \underline{8 \cdot 10^6 \text{ N} \cdot \text{m}^{-1}}. \tag{3.64}$$

The overall mass can be determined from the frequency ratio η and the spring constant c.

$$\eta^2 = \frac{\Omega^2}{\omega_0^2}; \qquad \omega_0^2 = \frac{c}{m}; \qquad \Omega = \frac{\pi n}{30} = 62.8 \text{ s}^{-1};$$

$$m = \frac{\eta^2 c}{\Omega^2} = \frac{21 \cdot 8 \cdot 10^6 \text{ N} \cdot \text{m}^{-1}}{62.8^2 \cdot \text{s}^{-2}} = 42.6 \cdot 10^3 \text{ kg}. \tag{3.65}$$

The static deflection of the replacement spring when installed is

$$\underline{\underline{x_{\text{st}}}} = \frac{mg}{c} = \frac{42.6 \cdot 10^3 \cdot \text{kg} \cdot 9.81 \cdot \text{m} \cdot \text{s}^{-2}}{8 \cdot 10^6 \cdot \text{N} \cdot \text{m}^{-1}} = \underline{0.052 \text{ m}}. \tag{3.66}$$

This value is also obtained from Fig. 3.3 if the natural frequency $f = n/(60\eta) = 2.18$ Hz is inserted.

S3.2 The equations of motion follow from the equilibrium of dynamic forces and moments for $\varphi_y \ll 1$, $\sin \varphi_y = \varphi_y$, $\cos \varphi_y = 1$, see Fig. 3.17:

$$m\ddot{x}_S - F_1 + F_2 - F_H = 0 \tag{3.67}$$

$$J_S \ddot{\varphi}_y - F_1 \zeta_1 + F_2 \zeta_2 - F_H h = 0. \tag{3.68}$$

The horizontal force applied to the springs is

$$F_H = 2c_x(x_S - h\varphi_y). \tag{3.69}$$

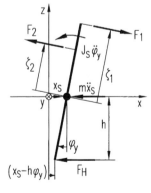

Fig. 3.17
Forces and coordinates at the compressor block ($\varphi_y \approx 0$)

No horizontal force is transmitted to the housing if

$$F_H = 0; \qquad x_S = h\varphi_y \tag{3.70}$$

is satisfied. After solving for the accelerations, it follows from (3.70) with (3.67) and (3.68)

$$\ddot{x}_S = \frac{F_1 - F_2}{m} = h\frac{F_1\zeta_1 - F_2\zeta_2}{J_S} = h\ddot{\varphi}_y. \tag{3.71}$$

This provides an equation for calculating the optimum distance h:

$$h = \frac{J_S(F_1 - F_2)}{m(F_1\zeta_1 - F_2\zeta_2)}. \tag{3.72}$$

Due to $F_2(t) = p \cdot F_1(t)$, the result with the given numerical values is

$$\underline{\underline{h}} = \frac{J_S(1-p)F_1}{m(F_1 l - pF_1 0.8l)} = \frac{ml^2(1-0.8)F_1}{ml F_1(1-0.8^2)} \underline{\underline{= 0.556l}} \tag{3.73}$$

This distance (which was part of the subject matter of a patent) is realistic since it is within the housing dimensions. In a more detailed analysis of the forced vibrations, the spatial motion, horizontal spring forces, damping, e. t.c. have to be taken into account. The outcome will be that the bearing forces are not exactly zero but that a minimum load occurs near the distance according to (3.72). It is remarkable that this solution applies to arbitrary time functions of the forces F_1 and F_2 as long as they are proportional to each other.

S3.3 The excitation force of the accelerated piston motion that acts in s direction is

$$F(t) = -m\ddot{s} = mr\Omega^2 \cos \Omega t. \tag{3.74}$$

It is considered to be independent of the motion of the rocker, and the total moment of inertia with respect to A is assumed to be constant according to (3.75):

$$J_A = J_G^A + m \cdot (l_2^2 + l^2) = (0.523 + 0.027) \text{ kg} \cdot \text{m}^2 = 0.55 \text{ kg} \cdot \text{m}^2. \tag{3.75}$$

Thus a differential equation with constant coefficients is derived from the equilibrium of moments about the pivot point A with $M = Fl_2$:

$$\underline{\underline{J_A \ddot{\psi} + cl_1^2 \psi = -mrl_2\Omega^2 \cos \Omega t.}} \tag{3.76}$$

If, instead of the angle ψ, the coordinate x_P of point P is introduced using the relationship $x_P = l_2 \psi$ for $\sin \psi \ll 1$, an equation of motion that is linear with respect to x_P and has the form of (3.8) is obtained:

$$\ddot{x}_P + \frac{cl_1^2}{J_A} x_P = -\frac{ml_2^2}{J_A} r\Omega^2 \cos \Omega t. \tag{3.77}$$

The solution for the steady-state vibration is $x_P = \hat{x}_P \cos \Omega t$. With $\omega_0^2 = cl_1^2/J_A$ and $\eta = \Omega/\omega_0$, the amplitude is, see (3.14):

$$\hat{x}_P = \frac{mrl_2^2}{J_A} \cdot \frac{\eta^2}{\eta^2 - 1}. \tag{3.78}$$

According to the problem statement, the requirement $|\hat{x}_P| \leq \hat{x}_{\text{zul}}$ has to be satisfied. If a low tuned system ($\eta > 1$) is assumed – since only in such a system the amplitude decreases with increasing speed, see the nondimensionalized amplitude function V_3 in Fig. 3.4 – it follows from (3.78) that

$$(\eta^2 - 1)x_{\text{zul}} = \frac{mrl_2^2}{J_A}\eta^2 \quad \text{and} \tag{3.79}$$

3.3 Foundations under Impact Loading

$$\eta^2 = \frac{\Omega^2}{\omega_0^2} = \frac{J_A \Omega^2}{c \cdot l_1^2} \geq \frac{1}{1 - \dfrac{mrl_2^2}{J_A \hat{x}_{zul}}}. \tag{3.80}$$

Using the parameter values from the problem statement, the stiffness is then ($\Omega = \Omega_{min} = \pi \cdot 1000/30 = 104.7\ \mathrm{s}^{-1}$):

$$c \leq \left(1 - \frac{ml_2^2}{J_A} \cdot \frac{r}{x_{zul}}\right) \cdot \frac{J_A \Omega^2}{l_1^2} = 9985\ \mathrm{N/m}. \tag{3.81}$$

3.3 Foundations under Impact Loading

3.3.1 Modeling Forging Hammers

A foundation is exposed to impact loading in the case of forging hammers, shears, presses, and punch presses. A designer of such equipment has to create a design that absorbs the forces (the concrete must not be crushed by impacts). The transmission of unacceptable shocks onto the environment and "settling" (vibrating itself into the ground) or tilting of the equipment must be prevented. None of the measures should impair any one of the technological requirements (such as good forging efficiency).

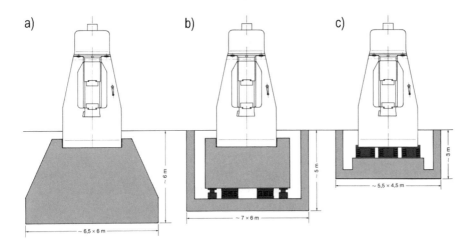

Fig. 3.18 Mounting of forging hammers (Source: GERB, Berlin); **a)** Firm grounding in the subsoil, **b)** Elastic foundation block, **c)** Direct spring suspension

Forging hammers typically require large reinforced concrete foundations (Fig. 3.18a) to absorb the high dynamic loads of the hammer. Figures 3.18b and c show alternative mountings of a forging hammer. By comparing the volumes of the concrete bodies, it can be deduced that the cost for a foundation can vary. It is very beneficial to provide a resilient support for such equipment. Using VISCO dampers (in parallel to the springs), according to Fig. 1.22a, can considerably reduce the mass of the spring-mounted foundation block as compared to a firm mounting. The system shown in Fig. 3.18c has proven its worth for hammers with a work capacity of up to 400 kJ.

The basic structure is shown in Fig. 3.19. The frame of the forging hammer is firmly mounted onto the foundation block. It houses the lifting or accelerating mechanism for the hammer (ram). One can distinguish between drop hammers and hammers that are accelerated by a drive system (such as compressed air). When the hammer hits the anvil, it has a specific amount of kinetic energy, which has to be converted into deformation energy, acting on the workpiece as efficiently as possible. The anvil bed, which supports the anvil, rests separately on top of a resilient intermediate layer on the foundation block. To protect the spring system of the foundation block against moisture, it is usually contained in a foundation trough, which rests on the subsoil. The foundation trough rests on the elastic subsoil.

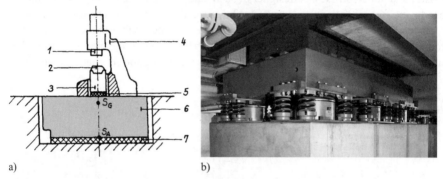

Fig. 3.19 Forging hammer; **a)** General view, *1* Ram; *2* Anvil; *3* Anvil bed; *4* Frame; *5* Spring; *6* Foundation block; *7* Spring; S_G Center of gravity of the total mass; S_A Center of gravity of the base area; **b)** View of a support (Source: GERB, Berlin)

The mass of the ram m_B is determined by the required impact energy. The masses of the anvil and anvil bed are on the order of magnitude of $m_A = 20 m_B$ to ensure a good forging efficiency. The foundation mass m_F is

$$m_F > 60 m_B \quad \text{or} \quad m_F = 2, 4 v_B^2 m_B. \tag{3.82}$$

v_B is the ram velocity immediately before the impact in m/s. (3.82), however, does not apply to double-acting or counterblow hammers. The foundation mass should be designed in such a way that the overall center of gravity S_G and the area center of the spring S_A are collinear with the direction of impact. This ensures uniform loading and avoids tilting motions.

The calculation model to be studied would have to be a three-mass system (anvil bed - foundation - foundation trough). Due to the differences in mass and the tremendous stiffness differences (the spring of the anvil bed is much stiffer than the foundation

3.3 Foundations under Impact Loading

spring), most rough calculations are based on a two-mass system. It is important to determine the natural frequencies and the maximum dynamic deflections. The consideration of damping is not necessary for that. However, the damping must be included if one wants to check if the vibrations decay sufficiently between two impacts, which is important for double-acting hammers, see Fig. 3.19.

3.3.2 Calculation Model with Two Degrees of Freedom

The model on which the calculation is based is shown in Fig. 3.20. The masses m_1, m_2 and the stiffnesses c_1, c_2 of the model can be selected depending on whether the task is a) or b). These depend on the system's stiffness and mass parameters. The origin of the coordinates x_1, x_2 is at the static equilibrium position before the ram mass hits. Since $m_B \ll m_1$ and $m_B \ll m_2$, the low static deflection and the dynamic influence of m_B after the impact are neglected.

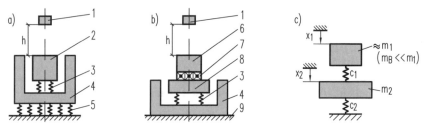

Fig. 3.20 Calculation model **c)** for the forging hammer for various applications **a)** and **b)**; *1* Ram; *2* Anvil and foundation; *3* Foundation spring; *4* Foundation trough; *5* Soil spring; *6* Anvil; *7* Framing (spring); *8* Foundation; *9* Rigid soil

The equations of motion as a special case of (3.50) are as follows:

$$m_1 \ddot{x}_1 + c_1(x_1 - x_2) = 0$$
$$m_2 \ddot{x}_2 - c_1(x_1 - x_2) + c_2 x_2 = 0. \tag{3.83}$$

They must be solved with the initial conditions:

$$t = 0: \quad x_1 = 0; \quad x_2 = 0; \quad \dot{x}_1 = u_1; \quad \dot{x}_2 = 0. \tag{3.84}$$

The initial velocity u is determined by the impact of the ram onto the anvil. Assuming a short impact, the principle of conservation of linear momentum applies to the two masses m_B and m_1:

$$m_B u_{v0} = m_B u_0 + m_1 u_1. \tag{3.85}$$

where u_{v0} is the velocity of m_B immediately before the impact and u_0; u_1 are the velocities after the impact. *Newton's impact hypothesis* is typically used as an additional equation:

$$k = -\frac{u_0 - u_1}{u_{v0}}. \tag{3.86}$$

Empirical values are available for the *coefficient of restitution* k:

$k = 0.2$ for slight hot straining
$k = 0.5$ for cold straining
$k = 0.8$ for heavy drop-forging works

The velocity u_{v0} can be determined for drop hammers: $u_{v0} = \sqrt{2gh}$ (h drop height). (3.85) and (3.86) provide the following for the velocity of the anvil after the impact:

$$u_1 = \frac{(1+k)u_{v0}m_B}{m_1 + m_B}. \tag{3.87}$$

The general solution is:

$$x_k(t) = \sum_{i=1}^{2} v_{ki}(a_i \cos \omega_i t + b_i \sin \omega_i t); \qquad k = 1,\, 2, \tag{3.88}$$

see also Sects. 4.2.1.1 and 4.4.2.

The two natural circular frequencies are obtained in analogy to (3.39) from

$$\omega_{1,2}^2 = \frac{1}{2}\left(\frac{c_1 + c_2}{m_2} + \frac{c_1}{m_1}\right) \mp \sqrt{\frac{1}{4}\left(\frac{c_1 + c_2}{m_2} + \frac{c_1}{m_1}\right)^2 - \frac{c_1 c_2}{m_1 m_2}}. \tag{3.89}$$

The dimensionless characteristic quantities for the hammer foundation are

$$\mu = \frac{m_1}{m_2} \ll 1; \qquad \gamma = \frac{c_1}{c_2} \gg 1. \tag{3.90}$$

The approximation

$$\omega_1^2 \approx \frac{\frac{c_1}{m_1}}{1 + \frac{c_1}{c_2}\left(1 + \frac{m_2}{m_1}\right)}; \qquad \omega_2^2 \approx \frac{c_2}{m_2}\left[1 + \frac{c_1}{c_2}\left(1 + \frac{m_2}{m_1}\right)\right], \tag{3.91}$$

applies, where the exact value of ω_1 is slightly larger, and that of ω_2 is slightly smaller.

The guideline for hammer foundations is $\mu \approx 0.3$; $\gamma > 5$, thus the two natural frequencies are far apart.

The amplitude ratios are $v_{11} = v_{12} = 1$ and analogously (4.29)

3.3 Foundations under Impact Loading

$$v_{21} = 1 - \frac{m_1 \omega_1^2}{c_1}; \quad v_{22} = 1 - \frac{m_1 \omega_2^2}{c_1} \tag{3.92}$$

and one finds $a_1 = a_2 = 0$ from the initial conditions (3.84)

$$b_1 = \frac{u_1}{\omega_1} \cdot \frac{c_1/m_1 - \omega_2^2}{\omega_1^2 - \omega_2^2}; \quad b_2 = -\frac{u_2}{\omega_2} \cdot \frac{c_1/m_1 - \omega_1^2}{\omega_1^2 - \omega_2^2}. \tag{3.93}$$

The solution of (3.83) results from (3.88), (3.92), and (3.93):

$$\begin{aligned} x_1 &= b_1 \sin \omega_1 t + b_2 \sin \omega_2 t \\ x_2 &= v_{21} b_1 \sin \omega_1 t + v_{22} b_2 \sin \omega_2 t. \end{aligned} \tag{3.94}$$

The oscillating motion of both masses is thus composed of two harmonic portions at their natural frequencies. Both masses typically perform non-periodic motions. Since $\omega_1 \ll \omega_2$ and $m_1 \omega_1^2/c_1 \ll 1$ apply to the hammer foundations, $v_{21} \approx 1$. For the relative displacement Δx between anvil bed and anvil, only the vibration at the second natural frequency where both masses vibrate in the opposite direction is of interest:

$$\Delta x = x_1 - x_2 \approx (1 - v_{22}) b_2 \sin \omega_2 t$$

$$\Delta x_{\max} = b_2 (1 - v_{22}) = \frac{b_2 m_1 \omega_2^2}{c_1}. \tag{3.95}$$

Table 3.4 Guidelines for permissible deflections in hammer foundations

Permissible maximum deflections of the anvil bed (m_1) on the foundation (models a, b)	Permissible maximum deflections of the foundation (m_2) (model a)
$\Delta x = 1$ mm ($m_B < 1$ t)	$x_{2\max} = 0.5 \ldots 2$ mm for straining hammers
$\Delta x = 2$ mm ($m_B = 1 \ldots 2$ t)	
$\Delta x = 3 \ldots 4$ mm ($m_0 > 3$ t)	$x_{2\max} = 3 \ldots 4$ mm for drop-forge hammers

Guidelines for the maximum deflections of the hammer foundations are listed in Table 3.4.

So far, the calculations were performed without taking damping into account, and thus yielded amplitude values that were above the actual values. Furthermore, the initial conditions (3.84) stipulated that only one initial velocity caused by the impact exists. Any vibrations caused by the previous impact would have to have decayed completely. An estimate of this process is performed using a damped system with one degree of freedom and based on the lowest natural frequency. First, the number of vibrations between two impacts is determined.

$$z = T_0/T \tag{3.96}$$

T_0 Cycle time, time between two impacts ($T_0 = 2\pi/\Omega$)
T Period of oscillation that corresponds to the lowest natural frequency ($T = 2\pi/\omega_1$)

The amplitude decline between two consecutive positive maxima for $D \ll 1$ is: $x_k/x_{k+1} = e^{2\pi D}$, see (1.97).

The value $x_{\min}$ to which the amplitude should have dropped at the next impact must be predetermined. There are z full oscillations between $x_{\max}$ and $x_{\min}$. Therefore,

$$\frac{x_{\max}}{x_{\min}} = e^{2\pi z D}; \qquad D = \frac{1}{2\pi}\frac{T}{T_0}\ln\left(\frac{x_{\max}}{x_{\min}}\right). \tag{3.97}$$

If, e. g., the following is given: $x_{\max}/x_{\min} = 10$; $T/T_0 = 1/3$, a damping ratio of $D = 0.12$ must be present. This value is not always achieved by natural damping.

3.3.3 Problems P3.4 to P3.6

P3.4 Hammer Foundation

Estimate the maximum oscillation amplitudes for mounting an air hammer. Use Fig. 3.20c as a calculation model.

Given: Ram mass: $m_B = 0.1$ t; Mass of anvil and anvil bed: $m_1 = 1.5$ t; Mass of hammer and foundation: $m_2 = 22.1$ t; Number of blows per minute: $n = 190$, Blow energy: $W = 1.6 \cdot 10^3$ N · m

Resilient intermediate layer between anvil bed and foundation: hammer felt $d = 40$ mm thick; dynamic modulus of elasticity of the hammer felt; $E = 8 \cdot 10^7$ N/m^2. Area between anvil bed and hammer: $A = 0.5$ m^2 The foundation rests on 6 spring elements. Their total spring constant is $c_2 = 4 \cdot 10^6$ N/m. The assumed coefficient of restitution is $k = 0.6$.

P3.5 Mounting of a Piston Compressor

A horizontal piston compressor with a speed $n = 258$ rpm is to be installed onto the subsoil with high tuning. The size of the foundation block was determined from the guideline $m_F = (5\ldots 10)m_M$ for $n < 300$ rpm. The system dimensions are shown in Fig. 3.21. The center of gravity of the entire system is above the area center of the base surface. The axes shown represent principal axes of inertia.

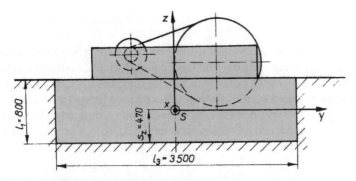

Fig. 3.21 Foundation of a piston compressor

3.3 Foundations under Impact Loading

The following parameter values are given:

Machine mass: $m_M = 1\,500$ kg
Foundation mass: $m_F = 12\,000$ kg

Principal moments of inertia:

$$J_x = 1.483 \cdot 10^4 \text{ kg} \cdot \text{m}^2; \quad J_y = 0.725 \cdot 10^4 \text{ kg} \cdot \text{m}^2; \quad J_z = 2.014 \cdot 10^4 \text{ kg} \cdot \text{m}^2$$

Distance of the center of gravity from the subsoil: $s_z = 0.47$ m
Outer foundation dimensions:
Height: $l_1 = 0.8$ m; Width: $l_2 = 2.4$ m; Length: $l_3 = 3.5$ m
Subsoil: Very soft clay with a foundation modulus of $C_z = 4 \cdot 10^4$ kN/m^3, see . (3.43)

Check if a mounting with high tuning has been achieved taking into consideration the second excitation harmonic.

P3.6 Periodic Impact Sequence (Hammer)

Blows whose impact period Δt is considerably smaller than the period of oscillation $T = 2\pi/\omega_0$ of the oscillator act at constant intervals ($T_0 = 2\pi/\Omega$) on a foundation that is assumed to be without damping. Calculate the foundation force that is generated in the steady-state condition. Compare the solution that can be obtained using a Fourier series with the solution resulting from treating it as a free vibration, if the periodicity condition and the velocity jump Δv as a result of the impact action are taken into account.

Given:
Oscillator with one degree of freedom, see figure in Table 3.3 with $s(t) \equiv 0$ and $b = 0$

Mass	m
Spring constant	c
Natural circular frequency	$\omega = \omega_0 = \sqrt{c/m}$
Cycle time of the impact sequence	$T_0 = 2\pi/\Omega > T$
Period of oscillation	$T = 2\pi\sqrt{m/c} \gg \Delta t$, see Fig. 3.22

Change in linear momentum by one impact $\Delta I = \int_0^{\Delta t} F(t)dt = F_S \cdot \Delta t = m \cdot \Delta v$

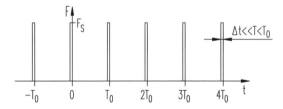

Fig. 3.22 Impact sequence

Find:
1. Fourier series of the excitation force
2. Steady-state solution for the foundation force
   2.1 using a Fourier series
   2.2 using the method of undetermined initial conditions
   2.3 using the program SimulationX® for the damped system with the following parameter values:
   Impact force $F_S = 10\,000$ N; cycle time $T_0 = 1$ s; impact time $\Delta t = 0.01$ s; mass $m = 400$ kg; damping ratio $D = 0.02$; spring constant $c = 2 \cdot 10^5$ N/m
3. Comparison of solutions and methods

3.3.4 Solutions S3.4 to S3.6

S3.4 The impact velocity of the ram must be calculated from the blow energy to be able to work with the calculation model in Fig. 3.20c. The following applies:

$$W = \frac{m_B u_{v0}^2}{2}; \qquad u_{v0} = \sqrt{\frac{2W}{m_B}}; \qquad u_{v0} = 5.65 \text{ m/s.} \qquad (3.98)$$

The initial velocity of the anvil and anvil bed is determined according to (3.87):

$$u_1 = (1 + 0.6)\, 5.65 \text{ m/s} \cdot \frac{0.1 \text{ t}}{1.6 \text{ t}} = 0.57 \text{ m/s.} \qquad (3.99)$$

The spring constant between anvil bed and foundation is calculated as follows:

$$c_1 = \frac{AE}{d} = \frac{0.5 \text{ m}^2 \cdot 8 \cdot 10^7 \cdot \text{N/m}^2}{0,04 \text{ m}} = 10^9 \text{ N/m.} \qquad (3.100)$$

The following is obtained for the natural frequencies from (3.89):

$$\omega_1 = 13.02 \text{ 1/s}; \qquad \omega_2 = 843.8 \text{ 1/s.} \qquad (3.101)$$

Using (3.93), one finds

$$b_1 = 2.78 \cdot 10^{-3} \text{m}; \qquad b_2 = 6.33 \cdot 10^{-4} \text{m} \qquad (3.102)$$

and according to (3.92)

$$v_{21} = 1; \qquad v_{22} = -0.068. \qquad (3.103)$$

Thus, according to (3.94):

$$x_1 = (2.78 \sin \omega_1 t + 0.63 \sin \omega_2 t) \text{ mm} \qquad (3.104)$$
$$x_2 = (2.78 \sin \omega_1 t - 0.04 \sin \omega_2 t) \text{ mm} \qquad (3.105)$$

There is almost no relative displacement between anvil bed and foundation at the first natural frequency. The displacement of the anvil bed, relative to the foundation, that occurs at the second natural frequency of $(0.63 + 0.04)$ mm is permissible according to Table 3.4. The amplitude of the foundation vibration at the fundamental frequency is also in the permissible range.

The tuning is still of interest. The first-order circular frequency of the excitation results from the number of blows $\Omega = \pi n/30 = 19.88$ 1/s. $\eta = \Omega/\omega_1 = 1.5$ applies for the fundamental vibration.

This means there is slight low tuning. Lowering the natural frequency is recommended to avoid getting too close to the resonance because of uncertainties in the model parameters.

S3.5 The foundation area and surface pressure are required for determining the foundation moduli. One finds from these data:

Foundation area: $A = 3.5 \text{ m} \cdot 2.4 \text{ m} = 8.4 \text{ m}^2$
Total mass: $m = m_M + m_F = 13\,500$ kg
Surface pressure: $p = mg/A = 1.58 \text{ N/cm}^2$

The following applies for the correction constants: Since $A < 10$ m^2: $a_1 = 1.09$, due to $p_{\text{stat}} < 5$ N/cm^2, the following applies according to (3.43):

$$C_z = 1.09 \cdot 4 \cdot 10^4 = 4.36 \cdot 10^4 \text{ kN/m}^3. \qquad (3.106)$$

3.3 Foundations under Impact Loading

The following are needed to calculate the elements of the stiffness matrix according to (3.46) bis (3.49):

$$I_x = \frac{l_2 l_3^3}{12} = 8.58 \text{ m}^4; \quad I_y = \frac{l_3 l_2^3}{12} = 4.03 \text{ m}^4; \quad I_z = I_x + I_y = 12.61 \text{ m}^4 \quad (3.107)$$

(3.39) ($f_i = \omega_i/2\pi$) results in:

$$\begin{aligned} f_1 &= 26.2 \text{ Hz}; & f_2 &= 26.9 \text{ Hz} \\ f_3 &= 38.9 \text{ Hz}; & f_4 &= 19.7 \text{ Hz} \\ f_5 &= 37.7 \text{ Hz}; & f_6 &= 20.8 \text{ Hz} \end{aligned} \quad (3.108)$$

The fundamental frequency of the excitation is $f_{\text{err}1} = n/60 = 4.3$ Hz. The frequency ratio, with respect to the second harmonic of the excitation, is $f_{\text{err}2}/f_{\min} = 8.6/19.7 = 0.44$, which represents high tuning. Only the 4th harmonic of the excitation would come close to resonance.

S3.6 First, the solution using the method discussed in Sect. 3.2.1.3 is determined. The Fourier coefficients of the excitation force must be known. They are given in Table 1.11 for case 6: $F_{kc} = 2\Delta I/T_0$, $F_{ks} = 0$. The excitation force function thus corresponds to the Fourier series

$$F(t) = \frac{\Delta I}{T_0}\left(1 + 2\sum_{k=1}^{\infty} \cos k\Omega t\right). \quad (3.109)$$

According to (3.24) to (3.26) for $D = 0$, the steady-state solution of the equation of motion (3.8) is:

$$x(t) = \frac{\Delta I}{cT_0}\left(1 + 2\sum_{k=1}^{\infty} \frac{\cos k\Omega t}{1 - k^2\eta^2}\right), \quad \eta = \frac{\Omega}{\omega} = \frac{2\pi}{\omega T_0} \quad (3.110)$$

To obtain a result with a precision of about three digits, the denominator would have to be about 10 000, that is, about $k^* = 100$ summands have to be taken into account to arrive at a usable result.

The results obtained for one cycle ($0 < t \leq T_0$) apply analogously to all other cycles in the ranges $jT_0 < t \leq (j+1)T_0$ with the cycle $j = 0, \pm 1, \pm 2, \ldots$ since, according to Fig. 3.22, the periodicity condition applies both to the excitation force $F(t) = F(t + jT_0)$ and to the foundation force $F_B(t) = c \cdot x(t) = c \cdot x(t + jT_0) = F_B(t + jT_0)$.

Within the cycle ($0 < t \leq T_0$) under consideration here, the oscillator moves according to (1.94) with $\delta = 0$ according to

$$\begin{aligned} x(t) &= x_0 \cos \omega t + \frac{v_0}{\omega} \sin \omega t \\ \dot{x}(t) &= -x_0 \omega \sin \omega t + v_0 \cos \omega t. \end{aligned} \quad (3.111)$$

The initial values x_0 and v_0 are, at first, unknown quantities that have to be determined in such a way that a periodic solution is obtained due to the periodic impact sequence. It is therefore required that the deflection x immediately before the impact coincides with that immediately thereafter, at each end point of an interval at which an impact occurs. Due to the principle of conservation of linear momentum, the velocity at these points has to increase abruptly by the amount $\Delta v = \frac{\Delta I}{m}$. Thus, the following conditions must be satisfied at the end points of the intervals:

$$\begin{aligned} x(T_0 - 0) &\stackrel{!}{=} x(T_0 + 0) = x(0) \\ \dot{x}(T_0 - 0) + \frac{\Delta I}{m} &\stackrel{!}{=} \dot{x}(T_0 + 0) = \dot{x}(0) \end{aligned} \quad (3.112)$$

This provides, in conjunction with the solution (3.111), two linear equations for the two unknown quantities x_0 and v_0:

$$x_0 \cos \omega T_0 + \frac{v_0}{\omega} \sin \omega T_0 = x_0$$
$$-x_0 \omega \sin \omega T_0 + v_0 \cos \omega T_0 = v_0 - \frac{\Delta I}{m}. \quad (3.113)$$

Due to $\sin \omega T_0/(1-\cos \omega T_0) = \cot(\omega T_0/2)$ their solution provides the sought-after initial values:

$$x_0 = \frac{\Delta I}{2m\omega} \cot\left(\frac{\omega T_0}{2}\right); \quad v_0 = \frac{\Delta I}{2m}. \quad (3.114)$$

The foundation force function in the interval $0 < t \leq T_0$ is thus:

$$F_B(t) = c \cdot x(t) = \frac{\omega \cdot \Delta I}{2}\left(\cot\left(\frac{\omega T_0}{2}\right) \cos \omega t + \sin \omega t\right). \quad (3.115)$$

The function of the foundation force and the velocity are shown for some values of $\omega T_0 = \omega \cdot 2\pi/\Omega = 2\pi/\eta$ in Fig. 3.23. One can see that there is a kink in the foundation force function at the end point of each interval. This is due to the jump in the velocity of the foundation mass.

The foundation force has the amplitude

$$\hat{F}_B = \frac{\omega \cdot \Delta I}{2}\sqrt{1+\left(\cot\left(\frac{\omega T_0}{2}\right)\right)^2} = F_S \pi \frac{\Delta t}{T} \frac{1}{|\sin(\pi/\eta)|} \quad (3.116)$$

Figure 3.24 shows the amplitude of the foundation force as a function of the frequency ratio, and one can recognize the resonance points at $\eta = 1/k$ for $k = 1, 2, 3, 4$.

(3.115) provides the same time function as the solution (3.110) multiplied by the stiffness c, although it looks completely different. One can see from (3.110) that the denominator of the kth summand becomes zero for $k\eta = 1$ (i. e. $\omega T_0 = 2k\pi$), and thus infinitely high resonance deflections (kth-order resonance) occur as described by solution (3.115), as a result of the cotangent function.

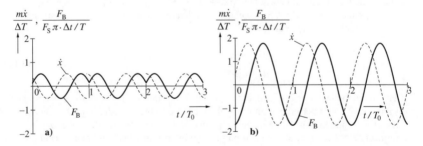

Fig. 3.23 Functions of foundation force and vibration velocity for two different frequency ratios; **a)** $\eta = 0.7$, **b)** $\eta = 1.1$

The following results from the numerical values given in paragraph 2.3 of the problem statement:
Shock pulse

$$\Delta I = F_S \Delta t = 100 \, \text{N} \cdot \text{s},$$

3.3 Foundations under Impact Loading

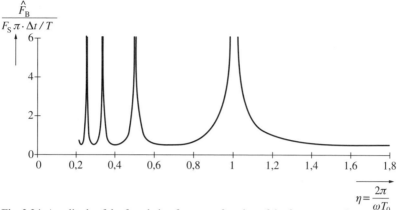

Fig. 3.24 Amplitude of the foundation force as a function of the frequency ratio $\eta = \dfrac{2\pi}{\omega T_0}$

Natural frequency

$$f = \frac{1}{2\pi}\sqrt{\frac{c}{m}} = \frac{\omega}{2\pi} = \frac{22.36\,\text{s}^{-1}}{2\pi} = 3.558\,\text{Hz},$$

Damping constant

$$b = 2D\sqrt{m\,c} = 358\,\text{N}\cdot\text{s/m}.$$

The maximum foundation force $F_{B\max} = 1117.8 \cdot 1.0173 = 1137\,\text{N} \ll F_S = 10^4\,\text{N}$ follows from (3.116) because of $1/|\sin(\omega T_0/2)| = 1/|\sin 11.18| = 1.0173$ and $F_S \pi \cdot \Delta t/T = 1117.8$, see Fig. 3.24.

The simulation provides the result shown in Fig. 3.25, taking into account the damping constant for the infinite sequence of short rectangular impulses.

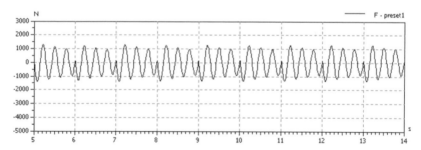

Fig. 3.25 Function of the foundation force as a result of the impact sequence for $\eta = 0.28$ and $D = 0.02$, calculated using SimulationX®

The maximum value approximately matches the value obtained above. If **viscous damping** is taken into account, the following solution for the foundation force in the range $0 < t < T_0$ results, with the decay rate $\delta = \omega_0 D$ and natural circular frequency $\omega = \omega_0\sqrt{1 - D^2}$, as well as with the abbreviations

$$E = \exp(-\delta T_0);\quad C = E\cos\omega T_0;\quad S = E\sin\omega T_0;\quad N = 1 - 2EC + E^2 \quad (3.117)$$

after elementary but extensive intermediate calculations, which for $D = 0$ matches the solution (3.115):

$$F_\text{B}(t) = c\,x(t) = \frac{\omega \cdot \Delta I}{N} \exp(-\delta t) \left[S\cos\omega t + \left(1 - C + S\frac{\delta}{\omega}\right) \sin\omega t \right]. \qquad (3.118)$$

The damping determines the decay between impacts and influences the number of cycles, and thus the material fatigue in all processes of this type.

Conclusion: A Fourier series converges slowly when short-time impacts and force jumps occur over time. It contains very many coefficients that decrease only slowly with increasing order. It is therefore more useful in cases of discontinuous periodic loads to use the method that provides the solution in intervals, rather than Fourier series. The time functions of the motion parameters and the resonance points that result from the Fourier series are the same as the ones from the calculation for a single cycle, taking into account the respective periodicity and transition conditions.

Another "method" to solve this problem is its numerical treatment, e. g. using the software SimulationX® [34]. Figure 3.25 shows a calculated function that does take viscous damping into account.

It follows from these results that outside the resonance ranges (because of $\Delta t/T \ll 1$) the foundation force is much smaller than the impact force F_S. This effect, where the inertia force at the impact site "neutralizes" the impact force, is utilized in many applications to provide relief for downstream sub-assemblies.

Chapter 4
Torsional Oscillators and Longitudinal Oscillators

4.1 Introduction

Torsional vibrations in reciprocating engines historically were among the first problems of machine dynamics. They first occurred in marine engines and were calculated and measured by O. FRAHM as early as 1902. That type of research received a major boost by the demand for lightweight designs for airships and airplanes. In particular, the review of the causes of engine damage, which resulted in enormous subsequent damage in the crash of the Zeppelin LZ 4 after an emergency landing (8/5/1908 near Echterdingen), sped up the development.

The discussion of engine vibrations in an electric locomotive by E. MEISSNER (1918) can be viewed as one of the first works in the field of parameter-excited vibrations within machine dynamics. An overview of pre-computer age reciprocating engine dynamics is contained in "Engineering Dynamics" by BIEZENO and GRAMMEL [1].

The reciprocating engine is still prominent, though it is now about studying not just the engine itself but the entire drive system. This leads to calculation models with a large number of degrees of freedom. Torsional vibrations in drive systems of other machine types are increasingly becoming of interest. For example, increasing demands for higher printing quality made it inevitable to dynamically model printing machines. Growing productivity demands require capturing the dynamics of drive systems as accurately as possible in many processing machines. One can generally say that torsional vibrations have to be considered for almost all machine types in which there is a rotating motion. Their mathematical treatment does not depend on whether the machine in question is a machine tool, cement mill, marine engine, or motorbike engine.

This section deals particularly with torsional vibrations in drive systems. Longitudinal and torsional oscillators are special cases of the general linear oscillators that are discussed in Chap. 6 (including the influence of damping). Longitudinal and torsional oscillators show similarities in their mathematical treatment. There are many drive systems that can be reduced to a torsional vibration model. Table 4.1 shows various basic forms of unconstrained torsional models.

The most important question that has to be answered prior to the dynamic analysis of a machine is the classification of the problem into the model types *rigid ma-*

chine or *vibration system*, see Chap. 1. The natural frequencies must be known and compared to the excitation frequencies. Since the damping of machines is typically weak, the natural frequencies can be calculated neglecting damping. The calculations also yield the mode shapes from which important conclusions can be drawn with regard to the influence on the natural frequencies. Mode shapes are also used to estimate forced vibrations and the behavior of a system under predefined initial conditions, which, for example, occurs during clutching processes.

The simplest calculation model for dynamic loads of drive systems is the **rigid machine model** (rigid-body system),, which provides *kinetostatic* loads. It applies to "slow" loads – see (1.1) or (1.2) – and forms an intermediate step for the calculation of the static internal forces. The torsional moments are found to be variable over time (without vibrations), and they depend on the distribution of inertia along the drive train.

The kinetostatic moments, which result from the rigid-body system model, provide **mean values of the dynamic load**. They are superposed onto the moments due to the vibrations of the elastic system (the "vibration moments"), that can be considerably larger, see, for example, (4.41) and (4.42).

Table 4.1 Basic forms of unconstrained torsional models

	Straight shafts. This model applies, for example, to an inline reciprocating engine without considering auxiliary drives.
	Shafts with speed transformation. This model applies when no torque split occurs.
	Drives with torque split. This model applies as long as an arbitrary number of torque splits occurs and the split portions are not recombined. Examples are vehicle drives with a load distribution mechanism for the front and rear drives, or a marine engine with multiple drive shafts.
	Meshed drives. This model applies when the torque is first split and then recombined on one shaft. Examples of this can be found in printing machine design, in pre-loaded mechanisms for gear testing and in load distribution mechanisms.

First, a rod moved by an input torque M_{an} at $x = 0$ is considered, see Fig. 4.1a. The torsion rod is accelerated with

$$\ddot{\varphi}(t) = \frac{M_{\text{an}}(t)}{J_{\text{p}}}, \qquad (4.1)$$

4.1 Introduction

where the moment of inertia is $J_\mathrm{p} = \varrho L I_\mathrm{p}$. I_p denotes the polar area moment of inertia. The rod is therefore twisted in alternate directions, and inside of it the **kinetostatic torsional moment**

$$M(x, t) = \varrho(L - x)I_\mathrm{p}\ddot{\varphi}(t) = \left(1 - \frac{x}{L}\right) J_\mathrm{p}\ddot{\varphi}(t) = \left(1 - \frac{x}{L}\right) M_\mathrm{an}(t) \quad (4.2)$$

develops at point x due to the inertia.

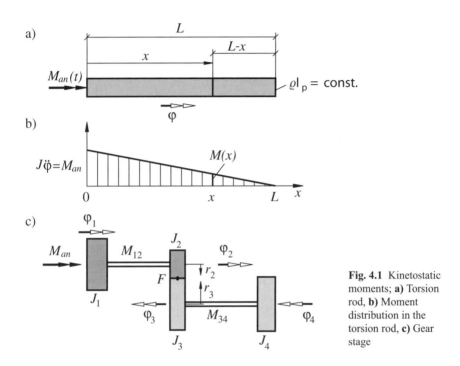

Fig. 4.1 Kinetostatic moments; **a)** Torsion rod, **b)** Moment distribution in the torsion rod, **c)** Gear stage

This moment distribution is shown in Fig. 4.1b. The rotational inertia of the rotating mass behind point x loads the rod with the **kinetostatic moment**. If the torsion rod is split at point x, where the moment of inertia is $J_1 = J_\mathrm{p}x/L$ on the left and $J_2 = J_\mathrm{p}(1 - x/L)$ on the right, the torsional moment at the interface is

$$M(x, t) = J_2\ddot{\varphi} = \frac{J_2 M_\mathrm{an}(t)}{J_1 + J_2}. \quad (4.3)$$

Another example for explaining the kinetostatic moments is the gear mechanism in Fig. 4.1c. The input torque causes an angular acceleration, which can be calculated using the reduced moment of inertia that follows from (2.199) (see (2.213))

$$J_\mathrm{red} = J_1 + J_2 + (J_3 + J_4)\left(\frac{r_2}{r_3}\right)^2. \quad (4.4)$$

Thus the angular accelerations are

$$\ddot{\varphi}_1 = \ddot{\varphi}_2 = \frac{M_{\mathrm{an}}}{J_{\mathrm{red}}}; \quad \ddot{\varphi}_4 = \ddot{\varphi}_3 = \left(\frac{r_2}{r_3}\right)\ddot{\varphi}_1 = \left(\frac{r_2}{r_3}\right)\frac{M_{\mathrm{an}}}{J_{\mathrm{red}}}. \quad (4.5)$$

The two shafts of the gear mechanism are exposed to different values for the torsional moments. These values are found by applying a free-body diagram:

$$M_{12} = (J_{\mathrm{red}} - J_1)\ddot{\varphi}_1 = \left[J_2 + (J_3 + J_4)\left(\frac{r_2}{r_3}\right)^2\right]\ddot{\varphi}_1$$

$$= \left[J_2 + (J_3 + J_4)\left(\frac{r_2}{r_3}\right)^2\right]\frac{M_{\mathrm{an}}}{J_{\mathrm{red}}} \quad (4.6)$$

$$M_{34} = J_4\ddot{\varphi}_4 = J_4\left(\frac{r_2}{r_3}\right)\frac{M_{\mathrm{an}}}{J_{\mathrm{red}}}.$$

Depending on whether one looks at the left or right side, the tangential force F at the gear teeth is:

$$F = \frac{(J_{\mathrm{red}} - J_1 - J_2)\ddot{\varphi}_1}{r_2} = \frac{(J_3 + J_4)\ddot{\varphi}_4}{r_3}. \quad (4.7)$$

One should see for oneself that the two results are identical.

Table 4.2 gives examples of reducing oscillator chains to an unbranched standard model with equations for calculating the free vibrations. These examples could represent the following systems: *a)* elastically coupled vehicles, *b)* hammer foundation, *c)* elevator with counterweight, *d)* drive system with gear ratio, *e)* crane with pendular load, *f)* rotary crane with pendular load. This reduction can only be performed assuming small swing angles for *e)* and *f)*.

Table 4.2 Examples for reductions to a torsional oscillator chain

a)	b)	c)	d)	e)	f)
φ_1	x_1	x_1	φ_4	x_M	φ_M
φ_2	x_2	$r\varphi$	φ_5	x_K	$i\varphi_K; i = r_K/r_M$
φ_3	0	x_3	$(r_6/r_5)\varphi_7$	$x_K + l\alpha$	$i[\varphi_K + \alpha(l/R)]$
J_1	m_1	m_1	J_4	m_K	J_M
J_2	m_2	J_T/r^2	$J_5 + J_6(r_5/r_6)^2$	m_L	J_K/i^2
J_3	$1/J_3 = 0$	m_3	$J_7(r_5/r_6)^2$	m_L	$m_L R^2/i^2$
c_{T1}	c_1	c_1	c_{T4}	c	c_T
c_{T2}	c_2	c_2	$c_{T6}(r_5/r_6)^2$	$m_L g/l$	$m_L g R^2/l i^2$

4.1 Introduction

The equations of motion of the torsional oscillator in Table 4.2, Case a, are

$$J_1\ddot{\varphi}_1 + c_{T1}(\varphi_1 - \varphi_2) = 0 \quad (4.8)$$
$$J_2\ddot{\varphi}_2 - c_{T1}(\varphi_1 - \varphi_2) + c_{T2}(\varphi_2 - \varphi_3) = 0 \quad (4.9)$$
$$J_3\ddot{\varphi}_3 \qquad\qquad - c_{T2}(\varphi_2 - \varphi_3) = 0 \quad (4.10)$$

The equations of motions for systems b) to f) in Table 4.2 have exactly the same structure, only the coordinates and parameters have the designations given in Table 4.2. The following equations for the natural frequencies and mode shapes apply to all these six systems:

Reference circular frequency:

$$\omega^{*2} = \frac{c_{T1}}{J_1} + \frac{c_{T1} + c_{T2}}{J_2} + \frac{c_{T2}}{J_3}, \quad (4.11)$$

Natural circular frequencies, see also Table 4.4, Case 1:

$$\omega_1^2 = 0; \quad \omega_{2,3}^2 = \frac{1}{2}\left[\omega^{*2} \mp \sqrt{\omega^{*4} - \frac{4c_{T1}c_{T2}}{J_1 J_2}\left(\frac{J_1 + J_2}{J_3} + 1\right)}\right]. \quad (4.12)$$

The mode shapes can be normalized in different ways, e.g. using

$$v_{1i} = 1; \quad v_{2i} = 1 - \frac{J_1\omega_i^2}{c_{T1}}; \quad v_{3i} = \frac{1 - \dfrac{J_1\omega_i^2}{c_{T1}}}{1 - \dfrac{J_3\omega_i^2}{c_{T2}}} \quad (4.13)$$

or

$$v_{1i} = \frac{1 - \dfrac{J_3\omega_i^2}{c_{T2}}}{1 - \dfrac{J_1\omega_i^2}{c_{T1}}}; \quad v_{2i} = 1 - \frac{J_3\omega_i^2}{c_{T2}}; \quad v_{3i} = 1. \quad (4.14)$$

Oscillator chains can be represented as so-called image shafts. An image shaft is a mechanically similar vibration system in which all parameters are referred to the same axis (as in Table 4.2) as moments of inertia and torsional spring constants and drawn to scale. The moments of inertia are then proportional to the radius and the compliance of the torsional springs (that is, the inverse of the torsional stiffness) is proportional to the lengths l_{red} of the sections between the disks, see Sect. 1.3 and Table 1.5. The reduction is performed such that the kinetic and potential energy of the original system and the image shaft match.

The representation of an oscillator chain as an image shaft is popularly used to illustrate the stiffness and mass distributions. This illustrative method (formerly used quite often) has lost significance, so it will not be described in detail here. One example, however, shall suffice. Figure 4.2a shows the calculation model of a two-step gear mechanism and Fig. 4.2b the associated image shaft. Given are the parameter values

$$i_{12} = \frac{r_1}{r_2} = \frac{1}{2}; \quad i_{34} = \frac{r_3}{r_4} = \frac{1}{3}; \quad i_{13} = \left(\frac{r_1}{r_2}\right)\left(\frac{r_3}{r_4}\right) = \frac{1}{6} \quad (4.15)$$

$$J_1 = 2J; \quad J_{21} = J; \quad J_{22} = 3J; \quad J_{23} = 2J; \quad J_{24} = 3J; \quad J_3 = 12J \quad (4.16)$$

$$c'_{T1} = 3c_T; \quad c''_{T1} = c_T; \quad c_{T2} = c_T. \quad (4.17)$$

One can show that these two systems are dynamically equivalent by performing the following transformations (using linear conversion of the angles and conversion of the torsional spring constants and moments of inertia with the squared gear ratios):

$$\varphi_{1r} = \varphi_1; \quad \varphi_{2r} = \varphi_2; \quad \varphi_{3r} = \varphi_3/i_{13} \quad (4.18)$$

$$J_1^r = J_1 = 2J; \quad J_2^r = J_{21} + i_{12}^2(J_{22} + J_{23}) + i_{13}^2 J_{24} = \frac{7J}{3}; \quad J_3^r = i_{13}^2 J_3 = \frac{J}{3} \quad (4.19)$$

$$c_{T1}^r = c_{T1} = \frac{c'_{T1} c''_{T1}}{c'_{T1} + c''_{T1}} = \frac{3}{4} c_T \sim \frac{1}{l_{1red}}; \quad c_{T2}^r = i_{13}^2 c_{T2} = \frac{c_T}{36} \sim \frac{1}{l_{2red}}. \quad (4.20)$$

The representation to scale makes it evident that the first shaft is considerably stiffer than the second one. For the purposes of a rough calculation, the gear mechanism can be treated as if the two disks on the left were rigidly connected and it behaves like a two-mass system in the lower frequency range.

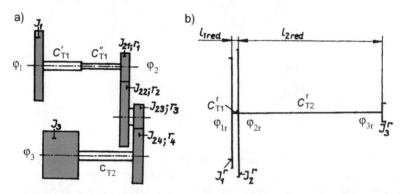

Fig. 4.2 Example of a image shaft; **a)** Original model of the gear mechanism, **b)** Image shaft

4.2 Free Vibrations of Torsional Oscillators

4.2.1 Models with Two Degrees of Freedom

4.2.1.1 Linear Torsional Oscillators with Two Degrees of Freedom

Torsional oscillators are divided into constrained and unconstrained oscillators. Constrained oscillators are used if either at least one spring is fixed and no free rotation of the rigid-body system can occur or if at least one spring is connected to a rotating mass, the motion of which is known. Torsional oscillators with angular motion excitation are thus considered constrained models (like the motion excitation of longitudinal oscillators), see Sect. 4.3.2.3. An unconstrained model becomes a constrained model if one disk is fixed (clamped), like during braking, see Fig. 4.3.

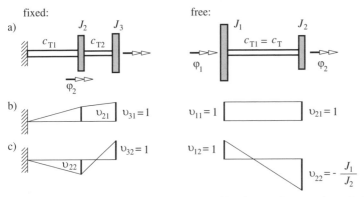

Fig. 4.3 Torsional oscillator with two degrees of freedom; **a)** System schematic, **b)** First mode shape, **c)** Second mode shape

The equations of motion for the constrained model are, see (4.9) and (4.10):

$$\begin{aligned} J_2\ddot{\varphi}_2 + c_{T2}(\varphi_2 - \varphi_3) + c_{T1}\varphi_2 &= 0 \\ J_3\ddot{\varphi}_3 - c_{T2}(\varphi_2 - \varphi_3) &= 0 \end{aligned} \quad (4.21)$$

(4.21) can be compared with the equation of motion (3.83) of the hammer foundation. Since they are structured alike, the results from Chap. 3 can be applied here. The solution according to (3.94), therefore, is:

$$\varphi_k(t) = \sum_{i=1}^{2} v_{ki}(a_i \cos \omega_i t + b_i \sin \omega_i t), \quad k = 2, 3 \quad (4.22)$$

with the amplitude ratios v_{ki} of the mode shapes from (4.13) or (4.14) and the natural circular frequencies ω_i. The unknown quantities a_i and b_i can be determined using the initial conditions, see, for example, solution S4.2 or S4.3. The equations of motion for the unconstrained model are a special case of (4.8) and (4.9) and can be written as follows:

$$\begin{aligned} J_1 \ddot{\varphi}_1 + c_T(\varphi_1 - \varphi_2) &= 0 \\ J_2 \ddot{\varphi}_2 - c_T(\varphi_1 - \varphi_2) &= 0 \end{aligned} \tag{4.23}$$

Assuming a solution of the form $\varphi_k = v_k \cdot \sin \omega t$, it follows (for $k = 1, 2$) after insertion:

$$\begin{aligned} (c_T - J_1\omega^2) v_1 - c_T v_2 &= 0 \\ -c_T v_1 + (c_T - J_2\omega^2) v_2 &= 0. \end{aligned} \tag{4.24}$$

The solution of each equation provides the amplitude ratio

$$\frac{v_2}{v_1} = \frac{c_T - J_1\omega^2}{c_T} = \frac{c_T}{c_T - J_2\omega^2} \tag{4.25}$$

and the frequency equation results from the two fractions on the right:

$$(c_T - J_1\omega^2)(c_T - J_2\omega^2) - c_T^2 = \omega^2 \left[J_1 J_2 \omega^2 - c_T(J_1 + J_2) \right] = 0. \tag{4.26}$$

Its roots are

$$\omega_1 = 0; \qquad \omega_2^2 = c_T \frac{J_1 + J_2}{J_1 J_2} = \frac{c_T}{J_1} + \frac{c_T}{J_2}. \tag{4.27}$$

$\omega_1 = 0$ is counted as the first natural circular frequency, not only for formal, but for physical reasons here. The solution of (4.23) is thus, unlike (4.22),

$$\begin{aligned} \varphi_1 &= v_{11}(\varphi_0 + \omega_0 t) + v_{12}(a_2 \cos \omega_2 t + b_2 \sin \omega_2 t) \\ \varphi_2 &= v_{21}(\varphi_0 + \omega_0 t) + v_{22}(a_2 \cos \omega_2 t + b_2 \sin \omega_2 t). \end{aligned} \tag{4.28}$$

φ_0, ω_0, a_2, and b_2 are constants that can be determined from the initial conditions. The amplitude ratios result from (4.25) when inserting ω_1 and ω_2 from (4.27) into this expression and using the normalization $v_{11} = v_{12} = 1$:

$$\left(\frac{v_2}{v_1}\right)_1 = \frac{v_{21}}{v_{11}} = v_{21} = 1 - \frac{\omega_1^2 J_1}{c_T} = 1; \qquad v_{22} = 1 - \frac{\omega_2^2 J_1}{c_T} = -\frac{J_1}{J_2}. \tag{4.29}$$

As follows from 4.3b, the first mode shape is the rigid-body rotation whose "oscillation period" is $T_1 = 2\pi/\omega_1 \to \infty$. The second mode shape has a node, the position of which depends on the ratio J_1/J_2.

The angular deflections of torsional oscillators are depicted at the respective rotating mass as lines perpendicular to the axis of rotation, see Figs. 4.3, 4.7, 4.9, 4.14, 4.20 et al. The first natural frequency in all unconstrained systems equals zero, see also (4.87) and (4.103). This is why the remarkable phenomenon occurs where a machine drive exhibits a lower fundamental frequency, in measurement results, after braking than during the preceding state of motion.

4.2.1.2 Drive System with Backlash

Impacts occur in drive systems during starting and braking processes due to backlash (e. g. in couplings or gear mechanisms). The dynamic forces generated in this way, in particular when the drive changes its direction of rotation (e. g. when reversing the input direction of the slewing gear of a crane or the drive of an excavator shovel) are often considerably larger than the kinetostatic loads. Overloading and damages may occur if the designer does not include the influence of gear backlash in the calculations.

The magnitude of the **rotational backlash** φ_S **of a gear mechanism** can often be felt when the drive moves back and forth, and it is quite common that the reduced backlash at the engine shaft in drives with large gear ratios amounts to dozens of degrees. Experience shows that the backlash increases with the length of the service period of a drive since wear and tear (e. g. of gears or couplings) increases with time. It gets particularly dangerous when the press fits loosen. Especially the components on the slow running shaft of a drive need to be analyzed for extreme impact loads since backlash has the most considerable effect there.

The high dynamic loads, as compared to drives without backlash, result from the fact that the driven rotating mass reaches a high angular velocity passing through the backlash and impacts on the opposite side. The minimal model for calculating the impact of backlash is shown in Fig. 4.4. It shows the longitudinal oscillator in addition to the torsional oscillator model, and the equations of motion of the former match those of the latter due to the analogy indicated in Table 4.2. The two phases of passing through the backlash can be better illustrated (the rotational backlash φ_S corresponds to the axial backlash δ).

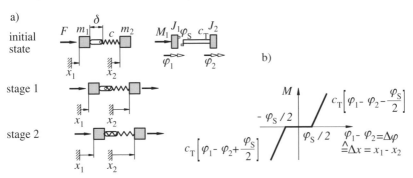

Fig. 4.4 Minimal model of a drive with gear backlash; **a)** System schematic with parameters, **b)** Characteristic with backlash

The extreme case where the input torque jumps immediately to the value M_{10} at the outset is assumed. A sequence of several moment jumps is studied for a drive without backlash in 4.3.3.1. The equations of motion are:

$$J_1 \ddot{\varphi}_1 + M = M_{\text{an}} = M_{10} \qquad (4.30)$$
$$J_2 \ddot{\varphi}_2 - M = 0. \qquad (4.31)$$

The torsional moment M develops in the torsional spring and, considering backlash, it amounts to:

$$M = \begin{cases} c_T(\varphi_1 - \varphi_2 + \varphi_S/2) & \text{for} & \varphi_1 - \varphi_2 \leq -\varphi_S/2 \\ 0 & \text{for} & -\varphi_S/2 \leq \varphi_1 - \varphi_2 \leq +\varphi_S/2 \\ c_T(\varphi_1 - \varphi_2 - \varphi_S/2) & \text{for} & \varphi_1 - \varphi_2 \geq \varphi_S/2 \end{cases}$$

$$= \frac{1}{2} c_T \left[\Delta\varphi - \frac{\varphi_S}{2}\text{sign}(\Delta\varphi) \right] \left[1 + \text{sign}\left(|\Delta\varphi| - \frac{\varphi_S}{2} \right) \right]. \tag{4.32}$$

If one divides (4.30) by J_1 and (4.31) by J_2, the following differential equation for the torsional moment in the angular range $|\varphi_1 - \varphi_2| \geq \varphi_S/2$ can be obtained from the difference of the angular accelerations and after multiplying by c_T, taking into account (4.32):

$$\ddot{M} + \omega^2 M = \frac{c_T M_{10}}{J_1}. \tag{4.33}$$

The square of the natural circular frequency is known from (4.27). During the *first phase*, the engine (J_1) is accelerated by the constant input torque M_{10} so that the entire backlash is passed through, see Fig. 4.4. The rotating mass J_2 remains at rest during this phase ($0 \leq t \leq t_1$), and $\varphi_2 \equiv 0$ applies. The initial conditions are

$$t = 0: \quad \varphi_1(0) = -\frac{\varphi_S}{2}, \quad \dot{\varphi}_1(0) = 0. \tag{4.34}$$

They express the fact that the left mass (rotating mass) rests against the left stop from a previous motion (other initial conditions would have to be formulated if one started in the middle of the backlash). During this phase $M \equiv 0$ according to (4.32), and it follows from the solution of the differential equation (4.30):

$$\varphi_1(t) = -\frac{\varphi_S}{2} + \frac{M_{10}}{2J_1} t^2, \quad \dot{\varphi}_1(t) = \frac{M_{10}}{J_1} t; \quad 0 \leq t \leq t_1. \tag{4.35}$$

The first phase is completed at the time t_1 when the rotating mass has passed through the backlash φ_S:

$$\varphi_1(t_1) = -\frac{\varphi_S}{2} + \frac{M_{10}}{2J_1} t_1^2 = \frac{\varphi_S}{2}; \quad \dot{\varphi}_1(t_1) = \frac{M_{10}}{J_1} t_1 = \sqrt{2\frac{M_{10}\varphi_S}{J_1}}$$
$$\varphi_2(t_1) = 0; \quad \dot{\varphi}_2(t_1) = 0. \tag{4.36}$$

The time for passing through the backlash follows from the first equation

$$t_1 = \sqrt{2\frac{J_1 \varphi_S}{M_{10}}}. \tag{4.37}$$

The end conditions of the first phase are at the same time the initial conditions of the *second phase* ($t_1 \leq t \leq t_2$). The two masses (rotating masses) are now connected by the spring. The following results from (4.30) to (4.32) for the angular range $\varphi_1 - \varphi_2 \geq \varphi_S/2$:

4.2 Free Vibrations of Torsional Oscillators

$$J_1\ddot{\varphi}_1 + c_T\left(\varphi_1 - \varphi_2 - \frac{\varphi_S}{2}\right) = M_{10}$$
$$J_2\ddot{\varphi}_2 - c_T\left(\varphi_1 - \varphi_2 - \frac{\varphi_S}{2}\right) = 0. \tag{4.38}$$

What is to be calculated here is not the time functions of the angles but that of the torsional moment. Using the initial conditions for the moment resulting from (4.36):

$$t = t_1: \quad M(t_1) = 0, \qquad \dot{M}(t_1) = c_T[\dot{\varphi}_1(t_1) - \dot{\varphi}_2(t_1)] = c_T\sqrt{2\frac{M_{10}\varphi_S}{J_1}} \tag{4.39}$$

the solution of (4.33) yields the variation of the moment in the shaft during this phase:

$$M(t) = M_{10}\frac{J_2}{J_1+J_2}\left[1 - \cos\omega(t-t_1) + \sqrt{2\frac{c_T(J_1+J_2)\varphi_S}{M_{10}J_2}}\sin\omega(t-t_1)\right]. \tag{4.40}$$

One can see that, in addition to backlash, the ratio of the moments of inertia is another major parameter influencing the shaft load. As can be seen, the "vibration load" of the free vibration is superposed onto the mean value of the kinetostatic moment from (4.3). The maximum moment results from (4.40):

$$M_{\max} = \frac{M_{10}J_2}{J_1 + J_2}\left(1 + \sqrt{1 + 2\frac{J_1 + J_2}{J_2}\cdot\frac{c_T\varphi_S}{M_{10}}}\right)$$
$$= \frac{M_{10}J_2}{J_1 + J_2}\left(1 + \sqrt{1 + \frac{2J_1\omega^2}{M_{10}}\cdot\varphi_S}\right). \tag{4.41}$$

It follows that the influence of the backlash on the magnitude of the dynamic load is the smaller, the lower the natural frequency of the elastic system is. Therefore, higher dynamic loads occur in "rigid" couplings than in "compliant" couplings. If one expresses this maximum moment using the angular velocity $\dot{\varphi}_1(t_1) = \Omega$ by replacing φ_S in (4.41) using (4.36), it takes the following form:

$$M_{\max} = M_{10}\frac{J_2}{J_1 + J_2}\left(1 + \sqrt{1 + \frac{J_1 + J_2}{J_2}\cdot\frac{c_T J_1 \Omega^2}{M_{10}^2}}\right). \tag{4.42}$$

The maximum moment for the loading case **"coupling impact"**, that results for $M_{10} = 0$ from a limiting process is

$$M_{\max} = \Omega\sqrt{\frac{J_1 J_2 c_T}{J_1 + J_2}} = \sqrt{\frac{\Omega\omega J_1 J_2}{J_1 + J_2}} = \frac{c_T\Omega}{\omega}. \tag{4.43}$$

(4.41) to (4.43) are suitable for calculating the maximum moments. They can also be applied to modal oscillators, see Sect. 6.3.3.

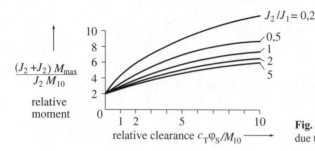

Fig. 4.5 Maximum moment due to backlash in the drive

Figure 4.5 illustrates the relationship of (4.41), i. e. the dependence of the maximum moment on the backlash. Its relationship with the kinetostatic moment is plotted along the ordinate. It already has a value of 2 without backlash but can reach much higher values with backlash in the drive, as these curves show.

It is sometimes incorrectly assumed that dynamic loads only reach double the value of the kinetostatic ones. Figure 4.5 shows that this is erroneous. Such underrating can result in incorrect load assumptions, unexpected overloading and considerable damages. Particularly large forces are generated in drives in which the ratio is $J_1/J_2 \ll 1$. This is often true for drives that do not have to transmit a large static moment, such as slewing gear drives of cranes or excavators. The ratio in drives that have to transmit large static forces, as compared to the dynamic forces, is mostly $J_1/J_2 \gg 1$ so that the impact of backlash on the loads in the drive is not that significant.

4.2.2 Oscillator Chains with Multiple Degrees of Freedom

Oscillator chains are calculation models of elements that are linked like chains, such as massless springs and rigid masses connected to each other, see Fig. 4.6. Longitudinal oscillator chains are characterized by the masses (m_k), spring constants (c_k), and excitation forces (F_k). The coordinate vector $x^\text{T} = (x_1, x_2, \ldots, x_n)$ contains the displacements x_k that correspond to the elongations from the position of static equilibrium. For torsional oscillators, the parameter values of the moments of inertia (J_k) and torsion spring constants ($c_{\text{T}k}$) are responsible for the dynamic behavior, and the absolute angles of rotation φ_k of the rotating masses are used as coordinates. Equations of motion for the torsional oscillator chain are derived below that similarly apply to longitudinal oscillators. The initial focus is on the natural vibrations. The moment equilibrium at the kth disk is stated in general for all rotating masses ($k = 1, 2, \ldots, n$). According to Fig. 4.6b, it is:

$$J_k \ddot{\varphi}_k - c_{\text{T}k-1}(\varphi_{k-1} - \varphi_k) + c_{\text{T}k}(\varphi_k - \varphi_{k+1}) = 0, \quad k = 1, 2, \ldots, n \quad (4.44)$$
$$\varphi_0 = \varphi_{n+1} \equiv 0.$$

4.2 Free Vibrations of Torsional Oscillators

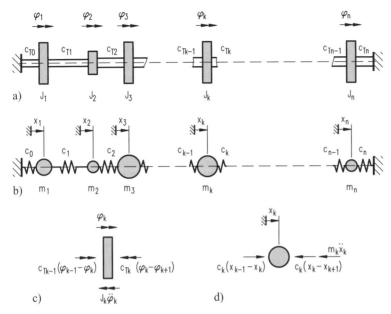

Fig. 4.6 Oscillator chain; **a)** Torsional oscillator with n disks (rotating masses), **b)** Longitudinal oscillator with n masses, **c)** free-body diagram for kth rotating mass, **d)** free-body diagram for kth mass

It represents n equations of motion for the torsional oscillator chain. It is $c_{T0} = c_{Tn} = 0$ if the input and output members can freely move (i.e. are not constrained). For clarity, one can express this set of equations in matrix notation (index "T" omitted for c_{Tk}), see the special cases (4.21), (4.23), and the last line of (4.45).

$$\begin{bmatrix} J_1 & 0 & 0 & \cdots & 0 & 0 \\ 0 & J_2 & 0 & \cdots & 0 & 0 \\ 0 & 0 & J_3 & \cdots & 0 & 0 \\ \vdots & & & \ddots & & \vdots \\ 0 & 0 & 0 & \cdots & J_{n-1} & 0 \\ 0 & 0 & 0 & \cdots & 0 & J_n \end{bmatrix} \begin{bmatrix} \ddot{\varphi}_1 \\ \ddot{\varphi}_2 \\ \ddot{\varphi}_3 \\ \vdots \\ \ddot{\varphi}_{n-1} \\ \ddot{\varphi}_n \end{bmatrix} + \begin{bmatrix} c_0+c_1 & -c_1 & 0 & \cdots & 0 & 0 \\ -c_1 & c_1+c_2 & -c_2 & \cdots & 0 & 0 \\ 0 & -c_2 & c_2+c_3 & \cdots & 0 & 0 \\ \vdots & & & \ddots & & \vdots \\ 0 & 0 & 0 & \cdots & c_{n-2}+c_{n-1} & -c_{n-1} \\ 0 & 0 & 0 & \cdots & -c_{n-1} & c_{n-1}+c_n \end{bmatrix} \begin{bmatrix} \varphi_1 \\ \varphi_2 \\ \varphi_3 \\ \vdots \\ \varphi_{n-1} \\ \varphi_n \end{bmatrix} = \begin{bmatrix} 0 \\ 0 \\ 0 \\ \vdots \\ 0 \\ 0 \end{bmatrix}$$

$$M \quad \cdot \quad \ddot{\varphi} \quad + \quad C \quad \cdot \quad \varphi = o$$

(4.45)

This defines the coordinate vector $\varphi^T = [\varphi_1, \varphi_2, \ldots, \varphi_k, \ldots, \varphi_n]$, which consists of all angles of vibration, the mass matrix M, and the stiffness matrix C. One can show that all natural circular frequencies follow from the characteristic equation

$$\det(C - \omega^2 M) = 0, \qquad (4.46)$$

see Chap. 6.

The matrix equation (4.45) can be transformed by the modal transformation described in 6.3.2 into a system of n equations for oscillators with one degree of freedom of the form

$$\mu_i \ddot{p}_i + \gamma_i p_i = 0, \qquad i = 1, 2, \ldots, n, \tag{4.47}$$

see (6.112) to (6.116). p_i represent the so-called modal coordinates or principal coordinates. They are related to the original angles of rotation by the transformation

$$\varphi_k(t) = \sum_{i=1}^{n} v_{ki} p_i(t) = v_{k1} p_1(t) + v_{k2} p_2(t) + \ldots + v_{kn} p_n(t), \tag{4.48}$$

see the special cases (4.22) and (4.28). The quantities v_{ki} (amplitude of the kth disk at the ith natural frequency) are proportional to a selected scale factor, i. e. they depend on a normalization condition, see (6.86). The totality of v_{ki} describes the ith mode shape, as is already known for special cases from Fig. 4.3, see also Fig. 4.7.

Section 6.3.2 describes the general relationship between the matrices $\boldsymbol{M}$, and $\boldsymbol{C}$ and the quantities v_{ki}, μ_i, γ_i, and ω_i. The modal masses μ_i for the torsional oscillator chain according to Fig. 4.6 result from (6.112)

$$\mu_i = \sum_{k=1}^{n} J_k v_{ki}^2, \qquad i = 1, 2, \ldots, n \tag{4.49}$$

see also (6.191). A **modal mass** μ_i characterizes the kinetic energy of the ith mode shape since

$$2W_{\text{kin}} = \sum_{k=1}^{n} J_k \dot{\varphi}_k^2 = \sum_{i=1}^{n} \mu_i \dot{p}_i^2, \tag{4.50}$$

because the following applies for $i \neq l$ as a special case of (6.103)

$$\sum_{k=1}^{n} J_k v_{ki} v_{kl} = 0. \tag{4.51}$$

In the sum of μ_i, those rotating masses J_k that vibrate at the greatest amplitudes (in the vicinity of the antinode) have the largest **sensitivity coefficients** μ_{ik} while the ones near the node only produce small summands μ_{ik}.

$\omega_1 = 0$ in the torsional oscillator chain that is unconstrained on both sides, and all "amplitudes" v_{k1} are equal because this first "mode shape" corresponds to a rigid-body rotation. If one normalizes using $v_{k1} = 1$, one gets

$$\mu_1 = \sum_{k=1}^{n} J_k v_{k1}^2 = \sum_{k=1}^{n} J_k \cdot 1^2 = J_1 + J_2 + \ldots + J_n, \tag{4.52}$$

i. e., the first modal mass equals the total moment of inertia. The modal spring constants γ_i that occur in (4.47) amount to

$$\gamma_i = \sum_{k=0}^{n} c_{Tk} \left(v_{ki} - v_{k+1,i} \right)^2. \tag{4.53}$$

4.2 Free Vibrations of Torsional Oscillators

in the torsional oscillator chain according to (6.112). The sensitivity coefficients γ_{ik} are dimensionless factors that one can use like μ_{ik} to calculate the influence of parameter changes on the natural frequencies, see (6.189) and (6.191). Their derivation is described in Sect. 6.4.2.

$$\mu_{ik} = \frac{J_k v_{ki}^2}{\sum_{j=1}^{n} J_j v_{ji}^2}; \qquad \gamma_{ik} = \frac{c_{Tk}(v_{ki} - v_{k+1,i})^2}{\sum_{j=0}^{n} c_{Tj}(v_{ji} - v_{j+1,i})^2}. \tag{4.54}$$

The change of the ith natural frequency for relatively small changes of the kth torsion spring and/or the lth rotating mass can be approximated using the sensitivity coefficients, see also (6.189):

$$\frac{\Delta f_i}{f_i} \approx \frac{1}{2}\left(\gamma_{ik} \cdot \frac{\Delta c_{Tk}}{c_{Tk}} - \mu_{il} \cdot \frac{\Delta J_l}{J_l}\right), \qquad i = 1, 2, \ldots, n. \tag{4.55}$$

To get a vivid impression of the distribution of the kinetic and potential energies at the ith natural frequency, it is useful to sketch the products $\mu_{ik} J_k$ and $\gamma_{ik} c_{Tk}$ to scale. This representation shows better than the mode shape which parameters (mass, stiffness) can most easily influence the respective natural frequency, see Fig. 4.14.

Table 4.3 shows closed-form equations for all natural frequencies and mode shapes for the three potential support options at the shaft ends for a homogeneous torsional oscillator chain in which all torsional spring constants and all n rotating masses are of equal size. The equations in Table 4.3 are suitable for studying the qualitative and quantitative relations for torsional oscillators with n degrees of freedom.

Figure 4.7 shows a torsional oscillator with $n = 5$ degrees of freedom as an example. According to (2) in Table 4.3, the following applies for the natural circular frequencies and for the mode shapes ($i = 1, \ldots, 5$):

$$\omega_i = 2\sqrt{\frac{c_T}{J}} \sin\frac{(2i-1)\pi}{22}; \qquad v_{ki} = \sin\frac{k(2i-1)\pi}{11}. \tag{4.56}$$

The number of nodes increases with the index i. One can see that the node is never located exactly at the location of a disk or at the unconstrained end of the oscillator chain for any mode shape.

The scope of the calculation model for real objects described by Figs. 4.6 and (4.45) is limited amongst others by the following assumptions:

- The division into massless elastic torsional springs and rigid disks with masses is a permissible simplification of the real continuum (would be violated by very wide disks);
- The ever present damping does not have a major influence on the natural frequencies and mode shapes;

Table 4.3 Natural frequencies and mode shapes of homogeneous torsional oscillator chains (Order of the natural frequency $i = 1, 2, \ldots, n$; Index of rotating mass $k = 1, 2, \ldots, n$; Number of rotating masses n)

Case	Torsional oscillator with n equal disks	Natural circular frequencies $\omega_i = 2\pi f_i$	Mode shapes v_{ki}	
1		$2\sqrt{\dfrac{c_T}{J}} \sin \dfrac{i\pi}{2(n+1)}$	$\sin \dfrac{ki\pi}{n+1}$	(1)
2		$2\sqrt{\dfrac{c_T}{J}} \sin \dfrac{(2i-1)\pi}{2(2n+1)}$	$\sin \dfrac{k(2i-1)\pi}{2n+1}$	(2)
3		$2\sqrt{\dfrac{c_T}{J}} \sin \dfrac{(i-1)\pi}{2n}$	$\cos \dfrac{(2k-1)(i-1)\pi}{2n}$	(3)

- Time-dependent or position-dependent changes of the rotating masses (such as the influence of a crank mechanism) or the torsional springs (such as gear contact ratio) can be neglected for calculating the behavior of the oscillator.

4.2.3 Evaluation of Natural Frequencies and Mode Shapes

Unlike in the past when the engineers had to become familiar with the respective calculation procedures in detail, today there are software products available so that engineers primarily have to know the model generation, the scope of the calculation model, and the applicability limits of the software and to ensure the accuracy of the input data as well as interpret the results.

After the representation of a real system by a calculation model has been found, the first analysis is usually that of the eigenmotions. This calculation provides the natural frequencies and the associated mode shapes. The required calculation model only consists of masses and springs, and thus it does not include values for damping and excitation, which are often difficult to determine. Once the natural frequencies are known, several conclusions regarding the model can be drawn. One first compares the lowest natural frequency with the highest excitation frequency of interest. If the lowest natural frequency is far above the highest excitation frequency, the calculation can be based on the rigid machine model (rigid-body mechanism). If the natural frequencies are in the range of the excitation frequencies, resonance frequencies can be stated.

4.2 Free Vibrations of Torsional Oscillators

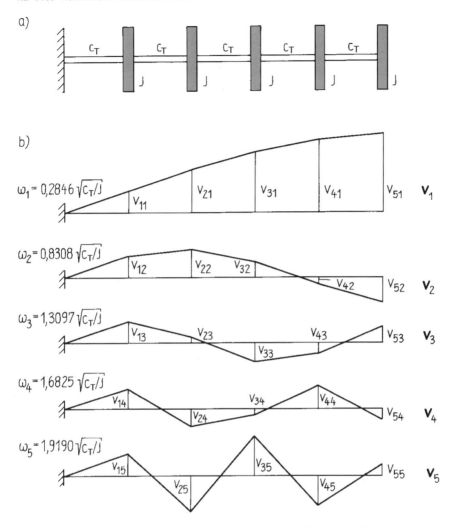

Fig. 4.7 Torsional oscillator chain with $n = 5$ disks; **a)** System schematic, **b)** Five mode shapes

If higher natural frequencies of the model are far above the range of excitation frequencies, the calculation model contains too many degrees of freedom. On the other hand, the model does not contain enough degrees of freedom if the highest calculated natural frequency is in the range of the excitation frequencies. Figure 4.8 illustrates these various cases. The calculated natural frequencies are also often used for model verification in that natural frequencies measured at the system are used for comparison. Since models with many degrees of freedom often have directly adjacent natural frequencies, the comparison is useful only when including the mode shapes. However, these mode shapes are difficult to measure.

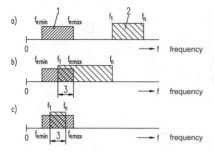

Fig. 4.8 Evaluation in the frequency domain; **a)** Rigid machine, **b)** Model with too many degrees of freedom, resonance range is relevant, **c)** Model with too few degrees of freedom, change required; *1* Excitation frequency range ($f_{e\,min} < f_e < f_{e\,max}$), *2* Natural frequency range ($f_1 < f_i < f_n$), *3* Resonance range

If the natural vibrations are known in the form of the natural frequencies and mode shapes, the free vibrations, i.e. the motions in each coordinate direction, depending on the initial conditions, can be calculated. This is particularly important for determining impact and step responses.

The mode shapes are used for the above mentioned analysis of the natural frequencies as well as for parameter discussions and for the calculation of the resonance amplitudes. It should first be noted that any increase in stiffness and any reduction in mass increases the natural frequencies. The effectiveness of such measures, however, depends mainly on the mode shape, see the sensitivity coefficients in Fig. 4.14 and Sect. 6.4.2.

The following applies: Changes in the stiffness are most effective in the shaft portion in which the largest difference of the relative deflections occurs. Changes in the mass are most effective in the antinode, i.e. at the site of the largest relative deflection of the mode shape. Mass changes at the node do not influence the natural frequency associated with the corresponding mode shape.

The *fundamental frequency* (the lowest frequency that is different from zero) of each oscillator is the most important characteristic of its dynamic behavior. The task is often just to determine the fundamental frequency for which it is useful to first estimate the fundamental frequency using a minimal model before analyzing a comprehensive calculation model for which many input data are required [4].

Approximation methods developed in the "computerless" past still have a certain relevance for some purposes since they allow the estimation of the natural frequencies as a function of a few essential parameters for a complex torsional vibration system. The rationale is to reduce the real object to a *system with a known characteristic equation*, such as to oscillator chains with two or three degrees of freedom or to continua, see Table 4.2, (4.27), Table 4.3, and Table 4.4.

The frequency equations given in Table 4.4, cases 2 to 4, result from a *continuum coupled with a discrete oscillator* that can be modeled using the torsional spring constant c_T and the moment of inertia J. (4) in Table 4.4 with $c_T = 0$ and $J = 0$ provides the following for the special case of the continuum ($\pi_1 = \pi_2 = 0$) with both ends free:

$$\lambda \sin \lambda = 0; \qquad \lambda_i = (i-1)\pi; \qquad i = 1, 2, 3, \ldots \qquad (4.57)$$

4.2 Free Vibrations of Torsional Oscillators 241

Table 4.4 Characteristic equations for torsional oscillators (analogy: longitudinal oscillators)

Case	System schematic	Frequency equation
1	c_{T1}, J_1 — c_{T2}, J_2 — J_3	$\omega_1 = 0;\quad \omega^4 - \left[\dfrac{c_{T2}}{J_3} + \dfrac{c_{T1}+c_{T2}}{J_2} + \dfrac{c_{T1}}{J_1}\right]\omega^2$ $+ \dfrac{c_{T1}\, c_{T2}}{J_1\, J_2\, J_3}(J_1+J_2+J_3) = 0 \quad (1)$
2	l; $GI_T, \rho I_p$; c_T, J	$(\pi_1 - \pi_2 \lambda^2)\cos\lambda - \pi_1 \pi_2 \lambda \sin\lambda = 0 \quad (2)$
3	l; $GI_T, \rho I_p$; c_T, J	$(\pi_1 - \pi_2 \lambda^2)\sin\lambda + \pi_1 \pi_2 \lambda \cos\lambda = 0 \quad (3)$
4	l; c_T, J; $GI_T, \rho I_p$	$(\pi_1 - \pi_2 \lambda^2)\cos\lambda - \lambda \sin\lambda = 0 \quad (4)$
5	n disks; c_{T0}, J_0, c_T, J, … c_T, J, c_T, J	$\omega_i = 2\sqrt{c_T/J}\,\sin(\kappa_i/2) \quad (5)$ $\sin\kappa\,\cot(n\kappa) - \dfrac{J}{J_0} + \left(2\dfrac{c_T}{c_{T0}} - 1\right)(1+\cos\kappa) = 0 \quad (6)$
	$\lambda = \omega l\sqrt{\dfrac{\rho I_p}{GI_t}} \cong \omega l\sqrt{\dfrac{\rho}{E}}\,,\quad \pi_1 = \dfrac{c_T l}{GI_T} \cong \dfrac{c l}{EA}\,,\quad \pi_2 = \dfrac{J}{\rho I_p l} \cong \dfrac{m}{\rho A l}$	(7)

Case 5 in Table 4.4 used to be popular for the "homogeneous machine with an additional mass" because it corresponds to a multicylinder engine with a flywheel coupled to it and allows easy calculation of all frequencies. To simplify the model, one can "spread out" the individual disks of the discrete torsional oscillator and thus change the model into a continuum model.

The natural frequencies of the continuum-type longitudinal and torsional oscillator depend on the boundary conditions and follow as special cases from the characteristic equations of Table 4.4:

- Unconstrained at both ends (from (4.57) or (4)):

$$f_i = \dfrac{i-1}{2}\sqrt{\dfrac{GI_t}{\varrho I_p l^2}} \cong \dfrac{i-1}{2}\sqrt{\dfrac{E}{\varrho l^2}} \qquad (4.58)$$

- Fixed at one end (from (2)):

$$f_i = \frac{2i-1}{4}\sqrt{\frac{GI_t}{\varrho I_p l^2}} \,\hat{=}\, \frac{2i-1}{4}\sqrt{\frac{E}{\varrho l^2}} \qquad (4.59)$$

- Fixed at both ends:

$$f_i = \frac{i}{2}\sqrt{\frac{GI_t}{\varrho I_p l^2}} \,\hat{=}\, \frac{i}{2}\sqrt{\frac{E}{\varrho l^2}}. \qquad (4.60)$$

An infinite number of dimensionless eigenvalues λ and corresponding natural frequencies, of which mostly only the lower ones are of interest, can be found from the characteristic equations for the continuum using a program for calculating the roots of equations.

The associated *mode shapes* are as important as the natural frequencies. One can write closed-form equations for simple models, see (4.13), Fig. 4.3, (4.29), Table 4.3, Fig. 4.7, and Fig. 4.14. It is important to know them in order to evaluate the effect of parameter changes on the natural frequencies and to analyze the excitability of forced vibrations [5].

An important aid are the sensitivity coefficients μ_{ik} and γ_{ik}, which can easily be calculated for torsional oscillators, see (4.54). They are explicitly provided by advanced programs and give engineers valuable clues as to how the vibration behavior can be influenced, see Figs. 4.10 and 4.14.

4.2.4 Examples

4.2.4.1 Four-Cylinder Engine

Figure 4.9 shows the model of a reciprocating engine with four cylinders, pulley and flywheel. The following parameter values are present in the initial state:

$J_1 = 4.611 \cdot 10^{-2}$ kg·m²; $c_{T1} = 16.19 \cdot 10^4$ N·m; $v_{12} = 1.000$
$J_2 = 1.350 \cdot 10^{-2}$ kg·m²; $c_{T2} = 76.03 \cdot 10^4$ N·m; $v_{22} = 0.4285$
$J_3 = 1.350 \cdot 10^{-2}$ kg·m²; $c_{T3} = 61.61 \cdot 10^4$ N·m; $v_{32} = 0.2916$
$J_4 = 1.350 \cdot 10^{-2}$ kg·m²; $c_{T4} = 76.03 \cdot 10^4$ N·m; $v_{42} = 0.1097$
$J_5 = 1.350 \cdot 10^{-2}$ kg·m²; $c_{T5} = 110.85 \cdot 10^4$ N·m; $v_{52} = -0.0415$
$J_6 = 39.34 \cdot 10^{-2}$ kg·m²; $v_{62} = -0.1443$

The first natural frequency of this torsional oscillator is $f_1 = 0$, the second and third natural frequencies are to be influenced. The natural frequency changes due to relatively small parameter changes can be calculated using the sensitivity coefficients, see (4.55):

4.2 Free Vibrations of Torsional Oscillators

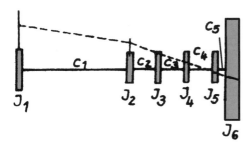

Fig. 4.9 Calculation model of a reciprocating engine with the mode shape of the fundamental frequency $(c_i \hat{=} c_{Ti})$

$$\frac{\Delta f_i}{f_{i0}} \approx \frac{1}{2}\left(\sum_{l=1}^{5}\gamma_{il}\frac{\Delta c_{Tl}}{c_{Tl}} - \sum_{k=1}^{6}\mu_{ik}\frac{\Delta J_k}{J_k}\right) \quad (4.61)$$

The printout from the SimulationX® program provides the system parameters *spring l* and *inertia k* and coefficients from which the sensitivity coefficients γ_{ik} and μ_{ik} can be calculated after division by the totals that are provided as well. The values computed for kinetic and potential energies in Fig. 4.10b can be understood as the numerators of the sensitivity coefficients μ_{ik} and γ_{ik} in (4.54) (for $i = 2$ in the left column and $i = 3$ in the right column; both for all k). The sum values reflect the denominators in (4.54). Thus the analysis results can be used for computing the sensitivity coefficients μ_{ik} and γ_{ik}.

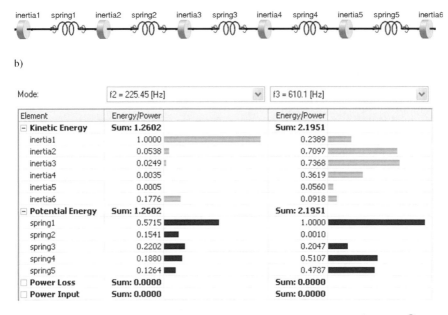

Fig. 4.10 Analyzing the torsional vibrations of a reciprocating engine using SimulationX®
a) Model representation in SimulationX®, **b)** Energy distribution for mode shapes 2 and 3

If only the parameters c_{T1}, c_{T5}, J_1 and J_6 are changed, the following linear approximation results for the influence of small parameter changes on the second and third natural frequencies, see the numerical values in Fig. 4.10:

$$\Delta f_2 \approx \frac{f_{20}}{2 \cdot 1.2602}\left(-1.0000\frac{\Delta J_1}{J_1} - 0.1776\frac{\Delta J_6}{J_6} + 0.5715\frac{\Delta c_{T1}}{c_{T1}} + 0.1264\frac{\Delta c_{T5}}{c_{T5}}\right)$$

$$\approx \left(-89.5\frac{\Delta J_1}{J_1} - 15.7\frac{\Delta J_6}{J_6} + 51.1\frac{\Delta c_{T1}}{c_{T1}} + 11.3\frac{\Delta c_{T5}}{c_{T5}}\right) \text{ Hz}$$

$$\Delta f_3 \approx \frac{f_{30}}{2 \cdot 2.1951}\left(-0.2389\frac{\Delta J_1}{J_1} - 0.0918\frac{\Delta J_6}{J_6} + 1.0000\frac{\Delta c_{T1}}{c_{T1}} + 0.4787\frac{\Delta c_{T5}}{c_{T5}}\right)$$

$$\approx \left(-33.2\frac{\Delta J_1}{J_1} - 25.5\frac{\Delta J_6}{J_6} + 277.9\frac{\Delta c_{T1}}{c_{T1}} + 133.0\frac{\Delta c_{T5}}{c_{T5}}\right) \text{ Hz} \quad (4.62)$$

Note that the relative parameter changes refer to the parameter values on which the calculations of f_{20} and f_{30} were based.

Changing the pulley J_1 and the spring constant c_{T1} appear to be most effective for influencing the second natural frequency since their sensitivity coefficients are the largest, as was to be expected based on the second mode shape, see Fig. 4.9. The spring parameters c_{T1} and c_{T5} have the largest influence on the third natural frequency. If one compares the two equations in (4.62), one can see that the second natural frequency, due to an increase of J_1, drops more than the third natural frequency so that one could increase the distance between the two natural frequencies in this way.

It should be noted that many manufacturing and design issues have to be taken into account, which limits the implementation options from the start.

4.2.4.2 Torsional Vibrations of a Printing Machine

The following example is to illustrate the general statements made in Sect. 4.2.4.

Drive and roller systems of printing mechanisms represent branched torsional oscillators. Since the transmission of printing ink responds quite sensitively to deviations from the setpoint motion of the cylinders, those natural frequencies of the machine are of interest whose mode shapes comprise antiphase relative rotations of associated cylinders.

The natural frequencies and mode shapes for the calculation model of the printing mechanism of an offset printing machine shown in Fig. 4.11 are to be determined. The shafts are modeled as massless torsional springs, while the cylinders are considered as rigid bodies that are not in contact with each other.
Given are the following parameter values:

$$J_1 = J_2 = J_3 = 2.65 \text{ kg} \cdot \text{m}^2 \quad \text{moments of inertia of plate, rubber and printing cylinders}$$

4.2 Free Vibrations of Torsional Oscillators

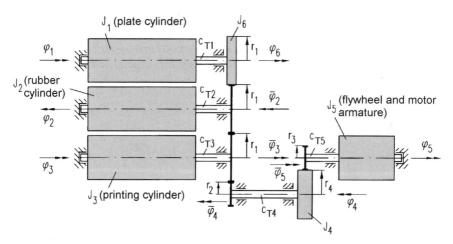

Fig. 4.11 Calculation model of a printing machine drive

$J_4 = 0.52 \text{ kg} \cdot \text{m}^2$	moment of inertia of the first gear step referred to angle φ_4
$J_5 = 1.98 \text{ kg} \cdot \text{m}^2$	moment of inertia of flywheel and motor armature
$J_6 = 1.10 \text{ kg} \cdot \text{m}^2$	moment of inertia of the intermeshing gears referred to angle φ_6
$r_4/r_3 = 4.88$	gear ratio of the first gear step
$r_1/r_2 = 2.26$	gear ratio of the second gear step
$c_{T1} = c_{T2} = c_{T3}$ $= 8.3 \cdot 10^5 \text{ N} \cdot \text{m}$	torsional spring constants of the cylinder shafts
$c_{T4} = 2.4 \cdot 10^5 \text{ N} \cdot \text{m}$	torsional spring constant of the intermediate shaft
$c_{T5} = 4.2 \cdot 10^5 \text{ N} \cdot \text{m}$	torsional spring constant of the input shaft

The kinetic and potential energies can be stated as follows, using the angular coordinates shown in Fig. 4.11:

$$2W_{\text{kin}} = \sum_{k=1}^{6} J_k \dot{\varphi}_k^2,$$

$$2W_{\text{pot}} = c_{T1} \cdot (\varphi_1 - \varphi_6)^2 + \sum_{k=2}^{5} c_{Tk} \cdot (\varphi_k - \bar{\varphi}_k)^2. \qquad (4.63)$$

The following constraints apply for the intermeshing gears:

$$\bar\varphi_2 = \varphi_6; \qquad \bar\varphi_3 = \bar\varphi_2; \qquad r_2\bar\varphi_4 = r_1\bar\varphi_3; \qquad r_3\bar\varphi_5 = r_4\varphi_4. \qquad (4.64)$$

If these equations are used to eliminate the angular coordinates with bars in (4.63), the following result is obtained after introducing the generalized coordinates $q_k = \varphi_k$ ($k = 1, \ldots, 6$):

$$2W_{\text{pot}} = c_{T1} \cdot (q_1 - q_6)^2 + c_{T2} \cdot (q_2 - q_6)^2 + c_{T3} \cdot (q_3 - q_6)^2$$

$$+ c_{T4} \cdot \left(q_4 - \frac{r_1}{r_2}q_6\right)^2 + c_{T5} \cdot \left(q_5 - \frac{r_4}{r_3}q_4\right)^2. \qquad (4.65)$$

Introducing the Lagrangian $L = W_{\text{kin}} - W_{\text{pot}}$, the equations of motion can now be established according to Lagrange's equations of the second kind. If these are written in matrix notation according to (4.45), one obtains the mass matrix $M = \text{diag}(J_k)$. Using the given parameter values, after factoring out the reference quantity J_1, results in:

$$M = 2.65 \text{ kg} \cdot \text{m}^2 \cdot \text{diag}\,(1;\ 1;\ 1;\ 0.196\,23;\ 0.747\,17;\ 0.415\,09) = J_1 \cdot \overline{M} \qquad (4.66)$$

and in the stiffness matrix

$$C = \begin{bmatrix} c_{T1} & 0 & 0 & 0 & 0 & -c_{T1} \\ & c_{T2} & 0 & 0 & 0 & -c_{T2} \\ & & c_{T3} & 0 & 0 & -c_{T3} \\ & & & c_{T4} + \left(\dfrac{r_4}{r_3}\right)^2 c_{T5} & -\dfrac{r_4}{r_3}c_{T5} & -\dfrac{r_1}{r_2}c_{T4} \\ & & & & c_{T5} & 0 \\ \text{symmetrical} & & & & & c_{T1} + c_{T2} + c_{T3} + \left(\dfrac{r_1}{r_2}\right)^2 c_{T4} \end{bmatrix}$$

$$= 8.3 \cdot 10^5 \text{ N} \cdot \text{m} \begin{bmatrix} 1 & 0 & 0 & 0 & 0 & -1 \\ & 1 & 0 & 0 & 0 & -1 \\ & & 1 & 0 & 0 & -1 \\ & & & 12.3398 & -2.4694 & -0.6535 \\ & & & & 0.506\,02 & 0 \\ \text{symmetrical} & & & & & 4.4769 \end{bmatrix} = c_{T1}\overline{C} \qquad (4.67)$$

Basic mathematical software for solving linear eigenvalue problems is used for calculating the natural frequencies and mode shapes.

The natural frequencies are:

$$\begin{aligned} f_1 &= 0 \text{ Hz}; & f_2 &= 50.1 \text{ Hz}; & f_3 &= 89.1 \text{ Hz} \\ f_4 &= 89.1 \text{ Hz}; & f_5 &= 300.6 \text{ Hz}; & f_6 &= 710.6 \text{ Hz}. \end{aligned} \qquad (4.68)$$

4.2 Free Vibrations of Torsional Oscillators

The eigenvectors v_i $(i = 1, \ldots, 6)$ are normalized in such a way that $v_i^T \cdot M \cdot v_i = 1$. They are combined in the modal matrix V, see (6.87) and (6.88):

$$V = \begin{bmatrix} 0.10244 & 0.54950 & 0.81651 & 0 & -0.14451 & 0.00107 \\ 0.10244 & 0.54950 & -0.40824 & 0.70711 & -0.14451 & 0.00107 \\ 0.10244 & 0.54950 & -0.40824 & -0.70711 & -0.14451 & 0.00107 \\ 0.23151 & -0.02378 & 0 & 0 & 0.09027 & 2.24360 \\ 1.12976 & -0.21770 & 0 & 0 & -0.02786 & -0.11776 \\ 0.10244 & 0.37577 & 0 & 0 & 1.50098 & -0.06698 \end{bmatrix}$$

$$= [v_{ki}]. \tag{4.69}$$

The first eigenvector reflects the rigid rotation of the printing mechanism. In order to assess the "danger" associated with the mode shapes, one should look at the relative rotations of the cylinders. The relative displacement at the contact points of the cylinders is a measure of the degree to which the transmission of the printing ink is impaired. Note the sign definitions of the angles in Fig. 4.11.

Figure 4.12 shows the distortion of the cylinders for two mode shapes (side view). It becomes evident that the doubly occurring natural frequency of $f_3 = f_4 = 89.1$ Hz can become a problem for the printing mechanism. Excitation with excitation frequencies that are in the vicinity, such as meshing frequencies or integer multiples of the rotational frequencies of the cylinders, can result in dangerously large resonance amplitudes that impair the printing quality.

This statement can be made without having to calculate the resonance amplitudes. As countermeasures, one can strive for a shift of the excitation and/or natural frequencies in opposite directions based on these results.

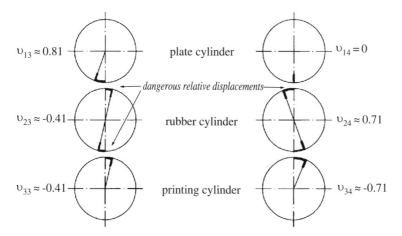

Fig. 4.12 Relative cylinder amplitudes for the third and fourth mode shapes

4.2.4.3 Vehicle Drive Train

A typical example of a complex drive system is the vehicle drive train, considering the engine as an elastic subsystem and the remainder of the drive train with clutch, transmission, articulated shaft, differential, axle shafts, and wheel/tire. Figure 4.13a shows a corresponding vibration model with all rotating masses and rotational stiffnesses as an unbranched unmeshed system (rear-wheel-driven vehicle with engaged 4th gear).

To minimize vibrations (and thus for noise reduction purposes), vehicles with mechanical transmissions are equipped with so-called "dual-mass flywheels" (DMF) that exert a major positive influence on the overall vibration behavior of the drive system, see Fig. 4.13b.

A modal analysis was performed using the parameter values given in the table in Fig. 4.13c; its results are shown in Fig. 4.14. The typical dynamical properties can be identified when looking at the first natural frequencies and mode shapes. By comparing the amplitude distribution of the mode shapes (v_{ki}) and the sensitivity coefficients multiplied by the parameter values, one can try to interpret the physical causes.

It is evident that the "engine" subsystem is dynamically decoupled from the remaining vibration system between flywheel and vehicle mass. This also justifies the reduction of the elastic "engine" substructure to a single mass as a "rigid" substructure if the drive system itself is not to be analyzed, since the engine behaves like a rotating rigid body in the lower mode shapes.

Other interesting points in terms of model generation that can be seen from Fig. 4.14b include that

- the vehicle mass (J_{15}) is not involved in the vibrations due to its large inertia and acts like a fixed point;
- particularly the second mode shape is strongly influenced by the dual-mass flywheel (relative motion between disks 7 and 8);
- all natural frequencies have been partially lowered due to the dual-mass flywheel (additional mass, smaller torsion spring), compare f_i of system a with f_{i+1} of system b
- the higher mode shapes are only influenced insignificantly by the dual-mass flywheel, compare mode shapes v_i of system a with v_{i+1} of system b
- the first mode shape is due to the most resilient springs (c_7, c_{13}) and the large flywheel mass (J_7)
- the vibration behavior of the four-cylinder engine is relevant only above about 290 Hz

Figure 4.14 does not show the rigid-body rotation (theoretical first mode shape). The figure also shows that the graphic representation depends on the normalization condition, for v_{k1} only looks different in Figs. 4.14a and 4.14b because the signs were reversed.

4.2 Free Vibrations of Torsional Oscillators

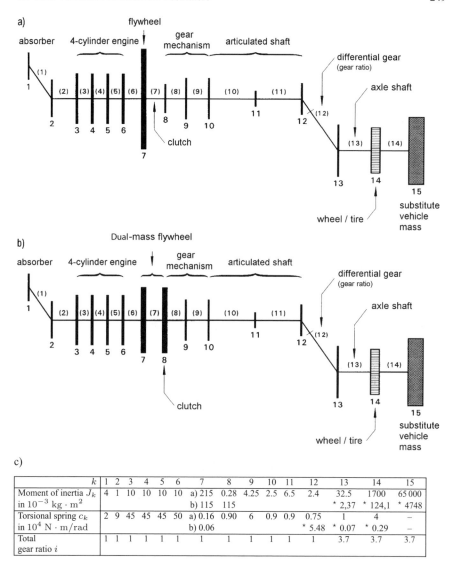

Fig. 4.13 Vehicle drive train; **a)** without dual-mass flywheel, **b)** with dual-mass flywheel, **c)** Table of parameter values (Source: ARLA); ⋆ values referred to principal shaft

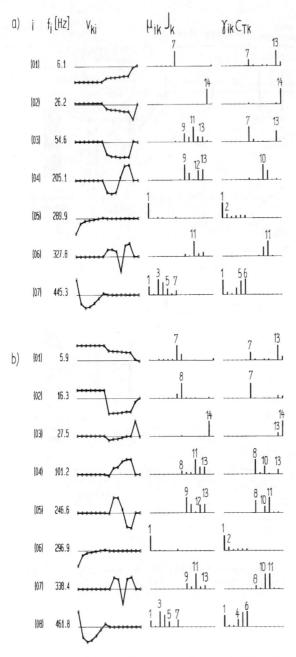

Fig. 4.14 Natural frequencies f_i, mode shapes v_{ki}, and products of parameter values with sensitivity coefficients of the kinetic (μ_{ik}) and potential energies (γ_{ik}) for Fig. 4.13, represented according to [21] (Source: ARLA); **a)** without dual-mass flywheel, **b)** with dual-mass flywheel

4.2 Free Vibrations of Torsional Oscillators

4.2.5 Problems P4.1 to P4.3

P4.1 Starting a Crane with Backlash in the Drive

A crane (see the schematic in Fig. 4.15) is accelerated with a constant motor torque from rest. Assuming that the rotating mass of motor and gear mechanism referred to the motor shaft has to traverse the entire reduced backlash before any force is transmitted to the rail, the torsional loads in the shafts of the driven wheels during the start-up process are to be determined. The crane is started without load, and kinetic resistances can be neglected.

Given:
$m = 98000$ kg	Translating mass of the crane
$J_M = 0.555$ kg $\cdot$ m^2	Rotating mass of gear mechanism and motor referred to the motor shaft
$\Delta \varphi_M = 0.35$ rad	Gear backlash referred to the motor shaft
$i = 25$	Gear ratio ($\dot\varphi_M = i \cdot \dot\varphi_W$)
$M_M = 91$ N $\cdot$ m	Constant motor torque
$R = 0.4$ m	Rolling radius of the wheels
$J_R = 16$ kg $\cdot$ m^2	Moment of inertia of a wheel
$G = 8.1 \cdot 10^{10}$ N/m^2	Shear modulus of the drive shafts
$\tau_{zul} = 1.33 \cdot 10^8$ N/m^2	Permissible torsional stress
$l_1 = 1.3$ m, $l_2 = 0.325$ m	Lengths of the drive shafts
$d_1 = 80$ mm, $d_2 = 65$ mm	Diameters of the drive shafts

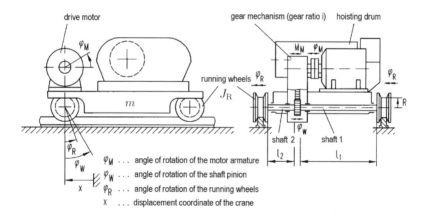

Fig. 4.15 Coordinates and parameters of the crane

Find:

1. Constraints for $\varphi_M, \varphi_W, \varphi_R$ and x
2. Equations of motion for the coordinates φ_W and φ_R defined in Fig. 4.15
3. Initial conditions after passing through the backlash
4. Torsional moments in the two drive shafts after the first pass through the backlash
5. Maximum torsional stress in both drive shafts

P4.2 Coupling Impact

A rotating mass J_3 that rotates at an angular velocity Ω_3 is suddenly coupled to a partial system at rest with the parameters J_1, c_{T1}, J_2 using an elastic pin and bush coupling with a torsional stiffness c_{T2} (Fig. 4.16).

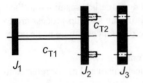

Fig. 4.16 Minimum model for the coupling process

Given:

Angular velocity J_3 upon impact: Ω_3
Moments of inertia $J_1 = J_2 = J_3 = J$
Torsional spring constants $c_{T1} = 10c_T$, $c_{T2} = c_T$

Find:

1. Natural frequencies and mode shapes
2. Time function for motion $\varphi_1(t)$
3. Elastic moment in the coupling for the case $\Omega_3 = \sqrt{c_T/J}$

Evaluate the results using plausibility checks that can be performed based on the principle of conservation of angular momentum and the principle of conservation of energy.

P4.3 Gear Blocking

For certain drive systems, a proof of safety in the event of shaft-breaking during a breakdown has to be provided. The dynamic moments that occur in the shafts in such a load case, especially their maximum values, must be known.

The drive system, which is initially rotating without vibrations, see Fig. 4.17, is suddenly blocked by engaging the engine brake. This excites torsional vibrations that apply a dynamic load to the shafts. The oscillator chain corresponds to case d in Table 4.2.

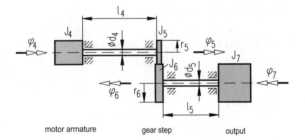

Fig. 4.17 Coordinates and parameters of the drive system

Assuming the extreme case in which the armature is braked instantly, the torsional stresses in both shafts are to be determined.

4.2 Free Vibrations of Torsional Oscillators

Given:

$l_4 = 400$ mm, $l_5 = 200$ mm	Lengths of the two shafts
$d_4 = 35$ mm, $d_5 = 32$ mm	Shaft diameters
$r_5 = 185$ mm, $r_6 = 250$ mm	Pitch radii of the mating gears
$J_4 = 0.8$ kg $\cdot$ m^2	Moment of inertia of the motor armature
$J_5 = 0.5$ kg $\cdot$ m^2, $J_6 = 1.7$ kg $\cdot$ m^2	Moments of inertia of the gears
$J_7 = 0.6$ kg $\cdot$ m^2	Moment of inertia of the drive
$G = 8.1 \cdot 10^{10}$ N/m^2	Shear modulus of the shafts
$n_M = 75$ rpm	Motor speed

Find:

1. Natural frequencies of the free and blocked (braked) system
2. Mode shapes of the braked system
3. Initial conditions for the sudden blocking of the motor armature
4. Torsional stresses (particularly their maximum values) that occur due to the sudden blocking in the two shafts

4.2.6 Solutions S4.1 to S4.3

S4.1 If there is no backlash, a constraint between φ_M and φ_W exists:

$$\varphi_M = i \cdot \varphi_W. \tag{4.70}$$

If there is stiction between the wheels and the rail, the following constraint applies:

$$x = R \cdot \varphi_R. \tag{4.71}$$

The angular speed upon impact follows from the fact that the kinetic energy of the motor armature after passing through the backlash and the work of the motor torque when passing through the backlash angle

$$\frac{1}{2} J_M \dot{\varphi}_M^2 = M_M \Delta \varphi_M$$

are equal:

$$\dot{\varphi}_M = \sqrt{\frac{2 M_M}{J_M} \cdot \Delta \varphi_M}. \tag{4.72}$$

After the backlash was traversed for the first time, power is transmitted between the gear mechanism and the pinion.

The kinetic and potential energies and the virtual work of the forces not covered by W_{pot} are needed to develop the equations of motion for the contact phase:

$$W_{\text{kin}} = \frac{1}{2} \cdot [m\dot{x}^2 + J_M \dot{\varphi}_M^2 + 4 J_R \dot{\varphi}_R^2] = \frac{1}{2} [(mR^2 + 4 J_R)\dot{\varphi}_R^2 + i^2 J_M \dot{\varphi}_W^2]$$

$$W_{\text{pot}} = \frac{1}{2} \cdot \left[c_{T1} \cdot (\varphi_W - \varphi_R)^2 + c_{T2} \cdot (\varphi_W - \varphi_R)^2 \right] \tag{4.73}$$

$$= \frac{1}{2} \cdot (c_{T1} + c_{T2}) \cdot (\varphi_W - \varphi_R)^2$$

with the torsional spring constants

$$c_{Tk} = \frac{G \cdot \pi d_k^4}{32 \cdot l_k}; \quad k = 1, 2. \tag{4.74}$$

The virtual work of the motor torque is

$$\delta W = M_M \cdot \delta\varphi_M. \tag{4.75}$$

Thus the equations of motion can be developed using Lagrange's equation of the second kind. They are:

$$\begin{bmatrix} i^2 J_M & 0 \\ 0 & mR^2 + 4J_R \end{bmatrix} \cdot \begin{bmatrix} \ddot{\varphi}_W \\ \ddot{\varphi}_R \end{bmatrix} + (c_{T1} + c_{T2}) \begin{bmatrix} 1 & -1 \\ -1 & 1 \end{bmatrix} \cdot \begin{bmatrix} \varphi_W \\ \varphi_R \end{bmatrix} = \begin{bmatrix} iM_M \\ 0 \end{bmatrix}. \tag{4.76}$$

The left side has the form of (4.23). The initial conditions after passing through the backlash are:

$t = 0$:

$$\varphi_W(0) = \frac{1}{i}\varphi_M(0) = 0; \qquad \dot{\varphi}_W(0) = \frac{1}{i}\dot{\varphi}_M(0) = \frac{1}{i} \cdot \sqrt{\frac{2M_M}{J_M}} \cdot \Delta\varphi_M \tag{4.77}$$

$$\varphi_R(0) = 0; \qquad \dot{\varphi}_R(0) = 0.$$

If one looks at the equations of motion (4.76), the problem can be modeled using the oscillator discussed in Sect. 4.2.1.2 according to Fig. 4.18 with two degrees of freedom and backlash.

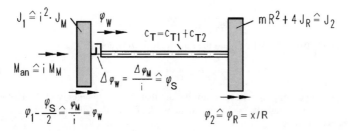

Fig. 4.18 Calculation model of crane and drive system with backlash

The equations (4.76) are structured in analogy with (4.38), and the initial conditions (4.77) are comparable to (4.36). Thus the solution obtained in Sect. 4.2.1.2 can be used. The moments in the two shafts are obtained with the square of the natural circular frequency, as in (4.27)

$$\omega^2 = (c_{T1} + c_{T2}) \cdot \frac{mR^2 + 4J_R + i^2 J_M}{(mR^2 + 4J_R) \cdot i^2 J_M} = 2025 \text{ s}^{-2} \tag{4.78}$$

and with the moment M known from (4.40) and (4.41):

$$\underline{\underline{M_k}} = c_{Tk}(\varphi_W - \varphi_R) = \frac{c_{Tk}}{c_{T1} + c_{T2}} \cdot M$$

$$= \frac{c_{Tk} M_M}{iJ_M \omega^2} \cdot \left[1 - \cos\omega t + \sqrt{\frac{2J_M \omega^2}{M_M}} \cdot \Delta\varphi_M \cdot \sin\omega t\right], \qquad k = 1, 2, \tag{4.79}$$

if one takes into account that the relationship

$$\frac{c_T(J_1 + J_2)}{M_{an} J_2} \hat{=} \frac{J_M \omega^2}{M_M} = \frac{(c_{T1} + c_{T2})(mR^2 + 4J_R + i^2 J_M)}{(mR^2 + 4J_R)i^2 M_M} \tag{4.80}$$

4.2 Free Vibrations of Torsional Oscillators

exists.

The maximum values are thus

$$M_{k\,\max} = c_{Tk}\frac{M_M}{iJ_M\omega_0^2}\left[1+\sqrt{1+\frac{2J_M\omega_0^2}{M_M}\Delta\varphi_M}\right], \quad k = 1,\,2. \tag{4.81}$$

The expression under the root enables the evaluation of the effect of the backlash (as compared to a drive without backlash). It is considerable in the case at hand, since it follows from the given parameter values:

$$\frac{2J_M\omega^2}{M_M}\Delta\varphi_M = \frac{2\cdot 0.555\cdot 2025.4}{91}\cdot 0.35 = 8.647, \tag{4.82}$$

see Fig. 4.5.

The resulting maximum torsional stresses are:

$$\tau_{k\,\max} = \frac{16M_{k\,\max}}{\pi\cdot d_k^3}; \quad k = 1,\,2. \tag{4.83}$$

The maximum values follow from (4.81) and (4.83) with (4.74):

$$\begin{array}{ll} \underline{M_{1\,\max} = 3.33\cdot 10^3\text{ N}\cdot\text{m};} & \underline{M_{2\,\max} = 5.81\cdot 10^3\text{ N}\cdot\text{m}} \\ \underline{\tau_{1\,\max} = 3.3\cdot 10^7\text{ N/m}^2;} & \underline{\tau_{2\,\max} = 10.8\cdot 10^7\text{ N/m}^2.} \end{array} \tag{4.84}$$

The stiffer shaft 2 ($c_{T2} = 4.37\cdot 10^5\text{ N}\cdot\text{m} > c_{T1} = 2.51\cdot 10^5\text{ N}\cdot\text{m}$) is stressed much more than the more flexible one.

Conclusion: The magnitude of the dynamic loads that occur in drive systems during start-up processes depends considerably on the magnitude of the backlash to be passed through, on the natural frequency, and on the motor torque.

S4.2 The problem leads towards an unconstrained model according to Table 4.2 (Case a), see (4.8), (4.9) and (4.10). The initial conditions are:

$$t = 0: \quad \begin{array}{lll} \varphi_1 = 0; & \varphi_2 = 0; & \varphi_3 = 0 \\ \dot{\varphi}_1 = 0; & \dot{\varphi}_2 = 0; & \dot{\varphi}_3 = \Omega_3. \end{array} \tag{4.85}$$

According to (4.12), the squares of the natural circular frequencies are

$$\omega_1^2 = 0; \quad \omega_2^2 = 1.4606\frac{c_T}{J}; \quad \omega_3^2 = 20.5394\frac{c_T}{J}, \tag{4.86}$$

i.e., the natural frequencies are

$$\underline{f_1 = 0}; \quad \underline{f_2 = 0.1923\sqrt{\frac{c_T}{J}}}; \quad \underline{f_3 = 0.7213\sqrt{\frac{c_T}{J}}}. \tag{4.87}$$

The mode shapes can be calculated from (4.13). The first one corresponds to a rigid body rotation.

$$v_{11} = 1; \quad v_{21} = 1; \quad v_{31} = 1. \tag{4.88}$$

The other mode shapes have the following amplitude ratios with $v_{1i} = 1$

$$\begin{array}{lll} v_{12} = 1; & v_{22} = 0.8539; & v_{32} = -1.8539 \\ v_{13} = 1; & v_{23} = -1.0539; & v_{33} = 0.0539. \end{array} \tag{4.89}$$

Figure 4.19 shows the second and third mode shapes. The complete solution for $k = 1, 2, 3$ in analogy with (4.28) is:

$$\varphi_k = \varphi_0 + \Omega t + v_{k2}(a_2 \cos \omega_2 t + b_2 \sin \omega_2 t) + v_{k3}(a_3 \cos \omega_3 t + b_3 \sin \omega_3 t). \quad (4.90)$$

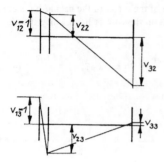

Fig. 4.19 Mode shapes of the torsional oscillator

If one satisfies the initial conditions (4.85), the following integration constants are obtained:

$$\varphi_0 = 0; \quad a_2 = 0; \quad a_3 = 0$$

$$b_2 = \frac{\Omega_3}{\omega_2} \cdot \frac{v_{23} - 1}{(v_{32} - 1)(v_{23} - 1) + (1 - v_{22})(v_{33} - 1)} = -0.2969\Omega_3 \sqrt{\frac{J}{c_T}}$$

$$b_3 = \frac{\Omega_3}{\omega_3} \cdot \frac{1 - v_{22}}{(v_{32} - 1)(v_{23} - 1) + (1 - v_{22})(v_{33} - 1)} = 0.0056\Omega_3 \sqrt{\frac{J}{c_T}} \quad (4.91)$$

$$\Omega = -(v_{22}\omega_2 b_2 + v_{23}\omega_3 b_3) = 0.3333\Omega_3.$$

According to (4.90), the disk *1* undergoes the following motion:

$$\varphi_1 = 0.3333\Omega_3 t - (0.2969 \sin \omega_2 t - 0.0056 \sin \omega_3 t)\Omega_3 \sqrt{\frac{J}{c_T}}. \quad (4.92)$$

The elastic moments generated in shaft 1 are as follows

$$M_{12} = c_{T1}(\varphi_1 - \varphi_2) = c_{T1}[b_2(v_{12} - v_{22}) \sin \omega_2 t + b_3(v_{13} - v_{23}) \sin \omega_3 t]$$
$$= c_T(-0.4337 \cdot \sin \omega_2 t + 0.1157 \cdot \sin \omega_3 t) \quad (4.93)$$

and in the coupling

$$M_{23} = c_{T2}(\varphi_3 - \varphi_2) = c_{T2}[b_2(v_{32} - v_{22}) \sin \omega_2 t + b_3(v_{33} - v_{23}) \sin \omega_3 t]$$
$$= c_T(0.8040 \cdot \sin \omega_2 t - 0.0060 \cdot \sin \omega_3 t). \quad (4.94)$$

As can be seen from the solution, time functions of the moments are composed of two harmonic vibrations, where one circular frequency is not an integer multiple of the other. This typically does not result in a periodic load.

The amplitude of the first component is much larger than that of the second, which was to be expected when looking back at the mode shapes.

The motion of the oscillator is composed of a rigid body rotation at an angular velocity Ω and a vibrational motion with the two natural frequencies. The angular velocity Ω results

4.2 Free Vibrations of Torsional Oscillators

from the condition that the angular momentum is conserved (i.e. $J_3\Omega_3 = (J_1+J_2+J_3)\Omega = 3J\Omega$), immediately as being $\Omega = \Omega_3/3$.

The initial kinetic energy introduced into the system by the disk *3* is distributed between the energies W_i of the three mode shapes, see Sect. 4.2.2 and (6.124):

$$W = W_1 + W_2 + W_3 = W_{\text{kin}\,0} = \frac{1}{2}J\Omega_3^2. \tag{4.95}$$

The rotational energy of the rigid-body system is the energy of the first "mode shape":

$$W_1 = \frac{1}{2}(J_1+J_2+J_3)\Omega^2 = \frac{1}{2}3J\left(\frac{\Omega_3}{3}\right)^2 = \frac{1}{6}J\Omega_3^2. \tag{4.96}$$

It does not contain any potential energy. The energy in the other two "proper" mode shapes is the **vibrational energy**, which can be calculated from the difference to the initial energy:

$$W_2 + W_3 = W - W_1 = \frac{1}{3}J\Omega_3^2. \tag{4.97}$$

It represents the sum total of the kinetic and potential energies that are contained in the two excited mode shapes. In an extreme case, the entire vibrational energy can be stored in one torsional spring during the vibration. This results in the two estimates

$$\frac{1}{2}c_{T1}(\varphi_1 - \varphi_2)_{\max}^2 \leqq \frac{1}{3}J\Omega_3^2; \qquad \frac{1}{2}c_{T2}(\varphi_2 - \varphi_3)_{\max}^2 \leqq \frac{1}{3}J\Omega_3^2. \tag{4.98}$$

This consideration thus allows one to state the maximum size of the maximum moments in the shafts (without performing the actual vibration calculation). It follows from (4.98), see also (6.129):

$$M_{12\,\max} = c_{T1}|\varphi_1 - \varphi_2|_{\max} \leqq \Omega_3\sqrt{\frac{2}{3}c_{T1}J} = 2.582 c_T$$
$$M_{23\,\max} = c_{T2}|\varphi_2 - \varphi_3|_{\max} \leqq \Omega_3\sqrt{\frac{2}{3}c_{T2}J} = 0.8165 c_T. \tag{4.99}$$

One can compare the resulting outcome with the results from (4.93) and (4.94) and try to determine why one estimate is more accurate than the other.

S4.3 The equations of motion of this vibration system are known from (4.8) to (4.10). Using the equations given in Table 4.2 (Case d), the parameter values of the problem are converted into those of the image shaft (Table 4.2, Case a):

$$J_1 \hateq J_4 = 0.8 \text{ kg} \cdot \text{m}^2;$$
$$J_2 \hateq J_5 + J_6\left(\frac{r_5}{r_6}\right)^2 = 1.4309 \text{ kg} \cdot \text{m}^2; \tag{4.100}$$
$$J_3 \hateq J_7\left(\frac{r_5}{r_6}\right)^2 = 0.3286 \text{ kg} \cdot \text{m}^2.$$

The torsional spring constants of the shafts result from

$$c_{Tj} = \frac{GI_{pj}}{l_j} = \frac{G\pi d_j^4}{32 l_j}; \qquad j = 4,\, 6 \tag{4.101}$$

and are converted for the coordinates of the image shaft of the system according to Fig. 4.2, Case a:

$$c_{T1} \hat{=} c_{T4} = 29\,833 \text{ N} \cdot \text{m}; \qquad c_{T2} \hat{=} c_{T6} \left(\frac{r_5}{r_6}\right)^2 = 22\,831 \text{ N} \cdot \text{m}. \tag{4.102}$$

The natural frequencies $f_i = \omega_i/(2\pi)$ of the unconstrained gear mechanism are derived with the data given in (4.100) and (4.102) from (4.12) or from (1) in Table 4.4:

$$f_1 = 0 \text{ Hz}; \qquad f_2 = 35.2 \text{ Hz}; \qquad f_3 = 48.9 \text{ Hz}. \tag{4.103}$$

The natural frequencies of the blocked (constrained) system are derived from the same equations by setting $J_1 = J_4 \to \infty$:

$$f_1 = 20.2 \text{ Hz}; \qquad f_2 = 47.8 \text{ Hz}. \tag{4.104}$$

All natural frequencies increase due to the braking, and $f_3 \to \infty$ applies for the third natural frequency. The mode shapes of the braked gear mechanism are first determined with respect to the coordinates φ_1 and φ_2 from (4.13):

$$v_{1i} = 0; \qquad v_{2i} = 1; \qquad v_{3i} = \frac{1}{1 - \dfrac{J_3 \omega_i^2}{c_{T2}}}; \qquad i = 1, 2 \tag{4.105}$$

$$v_{32} = -3.35; \qquad v_{31} = 1.30.$$

Conversion to the original coordinates using the transformation equations given in Table 4.2 provides the amplitude ratios depicted in Fig. 4.20, taking into account the gear ratio $r_5/r_6 = 0.74$.

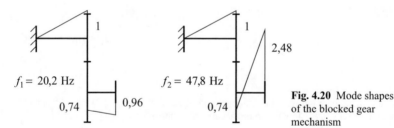

$f_1 = 20{,}2$ Hz $\quad 0{,}74 \quad 0{,}96$

$f_2 = 47{,}8$ Hz $\quad 0{,}74 \quad 2{,}48$

Fig. 4.20 Mode shapes of the blocked gear mechanism

The general solution of the equations of motion according to (4.22) is:

$$\left. \begin{aligned} \varphi_k(t) &= \sum_{i=1}^{2} v_{ki} \cdot (a_i \cos\omega_i t + b_i \sin\omega_i t) \\ \dot{\varphi}_k(t) &= \sum_{i=1}^{2} \omega_i v_{ki} \cdot (-a_i \sin\omega_i t + b_i \cos\omega_i t) \end{aligned} \right\}, \quad k = 2, 3. \tag{4.106}$$

Since all gears initially rotate without vibration, the initial conditions are:

$$t = 0: \qquad \varphi_2(0) = \varphi_3(0) = 0; \qquad \dot{\varphi}_2(0) = \dot{\varphi}_3(0) = \Omega = 7.854 \text{ s}^{-1}. \tag{4.107}$$

The coefficients a_i and b_i can be determined if one requires that the solutions (4.106) also satisfy the initial conditions (4.107). These requirements mean:

$$\begin{aligned} \varphi_2(0) &= v_{21}a_1 + v_{22}a_2 = 0; & \dot{\varphi}_2(0) &= \omega_1 v_{21} b_1 + \omega_2 v_{22} b_2 = \Omega \\ \varphi_3(0) &= v_{31}a_1 + v_{32}a_2 = 0; & \dot{\varphi}_3(0) &= \omega_1 v_{31} b_1 + \omega_2 v_{32} b_2 = \Omega. \end{aligned} \tag{4.108}$$

4.3 Forced Vibrations of Discrete Torsional Oscillators

These equations have the following solutions:

$$a_1 = a_2 = 0; \quad b_1 = 0.057\,98; \quad b_2 = 1.6897 \cdot 10^{-3}. \tag{4.109}$$

The motions of the blocked system can thus be stated:

$$\begin{aligned}\varphi_2(t) &= 0.057\,98 \cdot \sin \omega_1 t - 0.001\,69 \cdot \sin \omega_2 t \mathrel{\hat{=}} \varphi_5(t) \\ \varphi_3(t) &= 0.075\,40 \cdot \sin \omega_1 t - 0.005\,66 \cdot \sin \omega_2 t \mathrel{\hat{=}} \frac{r_6}{r_5}\varphi_7(t).\end{aligned} \tag{4.110}$$

The torsional stresses τ_{Tj} that occur at the outer rim of the shafts are proportional to the torsional moments, which themselves show a linear dependence on the vibration amplitudes. Using the section moduli $W_{Tj} = \pi d_j^3/16$, the result is:

$$\begin{aligned}\underline{\tau_{T4}} &= \frac{M_{T4}}{W_{T4}} = \frac{16 c_{T4}}{\pi d_4^3} \cdot \varphi_5(t) = \frac{G \cdot d_4}{2 \cdot l_4} \cdot \varphi_2(t) \\ &= \underline{(205.5 \cdot \sin \omega_1 t + 6.0 \cdot \sin \omega_2 t)\ \text{N}/\text{mm}^2}\end{aligned} \tag{4.111}$$

$$\begin{aligned}\underline{\tau_{T5}} &= \frac{M_{T5}}{W_{T5}} = \frac{16 c_{T5}}{\pi d_5^3} \cdot [\varphi_7(t) - \varphi_6(t)] \\ &= \frac{G d_5}{2 l_5} \cdot \frac{r_5}{r_6} \cdot [\varphi_3(t) - \varphi_2(t)] \\ &= \underline{(83.5 \cdot \sin \omega_1 t - 35.2 \cdot \sin \omega_2 t)\ \text{N}/\text{mm}^2}.\end{aligned} \tag{4.112}$$

4.3 Forced Vibrations of Discrete Torsional Oscillators

4.3.1 Periodic Excitation

The equations of motion for the forced damped vibrations of the torsional oscillator chain result from (4.44) if the viscous damping moments and excitation moments $M_k(t)$ are added (for $k = 1, 2, \ldots, n$) at the kth disk:

$$\begin{aligned}J_k \ddot{\varphi}_k &- b_{Tk-1}(\dot{\varphi}_{k-1} - \dot{\varphi}_k) + b_{Tk}(\dot{\varphi}_k - \dot{\varphi}_{k+1}) \\ &- c_{Tk-1}(\varphi_{k-1} - \varphi_k) + c_{Tk}(\varphi_k - \varphi_{k+1}) = M_k(t)\end{aligned} \tag{4.113}$$

The general solution of these equations is discussed in Chap. 6. This section discusses only the minimal models for introducing these problems.

It is useful to calculate forced vibrations in conjunction with damping values only, since otherwise no usable information about the amplitude values in the vicinity of resonance can be obtained. Since damping constants of real machines are often not available, it is common to introduce modal damping, see Sect. 6.6.1.

Forced vibrations are described by differential equations with constant coefficients, in which the excitation appears as an explicit time function. Periodic excitation functions occur in a large group of drive systems. These include reciprocating engines, which are excited by gas and inertia forces, reciprocating pumps and presses as well as drive systems with mechanisms with a variable transmission ratio, see Sect. 4.3.2.3.

In addition to these "external" excitation moments, excitation due to gear tooth impacts plays a part. The (mostly very high-frequency) vibrations thus produced impair the functional quality and result in additional tooth stresses, see Sect. 4.5.3.2.

The following sequence of calculations is useful for treating forced torsional vibrations with periodic excitation: after determining the model parameters, performing a modal analysis (ω_i, v_{ki}, γ_{ik}, μ_{ik}), and recording the excitation in the form of a *Fourier* series, potential resonance frequencies are determined using the resonance diagram. Then one can estimate the resonance amplitudes by comparing energies and assess the excitability of the resonance points due to the modal excitation forces. There are software packages that were specifically developed for torsional oscillators in drive systems. They can be used to calculate the time functions of all forces and motions of interest and to analyze parameter influences so that designs with favorable dynamic behavior can be developed even before construction and testing. VDI Guideline 3840 [36] gives hints as to which parameters should in general be considered when studying vibrations of shafts.

It is well known that any periodic excitation can be divided into a sum of harmonic terms using a Fourier series. For functions that are given not based on an analysis such as a measured indicator diagram, software provides the Fourier coefficients. The excitation moment at the disk k is thus obtained in the following form:

$$M_k = M_{k\,\text{st}} + \sum_{m=1}^{\infty} (M_{kmc} \cos m\Omega t + M_{kms} \sin m\Omega t) \qquad (4.114)$$
$$= M_{k\,\text{st}} + \sum_{m=1}^{\infty} \hat{M}_{km} \sin(m\Omega t + \beta_{km})$$

with the amplitude of the mth harmonic at k

$$\hat{M}_{km} = \sqrt{M_{kmc}^2 + M_{kms}^2} \qquad (4.115)$$

and the respective phase angle (and exact quadrant position)

$$\beta_{km} = \arccos\left(M_{kms}/\hat{M}_{km}\right) \cdot \text{sign}\left(M_{kmc}\right). \qquad (4.116)$$

In (4.114), $M_{k\,\text{st}}$ is the static moment that mostly represents the moment of the load. It causes the entire drive system to run under a preload. The circular frequency Ω of the periodic process corresponds to the angular velocity of the drive shaft. Note that a work cycle in four-stroke engines consists of two revolutions. If still reference is made to the rotational speed, the (mathematically unfavorable) effect of fractional orders will occur. It is therefore common to introduce fractional indices m.

Higher-order (mth-order) resonances exist for periodic excitation. Resonance can occur according to (6.363) if a circular frequency of an excitation $m\Omega$ coincides

4.3 Forced Vibrations of Discrete Torsional Oscillators

with a natural circular frequency ω_i:

$$\Omega = \frac{\omega_i}{m} \quad \text{or} \quad m\Omega = \omega_i \quad \text{or} \quad n_{im} = \frac{60 f_i}{m}. \tag{4.117}$$

The last equation refers to the rotational speeds n_{im} in rpm and the frequencies f_i in Hz.

This is a necessary (not a sufficient) condition, which can be conveniently represented in a resonance diagram, also known as *Campbell* diagram. The natural frequencies are plotted on the ordinate and rotational speeds on the abscissa. The lines of the orders then intersect with the natural frequencies at the critical rotational speeds that may occur there, see Figs. 4.24b, 4.27, and 6.31a.

The equations of motion for the minimal model in Fig. 4.21a result as a special case from (4.113) for $n = 2$ or as an extension of (4.23):

$$J_1 \ddot{\varphi}_1 + b_T(\dot{\varphi}_1 - \dot{\varphi}_2) + c_T(\varphi_1 - \varphi_2) = M_1(t) \tag{4.118}$$
$$J_2 \ddot{\varphi}_2 - b_T(\dot{\varphi}_1 - \dot{\varphi}_2) - c_T(\varphi_1 - \varphi_2) = M_2(t). \tag{4.119}$$

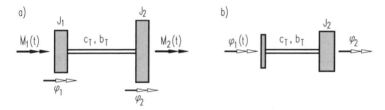

Fig. 4.21 Minimal models for forced vibrations; **a)** Two excitation moments, **b)** Motion excitation

If no driving or braking torques $M_k(t)$ are given, but instead a forced *motion excitation* $\varphi_1(t)$ is applied, the magnitude of J_1 is not relevant for the motion of $\varphi_2(t)$, and the calculation model of Fig. 4.21b is obtained, the equation of motion of which is derived as a special case from (4.119):

$$J_2 \ddot{\varphi}_2 + b_T \dot{\varphi}_2 + c_T \varphi_2 = b_T \dot{\varphi}_1(t) + c_T \varphi_1(t). \tag{4.120}$$

(4.118) can be used to determine the moment $M_1(t)$ that maintains the predefined motion excitation. It is sometimes helpful to use the relative angle $q = \varphi_1 - \varphi_2$ as a coordinate. From (4.120), one obtains the following for this angle:

$$\ddot{q} + \frac{b_T}{J_2}\dot{q} + \frac{c_T}{J_2}q = \ddot{\varphi}_1(t). \tag{4.121}$$

One can obtain a single equation of motion from (4.118) and (4.119) for the internal ("elastic") moment $M = c_T(\varphi_1 - \varphi_2)$ in the twisted shaft, see also (4.33). If one divides (4.118) by J_1 and (4.119) by J_2, the difference of the equations multiplied by c_T results in the following differential equation for the internal moment:

$$\ddot{M} + 2D\omega_0 \dot{M} + \omega_0^2 M = c_T \left(\frac{M_1(t)}{J_1} - \frac{M_2(t)}{J_2} \right) \quad (4.122)$$

where, in accordance with (1.88) and (1.92),

$$2D\omega_0 = 2\delta = b_T \frac{\omega_0^2}{c_T}; \quad \omega_0^2 = \frac{c_T}{J_1} + \frac{c_T}{J_2}, \quad (4.123)$$

see the special case (4.33). (4.120) and (4.122) basically have the same form as (4.121). If the right side of (4.121) is a periodic function, the periodically varying driving angle can be written as a Fourier series:

$$\varphi_1(t) = \sum_{m=1}^{\infty} (a_m \cos m\Omega t + b_m \sin m\Omega t). \quad (4.124)$$

If one inserts the second derivative into (4.121), the differential equation

$$\ddot{q} + 2D\omega_0 \dot{q} + \omega_0^2 q = \Omega^2 \sum_{m=1}^{\infty} m^2 (a_m \cos m\Omega t + b_m \sin m\Omega t) \quad (4.125)$$

is obtained. Its steady-state solution is

$$q(t) = \sum_{m=1}^{\infty} (A_m \cos m\Omega t + B_m \sin m\Omega t) = \sum_{m=1}^{\infty} C_m \cos(m\Omega t + \beta_m) \quad (4.126)$$

with the Fourier coefficients

$$\begin{aligned} A_m &= \frac{(1 - m^2\eta^2)a_m - 2Dm\eta b_m}{(1 - m^2\eta^2)^2 + (2Dm\eta)^2} m^2\eta^2; \\ B_m &= \frac{2Dm\eta a_m + (1 - m^2\eta^2)b_m}{(1 - m^2\eta^2)^2 + (2Dm\eta)^2} m^2\eta^2. \end{aligned} \quad (4.127)$$

The frequency ratio $\eta = \Omega/\omega_0$ was introduced here. The following applies for the amplitude of the mth harmonic:

$$C_m = \sqrt{A_m^2 + B_m^2} = \frac{m^2\eta^2 \sqrt{a_m^2 + b_m^2}}{\sqrt{(1 - m^2\eta^2)^2 + (2Dm\eta)^2}}. \quad (4.128)$$

If the circular frequency of the excitation $m\Omega$ of the mth harmonic coincides with the natural circular frequency ω_0, $m\eta = 1$ and the following estimate applies to the resonance amplitude:

$$|q|_{\max} \geq C_{m\,\max} \approx C_m(\eta = 1/m) = \frac{\sqrt{a_m^2 + b_m^2}}{2D}. \quad (4.129)$$

4.3.2 Examples

4.3.2.1 Motorbike Engine

Figure 4.22 shows the power unit of a motorbike engine and its calculation model. The engine is a two-stroke engine. Periodic excitation with four harmonics occurs:

$$M_2 = \sum_{m=1}^{4} \hat{M}_{2m} \sin(m\Omega t + \alpha_m). \quad (4.130)$$

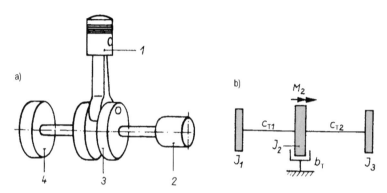

Fig. 4.22 Motorbike engine; **a)** Schematic of the power unit, *1* Piston, *2* Generator armature, *3* Crank mechanism, *4* Clutch, **b)** Vibration model

Based on the given mass and spring parameters

$$\begin{aligned} J_1 &= 1.027 \cdot 10^{-2} \text{ kg} \cdot \text{m}^2; & c_{T1} &= 25.9 \cdot 10^3 \text{ N} \cdot \text{m} \\ J_2 &= 0.835 \cdot 10^{-2} \text{ kg} \cdot \text{m}^2; & c_{T2} &= 20.6 \cdot 10^3 \text{ N} \cdot \text{m} \\ J_3 &= 0.079 \cdot 10^{-2} \text{ kg} \cdot \text{m}^2, & & \end{aligned} \quad (4.131)$$

the natural frequencies of the undamped system are derived from (4.12):

$$f_1 = 0; \quad f_2 = 366 \text{ Hz}; \quad f_3 = 853 \text{ Hz}. \quad (4.132)$$

With the help of known software, the steady-state periodic vibrations are calculated from (4.113) using the damping coefficient $b_T = 6$ N $\cdot$ m $\cdot$ s and the moment amplitudes

$$\begin{aligned} \hat{M}_{21} &= 53.8 \text{ N} \cdot \text{m}; & \hat{M}_{22} &= 43.6 \text{ N} \cdot \text{m} \\ \hat{M}_{23} &= 24.5 \text{ N} \cdot \text{m}; & \hat{M}_{24} &= 20.2 \text{ N} \cdot \text{m}. \end{aligned} \quad (4.133)$$

Figure 4.23a shows the second and third mode shapes of the undamped system. Figure 4.23b contains the resonance curve for the resulting moment in the shaft *1* with its components.

According to the general statements in Sect. 4.3.1, Fig. 4.23b shows several resonance points. One can calculate the resonance speeds from (4.117) using the natural frequencies given in (4.132). The resonance speeds in the speed range from $n = 5000$ to $15\,000$ rpm (Fig. 4.23b) are at $n_{im} = 60 f_i / m$, see (4.117):

$$n_{22} = 10\,990 \text{ rpm}; \qquad n_{23} = 7320 \text{ rpm}; \qquad n_{14} = 5490 \text{ rpm}. \qquad (4.134)$$

These peaks are clearly visible. The calculation also provides a smaller resonance peak at $n_{34} = 12\,800$ rpm. It corresponds to the resonance of the fourth harmonic with the third natural frequency.

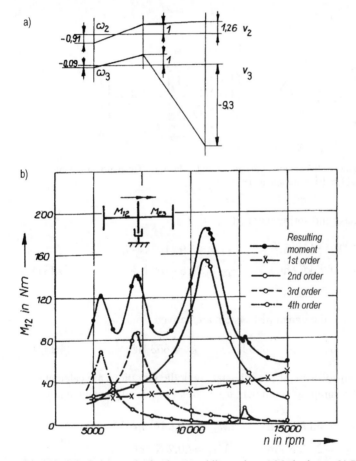

Fig. 4.23 Calculation results for the motorbike engine; **a)** Mode shapes, **b)** Resonance curve

4.3 Forced Vibrations of Discrete Torsional Oscillators

Figure 4.24a shows the *measured* resonance curve for this example. It is evident that there are many resonance speeds within the operating range. Since the excitation amplitudes of each order decline for increasing orders, the higher-order resonances result in smaller deflections. It is interesting, however, that they can still be measured without a problem. It is the engineer's job to design the crankshaft or to influence the vibration system in such a way that no damage occurs in the resonance range.

Figure 4.24b shows the Campbell diagram. One can see that the first natural frequency can result in 8 resonances (4th to 11th order) and the second natural frequency in 17 resonances (9th to 25th order) within the speed range. It cannot be avoided that resonances will occur in the operating speed range.

The resonance amplitudes can be limited by materials with strong damping, see (4.129). All resonance peaks in Fig. 4.24a can be explained using the Campbell diagram from Fig. 4.24b.

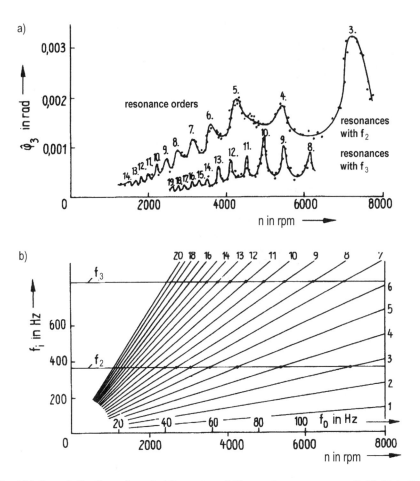

Fig. 4.24 Forced vibrations of a motorbike engine; **a)** Measured resonance curve divided into the two modal components, **b)** Campbell diagram with two natural frequencies and 20 harmonics

4.3.2.2 Vehicle Drive with Dual-Mass Flywheel

Dual-mass flywheels are sub-assemblies that are used in many passenger car drive trains. This complicated "flywheel" consists of a primary-side part that is connected to the crankshaft of the internal combustion engine, and a secondary-side part that provides the friction surface for the clutch disk. The primary side is connected to the secondary side by a set of springs that implements a progressive spring stiffness, that is, it shows a soft response for small deflections and a rigid response for great deflections, see Sect. 7.2. The dual-mass flywheel is operated supercritically and decouples the secondary side from the high-frequency vibrations of the primary side.

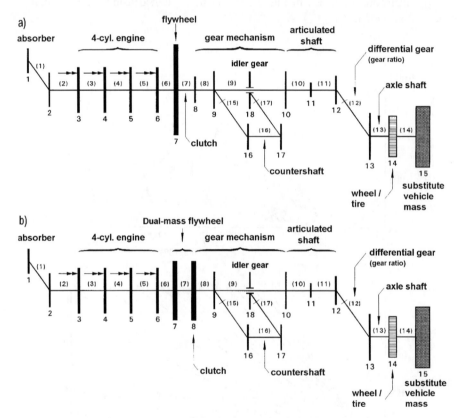

Fig. 4.25 Vehicle drive train with countershaft, loaded by periodic motor torques, see Fig. 4.13 (Source: ARLA); **a)** Version without dual-mass flywheel, **b)** Version with dual-mass flywheel

In analogy to the example presented in Sect. 4.2.4.3, the dynamic behavior of a torque-split vehicle drive will be studied, taking into account a gear step with backlash (in the unloaded drive branch). The torsional oscillator shown in Fig. 4.25 can be simulated despite the complexity of the system (branched structure, backlash

4.3 Forced Vibrations of Discrete Torsional Oscillators

in both gear steps) using advanced simulation programs [21]. This calculation model matches the one in Fig. 4.13, which was extended by periodic excitation moments at the four cylinders and by the countershaft with backlash.

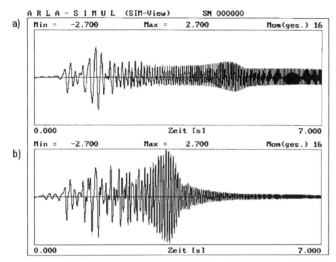

Fig. 4.26 Calculated time functions of the moment M_{16} (countershaft) according to Fig. 4.25 (Source: ARLA); **a)** without dual-mass flywheel, **b)** with dual-mass flywheel

The dynamic behavior in the time domain was simulated in the two versions without or with a dual-mass flywheel using the ARLA-SIMUL simulation software. Since the rotational speed increases monotonously in proportion with the time, time dependence also means rotational speed dependence. The results (time functions and amplitudes in the frequency domain) can be seen in Fig. 4.26 and Fig. 4.27.

A start-up process from zero speed to 1200 rpm in 7 seconds was simulated. Figure 4.26 shows that the vehicle drive train is capable of resonance in the speed range below the idle speed (especially the system *with dual-mass flywheel*). The dual-mass flywheel becomes advantageous only above the idle speed (minimization and "quasi-decoupling" of the forced engine excitation, no resonance). The system without dual-mass flywheel clearly shows an elevation in the moment curve due to the forced engine excitation at the base frequency of about 6 Hz. The system with the dual-mass flywheel also shows a resonance peak at the second natural frequency of 16 Hz, see Fig. 4.14 and Fig. 4.26b. In addition, the plots show another phenomenon: The dual-mass flywheel version is even more resonance-prone than the conventional drive train in the lower speed range (typically always below the idle speed); however, the amplitudes are getting considerably smaller in the higher speed ranges (i. e. a well tuned system from an acoustics point of view).

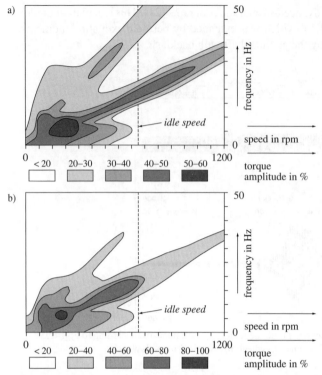

Fig. 4.27 Amplitude chart of the Campbell diagram for moment M_{16} in the vehicle model according to Fig. 4.25; Calculation results using ARLA-SIMUL; **a)** without dual-mass flywheel, **b)** with dual-mass flywheel

4.3.2.3 Stepping Mechanism with HS Curve Profile

The technological requirements are kept exactly within the manufacturing accuracy for conventional motion programs of cam mechanisms when the machine runs slowly. Real systems, however, comprise elasticity and backlash so that the actual motion of the driven link deviates from the desired motion program at higher speeds due to interfering vibrations.

The typical position function (VDI 2143)[36] of a stepping mechanism is composed of a perfectly straight dwell and a "Bestehorn sinoid" for each phase of the motion. This kinematic excitation corresponds to a Fourier series with an infinite number of summands (4.138). HS curve profiles do not use such segment-wise position functions. Instead, Fourier series with a minimum number of summands are applied [4].

The dynamic behavior for the calculation model of a stepping mechanism according to Fig. 4.28b is examined on the one hand for a motion program in the form of a "Bestehorn sinoid" and for an HS profile with $k = 4$ harmonics on the

4.3 Forced Vibrations of Discrete Torsional Oscillators

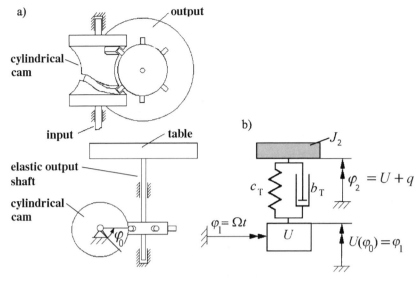

Fig. 4.28 Cam stepping mechanism; a) Schematic, b) Calculation model

other hand. To be determined is the speed up to which operation is possible if the permissible tolerance range is utilized.

The following parameter values are given:

$$\boldsymbol{x} = \begin{pmatrix} U_S \\ \varphi^* \\ n \\ J_2 \\ c_T \\ D \\ \Delta U \end{pmatrix} = \begin{pmatrix} 30° \\ 200° \\ 300 \text{ rpm} \\ 0.22 \text{ kg} \cdot \text{m}^2 \\ 5\,400 \text{ N} \cdot \text{m} \\ 0.02 \\ 0.3° \end{pmatrix} \begin{array}{l} \text{Pivoting angle, see Fig. 4.29} \\ \text{Input angle of rotation per step, see Fig. 4.29} \\ \text{Operating speed} \\ \text{Moment of inertia of the output link} \\ \text{Torsional spring constant of the output shaft} \\ \text{Damping ratio of the torsional oscillator} \\ \text{Permissible error of dwell ("tolerance range")} \end{array}$$
(4.135)

The equation of motion of the oscillator with motion excitation according to Fig. 4.28 results from the moment balance at the rotating output mass J_2. It is known from (4.120):

$$J_2 \ddot{\varphi}_2 + b_T \cdot \left(\dot{\varphi}_2 - \dot{U} \right) + c_T \cdot (\varphi_2 - U) = 0. \tag{4.136}$$

If the relative angle of the output shaft is used as the generalized coordinate

$$q = \varphi_2 - U; \qquad \ddot{q} = \ddot{\varphi}_2 - \ddot{U} = \ddot{\varphi}_2 - \Omega^2 U''; \qquad ()' = \frac{\text{d}()}{\text{d}\varphi}, \tag{4.137}$$

the equation of motion is obtained.

The zeroth-order position function of a stepping mechanism can be interpreted as superposition of a uniform motion and a periodic motion. The velocity and acceleration functions are then periodic functions of the input angle φ_0, see Fig. 4.29.

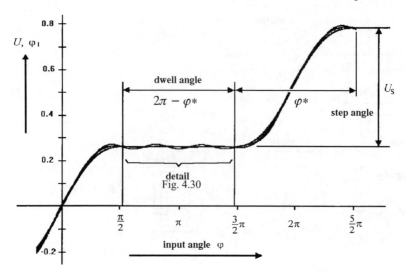

Fig. 4.29 Motion program of a stepping mechanism

The following applies to the motion programs:

$$U(\varphi_0) = \frac{U_S}{2\pi}\varphi_0 + \sum_{k=1}^{\infty} b_k \sin k\varphi_0. \qquad (4.138)$$

The cosine portion is missing in (4.138) since the periodic portion is an odd function, see Fig. 4.29. The equation of motion (4.136) thus looks as follows, see (4.123)::

$$\ddot{q} + 2D\omega_0\dot{q} + \omega_0^2 q = -\Omega^2 U'' = \Omega^2 \sum_{k=1}^{\infty} k^2 b_k \sin k\Omega t. \qquad (4.139)$$

If one compares the natural frequency $f = (1/2\pi)\sqrt{c_T(1-D^2)/J_2} = 24.9$ Hz to the fundamental excitation frequency $f_{\text{err}} = \Omega/(2\pi) = n/60 = 5$ Hz (operating speed), a frequency ratio of $\eta \approx 0.2$ is obtained. This means that there is a risk of resonance with the fifth harmonic of the position function. The Fourier coefficients b_k of the position function depend on the pivoting angle and the dwell width:

$$b_k = \frac{8\pi U_S}{\varphi^{*3}\left[\left(\dfrac{2\pi}{\varphi^*}\right)^2 - k^2\right]k^2} \cdot \sin\frac{k\varphi^*}{2} \qquad (4.140)$$

It follows for $k = 1, \ldots, 6$ with $\varphi^* = 200° \mathrel{\hat{=}} 10\pi/9$:

$$\begin{aligned}
b_1 &= 0.259\,789 \cdot U_S; & b_4 &= -0.001\,860 \cdot U_S \\
b_2 &= 0.066\,481 \cdot U_S; & b_5 &= -0.000\,698 \cdot U_S \\
b_3 &= 0.009\,872 \cdot U_S; & b_6 &= 0.000\,434 \cdot U_S.
\end{aligned} \qquad (4.141)$$

4.3 Forced Vibrations of Discrete Torsional Oscillators

Resonance peaks for the relative angle occur according to (4.129). Figure 4.35 shows the curve of the stepping motion for two different motion programs.

It is obvious that the dwell is not maintained exactly using an HS profile. This is permissible if the deviations stay within the tolerance range ($\Delta U = \pm 0,01 \cdot U_\mathrm{S}$).

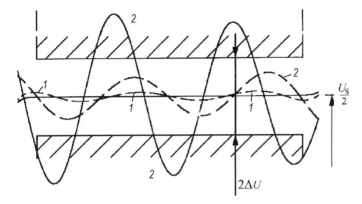

Fig. 4.30 Curve of the stepping motion; section from Fig. 4.29; (solid line: Bestehorn sinoid, dashed line: HS profile); curves 1: $\eta = 0$, curves 2: $\eta = 0,5$

The Fourier coefficients that determine the "genuine" HS profile are different from (4.141):
$$\begin{aligned} b_1 &= 0.249\,04 \cdot U_\mathrm{S}; & b_2 &= 0.054\,63 \cdot U_\mathrm{S} \\ b_3 &= 0.006\,88 \cdot U_\mathrm{S} & b_k &= 0 \quad \text{for} \quad k \geq 4. \end{aligned} \qquad (4.142)$$

Figure 4.30 shows the motion curves that occur for the available motion programs at the output link in the dwell area, as compared to the motion programs themselves. It is evident that strong vibrations are superposed onto the Bestehorn sinoid of the output motion while there are considerably smaller deviations at the HS profile. Figure 4.31 shows the maximum relative error of dwell as a function of the frequency ratio, which is proportional to the machine speed. The resonance points at
$$\eta = \frac{1}{k}, \qquad k = 4, 5, 6, 7, 8 \qquad (4.143)$$
are clearly visible for the Bestehorn sinoid and for the Bestehorn sinoid "truncated" above $k = 5$.

The magnitude of these resonance peaks is derived from (4.129) and exceeds the allowable value as early as from the fifth or sixth harmonic. The resonance with the fourth harmonic occurs for both motion programs and is of the same size for the "truncated" Bestehorn sinoid as for the normal Bestehorn sinoid.

The maximum obtainable frequency ratio can be increased by a "genuine" HS profile with $K = 3$ harmonics since it does not have any resonance point for $\eta = 1/4$, see curve 3 in Fig. 4.31.

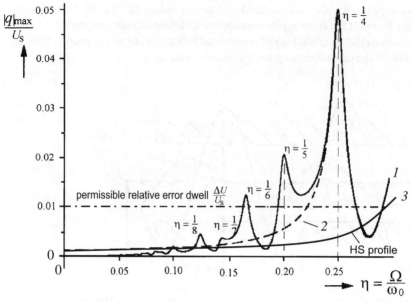

Fig. 4.31 Maximum error of dwell as a function of speed; Curve 1: Bestehorn sinoid, see (4.141), Curve 2: "truncated" Bestehorn sinoid, see (4.141) with $b_5 = b_6 = 0$, Curve 3: HS profile, see (4.142)

The strong vibrations caused by resonance are not present in the lower speed range for the HS profile. In addition to the positive effect that this has with regard to noise emission and wear and tear, it is therefore possible to increase the maximum achievable speed by about 80 % without violating the tolerance limits for the output motion.

Based on experience, the use of HS profiles for cam mechanisms allows vibration-free operation up to about *1.3 to 1.6 times the frequency ratio*, as compared to conventional cam profiles. In practice, this means an correspondingly higher operating speed without the occurrence of disturbing vibrations at the output link.

4.3.3 Transient Excitation

4.3.3.1 Multiple Torque Jumps

Transient excitations are non-periodic or short-time excitations that occur during transition processes. Such excitations can be found, for example, in starting, braking, accelerating or decelerating processes. In some drives or brakes, the maximum torques are not applied instantaneously, but rather in multiple steps. The dynamic loads in the vibration system depend on the magnitude and time sequence of such torque jumps.

4.3 Forced Vibrations of Discrete Torsional Oscillators

A torque jump is the harshest load case because the load jumps to its maximum value "in zero time". The other discussions in Sect. 4.3.3.2 will show the conditions under which this extremely high load case can be a useful approximation, even for loads applied over a finite time period.

The torsional moments that emerge after several torque jumps in the input shaft will be calculated for the torsional oscillator shown in Fig. 4.21a. The results can be transferred to the systems outlined in Table 4.2 and to any linear oscillator with multiple degrees of freedom if the torsional oscillator is viewed as a modal oscillator, see Chap. 6. The following considerations can therefore also be applied to the load oscillation of a crane, a stepping motor, an asynchronous motor that is controlled via several stages, multiple-step brakes and similar cases, which will not be discussed in detail here, see [4].

The internal moment for an undamped oscillator ($b_T = 0$, $D = 0$) with a sudden torque jump $M_{an} = M_{10}$ acting on the disk *1* is to be determined. The solution of (4.118) and (4.119) for the initial conditions

$$t = t_0 = 0: \quad \varphi_1(0) = \varphi_2(0) = 0; \quad \dot{\varphi}_1(0) = \dot{\varphi}_2(0) = 0 \qquad (4.144)$$

is as follows in the range $0 \leq t \leq t_1$:

$$\varphi_1(t) = \frac{M_{10}}{J_1 + J_2} \cdot \frac{t^2}{2} + \frac{M_{10} J_2 (1 - \cos \omega t)}{(J_1 + J_2)\omega^2 J_1} \qquad (4.145)$$

$$\varphi_2(t) = \frac{M_{10}}{J_1 + J_2} \cdot \frac{t^2}{2} - \frac{M_{10}(1 - \cos \omega t)}{(J_1 + J_2)\omega^2}. \qquad (4.146)$$

The torsional moment results from the solution of the equation of motion known from (4.122) with $D = 0$ and $M_2 = 0$

$$\ddot{M} + \omega^2 M = \frac{c_T M_{an}}{J_1} \qquad (4.147)$$

in the interval $0 \leq t \leq t_1$:

$$M(t) = M^{(1)}(t) = M_{10} \frac{J_2}{J_1 + J_2}(1 - \cos \omega t). \qquad (4.148)$$

This result is contained as a special case in (4.40). The exponent (1) indicates that this is the moment during the first interval (after the first jump at $t_0 = 0$). If the moment M_{11} begins to act on disk *1* at the time $t = t_1$, an additional moment is excited by the moment difference $(M_{11} - M_{10})$, and the following applies for $t > t_1$:

$$M^{(2)}(t) = \frac{J_2}{J_1 + J_2}[M_{10}(1 - \cos \omega t) + (M_{11} - M_{10})(1 - \cos \omega(t - t_1))]. \qquad (4.149)$$

A second moment function, which starts at the time t_1, is thus superimposed to the first moment function.

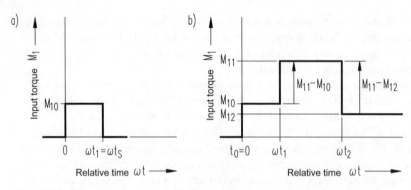

Fig. 4.32 Sequence of torque jumps; **a)** Rectangular impulse, **b)** three torque jumps

If the input torque jumps back to the value zero at the time $t_1 = t_s$, a so-called rectangular impulse occurs, see Fig. 4.32a. It is discussed in detail in Sect. 6.5.3 for oscillators with multiple degrees of freedom. Here, only the result will be provided, which follows from (4.149) if $M_{11} = 0$ is set. After transforming the expression using trigonometric identities, one obtains:

$$M = \frac{2M_{10}J_2}{J_1 + J_2} \cdot \sin\frac{\omega t_s}{2} \sin\left[\omega\left(t - \frac{t_s}{2}\right)\right]; \qquad t > t_s. \tag{4.150}$$

If the ratio t_s/T (with $T = 2\pi/\omega$) is introduced, the following expression is generated:

$$M = \frac{M_{10}J_2}{1+\mu} \cdot 2\sin\left(\pi\frac{t_s}{T}\right) \sin\left[\pi\left(\frac{2t}{T} - \frac{t_s}{T}\right)\right]. \tag{4.151}$$

One can see that there is no torque vibration for all even-numbered ratios of $t_s/T = n$. This, however, applies only to $t_s > T$ in the range $t > t_s$.

Both phases have to be considered when determining the maximum value:

$0 < t < t_s$:

$$\begin{aligned} M_{\max} &= 2\frac{M_{10}J_2}{J_1 + J_2} & \text{for} \quad & t_s \geq \frac{T}{2} \\ M_{\max} &= \frac{M_{10}J_2}{J_1 + J_2}\left[1 - \cos\left(2\pi\tfrac{t_s}{T}\right)\right] & \text{for} \quad & t_s < \frac{T}{2} \end{aligned} \tag{4.152}$$

$t > t_s$:

$$M_{\max} = 2\frac{M_{10}J_2}{J_1 + J_2}\sin\left(\pi\frac{t_s}{T}\right) \approx \frac{2M_{10}J_2}{J_1 + J_2}\left[\frac{\pi t_s}{T} - \frac{1}{6}\left(\frac{\pi t_s}{T}\right)^3 + \ldots\right] \tag{4.153}$$

The function $\sin(\pi t_s/T)$ can be expanded into the specified series for short impulses ($t_s \ll T/2$).

4.3 Forced Vibrations of Discrete Torsional Oscillators

If there are multiple torque jumps at the times t_j ($j = 1, 2, \ldots, J$) another summand is added for each j. After a total of J torque jumps, the torsional moment is

$$M^{(J)}(t) = \frac{J_2}{J_1 + J_2} \sum_{j=1}^{J}(M_{1j-1} - M_{1j-2})[1 - \cos\omega(t - t_{j-1})]; \quad t \geq t_J. \quad (4.154)$$

$M_{1,-1} = 0$ should be set. Since $\cos[\omega(t - t_j)] = \cos\omega t \cos\omega t_j + \sin\omega t \sin\omega t_j$, one can transform (4.154) in such a way that one can separate the mean value M_{J-1} from those torque components that change with the natural frequency:

$$M^{(J)}(t) = \frac{J_2}{J_1 + J_2} \left\{ M_{J-1} - \sum_{j=1}^{J}[(M_{1j-1} - M_{1j-2})\cos\omega(t - t_{j-1})] \right\}$$

$$= \frac{J_2}{J_1 + J_2}[M_{J-1} - \hat{M}^{(J)}\cos\omega(t - t^*)]; \quad t \geq t_J.$$

(4.155)

The formula for t^* has been omitted here. The torque amplitude after the Jth jump is obtained:

$$\hat{M}^{(J)} = \sqrt{[\sum_{j=1}^{J}(M_{1j-1} - M_{1j-2})\cos\omega t_{j-1}]^2 + [\sum_{j=1}^{J}(M_{1j-1} - M_{1j-2})\sin\omega t_{j-1}]^2}.$$

(4.156)

The maximum moment of the so-called *residual vibration* after the Jth torque jump is therefore:

$$M_{\max} = \frac{J_2}{J_1 + J_2}(|M_{J-1}| + \hat{M}^{(J)}). \quad (4.157)$$

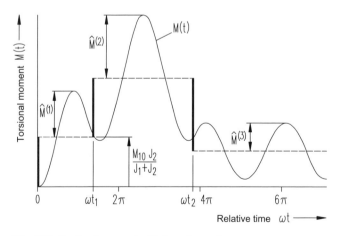

Fig. 4.33 Torsional moment with three torque jumps

Figure 4.33 shows a typical time function for the torsional moment as a result of three torque jumps. When designing drives, one task can be to work out an optimum distribution of the shifting times and torque jumps. The amplitude of the residual vibrations is of particular interest. There are two conditions under which the amplitude equals zero. They follow from (4.156):

$$\sum_{j=1}^{J}(M_{1j-1} - M_{1j-2})\cos\omega t_{j-1} = 0; \quad \sum_{j=1}^{J}(M_{1j-1} - M_{1j-2})\sin\omega t_{j-1} = 0.$$
(4.158)

They can be used to calculate favorable points in time and magnitudes for the torque jumps, see [4] and Problem P4.5.

4.3.3.2 Startup Functions

Loads are not really applied suddenly, as assumed in Sect. 4.3.3.1, but during a finite time that will be called *startup time* t_a here. This subsection is to answer the question of what influence the time function of the input torque $M_1 = M_{\mathrm{an}}(t)$ has on the torsional moment in the shaft if it increases during the startup time from zero to the maximum value M_{10}.

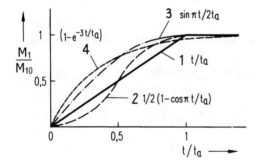

Fig. 4.34 Comparison of startup functions

Figure 4.34 shows four different moment curves and their associated equations. Like in Sect. 4.3.3.1, the torsional oscillator from Fig. 4.21a is considered for the torsional moment of which (4.147) applies. This section only exemplifies the calculation for Case 1 from Fig. 4.34. The solution of (4.147) for the initial conditions $t = 0$: $M = 0$, $\dot{M} = 0$ for the linearly increasing input torque

$$M_1(t) = M_{10}\frac{t}{t_a}; \quad 0 \leqq t \leqq t_a \tag{4.159}$$

is:

$$M(t) = \frac{M_{10}J_2}{J_1 + J_2}\left(\frac{t}{t_a} - \frac{\sin\omega t}{\omega t_a}\right); \quad 0 \leqq t \leqq t_a. \tag{4.160}$$

4.3 Forced Vibrations of Discrete Torsional Oscillators

Starting at the time $t = t_a$, the same function is excited once again (at a time offset of t_a) using the linearly decreasing moment $-M_{10}(t-t_a)/t_a$ so that an excitation at the constant value of M_{10} remains from this superposition:

$$\Delta M(t) = \frac{-M_{10}J_2}{J_1+J_2}\left(\frac{t-t_a}{t_a} - \frac{\sin\omega(t-t_a)}{\omega t_a}\right); \quad t \geq t_a. \quad (4.161)$$

Superposition of these two functions results in the curve

$$\begin{aligned}M(t) &= \frac{M_{10}J_2}{J_1+J_2}\left[1+\frac{\sin\omega(t-t_a)-\sin\omega t}{\omega t_a}\right]\\ &= \frac{M_{10}J_2}{J_1+J_2}\left[1-\frac{2}{\omega t_a}\sin\frac{\omega t_a}{2}\cos\omega(t-t_a/2)\right]; \quad t \geq t_a.\end{aligned} \quad (4.162)$$

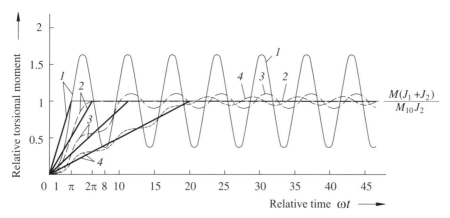

Fig. 4.35 Curve of the torsional moment as a function of the start-up time; Curve *1*: $t_a/T = 0.5$; Curve *2*: $t_a/T = 2$; Curve *3*: $t_a/T = 1.8$; Curve *4*: $t_a/T = 3.2$

Figure 4.35 shows the time function of the torsional moment that is composed of the solutions (4.160) and (4.162). No residual vibration remains for even-numbered ratios t_a/T.

Figure 4.36 shows the maximum values of the residual torsional moment as a function of relative startup time, not only for the case of the piecewise linear startup function calculated here, but rather for all four cases shown in Fig. 4.34. One can see that fast startup processes ($t_a/T \ll 1$) have an effect like a jump:

$$M_{\max} \approx \frac{M_{10}J_2}{J_1+J_2}\left[2-\frac{(\omega t_a)^2}{24}\right]; \quad t_a/T \leq 0.2. \quad (4.163)$$

The amplitude declines fast as the ratio t_a/T increases. There are only minor vibrations for slow startup processes ($t_a/T < 5$), and the results hardly differ for the

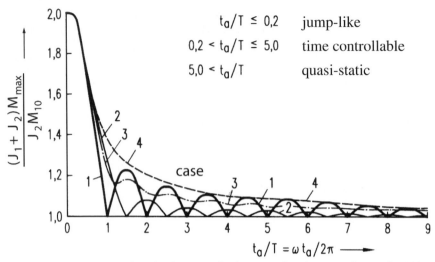

Fig. 4.36 Maximum residual torsional moment for the startup functions according to Fig. 4.34

various startup functions. The following asymptote applies:

$$M_{\max} \approx \frac{M_{10} J_2}{J_1 + J_2}\left(1 + \frac{2}{\omega t_a}\right); \qquad t_a/T \geqq 5. \qquad (4.164)$$

The realization that the form of the startup function only influences the *maximum torsional moment* in the approximate range from $0.2 \leq t_a/T \leq 5$ has practical relevance. The startup function has little influence in the other ranges, so it is futile to try to diminish the dynamic loads using controllers if $t_a/T < 0.2$ (one can calculate approximately using the step function).

4.3.4 Problems P4.4 to P4.6

P4.4 Moving a Resonance Point

Intense noises occurred at the operating speed in an already finished drive system, so that the assumption had to be that a critical speed with the gear meshing frequency had been reached and that damage would occur during permanent operation. It turned out that the tenth mode shape had an antinode near a gear pair. Design measures are to be taken on the existing object to move the resonance point. One option is the use of gears with another gear module, however, the gear ratio of the gear step, the center distance, the rotating masses and the torsional stiffnesses of the shafts can be changed only slightly for design reasons.

4.3 Forced Vibrations of Discrete Torsional Oscillators

Given:

Tooth numbers of existing gears	$z_1 = 19;\ z_2 = 43$
Input shaft speed	$n_1 = 1450$ rpm
Tenth natural frequency of the drive system	$f_{10} = 450$ Hz
Module of mating gears	$m = 6$
Permissible relative gear ratio change	$\pm 2.5\ \%$
Maximum achievable change in the natural frequency	$\pm 5\ \%$

Find:

1. Frequency ratio of the observed critical speed
2. Variants of tooth numbers for the two gears
3. Achievable change in the critical frequency ratio

P4.5 Load During a Three-Step Braking Process

A drive system, the model of which corresponds to the unconstrained torsional oscillator of Fig. 4.3b, moves without vibrations at angular velocity Ω_0. It is to be braked down to an angular velocity Ω_2 by three torque jumps within the time t_2, see Fig. 4.32b. At the beginning ($t = 0$), a constant braking torque M_{10} is acting at the disk *1*, and after a time t_1 the torque M_{11} is acting. After time t_2, the moment M_{12} acts with which the drive system is powered after the braking process. No residual vibrations should occur in the shaft after time t_2.

Given:

Angular velocity at the start	Ω_0
Angular velocity at the end	Ω_2
Moments of inertia	$J_1,\ J_2$
Natural circular frequency	$\omega = \sqrt{c_T(J_1 + J_2)/(J_1 J_2)}$
Brake time	t_2
Final torque	M_{12}

Find:

1. Condition that all moments M_{10}, M_{11} and the time t_1 have to satisfy
2. Angular velocities of disks J_1 and J_2 in the interval $0 \le t \le t_1$
3. Conditions for M_{10}, M_{11} and ωt_1 so that the residual vibrations vanish
4. M_{10}, M_{11} and ωt_1 for $M_{12} = 58.6$ N; $\omega t_2 = 9\pi/4$;
$(J_1 + J_2)(\Omega_2 - \Omega_0) = 400\pi/\omega$ N · m · s.

P4.6 Interpretation of Measurement Results (6-Cylinder Four-Stroke Engine)

Figure 4.37 shows the maximum values of the torsional moment in the shaft of a 6-cylinder four-stroke diesel engine, where resonances occurred with the first and second non-zero natural frequencies.

Given:
Natural frequencies $f_2 = 213$ Hz, $f_3 = 310$ Hz
Resonance curve according to Fig. 4.37 in the speed range $n = 1800$ to 4500 rpm.

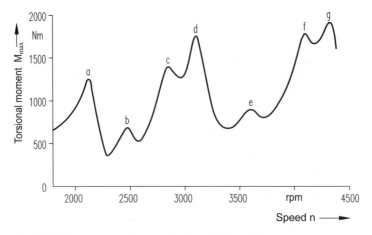

Fig. 4.37 Maximum moment as a function of the speed (resonance curve)

Find:

1. What resonance orders occur?
2. What are other critical speeds in the lower speed range?

4.3.5 Solutions S4.4 to S4.6

S4.4 The gear ratio of the gear step is

$$i = \frac{z_1}{z_2} = \frac{19}{43} = 0.441\,86. \tag{4.165}$$

The gear meshing frequency is

$$f_e = \frac{n_1 z_1}{60} = 1450\frac{19}{60} = 459.2 \text{ Hz}. \tag{4.166}$$

The frequency ratio at operating speed is therefore:

$$\eta_{10} = \frac{f_e}{f_{10}} = \frac{459.2}{450} = 1.100. \tag{4.167}$$

One can change the gear meshing frequency by achieving the required gear ratio with different tooth numbers of the two gears.

A secondary condition applies that the center distance s of the two shafts that carry the meshing gears should also be approximately retained (adjustment allowed):

$$s = \frac{1}{2}(z_1 + z_2)m = \frac{1}{2}(19 + 43)6 \text{ mm} = 186 \text{ mm}. \tag{4.168}$$

The existing gear ratio and center distance must be approximately retained for the new module m^* and the modified tooth numbers z_1^* and z_2^*. The following conditions apply:

4.3 Forced Vibrations of Discrete Torsional Oscillators

$$\frac{z_1^*}{z_2^*} \approx 0.441\,86 = i; \qquad \frac{1}{2}m^*(z_1^* + z_2^*) \approx \frac{1}{2}m(z_1 + z_2) = s. \tag{4.169}$$

This follows from the two "approximate" equations:

$$z_1^* \approx \frac{m(z_1 + z_2)i}{m^*(1+i)}; \qquad z_2^* \approx \frac{m(z_1 + z_2)}{m^*(1+i)}. \tag{4.170}$$

Tooth numbers $z_1^* \approx 22.8$ and $z_2^* \approx 51.6$ result for the module $m^* = 5$. The numbers are $z_1^* \approx 16.3$ and $z_2^* \approx 36.9$ for the module $m^* = 7$. Only integers can be selected, of course

Module $m^* = 5$: $\quad z_1^* = 23;\quad z_2^* = 51;\quad i^* = \dfrac{23}{51} = 0.450\,98;\quad s^* = 185$ mm

Module $m^{**} = 7$: $\quad z_1^* = 16;\quad z_2^* = 37;\quad i^* = \dfrac{16}{37} = 0.432\,43;\quad s^* = 185.5$ mm.

The center distance s can remain the same if the gear profile is shifted. The gear ratios change by less than 2.5 %. If one selected $z_1 = 17$ teeth for the smaller gear (since prime numbers are advantageous as tooth numbers) for module $m^* = 7$ the gear ratio change $(17/37 = 0.459\,45)$ would be unacceptably large (4 %).

With the *smaller module* ($m^* = 5$), the gear meshing frequency would be increased. In addition, it would be recommended to *reduce the natural frequency*, (e.g. by a larger rotating mass and/or a softer torsional spring constant). (According to the problem statement, a maximum of 5 % can be achieved by design changes.) The frequency ratio when implementing variant 1 is:

$$\underline{\underline{\eta_{10}^*}} = \frac{n_1 z_1^*}{0.95 \cdot 60 f_{10}} = \frac{1450 \cdot 23}{0.95 \cdot 60 \cdot 450} \underline{\underline{= 1.300}}. \tag{4.171}$$

The resonance point would be avoided, but operation above the resonance point requires passing through the resonance speed.

The gear meshing frequency would be decreased with the *larger module* ($m^* = 7$). According to the problem statement, one could *increase the natural frequency* to $1.05 f_{10}$, (e.g. by a smaller rotating mass and/or a harder torsional spring constant). The frequency ratio would then be

$$\underline{\underline{\eta_{10}^*}} = \frac{n_1 z_1^*}{1.05 \cdot 60 f_{10}} = \frac{1450 \cdot 16}{1.05 \cdot 60 \cdot 450} \underline{\underline{= 0.818}} \tag{4.172}$$

if variant 2 is implemented. The critical speed is above the operating speed and would therefore not be reached.

The final decision in favor of the appropriate variant only becomes possible if strength and service life of the gearing have been proven for the modified module (profile shift).

S4.5 A correlation with the total braking time t_2 can be found using the principle of conservation of angular momentum, see Sect. 2.3. The change in angular momentum equals the time integral of the applied moments. Therefore:

$$(J_1 + J_2)(\Omega_2 - \Omega_0) = M_{10} t_1 + M_{11}(t_2 - t_1) = \int_0^{t_2} M(t)\,dt = \Delta D. \tag{4.173}$$

For the elastic system according to Fig. 4.3, the initial conditions for the equations of motion known from (4.118) and (4.119) are as follows:

$$t = 0: \qquad \varphi_1(0) = \varphi_2(0) = 0; \qquad \dot{\varphi}_1(0) = \dot{\varphi}_2(0) = \Omega_0. \tag{4.174}$$

After determining the integration constants, one obtains the following solutions for the two angles in the interval $0 \leq t \leq t_1$ according to (4.145) and (4.146):

$$\varphi_1(t) = \Omega_0 t - \frac{M_{10}}{J_1 + J_2}\left[\frac{t^2}{2} + \frac{J_2(1-\cos\omega t)}{J_1 \omega^2}\right] \qquad (4.175)$$

$$\varphi_2(t) = \Omega_0 t - \frac{M_{10}}{J_1 + J_2}\left[\frac{t^2}{2} - \frac{1-\cos\omega t}{\omega^2}\right]. \qquad (4.176)$$

The resulting angular velocities are:

$$\dot\varphi_1(t) = \Omega_0 - \frac{M_{10}}{J_1 + J_2}\left[t + \frac{J_2 \sin\omega t}{J_1 \omega}\right] \qquad (4.177)$$

$$\dot\varphi_2(t) = \Omega_0 - \frac{M_{10}}{J_1 + J_2}\left[t - \frac{\sin\omega t}{\omega}\right]. \qquad (4.178)$$

The first two terms in these equations correspond to the rigid-body motion, while the third term arises due to the natural vibrations. Using a conversion, the term could also be written in such a way that the torsional spring constant appears in the denominator. A vibration where the two rotating masses move in opposite directions as for the second mode shape occurs around the rigid-body motion, see Fig. 4.3b. Both disks lose velocity.

In the first stage, a torsional moment, is generated in the shaft that is known from (4.148). The maximum moment in the first interval reaches double the kinetostatic value only if the time is $t_1 > T/2 = \pi/\omega$, that is, after at least half an oscillation:

$$M_{\max} = \frac{2 M_{10} J_2}{J_1 + J_2}. \qquad (4.179)$$

The moments at the other times result from (4.154). The problem can be solved using the equations given in 4.3.3. From (4.158), it follows for $J = 3$ jumps and initial time $t_0 = 0$:

$$M_{10} + (M_{11} - M_{10})\cos\omega t_1 + (M_{12} - M_{11})\cos\omega t_2 = 0 \qquad (4.180)$$

$$(M_{11} - M_{10})\sin\omega t_1 + (M_{12} - M_{11})\sin\omega t_2 = 0. \qquad (4.181)$$

Including (4.173), there are three equations for the three unknown quantities M_{10}, M_{11} and t_1. There are infinitely many solutions to these transcendental equations. The solutions for the "jump times" depend on the magnitude of the moments if these are predefined. Because of the following trigonometric identity,

$$(\sin\omega t_1)^2 + (\cos\omega t_1)^2 = 1 \qquad (4.182)$$

one can first obtain a single equation for $\sin\omega t_2$ and $\cos\omega t_2$ from (4.180) and (4.181):

$$[(M_{12}-M_{11})\sin\omega t_2]^2 + [(M_{12}-M_{11})\cos\omega t_2 - M_{10}]^2 = (M_{11}-M_{10})^2. \qquad (4.183)$$

If one eliminates $\cos\omega t_2$ from it, a quadratic equation for $\sin\omega t_2$ remains. This approach will not be discussed further, just a result for one numerical example will be given: One solution for the final moment ($M_{12} = 58.6$ N · m), the braking time ($\omega t_2 = 9\pi/4$), and the change in angular momentum $\Delta D = (J_1 + J_2)(\Omega_2 - \Omega_0) = 400\pi/\omega$ N · m · s is

$$\underline{M_{10} = 100 \text{ N} \cdot \text{m}}; \qquad \underline{M_{11} = 200 \text{ N} \cdot \text{m}}; \qquad \underline{\omega t_1 = \frac{\pi}{2}}. \qquad (4.184)$$

Check the validity by insertion into (4.173), (4.180), and (4.181).

S4.6 The speed $n_2 = 12\,780$ rpm corresponds to the fundamental frequency $f_2 = 213$ Hz, and the speed $n_3 = 18\,600$ rpm corresponds to the natural frequency $f_3 = 310$ Hz. The higher-

4.4 Absorbers and Dampers in Drive Systems

order critical speeds correspond to the local maxima of the resonance curve, see Fig. 4.37. They result from (4.117), see the numerical results in Table 4.5.

Table 4.5 Higher-order critical speeds n_{im} in rpm

Order m	3	3.5	4	4.5	5	6	7.5	9
n_2/m	**4260**	**3651**	3195	**2840**	2556	**2130**	1704	1420
n_3/m	6200	5314	4650	**4133**	3720	**3100**	**2480**	**2066**

The numbers in bold print correspond to points a through f in Fig. 4.37. The excitations with resonance orders m =3, 3.5, 4.5, 6, and 9 are dominant. The excitation orders not listed only have small amplitudes. At resonance peak a, the resonance of the 6th harmonic is superposed to the first natural frequency and that of the 9th harmonic is superposed to the second natural frequency. Other sizeable resonance peaks can be expected in the lower speed range at 1704 rpm and 1420 rpm.

4.4 Absorbers and Dampers in Drive Systems

4.4.1 Introduction

The first task when designing a dynamically loaded machine is to eliminate or reduce the causes of vibration excitations. This includes measures such as balancing, reduction of backlash, jumps, discontinuities, etc. If these measures are insufficient for keeping interfering vibrations within permissible limits, there are ways of vibration control using absorbers or dampers. This always requires an additional effort, and since these elements are only effective when there actually are vibrations, it is advisable to first exhaust the primary measures for reducing excitations.

Absorbers and dampers are components that are added to a vibration system to achieve a reduction of the vibration amplitudes or loads by

- Detuning the system through changing parameter values or
- Adding auxiliary partial oscillators ("*absorber*") so that the vibration amplitudes at a certain frequency become minimal at selected points of the system or
- Transforming the vibration energy into thermal energy using dampers.

Practical descriptions are contained in the VDI Guidelines for dampers (VDI 3833, Sheet 1) and absorbers (VDI 3833 Sheet 2) [36].

In drive trains, particularly in vehicle drive trains, frictionally attached auxiliary masses and permanent-slip clutches are frequently used for *vibration damping* since they reduce the noises of the drive train, especially the ones caused by gear rattle. Clutch disks with an integrated torsion damper are commonly used. Pendulums in the centrifugal force field allow the adjustment of the absorber frequency to the engine speed, see Sect. 4.4.5.2.

Various absorber designs are used to reduce vibrations of foundations due to excitation by machines, of bridges due to excitation by traffic loads, of towers and high-voltage power lines due to excitation by the wind and of large machines and structures (especially high-rise buildings) due to excitation by earthquakes. There are also absorbers that are based on the gyroscopic effect and have been suggested, for example, to absorb vibrations in large surface mining equipment.

4.4.2 Design of an Undamped Absorber

Figure 4.38 shows the model used for the discussion. Its free vibrations were discussed in 4.2.1.1. The equations of motion are, see (4.21):

$$J_1\ddot{\varphi}_1 + c_{T1}(\varphi_1 - \varphi_2) = 0 \\ J_2\ddot{\varphi}_2 - c_{T1}(\varphi_1 - \varphi_2) + c_{T2}\varphi_2 = M_2. \quad (4.185)$$

For a harmonic excitation moment $M_2 = \hat{M}_2 \cos \Omega t$, a particular solution of the undamped model of the form $\varphi_k = \hat{\varphi}_k \cos \Omega t$ ($k = 1, 2$) can be assumed.

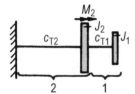

Fig. 4.38 Torsional oscillator used as absorber

The equations (4.185) formally correspond to (3.50) because the systems in Figs. 3.10 and 4.38 are similar.

Both the longitudinal oscillator in Fig. 3.10 and the torsional oscillator in Fig. 4.38 can be described using the dimensionless parameters

$$\mu = \frac{J_1}{J_2} = \frac{m_1}{m_2}; \quad \gamma = \frac{c_{T1}}{c_{T2}} = \frac{c_1}{c_2}; \quad \xi^2 = \frac{\gamma}{\mu}; \quad \eta^2 = \frac{\Omega^2}{c_{T2}/J_2} = \frac{\Omega^2}{c_2/m_2} \quad (4.186)$$

$$V_1 = \frac{\hat{\varphi}_1}{\varphi_{st}} = \frac{\hat{x}_1}{x_{st}}; \quad V_2 = \frac{\hat{\varphi}_2}{\varphi_{st}} = \frac{\hat{x}_2}{x_{st}}; \quad \varphi_{st} = \frac{\hat{M}_2}{c_{T2}}. \quad (4.187)$$

One can express the results obtained in Sect. 3.2.3 using the dimensionless parameters. The result of (3.54) is:

$$V_1 = \frac{\xi^2}{(1+\gamma-\eta^2)(\xi^2-\eta^2)-\gamma\xi^2}; \quad V_2 = \frac{\xi^2-\eta^2}{(1+\gamma-\eta^2)(\xi^2-\eta^2)-\gamma\xi^2}. \quad (4.188)$$

It is now to be determined which parameters would be useful to stabilize an oscillator with one degree of freedom (c_{T2}, J_2, M_2) that was originally operated in resonance by providing an absorber (c_{T1}, J_1) at the previous excitation frequency (Ω). The load in spring 2 vanishes for $\xi = \eta$ because $V_2 = 0$. If the original reso-

4.4 Absorbers and Dampers in Drive Systems

nance frequency matches the absorption frequency,

$$\Omega^2 = \frac{c_{T2}}{J_2} = \frac{c_{T1}}{J_1} = \omega_T^2; \qquad \mu = \gamma \qquad (4.189)$$

has to be valid. To find out what influence the mass ratio μ and the spring ratio γ have on the absorption effect, note that the moment of inertia of the absorber must not be too large for design and economic reasons, i.e. it has to be $J_1 < J_2$ (that is $\mu < 1$). If one expresses the known natural circular frequencies using the dimensionless parameters defined by (4.186), their relative magnitude can be calculated as a function of the mass ratio μ. The result is shown in Fig. 4.39a.

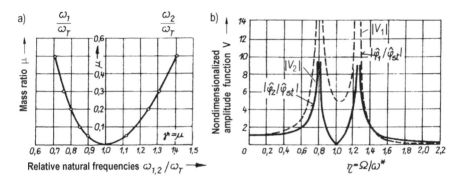

Fig. 4.39 Absorption effect; **a)** Natural frequencies as a function of the mass ratio, **b)** Non-dimensionalized amplitude functions for $\gamma = \mu = 0.2$

One can tell from Fig. 4.39a that when μ gets smaller (and only $\mu \ll 1$ is of engineering interest), the two natural frequencies move closer together. If one plots the relative amplitudes for $\mu = 0.2$ as functions of the frequency ratio (Fig. 4.39b), the following conclusions can be drawn:

A model extension by one degree of freedom is produced by coupling an absorber to it. The resonance frequency under consideration splits into two, which are located slightly above and below the original one. One can only achieve a large distance using a large absorber mass.

The source system remains at rest at the absorption frequency to which the absorber is tuned while the absorber oscillates with a large but finite amplitude. However, one resonance has to be passed through before the absorber frequency is reached.

The linear absorber discussed here can only be used in drive systems that run at a constant operating speed with little variation. Operation must also be permanent because the resonance range will be passed during starting and shutting down processes. The absorber has to be designed carefully because the absorber spring will be exposed to a large loads. In any case, it is advisable to provide the absorber with damping to diminish the resonance amplitudes. Such a component is the spring-constrained damper.

4.4.3 Design of a Spring-Constrained Damper

A damper differs from an absorber by the relative damping with a torsional damping constant b_T between the rotating masses of the model. The calculation model shown in Fig. 4.40 is obtained.

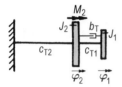

Fig. 4.40 Damped absorber

The equations of motion for a harmonic excitation moment are:

$$J_1\ddot{\varphi}_1 + b_T(\dot{\varphi}_1 - \dot{\varphi}_2) + c_{T1}(\varphi_1 - \varphi_2) = 0$$
$$J_2\ddot{\varphi}_2 - b_T(\dot{\varphi}_1 - \dot{\varphi}_2) - c_{T1}(\varphi_1 - \varphi_2) + c_{T2}\varphi_2 = \hat{M}_2 \cos \Omega t \quad (4.190)$$

Only the damping of the damper is considered. The following is obtained for the relative amplitude of rotating mass 2 from (4.190), see (4.187)

$$V_2 = \sqrt{\frac{c_{T2}^2\,[(c_{T1}-J_1\Omega^2)^2 + b_T^2\Omega^2]}{[(c_{T2}-J_2\Omega^2)(c_{T1}-J_1\Omega^2) - c_{T1}J_1\Omega^2]^2 + b_T^2\Omega^2[c_{T2}-J_2\Omega^2-J_1\Omega^2]^2}}. \quad (4.191)$$

If one introduces the damping ratio as

$$2D = \frac{b_T}{J_1}\sqrt{\frac{J_2}{c_{T2}}}, \quad (4.192)$$

one obtains

$$V_2 = \sqrt{\frac{(2D\eta)^2 + (\xi^2 - \eta^2)^2}{(2D\eta)^2\,[\eta^2(1+\mu) - 1]^2 + [(1+\gamma - \eta^2)(\xi^2 - \eta^2) - \gamma\xi^2]^2}}. \quad (4.193)$$

As can easily be seen, (4.193) transforms into (4.188) for $D = 0$.
The greatest effect of an absorber occurred at $\xi^2 = 1$. Figure 4.41 therefore shows $V = V(\eta; D)$ for $\xi^2 = 1$. The mass ratio is assumed to be $\mu = 0.2$ as in Fig. 4.39. The curves for $D = 0; 0.1; 0.3$ and $D \to \infty$ were plotted. The curve already shown in Fig. 4.39 is obtained for $D = 0$. For $D \to \infty$, the damper mass J_1 firmly "adheres" to the rotating mass J_2. An undamped one-mass model is created whose natural circular frequency is $\omega^2 = c/(J_1 + J_2)$. It should be noted for $D = 0.1; 0.3$ that both curves intersect with the curves for $D = 0; D \to \infty$ at the points P and

4.4 Absorbers and Dampers in Drive Systems

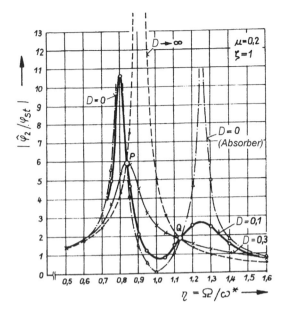

Fig. 4.41 Non-dimensionalized amplitude function for a model with a spring-constrained damper

Q. These points of intersection are independent of damping. They can be used for damper optimization, especially since they are easily determined from the curves for $D = 0$; $D \to \infty$. The following must apply with (4.188) and (4.193):

$$\frac{\xi^2 - \eta_{P,Q}^2}{(1 + \mu\eta^2 - \eta_{P,Q}^2)(\xi^2 - \eta_{P,Q}^2) - \mu\xi^4} = \frac{1}{\eta_{P,Q}^2(1 + \mu) - 1}. \tag{4.194}$$

Therefore

$$\eta_{P,Q}^2 = \frac{1}{2 + \mu}\left[\xi^2(1 + \mu) + 1 \mp \sqrt{\xi^4(1 + \mu)^2 - 2\xi^2 + 1}\right]. \tag{4.195}$$

$\eta_P = 0.836$, $\eta_Q = 1.141$ applies to the example in Fig. 4.41. A non-dimensionalized amplitude $V_2 = \hat{\varphi}_2/\varphi_{st}$ corresponds to each η value. The damping effect is greatest if points P and Q are at the same level, i.e. if $V_P = V_Q$. This allows the determination of an optimum value for the setting ξ. One finds

$$\xi_{opt} = \frac{1}{1 + \mu}. \tag{4.196}$$

The value of the mass ratio is determined by the installation conditions at the engine. This leads, in most cases, to $\mu \ll 1$, and ξ_{opt} is then located in the vicinity of $\xi = 1$. The optimum damping would result from the requirement that the non-dimensionalized amplitude function should have a horizontal tangent when P and Q are at the same level.

This fairly complex calculation results in the following for a horizontal tangent in points P and Q:

$$D_{\text{opt}}^2(P) = \frac{\mu\left(3 - \sqrt{\frac{\mu}{\mu+2}}\right)}{8(1+\mu)^3}; \quad D_{\text{opt}}^2(Q) = \frac{\mu\left(3 + \sqrt{\frac{\mu}{\mu+2}}\right)}{8(1+\mu)^3}. \quad (4.197)$$

The mean value is:

$$D_{\text{opt}}^2 = \frac{3\mu}{8(1+\mu)^3}; \quad V_{\text{opt}} = \left|\frac{\hat{\varphi}_1}{\varphi_{\text{st}}}\right|_{\text{opt}} = \frac{1}{\eta_{P,Q}^2(1+\mu) - 1} = \sqrt{1 + \frac{2}{\mu}}. \quad (4.198)$$

From this simple relationship, one can tell the importance of a large damper mass, which, however, requires a damping D_{opt} that is hard to implement. There are various designs of spring-constrained dampers. Often, dampers with rubber elements as spring-damper components are used.

Figure 4.42 shows various embodiments. Note that the damper spring is subjected to large loads. Its load can be calculated using the method specified for the overall system. Incipient tears in the rubber spring can only be identified after re-

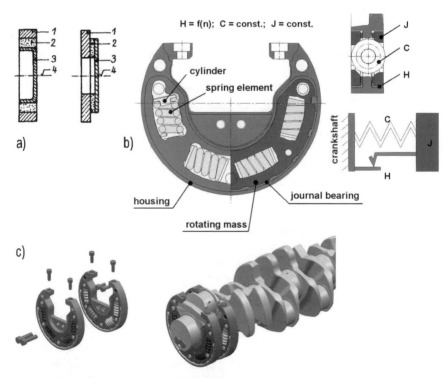

Fig. 4.42 Embodiments of a spring-constrained damper; **a)** Spring-constrained damper; *1* Damper mass, *2* Rubber spring, *3* Driving disk, *4* Axis of rotation, **b)** Internal crankshaft damper (ICD), **c)** Crankshaft with ICD (Source: LuK)

4.4 Absorbers and Dampers in Drive Systems

moving the damper. They have a major influence on the oscillation behavior as the damping effect is reduced by changing the frequency tuning. The load on the system is then much higher, which can result in breakage.

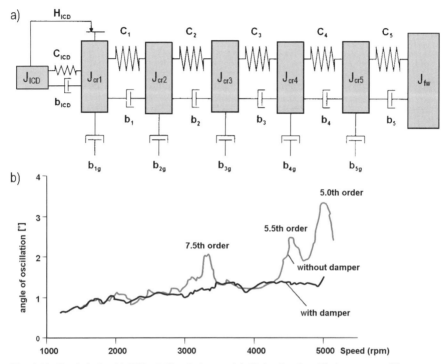

Fig. 4.43 Crankshaft with ICD; **a)** Simulation model, **b)** Angle of oscillation measured between the two crankshaft ends without and with ICD (Source: LuK)

Figure 4.43 shows the angle of rotation measured for a crankshaft angle of oscillation without damper and with ICD damper used on a 6-cylinder diesel engine for comparison. The results of measurements match those of the simulations and confirm that resonance peaks of the principal orders (here the 5th, 5.5th, and 7.5th order) were almost eliminated by the ICD.

4.4.4 Design of a Springless Damper

Springless dampers, the rotating damper masses of which are coupled to the system via a damper element only, are particularly important for the prevention of breakage of the damper spring and its consequences. Note, however, that the concept of a springless damper refers to the fact that a separate spring component is missing. Recent studies have shown that there is a spring effect that cannot be neglected, particularly in rotary viscosity dampers with highly viscous damper fluid. The following considerations in which $c_{T1} = 0$ was set can therefore be viewed as a limit evaluation only.

According to (4.193), the non-dimensionalized amplitude function for $\xi = 0$ is calculated from:

$$V_2 = \frac{\hat{\varphi}_2}{\varphi_{st}} = \sqrt{\frac{4D^2 + \eta^2}{4D^2(\eta^2(1+\mu)-1)^2 + \eta^2(\eta^2-1)^2}}. \tag{4.199}$$

The positions of the damping-independent points are found from (4.195)

$$\eta_{P,Q}^2 = \frac{1}{2+\mu}(1 \mp 1); \qquad \eta_P = 0; \qquad \eta_Q^2 = \frac{2}{2+\mu}. \tag{4.200}$$

One-mass models result for $D = 0$ and $D \to \infty$ so that the requirement of a minimum in the non-dimensionalized amplitude is satisfied if the non-dimensionalized amplitude function has its maximum at Q. This non-dimensionalized amplitude can be stated immediately when inserting (4.200) into the equation for $D \to \infty$, see (4.198). One finds

$$V_{opt} = 1 + \frac{2}{\mu}. \tag{4.201}$$

If one compares this result to (4.198), it becomes evident that the effect of a spring-constrained damper with the same rotating mass is much better or that a springless damper will require a larger mass for the same damping effect. The optimum damping at which the non-dimensionalized amplitude function has its maximum at Q, can be determined using the approach specified above. One finds:

$$D_{opt}^2 = \frac{1}{2(1+\mu)(2+\mu)}. \tag{4.202}$$

If one assumes that $\mu \ll 1$, $D_{opt}^2 = 1/4$ applies. The torsional damping constant of the damping element then amounts to:

$$\frac{b_T^2}{4J_1^2\omega^{*2}} = \frac{1}{4}; \qquad b_T = J_1\omega^* \tag{4.203}$$

The simple relationship (4.203), which only includes the rotating mass of the damper and the resonance frequency for multimass systems, is used frequently.

It is difficult to implement a specific damping constant in a damper due to the strong dependence on manufacturing and operating parameters.
Springless dampers are either based on friction damping or viscous damping. There is a frictional connection via the brake pad between the damper mass and the driving disk in the friction damper in Fig. 4.44, which is produced by the spring force.

The viscous damper (viscous rotary damper), Fig. 4.45, has a coupling of the damper mass and the housing that sits on the shaft by the viscosity of the damper oil. It can be controlled in wide limits for silicone oils. These dampers are primarily

4.4 Absorbers and Dampers in Drive Systems

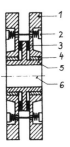

Fig. 4.44 Friction damper
1 Damper mass; *2* Clamping spring;
3 Brake pad; *4* Cylinder sleeve;
5 Driving disk; *6* Axis of rotation

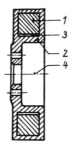

Fig. 4.45 Viscous damper
(viscous rotary damper)
1 Damper mass; *2* Housing
3 Damper fluid; *4* Axis of rotation

used in large-scale diesel engines. They are basically maintenance-free but take a great manufacturing effort. These dampers are produced up to a diameter of 2 m (mass approximately 6 000 kg).

4.4.5 Examples

4.4.5.1 Peculiarities of Viscous Rotary Dampers

If considerations are once again based on the spring-constrained damper, Fig. 4.40, the non-dimensionalized amplitude function (4.193) applies with the shortcuts (4.192). The spring and damper constants depend on several factors, particularly the gap coefficient Sp.

The gap coefficient depends on the geometry of the damper ring and the housing. If one neglects the flat spots on the ring, the following applies:

$$Sp = \pi \left[\frac{r_a^4 - r_i^4}{S_A} + 2B \left(\frac{r_a^3}{s_{ra}} + \frac{r_i^3}{s_{ri}} \right) \right] \quad (4.204)$$

r_a, r_i Outer and inner radius of the ring
s_{ra}, s_{ri} Radial gap outside and inside
B Ring width
S_A Axial gap width.

If the material parameters have been determined experimentally, the non-dimensionalized amplitude function $V = |\hat{\varphi}_2/\varphi_{st}|$ can be calculated as in Fig. 4.41. The non-dimensionalized amplitude function shown in Fig. 4.46 was obtained depending on the gap coefficient Sp. It is remarkable that there no longer is a damping-free point and that the values obtained with an optimum gap coefficient are below the optimum achieved at a damping-free point. The tests further showed that the required optimum gap coefficient increases as the natural frequency of the substitute model rises. Large gap numbers, however, require high manufacturing precision.

It should be noted that molecule decay can occur in high-viscosity oils after a specific service period that can result in the ring becoming jammed in the housing and thus in failure of the damper.

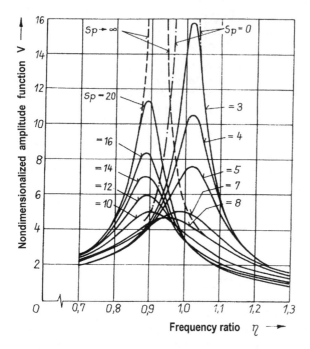

Fig. 4.46 Non-dimensionalized amplitude function of a viscous rotary damper, taking into account the gap coefficient Sp as a parameter. Sp is measured in m^3.

4.4.5.2 Pendulum Absorber

In many drive systems of machines, the excitation frequencies $k\Omega$ are proportional to the rotational speeds. Pendulum absorbers are an effective and adaptive means of eliminating hazardous resonances in rotating shafts. Several types of pendulum absorbers were proposed by B. SALOMON in the 1930s (Patent DRP 597 091), which either consist of swinging rings or of rollers and run on circular paths. The various design implementations that have been created since then all strive for the same goal of achieving a large mass and a short pendulum length, as these are required for high frequencies and large torques. Pendulum absorbers are attached to the crank web, if possible, to absorb the interfering variable torque at the location where it is initiated. Figure 4.47 shows various shapes of *pendulum absorbers by Salomon* [16]. A periodic moment $M_1(t) = \hat{M}_1 \sin k\Omega t$ acts on a rotating disk (rotating mass J_1) on the input shaft. Without the pendulum absorber, the rotation would be at a variable

4.4 Absorbers and Dampers in Drive Systems

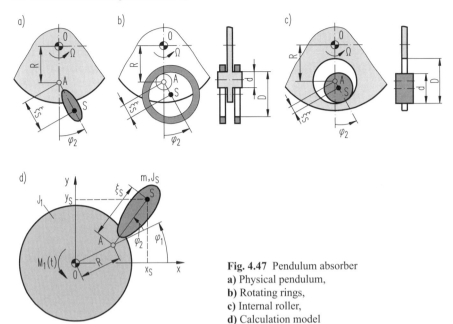

Fig. 4.47 Pendulum absorber
a) Physical pendulum,
b) Rotating rings,
c) Internal roller,
d) Calculation model

angular velocity ($\varphi_1 = \Omega t + \hat{\varphi}_1 \sin k\Omega t$). A physical pendulum (mass m, distance to the center of gravity ξ_S, moment of inertia J_S) is pivoted at the disk at point A (distance R), see Fig. 4.47a. In rolling pendulums (Figs. 4.47b and 4.47c), half the diameter difference $\xi_S = (D-d)/2$ corresponds to the distance to the center of gravity of the physical pendulum since both move as if they were supported at point A.

The system shown in Fig. 4.47d has two degrees of freedom (angle of rotation φ_1 and relative swing angle φ_2). The potential energy is not considered (horizontal position of the plane of rotation). The kinetic energy is the sum of the rotational energy of the rotating masses and the translational energy of the mass m:

$$2W_{\text{kin}} = J_1 \dot{\varphi}_1^2 + m(\dot{x}_S^2 + \dot{y}_S^2) + J_S(\dot{\varphi}_1 + \dot{\varphi}_2)^2. \tag{4.205}$$

The center of gravity of the pendulum absorber has the coordinates

$$x_S = R\cos\varphi_1 + \xi_S \cos(\varphi_1 + \varphi_2); \quad y_S = R\sin\varphi_1 + \xi_S \sin(\varphi_1 + \varphi_2). \tag{4.206}$$

The resulting velocities are

$$\begin{aligned}\dot{x}_S &= -\dot{\varphi}_1 R \sin\varphi_1 - (\dot{\varphi}_1 + \dot{\varphi}_2)\xi_S \sin(\varphi_1 + \varphi_2); \\ \dot{y}_S &= \dot{\varphi}_1 R \cos\varphi_1 + (\dot{\varphi}_1 + \dot{\varphi}_2)\xi_S \cos(\varphi_1 + \varphi_2).\end{aligned} \tag{4.207}$$

If one inserts these into (4.205), a brief conversion will yield the expression

$$2W_{\text{kin}} = m_{11}\dot{\varphi}_1^2 + 2m_{12}\dot{\varphi}_1\dot{\varphi}_2 + m_{22}\dot{\varphi}_2^2 \tag{4.208}$$

for the kinetic energy. This is a link to Sect. 2.4.1 where the same form occurs in (2.148). The partially varying generalized masses here are

$$\begin{aligned}m_{11} &= J_1 + mR^2 + J_S + m\xi_S^2 + 2mR\xi_S \cos\varphi_2 \\ m_{12} &= m_{21} = J_S + m\xi_S^2 + mR\xi_S \cos\varphi_2; \qquad m_{22} = J_S + m\xi_S^2.\end{aligned} \tag{4.209}$$

The equations of motion that correspond to the form of the rigid-body system with two drives, see (2.156) and (2.157) in Sect. 2.4.1.1, result from Lagrange's equations of the second kind. Each of these equations describes a moments balance:

$$\begin{aligned}m_{11}\ddot{\varphi}_1 + m_{12}\ddot{\varphi}_2 - 2mR\xi_S \sin\varphi_2 \dot{\varphi}_1\dot{\varphi}_2 - mR\xi_S \sin\varphi_2 \dot{\varphi}_2^2 &= M_1 \\ m_{21}\ddot{\varphi}_1 + m_{22}\ddot{\varphi}_2 + mR\xi_S \sin\varphi_2 \dot{\varphi}_1^2 &= 0.\end{aligned} \tag{4.210}$$

For small angles $|\varphi_2| \ll 1$, $\sin\varphi_2 = \varphi_2$ and $\cos\varphi_2 = 1$. This yields the two *linear equations of motion* with $k\hat{\varphi}_1 \ll 1$ and $\dot{\varphi}_1 \approx \Omega$:

$$J_{11} = J_1 + J_S + m(R+\xi_S)^2; \qquad J_{22} = J_A = J_S + m\xi_S^2 \tag{4.211}$$

in the form

$$\begin{aligned}J_{11}\ddot{\varphi}_1 + (J_{22} + mR\xi_S)\ddot{\varphi}_2 &= \hat{M}_1 \sin k\Omega t \\ (J_{22} + mR\xi_S)\ddot{\varphi}_1 + J_{22}\ddot{\varphi}_2 + mR\xi_S \Omega^2 \varphi_2 &= 0.\end{aligned} \tag{4.212}$$

The square of the natural circular frequencies can be read from the second equation if the term containing $\ddot{\varphi}_1$ is moved to the right side (centrifugal pendulum, forced vibration)

$$\omega_0^2 = \frac{mR\xi_S \Omega^2}{J_{22}} = \frac{mR\xi_S}{J_S + m\xi_S^2}\Omega^2. \tag{4.213}$$

This leads to an important finding: The *natural circular frequency of the physical pendulum in a centrifugal force field is proportional to the speed of rotation*.

The amplitudes for the steady-state case can be calculated using

$$\varphi_1 = \Omega t + \hat{\varphi}_1 \sin k\Omega t; \qquad \varphi_2 = \hat{\varphi}_2 \sin k\Omega t. \tag{4.214}$$

From the linear system of equations that is obtained when inserting (4.214) into (4.211) and (4.212), it follows that:

$$\hat{\varphi}_1 = -\frac{\hat{M}_1(mR\xi_S - J_{22}k^2)}{\Delta}; \qquad \hat{\varphi}_2 = -\frac{\hat{M}_1(J_{22} + mR\xi_S)}{\Delta} \tag{4.215}$$

with the denominator:

$$\Delta = \{[(J_{22} + mR\xi_S)^2 - J_{11}J_{22}]k^2 + J_{11}mR\xi_S\}(k\Omega)^2. \tag{4.216}$$

$\hat{\varphi}_1 = 0$ applies to the angular amplitude if the numerator is zero, that is, under the absorption condition

$$mR\xi_S = J_{22}k^2 = (J_S + m\xi_S^2)k^2. \tag{4.217}$$

For the case of absorption, the amplitude of the pendulum absorber follows from (4.215):

$$|\hat{\varphi}_2| = \frac{\hat{M}_1}{(J_{22} + mR\xi_S)(k\Omega)^2}. \tag{4.218}$$

Although the moment M_1 is variable, the disk remains at rest, regardless of the amplitude of the moment and at all speeds of rotation. The absorber acts as if the moment of inertia of the disk were infinitely large because the moment is directly absorbed by the pendulum.

For large deflections, the pendulum absorber represents a nonlinear oscillator whose natural frequency (when looked at closely) depends on the amplitude. The linearization is permissible and useful only up to angles of about $|\varphi_2| < \pi/6$. For greater angles, the swinging motion contains higher harmonics that also excite vibrations. There are design solutions with roller paths in the form of cycloids wherein the natural frequency of the rolling pendulum remains independent of the amplitude in the centrifugal force field.

4.5 Parameter-Excited Vibrations by Gear Mechanisms with Varying Transmission Ratio

4.5.1 Problem Formulation/Equation of Motion

Mechanisms with a varying transmission ratio are used in many processing machines and presses to achieve specific motions. It is known from Section 2 that mechanisms with a varying transmission ratio cause a variable moment of inertia with respect to the input shaft, which shows special dynamic effects even for the model of a rigid machine.

In addition to the concentration of mass (flywheel) at the drive motor, many machines exhibit a concentration of mass at the output of the mechanism with a varying transmission ratio, and the drive shafts or couplings located in between represent the essential elasticity that can be modeled as a torsion spring with the torsional spring constant c_T. This resilience *upstream* of the mechanism has a dynamic effect of a different quality than such a resilience *downstream* of the mechanism. This can produce parameter-excited vibrations, as opposed to forced vibrations that occur due to motion excitation at the output of a mechanism.

The following will show how the originally nonlinear equation of motion that is typical of such problems can be transformed into a linear differential equation. Subsequently, the specifics of the physical behavior of such torsional oscillators, methods of calculation, and ways to influence this behavior by means of design will be discussed.

The calculation model for the vibration system discussed here is shown in Fig. 4.48.

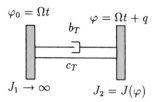

Fig. 4.48 Model of the torsional oscillator with a mechanism with a varying transmission ratio at the output

It consists of the input link rotating at a constant angular velocity Ω, the input shaft with the torsional spring constant c_T, the torsional damping constant b_T, and the moment of inertia $J(\varphi)$ that depends on the position of the input link, see (2.199). The equation of motion results from the equilibrium of moments between the moment from the inertia forces, see (2.209), and the restoring moment of the elastic and viscously damped torsion spring of the input shaft. The nonlinear equation of motion using the coordinates shown in Fig. 4.48 is:

$$J(\varphi)\ddot{\varphi} + \frac{1}{2} \cdot \frac{\mathrm{d}J(\varphi)}{\mathrm{d}\varphi}\dot{\varphi}^2 + c_T(\varphi - \varphi_0) + b_T(\dot{\varphi} - \dot{\varphi}_0) = 0. \tag{4.219}$$

The rotational motion φ is composed of a portion with the constant angular velocity Ω and a relative angle of rotation q:

$$\varphi = \varphi_0 + q = \Omega t + q; \qquad \dot{\varphi} = \Omega + \dot{q}; \qquad \ddot{\varphi} = \ddot{q}. \tag{4.220}$$

If one can assume that the angle q stays so small that linearization is permissible, the following expansion of $J(\varphi)$ into a Taylor series can be truncated after the linear term. If the derivative with respect to the angle of rotation of the rigid machine is denoted by a single prime, one obtains

$$\frac{\mathrm{d}(\ )}{\mathrm{d}(\Omega t)} = (\ )'; \qquad J(\varphi) = J(\Omega t) + J'(\Omega t)q + \cdots \tag{4.221}$$

$$\frac{\mathrm{d}J(\varphi)}{\mathrm{d}\varphi} = J'(\Omega t) + J''(\Omega t)q + \cdots \tag{4.222}$$

If these series are inserted into (4.219), the following *linear* differential equation with $J = J(\Omega t)$ results when neglecting all terms starting from the order q^2:

$$J\ddot{q} + (b_T + J'\Omega)\dot{q} + (c_T + J''\Omega^2/2)q = -\frac{1}{2}J'\Omega^2 \tag{4.223}$$

The reduced moment of inertia $J(\Omega t)$ is included in this equation as a time-dependent quantity. Thus, a rheolinear differential equation has arisen from the orig-

4.5 Parameter-Excited Vibrations

inally nonlinear differential equation (4.219). The kinetostatic moment of the rigid machine appears on the right-hand side as the excitation moment for the torsional vibrations. In addition, the variability of J also has an effect on the left-hand side as so-called parametric excitation. Time-dependent coefficients occur where the equations discussed so far had constant coefficients.

A "decay rate" and a "natural circular frequency" can be formally assigned to the equation of motion (4.223) of the torsional oscillator in analogy with (4.221), see (4.223).

$$\delta(t) = \frac{b_T + J'\Omega}{2J}; \qquad \omega_0^2(t) = \frac{c_T + J''\Omega^2/2}{J}. \qquad (4.224)$$

These quantities, however, typically depend on the position of the mechanism (and here especially on time).

4.5.2 Solution of the Equation of Motion, Stability Behavior

An engineer would expect that the behavior of a drive system where time-dependent reduced moment of inertia changes slightly or slowly is only slightly different from that of one with a constant moment of inertia. As (4.223) is linear, it must be possible to construct a fundamental system for its solution, and the following applies generally to (4.223): $q = q_\text{hom} + q_\text{part}$.

A homogeneous solution of the form

$$q_h = C_1 e^{\mu_1 t} q_1(t) + C_2 e^{\mu_2 t} q_2(t) \qquad (4.225)$$

is frequently assumed. C_1 and C_2 are the integration constants that can be determined from the initial conditions, $q_1(t)$ and $q_2(t)$ are periodic functions of time, and μ_1, μ_2 are the characteristic coefficients that determine the stability of the solution. Their dependence on parameters of the specific equation of motion leads to so-called stability charts that state the limits of parameter values for which stable or unstable dynamic behavior occurs.

In principle, concrete problems that include time-variable parameters can also be solved using the methods of numerical integration if a limited time interval is of interest only. However, it is also important for an engineer to obtain a general idea of the dynamic effects that can occur during parametric excitation. Parametric resonance is much rarer than forced resonance, but it should not be considered as an unessential or improbable side effect because, due to its specifics, it can result in dangerously large amplitudes. There are four fundamental differences as compared to forced resonance.

First, parametric resonance is a phenomenon of unstable equilibrium because the system can be stimulated by inevitable initial disturbances. It can happen as a result of an accidental disturbance at specific frequencies that the system absorbs energy amounts (proportional to the instantaneous amplitudes) at the rate of its natural frequency and thus becomes "agitated" fast.

A second peculiarity is that the vibration amplitudes increase in a *linear* manner over time for forced resonance without damping, but they increase *exponentially* for parametric resonance.

Third, the qualitative effect of damping is different. While viscous damping limits the resonance amplitudes in the event of forced resonance, the amplitudes can increase without bounds despite existing damping in the event of parametric resonance. However, parametric resonance only develops in specific frequency ranges only if the damping falls below a specific value.

A fourth particularity of parametric resonance is that it does not only occur at certain frequencies (like forced resonance), but in a certain frequency range in the vicinity of these specific frequencies. The principal parametric resonance occurs in the vicinity of that angular velocity that corresponds to double the averaged natural frequency of the oscillator, i.e. at $\Omega = 2\bar{\omega}$. The resonance points are generally located around the frequency ratios

$$\eta = \frac{\Omega}{\omega} = \frac{2}{k}; \qquad k = 1, 2, \cdots, \tag{4.226}$$

that is, at values that are by integral multiples smaller than the frequency of the principal resonance. The width of these critical regions depends on the pulse depth of the parametric excitation.

Another particularity of parameter-excited oscillators is that dynamic instabilities can occur in a finite time interval far away from the speed ranges of parametric resonance characterized by (4.226).

One can illustrate this effect using the definition of the "decay rate", as in (4.224). The motion is dynamically stable as long as

$$\delta(t) = \frac{b_{\mathrm{T}} + \Omega J'(\Omega t)}{2 J(\Omega t)} > 0, \tag{4.227}$$

because then there will be the decay according to $\exp(-\delta t)$ as known from the free vibrations of a linear oscillator.

Since $J'(\Omega t)$ is variable, it may happen, though, that this "decay rate" becomes negative in specific intervals and causes exponential stimulation of amplitudes. The dynamic instability manifests itself in amplitude-modulated vibrations at the natural frequency $\omega(t)$ that resembles a beat vibration. The interval of stimulation alternates with an interval of decay, which is why amplitudes do not increase without bounds in this case, as is otherwise common for parametric resonance.

The parameter value of b_{T} is often highly uncertain, whereas knowledge is available about the damping ratio D. Using $D = b_{\mathrm{T}}/(2\sqrt{c_{\mathrm{T}} J})$, the condition results from the requirement $\delta(t) > 0$:

$$D + \frac{\Omega J'(\Omega t)}{2\omega J(\Omega t)} > 0. \tag{4.228}$$

4.5 Parameter-Excited Vibrations

This condition may be violated in ranges where $J' < 0$. Since the inertia decreases, the mechanism is "automatically" accelerated in the absence of damping, and kinetic energy is transferred in these positions of the mechanism from the inert mass of the output link to the torsion spring.

One can also see from the parameters in (4.228) what countermeasures an engineer can take. Large damping is favorable but can only be achieved to a limited extent. One should always strive for a reduction of the variation of the moment of inertia J' because it is also the cause for the other interfering phenomena of parametric resonance. The dependence on Ω means that such instabilities will not occur at low speeds of a mechanism because they are suppressed by damping, but that they can occur at higher speeds. The natural circular frequency ω in the denominator of a term of (4.228) indicates that these instabilities will hardly occur in high-frequency (rigid) systems.

In cam mechanisms, instability ranges of torsional vibrations can be moved towards higher speeds, e. g. by influencing the cam profile in areas that are not completely predefined by engineering requirements. The time function of the moment of inertia can be smoothened in many mechanisms using dynamic compensators (compensatory mechanisms), similar to power balancing, see Sect. 2.4.2.

4.5.3 Examples

4.5.3.1 Transfer Manipulator

On modern presses, workpieces are transported using so-called transfer manipulators that operate in sync with the press stroke and have to keep the workpiece immobile during the forming process. Due to the large distance between the press drive and the output link (the gripper rails), which is bridged by an articulated shaft of several meters in length, the drive cannot have a torsionally stiff design. There is a risk with fast-running manipulators that the vibrations of the grippers result in inaccurate workpiece placement and collisions in the dwell phase.

Figure 4.49a shows the grossly simplified mechanism schematic and Fig. 4.49b the associated minimal model.

The cam mechanism, which is characterized by its motion program $U(\varphi)$, is a major element of the drive system. The motion program represents the connection between the input motion φ and the output motion, in the example on hand the motion of the gripper rails x; $x = U(\varphi)$. This motion is characterized by two dwell phases, the transitions of which are described by a "motion program of the 5th power" (VDI Guideline 2143 [36]). The following applies outside of the dwell phases:

$$x = U(\varphi) = U_{\max}(10z^3 - 15z^4 + 6z^5); \qquad z = z(\varphi) \qquad (4.229)$$

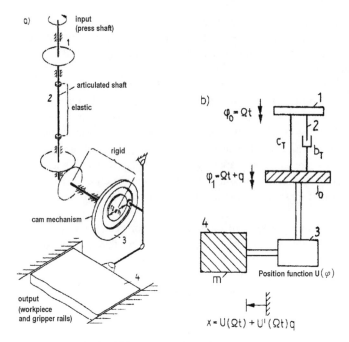

Fig. 4.49 Longitudinal lifting mechanism of a transfer manipulator; **a)** Mechanism schematic, **b)** Calculation model, *1* Drive from the press, *2* Articulated shaft (c_T; b_T), *3* Cam mechanism ($r_1 - r_2 = U_{\max}$), *4* Gripper rails and workpieces

The variable z is connected to the input angle φ in the following way:

$$-\frac{70\pi}{180} \leq \varphi \leq \frac{70\pi}{180}: \quad z = \frac{9}{7\pi}\varphi + \frac{1}{2} \quad \text{(rise)}; \qquad z' = \frac{9}{7\pi}$$

$$\frac{70\pi}{180} \leq \varphi \leq \frac{110\pi}{180}: \quad z = 1 \quad \text{(front dwell)} \qquad (4.230)$$

$$\frac{110\pi}{180} \leq \varphi \leq \frac{250\pi}{180}: \quad z = -\frac{9}{7\pi}\varphi + \frac{25}{14} \quad \text{(return)}; \qquad z' = -\frac{9}{7\pi}$$

$$\frac{250\pi}{180} \leq \varphi \leq -\frac{70\pi}{180}: \quad z = 0 \quad \text{(rear dwell)}.$$

The reduced moment of inertia is determined by the rotating masses of the drive (J_0), the mass m of the output and the motion program. The following applies according to (2.199)

$$J(\varphi) = J_0 \varphi'^2 + mx'^2 = J_0 + mU'^2. \qquad (4.231)$$

The equation of motion of the calculation model in Fig. 4.49b is derived from (4.223). One finds

4.5 Parameter-Excited Vibrations

$$(J_0 + mU'^2)\ddot{q} + (b_T + 2mU'U''\Omega)\dot{q} + [c_T + m\Omega^2(U''^2 + U'U''')]q \\ = -m\Omega^2 U'U''. \tag{4.232}$$

The following applies for the various derivatives of the motion program with $U' = \mathrm{d}U/\mathrm{d}(\Omega t)$ and z' according to (4.230):

$$\begin{aligned} \text{0th order: } U &= U_{\max}(10z^3 - 15z^4 + 6z^5) \\ \text{1st order: } U' &= U_{\max}30z'(z^2 - 2z^3 + z^4) \\ \text{2nd order: } U'' &= U_{\max}60z'^2(z - 3z^2 + 2z^3) \\ \text{3rd order: } U''' &= U_{\max}60z'^3(1 - 6z + 6z^2). \end{aligned} \tag{4.233}$$

Figure 4.50a shows the various derivatives of the motion program. This example is based on the following parameter values:

$$\begin{aligned} U_{\max} &= 0.4 \text{ m}; & J_0 &= 1.1 \text{ kg} \cdot \text{m}^2; & c_T &= 1.5 \cdot 10^3 \text{ N} \cdot \text{m} \\ \Omega &= 2\text{s}^{-1}; & m &= 136 \text{ kg}; & b_T &= 20 \text{ N} \cdot \text{m} \cdot \text{s} \end{aligned} \tag{4.234}$$

The "natural circular frequency" is $\omega(t) = \sqrt{c_T(t)/J(t)}$, see (4.224).
Since $c_T > m\Omega^2(U''^2 + U'U''')_{\max}$, ω_0 mainly depends on the mechanism position, but only to a minor extent on the speed of rotation. The following applies approximately

$$\omega_0^2(t) \approx \frac{c_T}{J_0 + mU'^2}. \tag{4.235}$$

Figure 4.50b shows the time function of the relative reduced moment of inertia and the relative natural circular frequency. The solution of (4.232) in this example was obtained using a numerical integration method. Figure 4.50c shows the resulting time function of the elastic moment in the input shaft $M_{\text{an}} = c_T q$, which is also a measure of the oscillatory motion of the output link as $U(\varphi) \approx U(\Omega t)$. Figure 4.50c also depicts the kinetostatic moment $M_{\text{st}} = m\Omega^2 U'U''$ for comparison. It would result for a rigid drive, see (4.232).

As can be expected based on the theory, intense vibrations occur during the decrease in the moment of inertia ($J'(\varphi) \leq 0$) that are superimposed on the kinetostatic moment. Intense vibrations are excited for a short time when $b_T < 2m\Omega U'U''$, which also becomes evident from (4.232). Figure 4.50d shows the time functions of $d = b_T/2m\Omega U'U''$ and η. If one looks at the zeros of the dynamic moment in Fig. 4.50c (the sum of kinetostatic moment and vibration moment), one can clearly see the change in $\omega(t)$, shown in Fig. 4.50d.

The moment curve becomes understandable from a qualitative perspective if one focuses on the motion of the gripper rails, for the major part of the (variable) kinetic energy is stored in them. They have the largest velocity at the positions $\varphi = 0$ and $\varphi = \pi$ and have to be decelerated by the input shaft during the rise in the range of $0° \leq \varphi \leq 70°$ and during the return in the range of $180° \leq \varphi \leq 250°$. This produces a negative kinetostatic moment that has a braking effect with respect to the input shaft and simultaneously excites vibrations. This twofold moment variation per crank revolution is typical for all drives with a motion reversal. It always results in a dominant second harmonic of the input torque. The vibrations occur only at

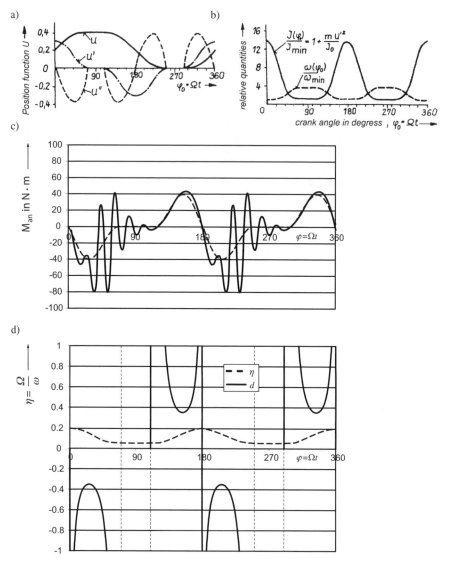

Fig. 4.50 Longitudinal lifting mechanism; **a)** Motion program and first two derivatives, **b)** Reduced moment of inertia and natural frequency as relative quantities, **c)** Input torque curve, **d)** Curve of the damping and frequency ratio

high speeds and only within the range of the negative J' (declining J). As can be seen from Fig. 4.50, no such excited vibrations can be expected for positive values of $d = b_T/2m\Omega U'U''$.

It is now the task of the designer to ensure through engineering changes that the interfering vibrations in the dwell phases vanish. All parameters of the calculation model were determined by experiment, and after finding that the results of measurement matched the calculated values well, the effect of the following measures was studied using the calculation model.

1. Reduction of the mass m of the output link. It results in a smaller variation of $J(\varphi)$ and thus reduces the parametric excitation. This change can be implemented by lightweight design (closed box section).
2. Increase of the torsional stiffness of the drive. The articulated shaft represented the greatest resilience but could not be stiffened considerably. But its reduced stiffness increases if it is operated at a higher speed than the press. This can be achieved by additional gear mechanisms instead of the simple bevel gears.
3. Reduction of the vibration excitation by minimizing the product $U'U''$ (kinetostatic moment $M_{\text{st}} = m\Omega^2 U'U''$) in the negative range. This can be achieved by an asymmetrical motion program (change of the point of inflection) of the cam mechanism.
4. Balancing of the kinetostatic moment in the negative range using a compensator. Inertia compensators increase the mean J and impact the braking time negatively. A spring compensator that is attached directly at the gripper rails has an optimum effect at a specific speed only, see VDI Guideline 2149 Part 1 [35].

4.5.3.2 Variable Tooth Stiffness as Vibration Excitation

In gear mechanisms, vibrations are excited in each gear mating, although a constant angular velocity of the two gears should occur kinematically based on the rolling conditions that were taken into account when calculating the tooth profiles. The inevitable deformation of the teeth of the meshing tooth pairs occurs as a result of the tooth forces required to transmit the torques of the gear mechanisms. Dynamic tooth forces that are much larger than the actual static forces to be transmitted can be excited as a result of constant moments. The significant influence of tooth backlash, which may lead to gear rattle, is discussed in [26] in more detail.

The internal excitation causes include variations of the gearing stiffness $c(t)$ that occur when the tooth flanks roll off and when switching from single to double meshing. Deviations from the optimum gearing geometry or pitch errors and radial runout can result in vibration excitation not only in the event of gear damage, but also in intact gear mechanisms. The periodic function of the gearing stiffness can be described as follows by the mean value c_{m}, the Fourier coefficients $\hat{c}_k$ and the meshing frequency f_z:

$$c(t) = c_m + \sum_k \hat{c}_k \cos(2\pi k f_z t + \varphi_k). \qquad (4.236)$$

Figure 1.12 shows a typical curve of the variable tooth stiffness, see Sect. 1.3. In a helical gear mechanism, the variations are small as compared to spur gear mechanisms, which means that a few Fourier coefficients suffice to describe them.

The meshing frequency is the product of the angular velocity and the number of teeth. The meshing frequency occurs at the contact point of two gear wheels (shaft

1: angular velocity Ω_1, number of teeth z_1 and shaft *2*: angular velocity Ω_2, number of teeth z_2)

$$f_z = zf = \frac{z_1 \Omega_1}{2\pi} = \frac{z_2 \Omega_2}{2\pi}. \qquad (4.237)$$

Note the standard work by Linke [22] for all problems relating to gear mechanisms.

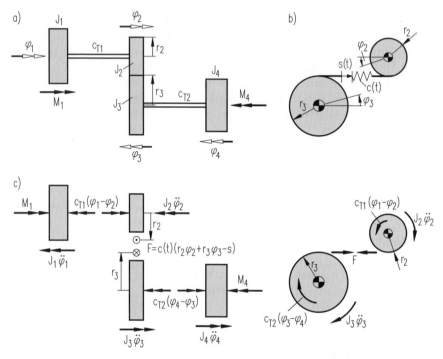

Fig. 4.51 Calculation model of a spur gear mechanism; **a)** System schematic, **b)** Side view, **c)** Free-body diagram

In the simplest case of a one-step spur gear mechanism (Fig. 4.51), the angles of rotation φ_1 and φ_4 of the two rotating masses, the angles φ_2 of the pinion and φ_3 of the gear as well as the torsional stiffnesses c_{T1} and c_{T3} between the disks are taken into account when modeling. The gearing errors that are modeled as motion excitation $s(t)$ and the gearing stiffness $c(t)$ are periodically variable.

The equations of motion for the calculation model result from the equilibrium of moments on each part of the free-body diagram, see Fig. 4.51

$$\begin{aligned}
J_1 \ddot{\varphi}_1 + c_{T1}(\varphi_1 - \varphi_2) &= M_1(t) & (4.238) \\
J_2 \ddot{\varphi}_2 - c_{T1}(\varphi_1 - \varphi_2) + r_2 c(t)[r_2 \varphi_2 + r_3 \varphi_3 - s(t)] &= 0 & (4.239) \\
J_3 \ddot{\varphi}_3 + r_3 c(t)[r_2 \varphi_2 + r_3 \varphi_3 - s(t)] + c_{T2}(\varphi_3 - \varphi_4) &= 0 & (4.240) \\
J_4 \ddot{\varphi}_4 - c_{T2}(\varphi_3 - \varphi_4) &= M_4(t). & (4.241)
\end{aligned}$$

4.5 Parameter-Excited Vibrations

(4.238) to (4.241) describe a forced vibration as a result of parametric excitation by $c(t)$ and the motion excitation due to $s(t)$. Such equations cannot be solved analytically in closed form. Many companies provide software for this. The torsional vibration system shown in Fig. 4.51 will be analyzed for given parameter values here. It will also be shown what major influence the selected calculation model has.

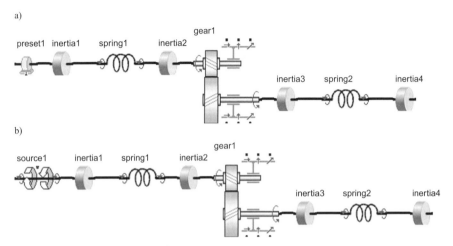

Fig. 4.52 Models in SimulationX® for the spur gear mechanism of Fig. 4.51; **a)** Motion excitation by a given angular acceleration, **b)** Excitation by a constant input torque

The following parameter values for the SimulationX® models shown in Fig. 4.52 are given:

Rotating masses $J_1 = 0.2\,\text{kg} \cdot \text{m}^2$, $J_2 = 0.2\,\text{kg} \cdot \text{m}^2$, $J_3 = 1\,\text{kg} \cdot \text{m}^2$,
$\qquad J_4 = 3\,\text{kg} \cdot \text{m}^2$,
Tooth numbers $z_1 = 23$, $z_2 = 57$, module $m = 3$, contact ratio $= 2.3$;
Mean tooth stiffness $c_m = 49 \cdot 10^6\,\text{N/m}$,
Damper constant of the gearing $b = 2000\,\text{N} \cdot \text{s/m}$;
Torsional spring constants $c_{T1} = 5 \cdot 10^5\,\text{N} \cdot \text{m}$; $c_{T2} = 3 \cdot 10^5\,\text{N} \cdot \text{m}$.

This drive system is accelerated so that it is rotated by an angle of 240 rad within 10 s. In variant a, this is done by a given angular acceleration of $4.756\,\text{rad/s}^2$ and in variant b by a moment of $M = 5\,\text{N} \cdot \text{m}$, which provides the same mean acceleration to the system that is considered rigid and has the reduced moment of inertia of $J_\text{red} = J_1 + J_2 + (J_3 + J_4)(z_1/z_2)^2 = 1.0513\,\text{kg} \cdot \text{m}^2$.

One can see from Fig. 4.53 that this drive system passes through a resonance point in both cases that must be excited by the meshing frequency because no other periodic excitations were applied. Why is the resonance speed in model a much lower than in model b? Note the significant statement at the end of Sect. 4.2.1.1.

The natural frequencies vary due to the variable contact ratio, which is also the cause of the parametric excitation. For model a, they are in the range of $f_1 = (34...38)\,\text{Hz}$, $f_2 = (130...140)\,\text{Hz}$ and $f_3 = (268...278)\,\text{Hz}$. For model

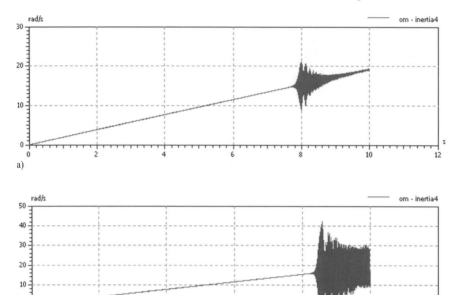

Fig. 4.53 Function of angular velocity of disk 4 for the models in Fig. 4.52; **a)** Motion excitation by a given angular acceleration, **b)** Excitation by a constant input torque

b, they are at $f_1 = (54\ldots59)$ Hz, $f_2 = (140\ldots158)$ Hz and $f_3 = (362\ldots365)$ Hz. Models a and b have different natural frequency ranges because the conditions at the left boundary differ. If an acceleration function is provided at the left boundary, it acts like a fixed point with respect to its natural frequencies, while the oscillator chain is free at the left boundary in the case of the predefined input torque.

The linearly varying angular velocity $\dot{\varphi}_1 = -\alpha t = -(M/J_{\text{red}})t$ enters a resonance area when the meshing frequency f_z is reached, which occurs earlier for model a than for model b. The meshing frequency according to (4.237) is $f_z = 23\Omega_1/(2\pi) = 23\alpha t/(2\pi) = (17.4t/\text{s})$ Hz for 23 teeth of the first gear step. This second frequency is strongly excited because the teeth of the gear step are severely deformed at the second natural frequency. The wider resonance range in case b is due to the fact that the third subharmonic is excited as well.

4.5.4 Problems P4.7 and P4.8

P4.7 Salomon Pendulum Absorber

The dimensions of a Salomon pendulum absorber according to Fig. 4.47c that absorbs the second excitation harmonic of an input torque are to be determined.

4.5 Parameter-Excited Vibrations

Given:
Distance of the center of rotation	$R = 100$ mm
Distance to the center of gravity	$\xi_S = 1$ mm
Mass of the roller	$m = 0.05$ kg
Moment amplitude of the second harmonic	$\hat{M}_1 = 15$ N·m at $k = 2$

Find:

1. Diameter of the roller d
2. Diameter of the hole D
3. Amplitude $\hat{\varphi}_2$ of the roller at angular velocity $\Omega = 300/\text{s}$
4. Discussion of the influence of the order of the harmonic on the magnitude of the absorber mass

P4.8 Natural Frequencies of a Drive System

The drive system to be examined consists of a straight plunger (constant cross-section A, length l), to which a rigid body of mass m is coupled using a spring assumed to be massless with stiffness c. The plunger is driven at the left boundary, which causes longitudinal vibrations.

In order to demonstrate the influence of the model generation on the natural frequencies to be calculated, different model variants will be considered, see Fig. 4.54. On the one hand, the drive is to be modeled using various boundary conditions (motion or force excitation), and on the other hand the plunger is to be captured either as a continuum or as a longitudinal oscillator chain with 5 point masses of equal size. 4 model variants are to be studied:

a) Plunger modeled as continuum with motion excitation (predefined displacement of the left plunger boundary)
b) Plunger modeled as oscillator chain with motion excitation
c) Plunger modeled as continuum with force excitation
d) Plunger modeled as oscillator chain with force excitation

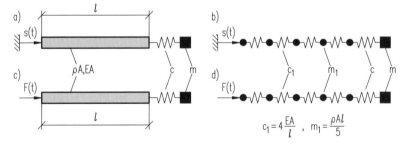

Fig. 4.54 Variants of calculation models of a drive system

Given:
Length and cross section of the plunger	$l = 0.2$ m, $A = 100$ mm² $= 10^{-4}$ m²
Density and Young's modulus of the plunger material	$\varrho = 7.85 \cdot 10^3$ kg/m³, $E = 2.1 \cdot 10^{11}$ N/m²
Spring constant of the coupling	$c = 3.0 \cdot 10^7$ N/m
End mass	$m = 0.6$ kg

Find:

1. Lowest four natural frequencies $(f_1, \ldots, f_4)$ for the respective model variants
2. Interpretation of the differences of these natural frequencies

4.5.5 Solutions S4.7 and S4.8

S4.7 The roller as a homogeneous circular cylinder has a moment of inertia of $J_S = md^2/8$, see Table 5.2. A correlation with the roller diameter results from the absorption condition (4.217):

$$mR\xi_S = (J_S + m\xi_S^2)k^2 = m\left(\frac{d^2}{8} + \xi_S^2\right)k^2, \qquad (4.242)$$

from which the diameter follows:

$$\underline{\underline{d}} = \sqrt{8\xi_S\left(\frac{R}{k^2} - \xi_S\right)} = \sqrt{8\left(\frac{100}{4} - 1\right)} = 13.8 \text{ mm}. \qquad (4.243)$$

The resulting outer diameter is $\underline{\underline{D = 2\xi_S + d = 15.856 \text{ mm}}}$.
The moment of inertia with respect to point A is:

$$J_{22} = (J_S + m\xi_S^2) = m\left(\frac{d^2}{8} + \xi_S^2\right) = 50\left(\frac{192}{8} + 1\right) \text{ g} \cdot \text{mm}^2$$
$$= 1250 \text{ g} \cdot \text{mm}^2. \qquad (4.244)$$

The pendulum amplitude at the specified angular velocity results from (4.218)

$$\underline{\underline{|\hat{\varphi}_2|}} = \frac{\hat{M}_1}{(J_{22} + mR\xi_S)(k\Omega)^2} = \frac{5 \cdot 10^9}{(1250 + 5000)(600)^2} \underline{\underline{= 0.222 \text{ rad}}} \qquad (4.245)$$

This angle satisfies the condition $|\hat{\varphi}_2| \ll 1$ so that one can expect linear behavior. One can see the following from this equation: The mass of the pendulum absorber is independent of the order of the harmonic. With a view to the pendulum amplitude, the lower order k of the harmonic is, the larger it has to be. The elongation of the pendulum absorber increases with decreasing speed.

S4.8 Since this problem is not about estimating forced vibrations but rather only about the natural frequencies of the drive system, the type of excitation only affects the boundary conditions to be prescribed.

Model a) with motion excitation corresponds to case 2 in Table 4.4, since the motion of the left boundary is given by a displacement-time function. Model c) with force excitation corresponds to case 3 in Table 4.4, since the left boundary can move freely.

The given parameter values result in the dimensionless characteristic parameters for the stiffness and mass as defined in Table 4.4:

$$\pi_1 = \frac{cl}{EA} = \frac{3 \cdot 10^7 \cdot 0.2}{2.1 \cdot 10^{11} \cdot 10^{-4}} = 0.2857;$$
$$\pi_2 = \frac{m}{\varrho Al} = \frac{0.6}{7.85 \cdot 10^3 \cdot 10^{-4} \cdot 0.2} = 3.8217 \qquad (4.246)$$

The lowest four natural frequencies of these two systems result, depending on the respective boundary conditions, from the transcendental equations (2) or (3) that are specified in Table 4.4 and are to be solved numerically.

According to (7) of Table 4.4, the resulting natural frequencies are:

$$f_i = \frac{\lambda_i}{2\pi l}\sqrt{\frac{E}{\varrho}} = 4115.9\lambda_i \text{ Hz} \qquad (4.247)$$

4.5 Parameter-Excited Vibrations

In the two discrete models b) and d), the drive is either subjected to motion excitation or force excitation. A system with 5 degrees of freedom results for b) and a system with 6 degrees of freedom results for d). The solution of the resulting eigenvalue problem according to (4.46) provides the respective natural frequencies.

Table 4.6 lists the results of the numerical analysis that was performed using a commercial software product.

Table 4.6 Natural frequencies for 4 model variants

i	\multicolumn{4}{c}{f_i in Hz}
	a)	b)	c)	d)
1	990.3	990.1	0	0
2	7 153	7 155	2 359	2 368
3	19 643	18 631	13 296	11 762
4	32 475	28 282	26 047	21 789

When comparing the variants, one can see that the natural frequencies of the models with motion excitation (variants a) and b)) clearly differ from those with force excitation (variants c) and d)). The first natural frequency is zero in the case of motion excitation because the plunger can move freely (the corresponding free translational rigid-body motion). The natural frequencies of the systems with motion excitation at the left boundary are generally higher than those of the systems that are free at the left boundary, since the prescribed motion excitation $s(t)$ produces an additional constraint [5].

A more accurate model would have to consider the characteristic of the drive motor, i.e. the reaction of the inertia force on the drive that was not taken into account here in all cases. The natural frequencies of the real system are between the two limiting cases of motion excitation or force excitation assumed here.

Since the end mass m in this problem is fairly resiliently coupled to the plunger($\pi_1 \approx 0.29$), the first nonzero natural frequency can be estimated. For the models with motion excitation, the plunger approximately acts like a rigid body with a predefined displacement against which the mass m can oscillate, i.e. the estimate

$$f_1 \approx \frac{1}{2\pi}\sqrt{\frac{c}{m}} \approx 1125\,\text{Hz}$$

applies. In the case of force excitation, there is an unconstrained two-mass oscillator as in Fig. 4.3a (right) so that the estimate according to (4.27)

$$f_1 \approx \frac{1}{2\pi}\sqrt{\frac{c}{m} + \frac{c}{\varrho Al}} \approx 2471\,\text{Hz}$$

can be applied to it. Such estimates of the order of magnitude of the results produced by the computer (here for the first natural frequencies, respectively) should be used whenever the opportunity arises to prevent gross errors.

Chapter 5
Bending Oscillators

5.1 Problem Development

In the 1870s, the rapid development of turbine manufacturing spawned the first theoretical works in England and Germany that dealt with the determination of critical speeds of shafts with variable cross sections. Critical speeds are such speeds at which a machine may not be operated for extended periods of time for reasons of operating safety because dangerous resonance conditions will occur.

The first studies of special issues of bending oscillations that were conducted by RANKINE (1869) and DE LAVAL (1889) were continued in the works by A. STODOLA who explained many important phenomena and presented his findings in a fundamental work in 1924. More progress in the field of calculating critical shaft speeds was made in the 1930s and 1940s by numerous researchers. They were summarized in classic form by BIEZENO and GRAMMEL [1].

The further development was characterized in that the parameters identified as essential were studied more closely in theory and in experiments. Furthermore, the calculation methods were perfected with the emergence of computers, and since about 1955 the most common method was that of transfer matrices [16], [32]. In addition to the transfer matrix method, the finite element method has influenced bending oscillator calculations since the mid 1960s. The current state of the art is characterized in that there are calculation programs available for most of the pertinent problems that have been solved theoretically. These programs are based on the finite element method (FEM).

The bending oscillations of (non-rotating) beams and beam systems, the calculation of which is closely linked to that of shafts, are relevant to structural dynamics and for machine frames. Excellent overviews of the vibrations of rods, beams and plates can be found in [10] and [7]. A modern introduction to the field of rotor dynamics is provided in [8].

In everyday engineering practice, many standards and regulations for calculations have to be observed. A few that are of international significance shall be pointed out here. The guidelines of the DIN ISO 7919 series are of major importance for monitoring turbomachines, e. g.: ISO 7919-5 (2005): Mechanical vibration – Evaluation of machine vibration by measurements on rotating shafts. Part 5: Machine sets in hydraulic power generating and pumping plants.

The guidelines of the DIN ISO 10 816 series are recommended for measurements of bearing vibrations: Mechanical vibration; Evaluation of machine vibration by measurements on non-rotating parts.

The problems discussed in the technical literature eventually serve the design of reliable frames, shafts, and rotors and can be divided into the following groups:

1. Selection of appropriate calculation models that capture the observed and experimentally studied phenomena with reasonable effort
2. Calculation of critical speeds (or natural frequencies, respectively) for given calculation models and parameter values
3. Calculation of actual motions and load parameters (bearing forces and deflections, spindle motions)
4. Selection of favorable parameter values of a design
5. Identification of causes for vibrations in real designs.

5.2 Fundamentals

5.2.1 Self-Centering in a Symmetrical Rotor

It is known from Sect. 2.6.2 that there is virtually always a residual unbalance in rotors because complete balancing cannot be achieved. This section will demonstrate the effect of an unbalance excitation in a simple rotor with a disk. For the sake of simplicity, it is assumed that the disk is located at the center of a shaft with two bearings and a constant bending stiffness EI so that only a symmetric mode shape occurs and the influences of the gyroscopic effect are eliminated.

Figure 5.1 shows the disk, which is displaced from the static equilibrium position, in a fixed coordinate system. The disk rotates at the angular velocity Ω. The deflections in the co-rotating reference system are denoted as r. The center of gravity of the disk S is away from the shaft center point W (eccentricity e). A torque M is applied at the disk. The disk mass is m, the moment of inertia about the axis of rotation is J_p. The massless shaft is isotropic, i.e., its spring constant c is the same in the x and y directions. Since there is no shaft inclination, no other spring constants are required.

The equilibrium of forces at the center of gravity of the disk rotating in the horizontal plane results in, see Fig. 5.1:

$$m\ddot{x}_\mathrm{S} + cx_\mathrm{W} = 0 \tag{5.1}$$

$$m\ddot{y}_\mathrm{S} + cy_\mathrm{W} = 0. \tag{5.2}$$

From the equilibrium of moments about the axis through the center of gravity, it follows that

5.2 Fundamentals

$$J_p \ddot{\varphi} + c x_W e \sin \varphi - c y_W e \cos \varphi = M. \tag{5.3}$$

If one considers the constraints, see Fig. 5.1,

$$x_W = x_S - e \cos \varphi, \qquad y_W = y_S - e \sin \varphi \tag{5.4}$$

the following equations of motion result:

$$m\ddot{x}_S + c x_S = c e \cos \varphi \tag{5.5}$$
$$m\ddot{y}_S + c y_S = c e \sin \varphi \tag{5.6}$$
$$J_p \ddot{\varphi} - c e (y_S \cos \varphi - x_S \sin \varphi) = M. \tag{5.7}$$

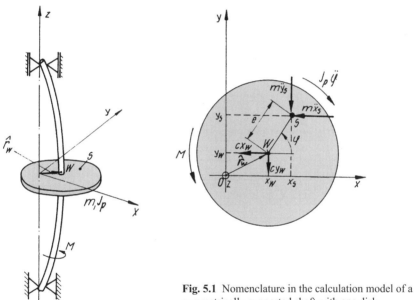

Fig. 5.1 Nomenclature in the calculation model of a symmetrically supported shaft with one disk

The natural circular frequency in this minimal model can be calculated from the static deflection ($f = mg/c$) at the mass m due to its weight (mg):

$$\omega_1 = \sqrt{\frac{c}{m}} = \sqrt{\frac{cg}{mg}} = \sqrt{\frac{g}{f}}. \tag{5.8}$$

If one denotes the ratio of angular frequency to natural circular frequency $\eta = \Omega/\omega_1$, the particular solutions of (5.5) and (5.6) with $\varphi = \Omega t$ (harmonic excitation) are:

$$x_S = \frac{e\omega_1^2}{\omega_1^2 - \Omega^2} \cos \Omega t = \frac{e}{1-\eta^2} \cos \Omega t = \hat{r}_S \cos \Omega t$$
$$y_S = \frac{e\omega_1^2}{\omega_1^2 - \Omega^2} \sin \Omega t = \frac{e}{1-\eta^2} \sin \Omega t = \hat{r}_S \sin \Omega t. \tag{5.9}$$

The input torque for the constant speed assumed here ($\dot{\varphi} = \Omega =$ const., $\ddot{\varphi} = 0$) for $\Omega \neq \omega_1$ is $M = 0$. Resonance occurs for $\Omega = \omega_1$ ($\eta = 1$), and the forced vibration amplitudes grow indefinitely (if the calculation is performed without damping), see Fig. 6.24.

(5.9) shows that the center of gravity moves along a circular path with angular velocity Ω. The radius of this circle is

$$\hat{r}_S = \sqrt{x_S^2 + y_S^2} = \frac{e\omega_1^2}{\omega_1^2 - \Omega^2} = \frac{e}{1-\eta^2}. \tag{5.10}$$

The motion of the disk center W results from (5.4) with (5.9):

$$x_W = \hat{r}_W \cos \Omega t; \qquad y_W = \hat{r}_W \sin \Omega t; \qquad \hat{r}_W = \frac{e\eta^2}{1-\eta^2}. \tag{5.11}$$

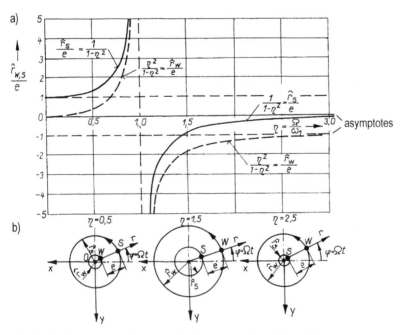

Fig. 5.2 Shaft motion as a function of the frequency ratio; **a)** Amplitude response curve, **b)** Center-of-mass trajectory and shaft center trajectory for $\eta = 0.5$; 1.5; 2.5

5.2 Fundamentals

The radius $\hat{r}_W$ corresponds to the deflection in the rotating system with the polar coordinates $(r; \varphi)$. Unlike $\hat{r}_S$, it can be measured.

The amplitudes of the motions of S and W and the trajectories of these points as a function of the frequency ratio η are shown in Fig. 5.2. Points O, S and W are located on a straight line (outside of the resonance range). The shaft rotates in bent condition and is under a quasi-static load. It is important for the strength calculation of such rotating shafts that no reversed bending occurs and thus no material damping by the shaft is acting.

Figure 5.2 shows that the shaft centers itself at high rotating speeds. The disk rotates about its axis through the center of gravity at speeds above the critical $(\Omega \gg \omega_1)$. This fact is utilized in centrifuges and other machines in which balancing is not an option due to the eccentric stock (Fig. 5.9). The system is tuned to run above the critical speed using soft springs in neck bearings. The disk rotates about its axis through the center of gravity, i. e. the springs in the neck bearings are deflected at the rotational frequency by the magnitude of eccentricity and apply a corresponding force to the frame. It is considerably smaller for a soft neck bearing than the respective unbalancing force that would occur with a rigid bearing.

5.2.2 Passing through the Resonance Point

Many drives are operated at speeds above the critical, i.e. the rotors have to pass through one or several resonance points during start-up until they reach their operating speed, and the same is true for coasting down and braking. This applies to the bending vibrations of rotors (such as turbomachines, textile mandrils, spin-driers, centrifuges), but also to machine foundations (see, for example, Fig. 3.14), torsional oscillators (such as vehicle drive trains, fans), and coupled-mass oscillators (such as screens, belt drives). Extreme loads are reached during this passing through resonance. Of primary interest is the behavior of the rotor near resonance since the largest dynamic deflections occur there.

If one can assume that the rotor starts with a *constant angular acceleration* α, the resulting amplitudes are as shown in Fig. 5.3.

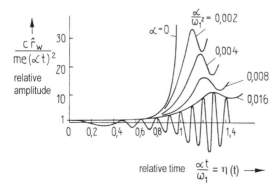

Fig. 5.3 Amplitude when passing through resonance as a function of acceleration

This is to show the result of the calculation only, see Fig. 5.3. The maximum amplitude is not reached when excitation frequency and natural frequency coincide, but slightly later. The maxima shift towards higher speeds during acceleration and towards lower speeds during deceleration processes. The faster the passing through resonance the smaller are the amplitudes. The dimensionless characteristic parameter α/ω_1^2 serves as a measure. If one simulates the start-up process considering the motor characteristic, it turns out that the speed increases at a slower rate near resonance because the drive has to generate energy for moving the resonating foundation. The rotor puts up a larger resistance than the one that would match its own rotational inertia when passing through the resonance point.

The reverse effect occurs when decelerating, i. e. at declining speed: a moment is exerted by the foundation onto the rotor so that the latter accelerates in excess of what the input torque would cause. The excess energy of the vibrating foundation is then transferred onto the rotor. The effect that the foundation acts as an additional drive can be utilized for the efficient operation of vibrating machines as oscillators, e. g. in a vibrating compactor, see Sect. 5.4.5 "Passing through resonance" in [4].

5.2.3 Rotating Shaft with Disk (Gyroscopic Effect)

The oscillation behavior of a rotating shaft is, in principle, different from that of a non-rotating beam since the dynamic behavior is influenced at high speeds by so-called gyroscopic moments, which act on the spatially extended disks. Gyroscopic moments occur when the principal axis of inertia of the disk does not coincide with the axis of rotation or if the disks tilt about the x or y axis when oscillating, see Fig. 5.4.

It is assumed that the disk rotates about the shaft axis at a constant angular velocity $\dot{\varphi} = \Omega$ ($\varphi = \Omega t$), sags in the x and y directions, and tilts at the same time by small angles ψ_x and ψ_y. Moments and forces, which are identified here as F_x, F_y, M_x, M_y, and M_z and are defined in Fig. 5.4, then act between rotor and shaft. The following moments result for unsymmetrical rotors:

$$\begin{bmatrix} M_x \\ M_y \\ M_z \end{bmatrix} = - \begin{bmatrix} J^S_{\xi\xi}\cos^2\varphi + J^S_{\eta\eta}\sin^2\varphi - J^S_{\xi\eta}\sin 2\varphi & \frac{1}{2}(J^S_{\xi\xi}-J^S_{\eta\eta})\sin 2\varphi + J^S_{\xi\eta}\cos 2\varphi \\ \frac{1}{2}(J^S_{\xi\xi}-J^S_{\eta\eta})\sin 2\varphi + J^S_{\xi\eta}\cos 2\varphi & J^S_{\xi\xi}\sin^2\varphi + J^S_{\eta\eta}\cos^2\varphi + J^S_{\xi\eta}\sin 2\varphi \\ J^S_{\eta\zeta}\cos\varphi - J^S_{\eta\zeta}\sin\varphi & J^S_{\xi\zeta}\sin\varphi + J^S_{\eta\zeta}\cos\varphi \end{bmatrix} \cdot \begin{bmatrix} \ddot{\psi}_x \\ \ddot{\psi}_y \end{bmatrix}$$

$$- \Omega \begin{bmatrix} -[(J^S_{\xi\xi}-J^S_{\eta\eta})\sin 2\varphi + 2J^S_{\xi\eta}\cos 2\varphi] & (J^S_{\xi\xi}-J^S_{\eta\eta})\cos 2\varphi - 2J^S_{\xi\eta}\sin 2\varphi + J^S_{\zeta\zeta} \\ (J^S_{\xi\xi}-J^S_{\eta\eta})\cos 2\varphi - 2J^S_{\xi\eta}\sin 2\varphi - J^S_{\zeta\zeta} & (J^S_{\xi\xi}-J^S_{\eta\eta})\sin 2\varphi + 2J^S_{\xi\eta}\cos 2\varphi \\ 0 & 0 \end{bmatrix} \cdot \begin{bmatrix} \dot{\psi}_x \\ \dot{\psi}_y \end{bmatrix}$$

$$- \Omega^2 \begin{bmatrix} 0 & 0 \\ 0 & 0 \\ J^S_{\xi\zeta}\cos\varphi - J^S_{\eta\zeta}\sin\varphi & J^S_{\xi\zeta}\sin\varphi + J^S_{\eta\zeta}\cos\varphi \end{bmatrix} \cdot \begin{bmatrix} \psi_x \\ \psi_y \end{bmatrix} + \Omega^2 \begin{bmatrix} J^S_{\xi\zeta}\sin\varphi + J^S_{\eta\zeta}\cos\varphi \\ -J^S_{\xi\zeta}\cos\varphi + J^S_{\eta\zeta}\sin\varphi \\ 0 \end{bmatrix}$$

(5.12)

5.2 Fundamentals

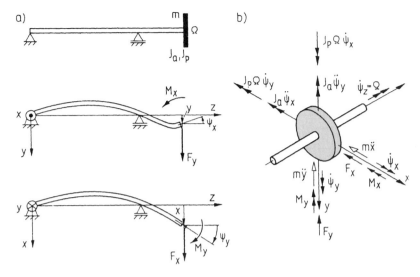

Fig. 5.4 Shaft equipped with a disk; a) Coordinates and load parameters, b) Forces and moments in the free-body diagram of the rotor

Taking into account the rotational transformation (2.15), these expressions result from the principle of conservation of momentum (2.87), according to

$$\boldsymbol{M}^S = \boldsymbol{A}\overline{\boldsymbol{M}}^S = \boldsymbol{A} \cdot \left(\overline{\boldsymbol{\omega}}\,\overline{\boldsymbol{J}}^S\overline{\boldsymbol{\omega}} + \overline{\boldsymbol{J}}^S\dot{\overline{\boldsymbol{\omega}}}\right) \equiv \frac{\mathrm{d}}{\mathrm{d}t}\left(\boldsymbol{A}\overline{\boldsymbol{J}}^S\overline{\boldsymbol{\omega}}\right), \qquad (5.13)$$

if a linearization is performed for small angles

$$|\psi_x| \ll 1, \qquad |\psi_y| \ll 1 \qquad (5.14)$$

and $\boldsymbol{M}^S = [-M_x, -M_y, -M_z]^T$ is set. The following approximations result from (2.14) and (2.30):

$$\boldsymbol{A} \approx \begin{bmatrix} \cos\varphi & -\sin\varphi & \psi_y \\ \sin\varphi & \cos\varphi & -\psi_x \\ \begin{pmatrix}-\psi_y\cos\varphi \\ +\psi_x\sin\varphi\end{pmatrix} & \begin{pmatrix}\psi_x\cos\varphi \\ +\psi_y\sin\varphi\end{pmatrix} & 1 \end{bmatrix}; \quad \overline{\boldsymbol{\omega}} = \begin{bmatrix}\omega_\xi \\ \omega_\eta \\ \omega_\zeta\end{bmatrix} \approx \begin{bmatrix}\begin{pmatrix}\dot\psi_x\cos\varphi \\ +\dot\psi_y\sin\varphi\end{pmatrix} \\ \begin{pmatrix}-\dot\psi_x\sin\varphi \\ +\dot\psi_y\cos\varphi\end{pmatrix} \\ \Omega\end{bmatrix} \qquad (5.15)$$

Insertion in (5.13) yields the moments specified in (5.12) if one neglects all nonlinear terms. (5.17) applies to the forces that immediately result from the equilibrium of forces in the free-body diagram of the rotor. The expressions in (5.12) become considerably simpler in the special case of a symmetrical rotor, that is for

$$J^S_{\xi\xi} = J^S_{\eta\eta} = J_a, \qquad J^S_{\zeta\zeta} = J_p, \qquad J^S_{\xi\eta} = J^S_{\xi\zeta} = J^S_{\eta\zeta} = 0. \qquad (5.16)$$

Table 5.1 Influence coefficients of beams supported by two bearings

No.	Systems schematic	d_{11}
1	(schematic with l_1, l_2, l_3, F_1, F_2, x_1, x_2, c_1, c_2, $l=l_1+l_2+l_3$, EI=const.)	$\dfrac{(l-l_1)^2}{c_1 l^2} + \dfrac{l_1^2}{c_2 l^2} + \dfrac{l_1^2(l-l_1)^2}{3EIl}$
2	(schematic with l_1, l_2, l_3, F_1, F_2, x_1, x_2, c_1, c_2, $(EI)_1, (EI)_2$)	$\dfrac{l_2^2}{c_1(l_1+l_2)^2} + \dfrac{l_1^2}{c_2(l_1+l_2)^2}$ $+\dfrac{l_1^2 l_2^2}{3EI_1(l_1+l_2)}$
3	(schematic with l_1, l_2, F_1, M_2, x_1, ψ_2, c_1, c_2, EI=const.)	$\dfrac{l_2^2}{c_1(l_1+l_2)^2} + \dfrac{l_1^2}{c_2(l_1+l_2)^2}$ $+\dfrac{l_1^2 l_2^2}{3EI(l_1+l_2)}$
4	(schematic with l_1, l_2, l_3, F_1, M_2, rigid, x_1, ψ_2, c_1, c_2, $(EI)_1, (EI)_2$)	$\dfrac{l_2^2}{c_1 l_1^2} + \dfrac{(l_1+l_2)^2}{c_2 l_1^2}$ $+\dfrac{l_1 l_2^2}{3EI_1} + \dfrac{l_3^3 - l_3^3}{3EI_2}$
5	(schematic with l_1, l_2, l_3, F_1, M_2, x_1, ψ_2, c_1, c_2, $l=l_1+l_2+l_3$, EI=const.)	$\dfrac{(l-l_1)^2}{c_1 l^2} + \dfrac{l_1^2}{c_2 l^2} + \dfrac{l_1^2(l_2+l_3)^2}{3EIl}$
6	(schematic with l_1, l_2, l_3, F_1, M_2, x_1, ψ_2, c_1, c_2, $(EI)_1, (EI)_2$)	$\dfrac{l_2^2}{c_1(l_1+l_2)^2} + \dfrac{l_1^2}{c_2(l_1+l_2)^2}$ $+\dfrac{l_1^2 l_2^2}{3EI_1(l_1+l_2)}$

(5.18) result, see also Fig. 5.4 and Sect. 2.3.2.

Equations for calculating the masses and moments of inertia of common rotor designs are specified in Table 5.2 (x; y are the body-fixed axes through the center of gravity).

$$F_x = -m\ddot{x}; \qquad F_y = -m\ddot{y} \qquad (5.17)$$
$$M_x = -J_a\ddot{\psi}_x - J_p\Omega\dot{\psi}_y; \qquad M_y = -J_a\ddot{\psi}_y + J_p\Omega\dot{\psi}_x; \qquad M_z = 0 \quad (5.18)$$

5.2 Fundamentals

Table 5.1 (continued)

d_{12}	d_{22}
$\dfrac{(l_2+l_3)l_3}{c_1 l^2} + \dfrac{l_1(l_1+l_2)}{c_2 l^2} + \dfrac{l_1 l_3(l^2 - l_1^2 - l_3^2)}{6EIl}$	$\dfrac{l_3^2}{c_1 l^2} + \dfrac{(l-l_3)^2}{c_2 l^2} + \dfrac{l_3^2(l_1+l_2)^2}{3EIl}$
$-\dfrac{l_2 l_3}{c_1(l_1+l_2)^2} + \dfrac{(l_1+l_2+l_3)l_1}{c_2(l_1+l_2)^2}$ $-\dfrac{l_1 l_2 l_3(2l_1+l_2)}{6EI_1(l_1+l_2)}$	$\dfrac{l_3^2}{c_1(l_1+l_2)^2} + \dfrac{(l_1+l_2+l_3)^2}{c_2(l_1+l_2)^2}$ $+\dfrac{(l_1+l_2)l_3^2}{3EI_1} + \dfrac{l_3^3}{3EI_2}$
$-\dfrac{l_2}{c_1(l_1+l_2)^2} + \dfrac{l_1}{c_2(l_1+l_2)^2} + \dfrac{l_1 l_2(l_2-l_1)}{3EI(l_1+l_2)}$	$\left(\dfrac{1}{c_1}+\dfrac{1}{c_2}\right)\dfrac{1}{(l_1+l_2)^2} + \dfrac{l_1^3+l_2^3}{3EI(l_1+l_2)^2}$
$\dfrac{l_2}{c_1 l_1^2} + \dfrac{l_1+l_2}{c_2 l_1^2} + \dfrac{l_1 l_2}{3EI_1} + \dfrac{l_2^2 - l_3^2}{2EI_2}$	$\left(\dfrac{1}{c_1}+\dfrac{1}{c_2}\right)\dfrac{1}{l_1^2} + \dfrac{l_1}{3EI_1} + \dfrac{l_2-l_3}{EI_2}$
$\dfrac{l_2+l_3}{c_1 l^2} + \dfrac{l_1}{c_2 l^2} - \dfrac{l_1(l^2 - l_1^2 - 3l_3^2)}{6EIl}$	$\left(\dfrac{1}{c_1}+\dfrac{1}{c_2}\right)\dfrac{1}{l^2} + \dfrac{(l_1+l_2)^3 + l_3^3}{3EIl^2}$
$-\dfrac{l_2}{c_1(l_1+l_2)^2} + \dfrac{l_1}{c_2(l_1+l_2)^2} - \dfrac{l_1 l_2(2l_1+l_2)}{6EI_1(l_1+l_2)}$	$\left(\dfrac{1}{c_1}+\dfrac{1}{c_2}\right)\dfrac{1}{(l_1+l_2)^2} + \dfrac{l_1+l_2}{3EI_1} + \dfrac{l_3}{EI_2}$

The correlations between the load parameters $(F_x; F_y; M_x; M_y)$ and deformation parameters $(x; y; \psi_x; \psi_y)$ are assumed to be linear:

$$y = +\alpha_y F_y - \gamma_y M_x; \qquad x = \alpha_x F_x + \gamma_x M_y \quad (5.19)$$
$$\psi_x = -\delta_y F_y + \beta_y M_x; \qquad \psi_y = \delta_x F_x + \beta_x M_y. \quad (5.20)$$

The different signs in the two planes can be explained by the different force and moment directions, see Fig. 5.4a.

For non-isotropic bearings, the influence coefficients α, β and $\gamma = \delta$ known from the mechanics of materials are dependent on the coordinate direction and were therefore given the indices x and y. If the support is isotropic, the influence coefficients in both planes are equal. Equations for calculating the influence coefficients for common types of shaft supports are included in Table 5.1. For isotropic support, it follows from (5.17) to (5.20):

$$y = +\alpha F_y - \gamma M_x = -\alpha m \ddot{y} - \gamma(-J_a \ddot{\psi}_x - J_p \Omega \dot{\psi}_y) \quad (5.21)$$
$$\psi_x = -\delta F_y + \beta M_x = +\delta m \ddot{y} + \beta(-J_a \ddot{\psi}_x - J_p \Omega \dot{\psi}_y) \quad (5.22)$$
$$x = +\alpha F_x + \gamma M_y = -\alpha m \ddot{x} + \gamma(-J_a \ddot{\psi}_y + J_p \Omega \dot{\psi}_x) \quad (5.23)$$
$$\psi_y = +\delta F_x + \beta M_y = -\delta m \ddot{x} + \beta(-J_a \ddot{\psi}_y + J_p \Omega \dot{\psi}_x). \quad (5.24)$$

The coupling that exists between the 4 equations is caused by the gyroscopic moment. One can see that decoupling occurs for $\Omega = 0$ (non-rotating shaft) so that only the displacement and tilt in one plane influence each other (y and ψ_x or x and ψ_y, respectively). The vibrations in the planes that are offset by 90° can then be examined separately. The gyroscopic moments for rotors in which $J_p \ll J_a$ are relatively small as compared to other moments so that they can be neglected without any substantial loss in accuracy (e. g. textile mandrils), see Fig. 5.5a and Fig. 6.30.

In the case of isotropic support, it is permissible and useful to use the rotating plane ($r; z$) formed by the z axis and the elastic line rather than the two fixed planes ($x, z; y, z$). The radial displacement r of the shaft center and the angle ψ of the cone formed by the rotating tangent can be used as coordinates.

Fig. 5.5 Rotor shapes
a) $J_a \gg J_p$ (Roller shape), gyroscopic effect negligible
b) $J_a < J_p$ (Disk shape), gyroscopic effect substantial

The shaft shown in Fig. 5.4 corresponds to case 6 in Table 5.1, the influence coefficients being $\alpha = d_{11}$, $\beta = d_{22}$, $\gamma = \delta = d_{12}$. The following equations apply likewise to cases 3 to 5.

(5.21) to (5.24) can be simplified if one introduces the complex variables $\tilde{r} = x + jy$, $\tilde{\psi} = \psi_y - j\psi_x$, $\tilde{F} = F_x + jF_y$, and $\tilde{M} = M_y - jM_x$ using $j = \sqrt{-1}$. Two equations of motion result if one combines (5.23) with (5.21) and (5.22) with (5.24):

$$\tilde{r} = \alpha \tilde{F} + \gamma \tilde{M} = -\alpha m \ddot{\tilde{r}} - \gamma(J_a \ddot{\tilde{\psi}} - jJ_p \Omega \dot{\tilde{\psi}})$$
$$\tilde{\psi} = \delta \tilde{F} + \beta \tilde{M} = -\delta m \ddot{\tilde{r}} - \beta(J_a \ddot{\tilde{\psi}} - jJ_p \Omega \dot{\tilde{\psi}}). \quad (5.25)$$

If one assumes the mode shapes of interest to be of the form

$$\tilde{r} = \hat{r}e^{j\omega t}; \qquad \tilde{\psi} = \hat{\psi}e^{j\omega t}, \quad (5.26)$$

a homogeneous linear system of equations results for the amplitudes $\hat{r}$ and $\hat{\psi}$:

$$\hat{r} = \alpha m \omega^2 \hat{r} + \gamma(J_a \omega^2 - J_p \Omega \omega)\hat{\psi}$$
$$\hat{\psi} = \delta m \omega^2 \hat{r} + \beta(J_a \omega^2 - J_p \Omega \omega)\hat{\psi}. \quad (5.27)$$

5.2 Fundamentals

Table 5.2 Masses, moments of inertia, and distances to the center of gravity of symmetrical rotors

Component	Distance t to the center of gravity mass m	Polar moment of inertia $J_{zz} = J_p$	Axial moment of inertia $J_{yy} = J_{xx} = J_a$
Solid cylinder	$t = \frac{h}{2}$ $m = \rho \frac{\pi}{4} \cdot d^2 \cdot h$	$m \frac{d^2}{8}$	$m \frac{3d^2 + 4h^2}{48}$
Hollow cylinder	$t = \frac{h}{2}$ $m = \rho \cdot \frac{\pi}{4}(d_1^2 - d_2^2) \cdot h$	$m \frac{d_1^2 + d_2^2}{8}$	$m \frac{3d_1^2 + 3d_2^2 + 4h^2}{48}$
Thin-walled hollow cyl.	$(d_1 \approx d_2 = d)$ $\left(\frac{d_1 - d_2}{2} = s\right)$ $t = \frac{h}{2}$ $m = \rho \cdot \pi \cdot d \cdot s \cdot h$	$m \frac{d^2}{4}$	$m \frac{3d^2 + 2h^2}{24}$
Thin-walled truncated cone	$t = \frac{h}{3} \cdot \frac{d_1 + 2d_2}{d_1 + d_2}$ $m = \rho \frac{\pi}{4}(d_1 + d_2) s \cdot$ $\sqrt{4h^2 + (d_1 - d_2)^2}$	$m \frac{d_1^2 + d_2^2}{8}$	$\frac{m}{18} \left[\frac{9}{8}(d_1^2 + d_2^2) + h^2 + 2 \frac{d_1 d_2 h^2}{(d_1 + d_2)^2} \right]$
Hub spider (n ribs)	$t = \frac{h}{2} + \frac{d_1 - d_2}{4} \tan\alpha$ $m = \rho \cdot h \cdot s \cdot \frac{d_1 - d_2}{2} \cdot n$	$\frac{m}{12} \frac{d_1^3 - d_2^3}{d_1 - d_2}$	$a = (d_1 - d_2) \cdot \tan\alpha$ $b = 4 - 3 \cdot \cos\alpha$ $m \cdot \frac{b}{3 \cdot \cos\alpha} \left[\frac{h^2}{4} + \frac{a^2}{16} + \frac{3}{4} \frac{ah(1 - \cos\alpha)}{b} \right]$

The natural frequencies result from setting the coefficient determinant to zero:

$$\begin{vmatrix} \alpha m\omega^2 - 1 & \gamma(J_a - J_p \Omega/\omega)\omega^2 \\ \delta m\omega^2 & \beta(J_a - J_p \Omega/\omega)\omega^2 - 1 \end{vmatrix} = 0. \quad (5.28)$$

The amplitude ratio results from (5.27) and characterizes the mode shape:

$$\left(\frac{\hat{r}}{\hat{\psi}}\right)_i = \frac{\gamma\omega_i^2(J_\mathrm{a} - J_\mathrm{p}\Omega/\omega_i)}{1 - \alpha m \omega_i^2} = \frac{1 - \beta\omega_i^2(J_\mathrm{a} - J_\mathrm{p}\Omega/\omega_i)}{\delta m \omega_i^2}. \tag{5.29}$$

It is interesting that the variable

$$J_\mathrm{R} = J_\mathrm{a} - J_\mathrm{p}\frac{\Omega}{\omega} \tag{5.30}$$

appears in (5.28) and (5.29) at the place where only the moment of inertia J_a occurs for the non-rotating beam ($\Omega = 0$). Therefore, if the ratio Ω/ω is given, one can calculate using J_R as for a non-rotating rigid body. Due to $\delta = \gamma$, the characteristic equation

$$\begin{aligned}\omega^4 m J_\mathrm{a}(\alpha\beta-\gamma^2) - \omega^3 m J_\mathrm{p}\Omega(\alpha\beta-\gamma^2) - \omega^2(\alpha m+\beta J_\mathrm{a}) + \omega\beta J_\mathrm{p}\Omega + 1 = \\ \omega^4 m J_\mathrm{R}(\alpha\beta - \gamma^2) - \omega^2(\alpha m + \beta J_\mathrm{R}) + 1 = 0 \end{aligned} \tag{5.31}$$

is obtained from the determinant (5.28) after solving and performing brief conversions. Thus, there is a total of four (signed) natural circular frequencies ω_i, which can be determined by solving (5.31). They depend on the angular velocity Ω of the shaft. Figure 5.6 shows the general curve. The natural circular frequencies are independent of J_p for $\Omega = 0$ because there is no gyroscopic effect for a non-rotating shaft. (5.31) becomes simpler and turns into a quadratic equation that yields the natural circular frequencies:

$$\omega_{10,20} = \sqrt{\frac{\alpha m + \beta J_\mathrm{a}}{2 m J_\mathrm{a}(\alpha\beta - \gamma^2)}\left[1 \mp \sqrt{1 - \frac{4 m J_\mathrm{a}(\alpha\beta - \gamma^2)}{(\alpha m + \beta J_\mathrm{a})^2}}\right]}. \tag{5.32}$$

The influence of the gyroscopic effect is mainly determined by $J_\mathrm{p}/J_\mathrm{a}$ and by Ω. (5.31) is solved for Ω to represent the dependence transparently for any speed. The inverse function

$$\Omega = \frac{J_\mathrm{a}}{J_\mathrm{p}}\omega + \frac{-1 + \omega^2 \alpha m}{[\beta - (\alpha\beta - \gamma^2) m \omega^2] J_\mathrm{p}\omega} \quad (-\infty < \omega < \infty) \tag{5.33}$$

is obtained after a few conversions.

One can determine the following asymptotes for the 4 natural circular frequencies from the frequency equation in the form of (5.31) or (5.33) by means of the limiting process $\Omega \to \infty$:

$$\omega_1 = -\frac{1}{\beta J_\mathrm{p}\Omega} \qquad \omega_3 = -\sqrt{\frac{\beta}{m(\alpha\beta - \gamma^2)}} = -\omega_\infty$$

$$\omega_2 = \sqrt{\frac{\beta}{m(\alpha\beta - \gamma^2)}} = \omega_\infty \qquad \omega_4 = \frac{J_\mathrm{p}}{J_\mathrm{a}}\Omega \tag{5.34}$$

5.2 Fundamentals

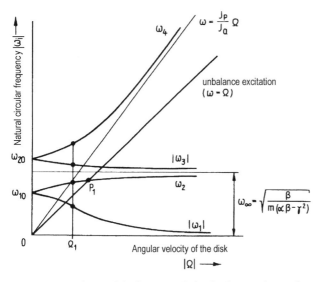

Fig. 5.6 Dependence of the four natural circular frequencies on the angular velocity Ω

It is not common to have negative natural circular frequencies, which is why only the magnitudes $|\Omega|$ have been plotted in Fig. 5.6 (natural circular frequencies are independent of the direction of rotation of the rotor). The negative sign in the case presented here indicates that the shaft center rotates in the opposite direction of the shaft rotating at Ω during the vibration at ω_1 or ω_3, see (5.26). This is called *rotation in opposite direction*.

The direction of rotation of the shaft vibration and the direction of rotation of the rotating shaft are the same for the positive values of the natural circular frequencies ω_2 and ω_4. This is called *rotation in the same direction*, see Fig. 5.7. The four natural circular frequencies that the shaft has at an angular velocity Ω_1 are depicted in Fig. 5.6 as small full circles. The associated mode shapes, which indicate the ratio of radial to angular amplitude at the respective natural frequency, are obtained by inserting the calculated ω_i in (5.29).

One can see from Fig. 5.6 that both for opposite and same directions of rotation there are two natural frequencies of the rotating shaft, the magnitude of which depends on the angular velocity Ω. Furthermore, it follows from this figure that for the same direction of rotation the natural frequencies are increased as compared to the non-rotating shaft due to the gyroscopic effect. The upper branch of the curve asymptotically approaches the straight line that results from (5.34). Critical speeds occur when the rotating shaft is excited by one of its natural frequencies. Principal and secondary excitations have to be distinguished.

The *principal excitation* of a rotating shaft is represented by its *unbalance excitation*. As in this case, a natural frequency ω_i coincides with the angular velocity Ω of the shaft, the resonance points are found as intersections of the straight line $\omega = \Omega$ and the curves according to (5.33). The *critical speeds of rotation in the same direction* result from (5.31) where $J_R = J_a - J_p$:

$$\frac{1}{\omega_{2,4}^2} = \frac{1}{2}\left\{\alpha m + \beta(J_a - J_p) \pm \sqrt{[\alpha m + \beta(J_a - J_p)]^2 - 4(\alpha\beta - \gamma^2)m(J_a - J_p)}\right\}. \tag{5.35}$$

For all rotors where $J_p/J_a > 1$, see Fig. 5.5 and Table 5.2 where $h \ll d : J_p/J_a = 2$, the root expression becomes larger than the preceding terms, and a negative ω^2 emerges. In other words: There is only one resonance of rotation in the same direction for disk-shaped rotors (Fig. 5.6, point P_1). $J_a > J_p$ can apply to drum-shaped rotors. Then there are two resonance points also during rotation in the same direction because the curve of $\omega_4(\Omega)$ is flatter, see Fig. 5.6.

The special case $J_a \approx J_p$ is worth noting because for high speeds the frequency of unbalance excitation approaches asymptotically the second natural frequency for rotation in the same direction. This means that for such rotors (e. g. milk centrifuges and spin driers), for which J_a/J_p depends on the load status and may come close to one, resonance occurs at all higher speeds, and passing through the second critical speed becomes impossible. The designer should avoid such phenomena by selecting favorable system parameters. *Resonance during rotation in the same direction is dangerous because the shaft deformation does not produce damping.* The shaft rotates in bent condition and is, as it were, exposed to static loading only. Only bearing damping is effective. The mode shapes of opposite rotation are exposed to stronger damping because material damping becomes active due to the alternating deformation of the shaft that occurs, see Fig. 5.7.

If the rotating shaft is not excited by the unbalance, but from outside by other forces or motions, it is called *secondary* or *external excitation*. Secondary excitations are, for example, motions of the mounting site that act onto the shaft as support motion(such as a shaft on a vibrating frame, centrifuges or fans in vibrating vehicles). Secondary excitations can also be caused by the input or output forces of a rotor, e. g. the forces with the meshing frequency of gear teeth or of the drive chain, the inertia forces of a linkage (crankshaft in reciprocating engines) et. al. In the case of secondary or external excitation, the rotor can be excited in resonance both in the natural frequencies of rotation in the opposite or in the same directions. Unlike with unbalance excitation, higher harmonics of excitation can be expected.

After calculating the natural circular frequencies ω_i for the fixed angular velocity Ω, the design engineer will have to check if it coincides with an integral multiple of a secondary circular frequency of the excitation ν that may exist independently of Ω.

$$\omega_i(\Omega) = \pm k \cdot \nu; \qquad k = 1, 2, \ldots \tag{5.36}$$

has to be avoided.

The influence of the gyroscopic effect on the position of the first critical speed depends on the ratio J_p/J_a and on whether the rotor tilts strongly (e. g. when overhung) or weakly (e. g. when centered between bearings) perpendicular to the axis of the shaft. This is determined by the influence coefficients. If it is positioned exactly symmetrically between the bearings, $\gamma = \delta = 0$ and thus the gyroscopic effect exerts no influence on ω_1 at any J_p.

5.2 Fundamentals

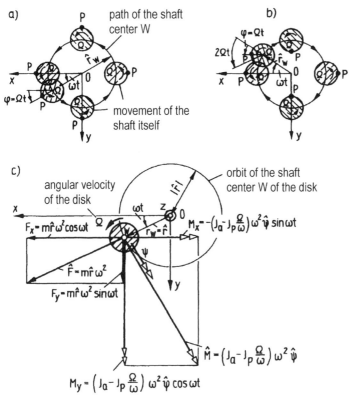

Fig. 5.7 Disk motion; **a)** Synchronous rotation in the same direction ($\omega = \Omega$), **b)** Synchronous rotation in opposite direction ($\omega = -\Omega$), **c)** Loading of the shaft rotating at angular velocity $\Omega = \pi n/30$ and vibrating circular frequency ω

It helps to imagine the forces and moments acting onto the disk that result from (5.17) and (5.18) at the natural frequencies according to (5.26). They are depicted in Fig. 5.7c.

The centrifugal force then acts in the direction of the displacement

$$\tilde{F} = F_x + jF_y = m\omega^2 \hat{r} e^{j\omega t} = m\omega^2 \hat{r}(\cos\omega t + j\sin\omega t) \quad (5.37)$$

and a moment that depends on the ratio Ω/ω, see (5.30), acts in the perpendicular direction:

$$\tilde{M} = M_y - jM_x = J_R\omega^2 \hat{\psi} e^{j\omega t} = J_R\omega^2 \hat{\psi}(\cos\omega t + j\sin\omega t). \quad (5.38)$$

Both rotate with the bent shaft. The center of the disk rotates at an angular velocity ω, the disk at an angular velocity Ω about its axis through the center of gravity.

5.2.4 Bending Oscillators with a Finite Number of Degrees of Freedom

The basis for the following is a calculation model that consists of massless shaft sections with n disks mounted on them, see Fig. 5.8. The moments of inertia J_{Rk} are with respect to the axis through the center of gravity that is perpendicular to the bending plane, see (5.30). Since the relationships between load and deformation parameters are linear for small deformations, the following results for $j = 1, 2, \ldots, n$, see (5.25):

$$r_j = \sum_k (\alpha_{jk} F_k + \gamma_{jk} M_k) \qquad (5.39)$$

$$\psi_j = \sum_k (\delta_{jk} F_k + \beta_{jk} M_k). \qquad (5.40)$$

If one summarizes all coordinates by $q_j = (r_j, \psi_j)$, all load parameters by $f_k = (F_k, M_k)$, and all inertia parameters by $m_{kk} = (m_k, J_{kk})$, one can introduce a uniform notation with generalized coordinates that is also used in Chap. 6. (5.39) and (5.40) can thus be written as

$$q_j = \sum_{k=1}^{n} d_{jk} f_k; \qquad j = 1, 2, \ldots, n. \qquad (5.41)$$

The influence coefficients $\alpha_{jk}, \beta_{jk}, \gamma_{jk}$ and δ_{jk} are then uniformly denoted as d_{jk} and arranged in the compliance matrix $\boldsymbol{D} = [d_{jk}]$.

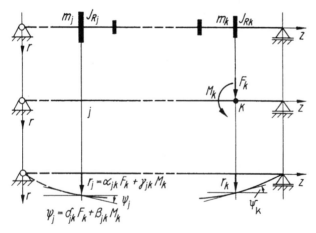

Fig. 5.8 Calculation model of the massless shaft with disks

(5.41) then becomes $\boldsymbol{q} = \boldsymbol{D}\boldsymbol{f}$ with the coordinate vector $\boldsymbol{q}$ and the force vector $\boldsymbol{f}$ according to

5.2 Fundamentals

$$q = [q_1, q_2, \ldots, q_n]^T; \qquad f = [F_1, F_2, \ldots, F_n]^T. \tag{5.42}$$

In the case of *non-rotating beams*, the inertia forces are $f = -M\ddot{q}$ with the diagonal mass matrix $M = \mathrm{diag}(m_{kk})$. The equations of motion (5.21) to (5.24) simplify for the free vibrations of a bending oscillator:

$$q = Df = -DM\ddot{q}. \tag{5.43}$$

Bending oscillators belong to the oscillators with n degrees of freedom, which are discussed in detail in Chap. 6. Only the general approach is indicated here.

The solution of (5.43) is assumed to be of the form $q = v\exp(\mathrm{j}\omega t)$. The result is n linearly independent equations for the components of the amplitude vector $v^T = [v_1, v_2, \ldots, v_n]$. The following applies in matrix notation

$$(E - \omega^2 DM)v = o. \tag{5.44}$$

This system of equations has non-trivial solutions if

$$\det(E - \omega^2 DM) = 0. \tag{5.45}$$

An nth-degree polynomial results for ω^2 from the solution of the frequency determinant. It has n real roots, from which one can calculate the natural frequencies $f_i = \omega_i/(2\pi)$ of this oscillator with n degrees of freedom.

Free vibrations are generated if a bending oscillator is displaced from its static equilibrium position and then left to itself. This displacement can be described by the initial conditions

$$t = 0: \qquad q(0) = q_0; \qquad \dot{q}(0) = u_0. \tag{5.46}$$

The solution of the equations of motion (5.43), taking into account (5.46), is discussed in detail in 6.3, see also the example in P5.6.

Only bending oscillators with 2 degrees of freedom that do not rotate will be discussed here. If a beam is equipped with $n = 2$ point masses, it follows from (5.44) with $\hat{r}_k = v_k$ and $\alpha_{jk} = d_{jk}$:

$$\begin{aligned} j = 1: &\quad (1 - \omega^2 d_{11} m_1)v_1 - \omega^2 d_{12} m_2 v_2 = 0 \\ j = 2: &\quad -\omega^2 d_{21} m_1 v_1 + (1 - \omega^2 d_{22} m_2)v_2 = 0. \end{aligned} \tag{5.47}$$

Setting the coefficient determinant to zero provides the characteristic equation (with $d_{12} = d_{21}$)

$$\begin{aligned} \det &= \begin{vmatrix} 1 - \omega^2 d_{11} m_1 & -\omega^2 d_{12} m_2 \\ -\omega^2 d_{21} m_1 & 1 - \omega^2 d_{22} m_2 \end{vmatrix} \\ &= \omega^4 m_1 m_2 (d_{11} d_{22} - d_{12}^2) - \omega^2 (d_{11} m_1 + d_{22} m_2) + 1 = 0. \end{aligned} \tag{5.48}$$

The two natural circular frequencies are found by solving this quadratic equation:

$$\omega_{1,2}^2 = \frac{d_{11}m_1 + d_{22}m_2}{2m_1 m_2 (d_{11}d_{22} - d_{12}^2)} \left(1 \mp \sqrt{1 - \frac{4m_1 m_2 (d_{11}d_{22} - d_{12}^2)}{(d_{11}m_1 + d_{22}m_2)^2}}\right). \quad (5.49)$$

The amplitude ratios that are brought about by the natural vibration of the beam with 2 point masses result from (5.47) if $\omega = \omega_i$ is set there:

$$\left(\frac{v_2}{v_1}\right)_i = \frac{1 - \omega_i^2 d_{11}m_1}{\omega_i^2 d_{12}m_2} = \frac{\omega_i^2 d_{21}m_1}{1 - \omega_i^2 d_{22}m_2}; \qquad i = 1, 2. \quad (5.50)$$

If the bending oscillator vibrates at one of its natural frequencies, the so-called mode shape is obtained at which the amplitudes have the fixed numerical ratio given by (5.50).

In the case of bending oscillators, the natural frequencies and mode shapes are calculated based on the influence coefficients. Table 5.1 lists the influence coefficients for shafts in two bearings, taking into account the bearing elasticity. Rigid support results as a special case ($1/c_1 = 0$ or $1/c_2 = 0$, respectively). Case 4 in Table 5.1 takes into account that the rotor-shaft joint does not coincide with the center of gravity of the rotor, as is the case, for example, for centrifuge drums and mandrils.

5.2.5 Examples

5.2.5.1 Natural Frequencies of a Milk Centrifuge

The milk centrifuge shown in Fig. 5.9 is used to separate liquids. The elastic neck bearing was introduced in 1889 for steam turbines, textile mandrils, and centrifuges after realizing that supercritical operation is feasible and beneficial. Questions relating to the set-up and sizing of mandril, neck bearing, and drum play an essential part in the optimal sizing of the design.

The shaft can be considered perfectly rigid and massless (bearing stiffness $\ll$ bending stiffness, shaft mass $\ll$ rotor mass). The minimal model is a hinged rigid body with a single elastic isotropic bearing, see Fig. 5.9b.

If the rotor is displaced from its static equilibrium position at $\psi_x = \psi_y = 0$, the forces and moments shown in Fig. 5.10 occur due to inertia, see Fig. 5.4. The positive directions of the small angles ψ_x and ψ_y have been specified here so that the respective angular velocities point in the positive axis directions (right-hand system).

The force components shown represent

- the restoring forces of the elastic bearing ($cl\psi_x$, $cl\psi_y$)
- the inertia forces from translating the center of gravity ($mL\ddot{\psi}_x$, $mL\ddot{\psi}_y$) that should be plotted in opposite direction of the positive coordinate direction

5.2 Fundamentals

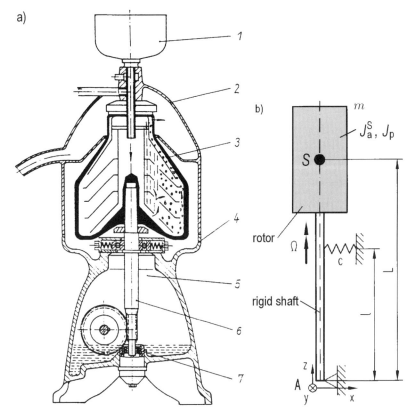

Fig. 5.9 Milk centrifuge; **a)** Drawing; *1* Inlet device, *2* Drum cover, *3* Drum, *4* Frame, *5* Neck bearing, *6* Mandril, *7* Footstep bearing; **b)** Calculation model

- the inertia moments from the rotational inertia when rotating about the axes through the center of gravity perpendicular to the image plane ($J_a^S \ddot{\psi}_x$, $J_a^S \ddot{\psi}_y$)
- the gyroscopic moments as a result of the rotation about the z axis and simultaneous rotation ψ_x or ψ_y ($J_p \Omega \dot{\psi}_x$, $J_p \Omega \dot{\psi}_y$), respectively.

The directions of the gyroscopic moments can also be explained as a consequence of the change of direction of the angular momentum vector, see also (5.18). It follows from the equilibrium of moments about point A (note the respective leverages):

$$\begin{aligned} J_a^S \ddot{\psi}_x + J_p \Omega \dot{\psi}_y + mL^2 \ddot{\psi}_x + cl^2 \psi_x = 0 \\ J_a^S \ddot{\psi}_y - J_p \Omega \dot{\psi}_x + mL^2 \ddot{\psi}_y + cl^2 \psi_y = 0. \end{aligned} \quad (5.51)$$

The following equations of motion are obtained after rearranging the terms and using the moment of inertia with respect to the bearing A ($J_A = J_a^S + mL^2$):

$$J_A \ddot{\psi}_x + J_p \Omega \dot{\psi}_y + cl^2 \psi_x = 0; \qquad J_A \ddot{\psi}_y - J_p \Omega \dot{\psi}_x + cl^2 \psi_y = 0. \quad (5.52)$$

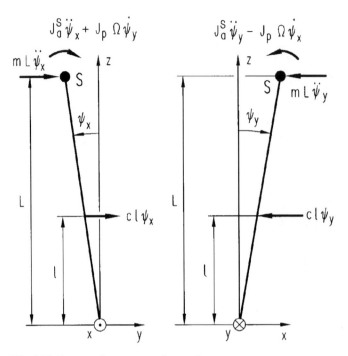

Fig. 5.10 Forces and moments acting on the rotor

Without the gyroscopic effect, the system of equations (5.52) is decoupled, and the following natural circular frequency is obtained from each equation (for $\Omega = 0$)

$$\omega_0 = \sqrt{\frac{cl^2}{J_A}}. \tag{5.53}$$

The motions in both planes are coupled due to the gyroscopic effect. Since bearing and shaft are isotropic, the center of gravity of the rotor moves on a circular path about the z axis. The equations

$$\psi_x = \hat{\psi} \cos \omega t; \qquad \psi_y = \hat{\psi} \sin \omega t \tag{5.54}$$

describe such a circular path with an initially unknown circular frequency ω. The amplitude $r = L\hat{\psi}$ is assumed for the radius. If one inserts the angular velocities $(\dot{\psi}_x, \dot{\psi}_y)$ and angular accelerations $(\ddot{\psi}_x, \ddot{\psi}_y)$ in (5.52), one obtains

$$\begin{aligned}
(-\omega^2 J_A + J_p \Omega \omega + cl^2)\, \hat{\psi} \cos \omega t &= 0 \\
(-\omega^2 J_A + J_p \Omega \omega + cl^2)\, \hat{\psi} \sin \omega t &= 0.
\end{aligned} \tag{5.55}$$

The characteristic equation is derived from both equations by setting the expressions in parentheses to zero:

5.2 Fundamentals

$$\omega^2 - \frac{J_p\Omega}{J_A}\omega - \frac{cl^2}{J_A} = 0. \tag{5.56}$$

From this follow the natural circular frequencies

$$\omega_{1,2} = \frac{J_p\Omega}{2J_A} \mp \sqrt{\frac{cl^2}{J_A} + \left(\frac{J_p\Omega}{2J_A}\right)^2}. \tag{5.57}$$

The influence of the gyroscopic effect can be evaluated using the characteristic parameter

$$\varepsilon = \frac{J_p\Omega}{2J_A\omega_0} = \frac{J_p\Omega}{2l\sqrt{cJ_A}}. \tag{5.58}$$

It is virtually negligible if $\varepsilon < 0.05$. The influence of the velocity of rotation Ω on the natural frequencies is smaller for cylindrical rotors than for disk-shaped rotors, see Fig. 5.5.

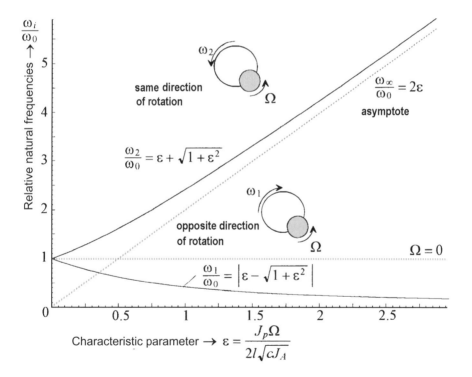

Fig. 5.11 Natural frequencies as a function of angular velocity

Note that according to (5.57) there are a "positive" and a "negative natural circular frequency" since the root expression is larger than the first summand. If one inserts the negative root in (5.54), one will see that it corresponds to another direction of rotation ("opposite direction of rotation": rotation of the rotor in opposite direction of the vibration direction) than the positive root ("same direction of rotation": rotation of the rotor in the same direction as the vibration direction).

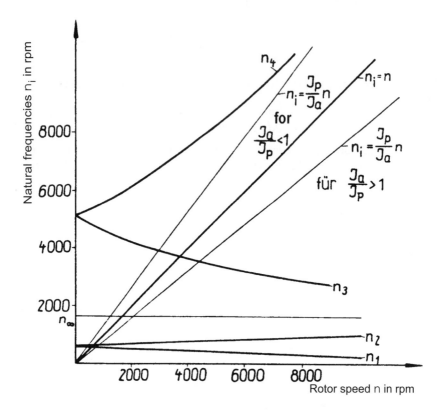

Fig. 5.12 Speed-dependent natural frequencies of a milk centrifuge

For the milk centrifuge shown in Fig. 5.9a, the speed-dependent natural frequencies shown in Fig. 5.12 were calculated using the model from Table 5.1, Case 4. This model with four degrees of freedom has four natural frequencies. The intersection of the straight line n_i with the curve n_2 provides the critical speed of synchronous rotation in the opposite direction that is generated in the event of unbalance excitation. Curves n_1 and n_3 correspond to the mode shapes of rotation in opposite direction.

However, not all dynamic phenomena found in experiments on such centrifuges can be explained with this simple model. Resonance points of the non-rotating drum ($\Omega = 0$) for $f_1 = 11.5$ Hz ($\widehat{=}690$ rpm), $f_2 = 41$ Hz ($\widehat{=}2460$ rpm) and $f_3 = 62$ Hz ($\widehat{=}3720$ rpm) were measured. The mode shapes of the shaft correspond to the natural frequencies f_1 and f_3, whereas f_2 was caused by the elastic support of the housing. Moreover, a rotation of the entire frame was determined as a mode shape that had as yet been unknown. A calculation model that takes into account the

5.2 Fundamentals

elastic set-up of the housing, the resilience of the drum and other effects is required to accurately clarify all dynamic influences.

5.2.5.2 Impact of a Moving Beam

A flap with a horizontal axis of rotation is slammed shut. The first three natural frequencies during free rotation ($c = 0$) and after the impact ($c \neq 0$) are to be calculated and represented as a function of the characteristic parameter $\bar{c} = 2EI/(cl^3)$. The flap is modeled as a bending oscillator with 4 degrees of freedom and the point of impact is modeled as a massless spring, see Fig. 5.13.

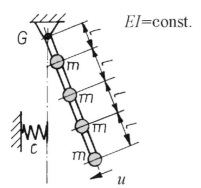

Fig. 5.13 Impact of a beam; Model parameters: c, m, l, EI

Figure 5.13 shows the bending oscillator prior to the impact. The absolute displacements can be combined in the coordinate vector

$$q = [q_1;\ q_2;\ q_3;\ q_4]^{\mathrm{T}}. \tag{5.59}$$

The equation of motion corresponds to (5.43).

The elements of the compliance matrix D are determined from Table 5.1. For example, the influence coefficients according to case 1 there for $l_1 = l_2 = l_3 = l$; $c_1 \to \infty$; $c_2 = c$ are:

$$d_{11} = \frac{1}{9c} + \frac{4l^3}{9EI}; \quad d_{12} = \frac{2}{9c} + \frac{7l^3}{18EI}; \quad d_{22} = \frac{4}{9c} + \frac{4l^3}{9EI}. \tag{5.60}$$

Case 2 provides the influence coefficients for $l_1 = l_3 = l$; $l_2 = 2l$; $c_1 \to \infty$ and $c_2 = c$:

$$d_{14} = \frac{4}{9c} - \frac{4l^3}{9EI}; \quad d_{44} = \frac{16}{9c} + \frac{4l^3}{3EI}. \tag{5.61}$$

All elements of the compliance matrix can be calculated in this way:

$$\boldsymbol{D} = \frac{1}{9c}\begin{bmatrix} 1 & 2 & 3 & 4 \\ 2 & 4 & 6 & 8 \\ 3 & 6 & 9 & 12 \\ 4 & 8 & 12 & 16 \end{bmatrix} + \frac{l^3}{18EI}\begin{bmatrix} 8 & 7 & 0 & -8 \\ 7 & 8 & 0 & -10 \\ 0 & 0 & 0 & 0 \\ -8 & -10 & 0 & 24 \end{bmatrix}. \qquad (5.62)$$

The mass matrix is

$$\boldsymbol{M} = \mathrm{diag}[m, m, m, m] = m\boldsymbol{E}. \qquad (5.63)$$

The natural circular frequencies result from solving the eigenvalue problem (5.45). The numerical solution was calculated using software. The results for the three lowest natural circular frequencies are presented in Fig. 5.14.

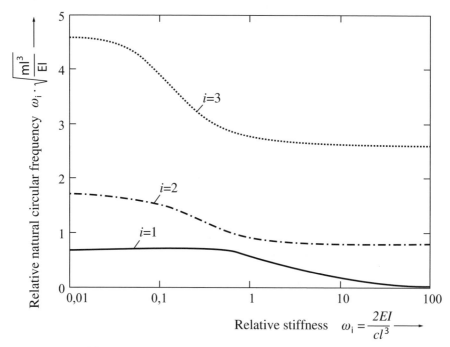

Fig. 5.14 Natural frequencies of the supported beam from Fig. 5.13 as a function of the bearing stiffness

Small values of the characteristic parameter $\bar{c}$ indicate a beam with relatively low bending resistance, and large $\bar{c}$ values indicate a very soft spring (limiting case: no spring), cf. see the asymptotes in Fig. 5.14. They approach asymptotically the values of the "pinned – free" case in Table 5.7 for $\bar{c} \to \infty$ ($ml^3 = \varrho AL^4$ where $L = 4l$) of the continuum.

The loads are calculated in Sect. 6.4.7, see Solution S6.7.

5.2 Fundamentals

5.2.6 Problems P5.1 to P5.3

P5.1 Evaluation of the Dynamic Bearing Force

A rigid unbalanced rotor with an elastic bearing (see Figs. 5.1 and 5.9) causes an eight times smaller dynamic bearing load at a speed of $n = 30\,000$ rpm than one in a rigid bearing. Explain this fact and calculate the natural frequency of the elastically supported rotor assuming that it behaves like an oscillator with one degree of freedom.

P5.2 Free Vibrations of a Bending Oscillator

Calculate the critical speeds and bending mode shapes of the calculation model of a machine shaft shown in Fig. 5.15a.

Given:

$l_1 = 300$ mm $\quad d = 25$ mm $\quad E = 2.1 \cdot 10^5$ N/mm^2
$l_2 = 200$ mm $\quad m_1 = 6$ kg
$l_3 = 500$ mm $\quad m_2 = 4$ kg

The shaft mass of 3.8 kg is distributed across the point masses.

Find: Influence coefficients, the two critical speeds, mode shapes

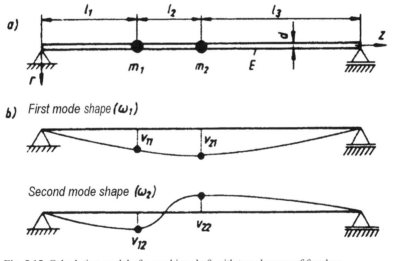

Fig. 5.15 Calculation model of a machine shaft with two degrees of freedom

P5.3 Influence of the Gyroscopic Effect on Natural Frequencies

To compare the elastic and gyroscopic influences, calculate the two natural circular frequencies using the models shown in Fig. 5.16 for the rotating and non-rotating shaft. Use the parameters m, l, EI and the numerical values specified in Table 5.3. Assume synchronous rotation in the same direction for the rotating shaft.

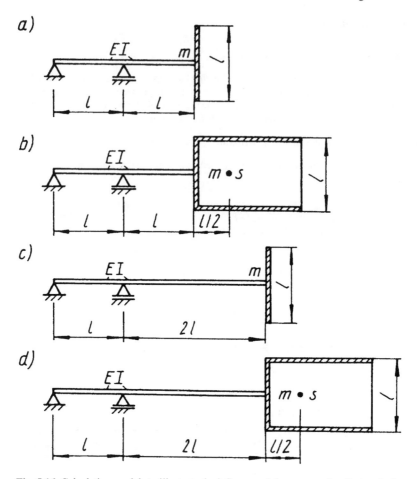

Fig. 5.16 Calculation models to illustrate the influence of the gyroscopic effect and of compliance

Table 5.3 Numerical values for the calculation models of Fig. 5.16

Model	$\alpha\dfrac{EI}{l^3}$	$\beta\dfrac{EI}{l}$	$\gamma\dfrac{EI}{l^2}$	$\dfrac{J_a}{ml^2}$	$\dfrac{J_p}{J_a}$	$\dfrac{\beta J_a}{\alpha m}$	$\alpha m\omega_\infty^2$
Figure 5.16a	0.6667	1.3333	0.8333	0.2500	2	0.5000	0.2188
Figure 5.16b	0.6250	0.8333	0.7083	0.3410	0.857	0.4547	0.0368
Figure 5.16c	4.0000	2.3333	2.6667	0.2500	2	0.1458	0.2381
Figure 5.16d	3.9583	1.8333	2.5417	0.3410	0.857	0.1579	0.1098

5.2.7 Solutions S5.1 to S5.3

S5.1 For a rigid rotor in rigid bearings, the dynamic bearing forces increase with the square of the speed due to the centrifugal forces of the unbalances, see (2.328). In a rigid rotor with

5.2 Fundamentals

elastic support, the bearing forces result from the centrifugal force and the inertia force from the free vibration that acts in opposite direction (in the speed range above the critical).

The ratio of the dynamic bearing force on the elastically supported rotor to that of the rigidly supported one, see (5.10) and (2.322), is

$$\left|\frac{F_{el}}{F_{st}}\right| = \frac{m\sqrt{\ddot{x}_S^2 + \ddot{y}_S^2}}{me\Omega^2} = \frac{me\Omega^2\omega_1^2}{me\Omega^2(\omega_1^2 - \Omega_1^2)} = \frac{1}{|1-\eta^2|} = 0.125. \tag{5.64}$$

From that it follows that $\Omega/\omega_1 = \eta = 3$, and thus the natural frequency is

$$\underline{f_1} = \frac{\omega_1}{2\pi} = \frac{\Omega}{2\pi \cdot 3} = \frac{\pi n}{30 \cdot 2\pi \cdot 3} = \underline{166.7 \text{Hz}} \tag{5.65}$$

S5.2 The influence coefficients can be taken from Table 5.1(Model 1). The following results for $c_1 = c_2 \to \infty$ and $l = l_1 + l_2 + l_3$ where $I = \pi d^4/64$:

$$d_{11} = \frac{l_1^2(l-l_1)^2}{3EIl} = 3.651 \cdot 10^{-3} \frac{\text{mm}}{\text{N}}$$

$$d_{22} = \frac{l_3^2(l_1+l_2)^2}{3EIl} = 5.174 \cdot 10^{-3} \frac{\text{mm}}{\text{N}} \tag{5.66}$$

$$d_{12} = \frac{l_1 l_3 (l^2 - l_1^2 - l_3^2)}{6EIl} = 4.098 \cdot 10^{-3} \frac{\text{mm}}{\text{N}}$$

The natural circular frequencies are calculated as $\omega_1 = 155.4 \text{ s}^{-1}$, $\omega_2 = 906.8 \text{ s}^{-1}$, see (5.49), i.e., the critical speeds are $\underline{n_1 = 1484 \text{rpm}}$ and $\underline{n_2 = 8659 \text{rpm}}$. The amplitude ratios result from (5.50)

$$\left(\frac{v_2}{v_1}\right)_1 = \left(\frac{v_{21}}{v_{11}}\right) = \underline{1.188}, \qquad \left(\frac{v_2}{v_1}\right)_2 = \left(\frac{v_{22}}{v_{12}}\right) = \underline{-1.262} \tag{5.67}$$

and correspond to the mode shapes shown in Fig. 5.15b.

S5.3 The natural circular frequencies of the non-rotating shafts ($\Omega = 0$) are obtained with $J_R = J_a$, and the circular frequencies of synchronous rotation in the same direction ($\Omega = \omega$) are derived from (5.31). The numerical values are contained in Table 5.4. One can see that ω_1 decreases with increasing distance of the mass from the bearings and that the gyroscopic effect increases the natural frequencies for synchronous rotation in the same direction, see also Fig. 5.6. The gyroscopic effect changes the second natural frequency considerably because the rotor has a larger tilt in the second mode shape. Since $J_p > J_a$ for cases a) and c), there is no second resonance point for synchronous rotation in the same direction.

Table 5.4 Natural frequencies of the calculation models of Fig. 5.16

Model	$\omega_1/\sqrt{EI/ml^3}$		$\omega_2/\sqrt{EI/ml^3}$	
	$\Omega = 0$	$\Omega = \omega$	$\Omega = 0$	$\Omega = \omega$
Figure 5.16a	1.027	1.502	4.417	imaginary
Figure 5.16b	1.053	1.227	11.760	26.707
Figure 5.16c	0.474	0.529	2.833	imaginary
Figure 5.16d	0.470	0.498	4.081	10.196

5.3 Beam with Distributed Mass

5.3.1 General Perspective

The calculation model of the continuum is an alternative to the calculation model with multiple degrees of freedom. This model features continuous distributions of the mass and elasticity. The vibration behavior of the beam is mainly determined by the bending stiffness and mass distribution as well as the bearing conditions.

The general assumption below is that the vibrating beam has a double symmetrical cross-section such as a circle, rectangle, or double T. Then one can limit the analysis to vibrations in one plane. Beams with a non-symmetrical cross-section typically do not vibrate in one plane. Coupled bending and torsional vibrations may occur because the shear center does not coincide with the center of gravity. Warped profiles, as found in turbine blades, are also excluded from the considerations below.

The differential equation of the bending line of a beam is known from the mechanics of materials:

$$(EIr'')'' = q. \tag{5.68}$$

If one replaces the distributed load q by the specific inertia forces that act on the beam when vibrations occur

$$q = -\varrho A \ddot{r}. \tag{5.69}$$

one obtains the partial differential equation for the free bending vibrations of the beam:

$$(EIr'')'' + \varrho A \ddot{r} = 0, \tag{5.70}$$

where:

$EI(z)$ Bending stiffness of the beam in the vibration plane
$\varrho A(z)$ Mass per unit length
$r(z, t)$ Time-dependent radial deflection
$(\ )'$ Derivative with respect to the z coordinate.

The approach of separation of variables separates the time and position functions for the free vibrations:

$$r(z, t) = v(z)\sin(\omega t + \beta). \tag{5.71}$$

$v(z)$ is the amplitude function and ω the circular frequency of the vibration. After inserting this function into (5.70), one obtains the ordinary differential equation

$$(EIv'')'' - \varrho A \omega^2 v = 0. \tag{5.72}$$

The natural circular frequencies ω_i and mode shapes $v_i(z)$ of beams that satisfy the conditions of the elementary beam theory can be calculated from this differential equation in conjunction with the boundary conditions for the respective case on hand.

5.3 Beam with Distributed Mass

Table 5.5 Parameter influences on the continuum beam (natural circular frequencies for hinged supports on both sides)

Case	Parameter	Differential equation of free oscillation	Natural circular frequencies $\omega_i \ (i=1,2,\ldots)$
1	Longitudinal force F	$EIr'''' + \varrho A \ddot{r} - F r'' = 0$ (1) (tension: $F>0$, compression: $F<0$)	$\omega_i = \dfrac{\pi^2 i^2}{l^2}\sqrt{\dfrac{EI}{\varrho A}}\sqrt{1+\dfrac{Fl^2}{\pi^2 i^2 EI}}$ (2)
2	Rotational inertia ϱI and transverse shear κG $\varepsilon = \dfrac{\pi^2 i^2 I}{Al^2}$, $\varepsilon \ll 1$	$EIr'''' + \varrho A \ddot{r} - \varrho I\left(1+\dfrac{E}{\kappa G}\right)\ddot{r}'' + \dfrac{\varrho^2 I}{\kappa G}\overset{....}{r} = 0$ (3) (Timoshenko beam, rectangle): $\kappa = 5/6$	$\omega_i = \dfrac{\pi^2 i^2}{l^2}\sqrt{\dfrac{EI}{\varrho A}}\left[1 - \dfrac{\varepsilon}{2}\left(1+\dfrac{E}{\kappa G}\right)\right]$ (4)
3	Angular velocity Ω, $\varepsilon \ll 1$	$EIr'''' + \varrho A \ddot{r} - \varrho I \ddot{r}'' - 2\varrho I \Omega \dot{w}'' = 0$ $EIw'''' + \varrho A \ddot{w} - \varrho I \ddot{w}'' + 2\varrho I \Omega \dot{r}'' = 0$ (5) Gyroscopic effect of rotating shafts	$\omega_i = \dfrac{\pi^2 i^2}{l^2}(1-\varepsilon)\left[\dfrac{I\Omega}{A} \pm \sqrt{\left(\dfrac{I\Omega}{A}\right)^2 + \dfrac{EI}{\varrho A}(1+\varepsilon)}\right]$ (6)
4	Velocity v_0	$EIr'''' + \varrho A \ddot{r} - (F - \varrho A v_0^2) r'' + 2\varrho A v_0 \dot{r}' = 0$ (7) (moving belts, straps, filaments, ...)	$\omega_i = \dfrac{\pi i}{l}\left[\sqrt{\dfrac{F}{\varrho A} + \dfrac{\pi^2 i^2}{\varrho A l^2}\dfrac{EI}{} - v_0^2}\sqrt{\dfrac{\varrho A}{F}}\right]$ (8)
5	Mass of liquid per unit length $\mu = \varrho_F A_F$, liquid velocity c, internal pressure p, flow cross sectional area A_F	$EIr'''' + (\varrho A + \mu)\ddot{r} + (\mu c^2 + pA_F - F)r'' + 2c\mu \dot{r}' = 0$ (9) (liquid-flow pipe)	$\omega_i = \dfrac{\pi i}{l}\left[\sqrt{\dfrac{F - pA_F}{\varrho A + \mu} + \dfrac{\pi^2 i^2}{(\varrho A + \mu) l^2}\dfrac{EI}{} - c}\right]$ (10)

Practical mechanical engineering tasks may require consideration of other influencing variables such as the axial force, angular velocity, shear deformation and others. The second column of Table 5.5 shows the respective differential equations that include the parameters mentioned. They are derived like in (5.68) to (5.70).

One can conclude from (2), Table 5.5 that tensile forces increase the natural frequencies and how. In the limiting case $EI \to 0$, the beam transforms into a taut string, the natural frequencies of which increase with the root of the tensile force: $\omega_1 = \pi\sqrt{F/\varrho Al^2}$. The natural frequencies decrease if compressive forces are applied, and when the critical buckling force $F_k = -\pi^2 EI/l^2$ is reached the vibrating beam becomes unstable because $\omega_1 = 0$.

The second row shows that the natural frequencies drop somewhat when the rotational inertia and transverse shear are taken into account. The physical cause of this is a reduced stiffness and increased inertia. This effect has a major impact only for higher mode shapes of short beams with a relatively high cross-sectional height and for hollow profiles.

The influence of the gyroscopic effect results in a splitting of the natural frequencies as is known from Sect. 5.2.2, see Figs. 5.6, 5.11 and 5.12. It is often negligibly small for small diameters. The critical speed of rotation in the same direction results from the condition $\omega_1 = \Omega$.

The influence of the velocity can result in undesirable instabilities for moving V-belts, paper sheets, textile sheets, belt conveyors and others. The vibration becomes unstable at $\omega_1 = 0$. This can be used to calculate the magnitude of the first critical speed:

$$v_{\text{krit}} = \sqrt[4]{\left(\frac{F}{\varrho A}\right)^2 \left(1 + \pi^2 \frac{EI}{Fl^2}\right)}. \tag{5.73}$$

This velocity would generate such large deflections that the respective motion would be subject to an unacceptable disturbance. This critical state can be avoided by increasing the pretension, which can also be seen from (8), Table 5.5. The destabilizing influence of internal pressure in fluid-filled pipes is worth noting: it acts like a axial compressive force. The critical conveyor speed v corresponds to a critical flow rate c analog, see Case 5 in Table 5.5.

The following consideration is to point to the important connection between the discrete model and the continuum. A beam with discrete masses m_k is examined while the rotational inertia is neglected, see Table 5.6. If one decreases the individual masses and at the same time increases their number, the final result will be a beam with continuous mass distribution consisting of an infinite number of mass elements $dm = \varrho A dz$.

The load is then transformed from individual forces F_k to the force elements $dF = q(z) \cdot dz$ that result from the line load $q(\zeta)$. The influence function $d(z, \zeta)$ that indicates the deflection at point z due to a force at point ζ corresponds to the influence coefficients $\alpha_{ik} = d_{ik}$. When calculating the deflection as due to a load, the sum that occurs in the discrete system is transformed into an integral (over the beam length l) due to the limiting process into an indefinite number of elements:

5.3 Beam with Distributed Mass

$$r_i(t) = \sum_{k=1}^{n} d_{ik} F_k(t) \hat{=} r(z, t) = \int_0^l d(z, \zeta) q(\zeta, t) d\zeta. \tag{5.74}$$

For free vibrations, the load corresponds to the inertia forces:

$$F_k(t) = -m_k \ddot{r}_k(t) \hat{=} q(\zeta, t) = -\varrho A(\zeta) \ddot{r}(\zeta, t). \tag{5.75}$$

$A(\zeta)$ is the cross-sectional area that can vary along the beam. The equations of motion are obtained after insertion:

$$r_i(t) = -\sum_{k=1}^{n} d_{ik} m_k \ddot{r}_k(t) \hat{=} r(z, t) = -\int_0^l d(z, \zeta) \varrho A(\zeta) \ddot{r}(\zeta, t) d\zeta. \tag{5.76}$$

The linear system of differential equations (5), Table 5.6 in the discrete model corresponds to the integro-differential equation (6), Table 5.6 of the continuum. Instead of the deflections $r_k(t)$ of individual masses ($k = 1, \ldots, n$) the continuum has a bend line $r(z, t)$ that is position- and time-dependent.

The approach known from (5.26) for determining the natural vibrations takes equivalent forms:

$$r_i(t) = v_i \exp(j\omega t) \hat{=} r(z, t) = v(z) \exp(j\omega t). \tag{5.77}$$

If one combines (7) and (8) with (5) and (6), Table 5.6, the following is obtained after cancelling $\exp(j\omega t)$

$$v_{ji} = \omega_i^2 \sum_{k=1}^{n} d_{jk} m_k v_{ki} \hat{=} v_i(z) = \omega_i^2 \int_0^l d(z, \zeta) \varrho A(\zeta) v_i(\zeta) d\zeta. \tag{5.78}$$

The far-reaching similarities of the equations for the discrete and continuous calculation models can be seen from Table 5.6, which lists essential formulae, the significance of some of which will only be explained in later sections, see Sects. 6.2.1, 6.3.2 and 6.4.1.

(10) in Table 5.6 is called a homogeneous *Fredholm integral equation* equation of the second kind. It represents the continuous analogon to the homogeneous linear system of equations (9), Table 5.6. This is to highlight the close connection between the theory of linear equation systems and that of integral equations, which is of interest to the theoretician and for comprehending advanced literature on the topic.

The physical meaning of integral equation (10), Table 5.6, partially coincides with that of differential equation (5.72) inasmuch as it describes a relationship between inertia forces and displacements. The major difference is that the influence function $d(z, \zeta)$ already considers the boundary conditions that capture the support situation of the respective beam. The differential equation (5.72) only contains the equilibrium on one beam element and has to be solved in conjunction with the

Table 5.6 Correlations between the discrete system and the continuous beam

	Discrete system	Continuum
System schematic	(diagram: discrete masses m_j, m_k at positions z_j, z_k with forces $F_j(t)$, $F_k(t)$ and deflections r_j, r_k)	(diagram: continuous beam with $r(\zeta,t)$, $\rho A(\zeta)$, $EI(\zeta)$ and $r(z,t)$, $\rho A(z)$, $EI(z)$, loading $q(\zeta,t)$)
Mechanical quantities	Discrete system	Continuum
Deflection	$r_j(z_j,t) = \sum_k d_{jk} F_k(t)$ (1)	$r(z,t) = \int_0^l d(z,\zeta) q(\zeta,t)\,d\zeta$ (2)
Inertia forces	$F_k(t) = -m_k \ddot{r}_k(t)$ (3)	$q(z,t) = -\rho A(z) \ddot{r}(z,t)$ (4)
Diff. equ. of motion	$r_j(t) + \sum_k d_{jk} m_k \ddot{r}_k(t) = 0$ (5)	$r(z,t) + \int_0^l d(z,\zeta) \rho A(\zeta) \ddot{r}(\zeta,t)\,d\zeta = 0$ (6)
Separation	$r_j(z_j,t) = v_j(z_j) \exp(j\omega t)$ (7)	$r(z,t) = v(z) \exp(j\omega t)$ (8)
Mode shape No. i	$v_{ji} - \omega_i^2 \sum_k d_{jk} m_k v_{ki} = 0$ (9)	$v_i(z) - \omega_i^2 \int_0^l d(z,\zeta) \rho A(\zeta) v_i(\zeta)\,d\zeta = 0$ (10)
Kinetic energy T	$T = \tfrac{1}{2} \sum_k m_k \dot{r}_k^2(t)$ (11)	$T = \tfrac{1}{2} \int_0^l \rho A(z) \dot{r}^2(z,t)\,dz$ (12)
Potential energy U	$U = \tfrac{1}{2} \sum_j \sum_k d_{jk} F_j F_k$ (13) $U = \tfrac{1}{2} \sum_j \sum_k c_{jk} r_j r_k$ (15)	$U = \tfrac{1}{2} \int_0^l \dfrac{M^2(z)}{EI(z)}\,dz$ (14) $U = \tfrac{1}{2} \int_0^l EI\, r''^2(z,t)\,dz$ (16)
Generalized orthogonality of mode shapes	$\sum_i m_i v_{ji} v_{ki} = \delta_{jk}$ (17)	$\int_0^l \rho A(z) v_j(z) v_k(z)\,dz = \delta_{jk}$ (18)
Summation (estimate)	$\sum_i 1/\omega_i^2 = B_1 = \sum_i m_i d_{ii} > \dfrac{1}{\omega_1^2}$ (19) $\sum_i 1/\omega_i^4 = B_2 = \sum_j \sum_k m_j m_k d_{jk}^2 > \dfrac{1}{\omega_1^4}$ (21)	$B_1 = \int_0^l \rho A(z) d(z,z)\,dz > 1/\omega_1^2$ (20) $B_2 = 2 \int_0^l \!\!\int_0^z \rho^2 A^2(\zeta) d^2(\zeta,z)\,d\zeta\,dz > 1/\omega_1^4$ (22)
Rayleigh quotient	$\omega_R^2 = \dfrac{\sum_j \sum_k c_{jk} v_j v_k}{\sum_k m_k v_k^2} \geq \omega_1^2$ (23)	$\omega_R^2 = \dfrac{\int_0^l EI(z) v''^2(z)\,dz}{\int_0^l \rho A(z) v^2(z)\,dz} \geq \omega_1^2$ (24)
Modal excitation force	$h_i(t) = \sum_{k=1}^n v_{ki} F_k(t)$ (25)	$h_i(t) = \int_0^l v_i(z) q(z,t)\,dz$ (26)

Note: In (17) and (18), δ_{jk} is the Kronecker delta, and $T = W_{\text{kin}}$ and $U = W_{\text{pot}}$ in (11) to (16)

5.3 Beam with Distributed Mass

boundary conditions to obtain mode shapes and natural frequencies. On the other hand, solving the integral equation is sufficient for calculating these parameters.

The actual motion of the beam in the case of free vibrations results by superposition, see (5.71), (6.126) and (6.127), of the mode shapes as follows:

$$r(z,\,t) = \sum_{i=1}^{\infty} \hat{p}_i v_i(z) \sin(\omega_i t + \beta_i). \tag{5.92}$$

A beam performs free vibrations if it is deflected (e. g. at time $t = 0$) from its static equilibrium position $r(z) = 0$ in the form of a bending line $r(z, 0) = r_0(z)$ and left to itself and/or if a velocity distribution $\dot{r}(z, 0) = u_0(z)$ was initially transferred to it, such as by a sudden impulse load distributed over its length.

The constants $\hat{p}_i$ and β_i result from the respective functions $r_0(z)$ and $u_0(z)$, see the analogy with (3.88), (4.22), (4.48), and (6.125) to (6.127):

$$\hat{p}_i \sin \beta_i = \int_0^l \varrho A(z) \cdot v_i(z) \cdot r_0(z) \mathrm{d}z$$

$$\hat{p}_i \cos \beta_i = \int_0^l \frac{\varrho A(z) \cdot v_i(z) \cdot u_0(z)}{\omega_i} \mathrm{d}z \tag{5.93}$$

$$\int_0^l \varrho A(z) \cdot v_i^2(z) \mathrm{d}z = 1$$

These equations follow from (5.92) when using the orthogonality relations (18), Table 5.6.

5.3.2 Straight Beam on Two Supports

Figure 5.17a shows a beam that is delimited on both sides by individual masses m or discrete springs c. The spring constants c represent the spring stiffness of the bearings that is practically always present. Figure 5.17b shows for both cases the dependence of the natural frequencies ($f_i = \omega_i/2\pi$) on these parameters, the natural circular frequencies resulting from equations of the form

$$\omega_i = \lambda_i^2 \sqrt{\frac{EI}{\varrho A l^4}}; \quad (i = 1,\,2,\,\ldots) \tag{5.94}$$

for all boundary conditions. Note the influence of the bending stiffness EI and the mass distribution ϱA in this equation: Like for a simple oscillator, the stiffness appears in the numerator and the mass appears in the denominator. The influence of other parameters on the natural circular frequencies is shown in the third column of

Table 5.7 Eigenvalues λ_i^2 and mode shapes $v_i(z)$ of cylindrical beams under various boundary conditions ($\omega_i = \lambda_i^2 \sqrt{EI/\varrho A l^4}$)

Case	$i=1$	$i=2$	$i=3$	$i=4$	$i=5$	$i \geq 6$
free - free	$\lambda_1^2 = 0$	$\lambda_2^2 = 0$	$\lambda_3^2 = 22{,}4$	$\lambda_4^2 = 61{,}7$	$\lambda_5^2 = 121$	$\lambda_i^2 = (i - \frac{3}{2})^2 \pi^2$
pinned - free	$\lambda_1^2 = 0$	$\lambda_2^2 = 15{,}4$	$\lambda_3^2 = 50{,}0$	$\lambda_4^2 = 104$	$\lambda_5^2 = 178$	$\lambda_i^2 = (i - \frac{3}{4})^2 \pi^2$
fixed - free	$\lambda_1^2 = 3{,}52$	$\lambda_2^2 = 22{,}0$	$\lambda_3^2 = 61{,}7$	$\lambda_4^2 = 121$	$\lambda_5^2 = 200$	$\lambda_i^2 = (i - \frac{1}{2})^2 \pi^2$
pinned - pinned	$\lambda_1^2 = 9{,}87$	$\lambda_2^2 = 39{,}5$	$\lambda_3^2 = 88{,}8$	$\lambda_4^2 = 158$	$\lambda_5^2 = 247$	$\lambda_i^2 = i^2 \pi^2$
fixed - pinned	$\lambda_1^2 = 15{,}4$	$\lambda_2^2 = 50$	$\lambda_3^2 = 104$	$\lambda_4^2 = 178$	$\lambda_5^2 = 272$	$\lambda_i^2 = (i + \frac{1}{4})^2 \pi^2$
fixed - fixed	$\lambda_1^2 = 22{,}4$	$\lambda_2^2 = 61{,}7$	$\lambda_3^2 = 121$	$\lambda_4^2 = 200$	$\lambda_5^2 = 298$	$\lambda_i^2 = (i + \frac{1}{2})^2 \pi^2$

5.3 Beam with Distributed Mass

Table 5.5. Compare the limiting cases $c \to 0$ and $c \to \infty$ with the data in Table 5.7 (λ_i^2 specified there, λ_i here). Very large masses act like rigid bearings since they also hamper the motion of the beam ends (due to their inertia). The limits for $m \to \infty$ therefore correspond to those of $c \to \infty$, however only for eigenvalues that are 2 orders lower.

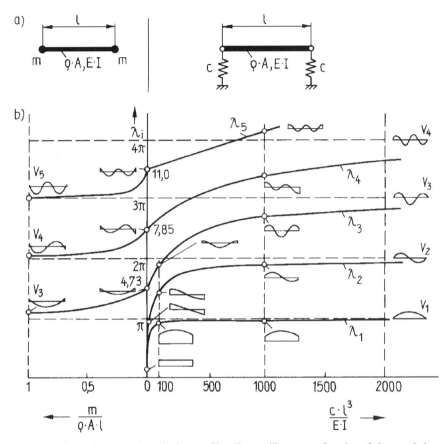

Fig. 5.17 Eigenvalues λ_i and mode shapes of bending oscillators as a function of characteristic parameters of the relative bearing stiffness cl^3/EI and of the mass ratio $m/\varrho Al$

It is important to know the respective mode shape $v_i(z)$ for each natural frequency f_i. Figure 5.17b makes it clear how strong the influence of the bearing stiffness is on the natural frequencies. One should take the time to imagine and memorize the effects of parameter changes on the natural frequencies and mode shapes.

5.3.3 Estimates by Dunkerley and Neuber

There are quite a few practical problems in which the approximate magnitude of the fundamental frequency is of interest. In this case, one can forego a comprehensive exact determination of all natural frequencies and use a method that estimates the natural frequencies. Estimating means to calculate the upper and lower limits for the eigenvalues using a relatively simple procedure.

The approximations and estimates described below were used by engineers in the past for an advance calculation in the case of simple models.

They have retained a certain significance since they are suitable for plausibility considerations and rough calculations.

The fundamental frequency is of special significance for all sub-assemblies of machines. It is the most important and sometimes the only significant natural frequency. It provides, for example, a criterion for deciding whether an object may be treated using the "rigid machine" model or not. When making design changes in a machine (such as to increase speed), one must check the shift of the fundamental frequency. The fundamental frequency is responsible for the lowest resonance point, e. g. it has to be "passed through" if a machine is operated at speeds above critical.

There are several ways to estimate the fundamental frequency, which are derived in detail in [6]. A *lower limit for the fundamental frequency* (assuming a regular system in which the lowest natural frequency is not zero) is provided by the inequality named after DUNKERLEY:

$$\frac{1}{\omega_1^2} < \sum_{k=1}^{n} \frac{1}{\omega_{(k)}^2}; \qquad \omega_1^2 > \omega_D^2 = \frac{1}{\sum_{k=1}^{n} \frac{1}{\omega_{(k)}^2}}. \tag{5.95}$$

The circular frequencies $\omega_{(k)}$ are those of the n subsystems of the *vibration system with n degrees of freedom* that each have a single degree of freedom. The subdivision is arbitrary to some extent, and one chooses subsystems that have at least one "parameter of their own" and can be calculated easily. These subsystems can, for example, be obtained from the original system if only a single inertia parameter (e. g. mass m_k) and all spring parameters are taken into account (Dunkerley). The other inertia parameters are set to zero. Another option is to look at n subsystems with just a single spring each (e. g. c_k), to take all inertia parameters into account and to assume all other springs to be perfectly rigid (infinitely stiff) (Neuber).

This estimate can also be used if the overall system is not divided into n subsystems but only into $n1$ *subsystems ($n1 < n$) with known first natural circular frequencies* $\omega_{(k)}$. These may be large systems, the fundamental frequencies of which are provided by a computer program. In this case, different inertia parameters (Dunkerley) or spring parameters (Neuber) have to be included so that these parameters do not form a "intersection". This is also a way of checking the results of computer programs and to analyze the influence of the subsystems on the overall system. The following estimate results for such subsystems:

5.4 Model Generation for Rotors

$$\omega_1^2 > \frac{1}{\sum_{k=1}^{n1} \frac{1}{\omega_{(k)}^2}}. \qquad (5.96)$$

The lowest fundamental frequency of a continuum can, therefore, be included in such an estimate, see Problem P5.4. For example, the lower limit of $(2\pi f_i = \omega_i)$ results from (5.96) for the fundamental frequency of a large system that is divided into two subsystems with the lowest natural frequencies $f_{(1)}$ and $f_{(2)}$:

$$f_1 > \frac{f_{(1)} f_{(2)}}{\sqrt{f_{(1)}^2 + f_{(2)}^2}}. \qquad (5.97)$$

This can be used for calculating, e. g. when coupling two systems whose lowest natural frequencies coincide $(f_{(1)} = f_{(2)})$ that the lowest natural frequency decreases after coupling but is higher than $f_{(1)}/\sqrt{2}$.

5.4 Model Generation for Rotors

5.4.1 General Considerations

The determination of an appropriate calculation model is the first theoretical task also when studying bending oscillations of shafts. The calculation model should be "as detailed as necessary and as rough as possible" since its only purpose is to facilitate a prediction of the dynamic behavior of a real system.

The calculation model should allow the determination of the future physical behavior of a real system when varying the parameters. In the case of bending vibrations, it is, for example, important to know how the magnitudes of the *natural frequencies* and the mode shapes depend on the parameters of the real system. The calculation model has to contain all essential parameters for this task. The calculation models discussed so far enable an engineer to take into account such parameters as masses m_i, moments of inertia J_{ai}, J_{pi}, lengths l_i, bearing spring constants c_i, density ϱ, modulus of elasticity E, variability of the cross-sectional area $A(z)$, variability of the area moment of inertia $I(z)$, angular velocity Ω, rotor dimensions, etc. when calculating the natural frequencies of isotropically supported shafts with a symmetrical cross-section, see Tables 5.2 and 5.1.

Engineers will have to estimate on a case by case basis which parameters are essential and include these accordingly in a refined calculation model.

Table 5.8 shows what other parameters have to be taken into account for which types of machines. This compilation is meant as a suggestion within this introductory overview. It is not possible at this point to give a comprehensive description of the parameter influences that are of technological significance because research in

Table 5.8 Overview of parameter influences [8]

Parameter	Critical speed n_k	Examples
Axial force	tension increases, pressure decreases n_k	truss member, coupler in crank mechanisms
Torque variations coefficient of speed fluctuation	kth order critical $\Omega_k = \omega/(k \pm 1)$, k order of the harmonic of the variation	reciprocating engines, mechanisms with variable transmission ratio
Magnetic tension (radial)	drops	electric motors, generators
Own weight mg (horizontal shaft)	$\omega_k = \omega/2$	turbine rotor
Meshing frequency	$\Omega_k = \omega \cdot z$	gear mechanism z number of teeth, ω angular velocity of the shaft
Anisotropy of the disks or the shaft	splitting of the spectrum into instability regions ω_1 to ω_2	shaft with groove, fan with two blades, two-pole rotors of synchronous machines
Bearing damping	increases	textile mandrils
Internal damping	self-excitation	material damping, shaft-hub joint
Disk elasticity	drops	blades of fans and turbines, saw blades
Coupling with other units	rises or drops, new spectrum	motor pump, compressor, turbine generator, rollers in opposite directions
Oil film in journal bearings	self-excitation	turbines, pumps, compressors
Speed (gyroscopic effect of the disks)	splitting	centrifuges
Shear deformation	drops	short beams, ship hulls

this field is still ongoing and new aspects keep emerging as mechanical engineering develops.

An important step of the generation is the specification of the degrees of freedom for a calculation model. It is common to apply a very fine FEM discretization with $n = 10^3$ to 10^5 elements.

The accuracy of all calculated natural frequencies increases with the number n of degrees of freedom, however the higher natural frequencies become less accurate with their increasing order i.

It is worth noting that the FEM discretization of the beam always yields larger natural frequencies than the continuum, while the natural frequencies are lower limits in the discretization with mass distribution that approach the exact value as the degrees of freedom increase, see Chap. 6.

A major component of the calculation models of rotating shafts and other bending oscillators are the joints to the mounting site. This does not just include the roller or journal bearings that connect the rotating machine part to the frame. The frame in which the shaft is supported often represents a vibration system in itself. This section has only discussed such bending oscillators in the narrower sense of

5.4 Model Generation for Rotors

the term that are supported in a fixed mounting site via elastic bearings. The general case of a bending oscillator contained in a vibration system is discussed in Chapt. 6, see for example Figs. 6.3, 6.4, and 6.7. It shall only be noted here that it is usually inevitable to treat the roller bearings as elastic supports.

Journal bearings pose particular problems. The resilience of the oil cushion in hydrodynamic bearings has a considerable influence on the dynamic behavior of rotating shafts, which STODOLA realized as early as 1925. The reason for this lies mainly in the anisotropic elastic and damping properties of the oil film. If in a journal bearing a force is applied to a shaft journal that in steady-state operation takes a specific equilibrium position, this force causes a displacement that does not coincide with the direction of the force.

This means in physical terms that certain motions of the journal in the bearing can transmit (torsional) energy from the energy source of the shaft drive to the transverse motion of the shaft. This may cause self-excited bending oscillations. The phenomenon of self-excitation is discussed comprehensively in the technical literature, for example in [8].

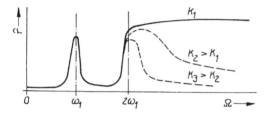

Fig. 5.18 Effect of the lubricating film in a journal bearing

Figure 5.18 shows an amplitude-frequency diagram for self-excited vibrations. The shaft initially behaves like an oscillator in which forced vibrations are excited. The system moves at the excitation frequency, and the known amplitude response curve occurs (resonance if the excitation frequency equals the natural frequency).

At a specific speed above critical, vibrations occur at the system's natural frequency that have very high amplitudes. These are self-excited vibrations the amplitude of which is only limited by the shaft bounding against the bearing (or the rotor hitting the housing).

It was found that these vibrations occur at virtually all speeds above this critical speed so that there is a resonance area without an upper limit. The behavior in this area above critical speed depends on a dimensionless characteristic parameter $K = \frac{2gB\eta}{(\Omega F\psi^3)}$. The amplitude of the self-excited vibrations changes less with speed for small K values (large bearing force F, large relative bearing play ψ, low oil viscosity η and small bearing width B) than for large K values, see Fig. 5.18.

5.4.2 Example: Grinding Spindle

The calculation models shown in Figs. 5.19b to f were used for examining the dynamic behavior of a grinding spindle to illustrate the influence of the various models on the magnitude of the critical speeds and on the mode shapes.

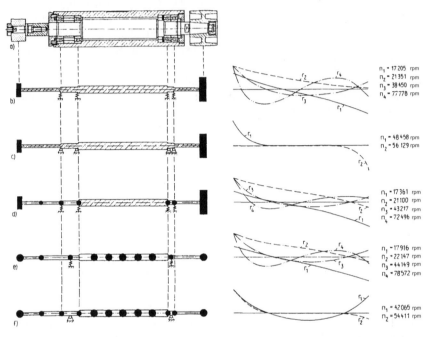

Fig. 5.19 Grinding spindle; **a)** Technical drawing; **b)** to **f)** Calculation models of a grinding spindle (hatching in the model indicates a continuum.) Right: Mode shapes and critical speeds of the grinding spindle models from b) to f)

Figure 5.19 shows the calculated critical speeds and mode shapes. If one compares the resulting numerical values, it becomes evident that the spring constant of the roller bearings has a significant influence. The lowest critical speed is at about $17\,200 \cdot$ rpm (model b). The values obtained using models b, d, and e are within the technologically justifiable tolerance limit of 5 % for the first two critical speeds. The values obtained using models c and f approximately correspond to the third natural frequency of the models with elastic support and provide an incorrect picture since the rigid-body motion is missing, see also Fig. 6.12.

5.5 Problems P5.4 to P5.6

P5.4 Estimate of the Influence of Elastic Bearings on the Lowest Natural Frequency

A computer program was used to calculate the lowest natural frequency for an input shaft with an arbitrary shape and mass distribution that is supported in two bearings without taking the bearing elasticity into account. Use Dunkerley's estimate to state an approximate formula for the lower limit of the lowest natural frequency, taking into account the bearing elasticity values.

Given:

Bearing span	l
Mass of the machine shaft	m
Moment of inertia of the shaft as a rigid body with respect to the bearings	J_A, J_B
Spring constants of the bearings	c_A, c_B
Lowest natural frequency of the shaft with perfectly rigid bearings	$f_{(1)}$
Bending stiffness of the shaft with a constant cross-section	EI

Find:
1. Natural frequencies of the input shaft subdivided into three subsystems
2. Estimate of the fundamental frequency according to Dunkerley
3. Estimate of the fundamental frequency for a homogeneous shaft with a constant cross-section

P5.5 Critical Speeds for the same Direction of Rotation

The parameters specified in the model in Fig. 5.20 are given for a rotating shaft (dimensions in mm). The material density is $\varrho = 7.85 \cdot 10^{-6}$ kg/mm^3, the modulus of elasticity is $E = 2.1 \cdot 10^5$ N/mm^2. Neglecting the mass of the shaft, calculate the diagram of speed-dependent natural frequencies and the critical speed for synchronous rotation in the same direction.

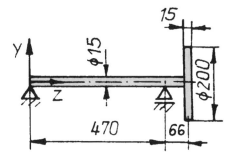

Fig. 5.20 Dimensions of a shaft

P5.6 Comparison of the "Continuous" and "Discrete Bending Oscillator" models

In Sect. 5.2.5.2, the model of the rotating flap was a bending oscillator with four masses, see Fig. 5.13. Using the continuous model, calculate the first three natural frequencies for this flap during free rotation and compare them with those of the discrete bending oscillator. Compare the result in Fig. 5.14 for the rigid flap as well.

5.6 Solutions S5.4 to S5.6

S5.4 The input shaft is divided into three subsystems, the first of which is the rigidly supported shaft itself (Fig. 5.21b), the second and third being the rigid body supported in an elastic bearing (Fig. 5.21c and d). This meets the stipulation specified in Sect. 5.3.3 that the limiting case of infinite stiffness is assumed for all members but one to calculate a natural circular frequency $\omega_{(k)}$.

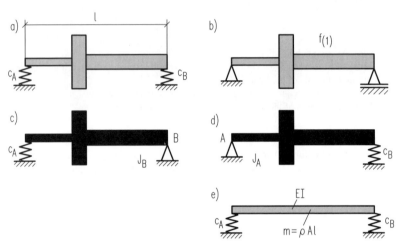

Fig. 5.21 Input shaft on elastic bearings; **a)** Original shaft, **b)** Elastic beam, **c)** Elastic left bearing, **d)** Elastic right bearing, **e)** Shaft with constant cross-section

The estimate according to (5.95) is valid:

$$\frac{1}{\omega_1^2} < \frac{1}{\omega_{(1)}^2} + \frac{1}{\omega_{(2)}^2} + \frac{1}{\omega_{(3)}^2}. \tag{5.98}$$

$\omega_{(1)} = 2\pi f_{(1)}$ of the first subsystem is given, the "shaft with rigid support". The two other subsystems have the natural circular frequency of the rigid body with an elastic bearing on one side that results from:

$$\omega_{(2)}^2 = \frac{c_A l^2}{J_B}; \quad \omega_{(3)}^2 = \frac{c_B l^2}{J_A}. \tag{5.99}$$

It first follows from (5.98) and (5.99)

$$\frac{1}{\omega_1^2} < \frac{1}{\omega_{(1)}^2} + \frac{J_B}{c_A l^2} + \frac{J_A}{c_B l^2}. \tag{5.100}$$

The sought-after estimate according to DUNKERLEY for the lowest natural frequency ($f_1 = \omega_1/2\pi$) of the machine shaft with elastic support is found after solving the expression:

$$f_1 > \frac{f_{(1)}}{\sqrt{1 + \left(\dfrac{J_A}{c_B} + \dfrac{J_B}{c_A}\right)\left(\dfrac{\omega_{(1)}}{l}\right)^2}}. \tag{5.101}$$

5.6 Solutions S5.4 to S5.6

This provides an easy way to calculate the influence of the two elastic bearings. $J_A = J_B = ml^2/3$ applies to the moments of inertia with respect to the bearing points for the special case of a homogeneous shaft with a constant cross-section (Fig. 5.21e). $\omega_{(1)}^2 = \pi^4 EI/ml^3$ is known from Sect. 5.3.2 and from Table 5.7 for the continuous beam hinged on two supports. This results in the estimate as a special case of (5.101)

$$f_1 > \frac{\frac{1}{2\pi}}{\sqrt{m\left[\dfrac{1}{3c_B} + \dfrac{1}{3c_A} + \dfrac{l^3}{\pi^4 EI}\right]}}. \tag{5.102}$$

S5.5 The shaft shown in Fig. 5.20 corresponds to calculation model 4 in Table 5.1.

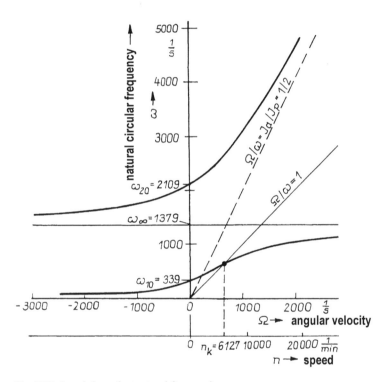

Fig. 5.22 Speed-dependent natural frequencies

The resulting parameter values (see Table 5.2) are

$$m_1 = \frac{\pi D^2 h}{4}\varrho = \frac{3.14 \cdot 200^2 \cdot 15}{4} \cdot 7.85 \cdot 10^{-6} \text{ kg} = 3.70 \text{ kg}$$

$$J_a = \frac{m_1}{16}\left(D^2 + \frac{4}{3}h^2\right) \approx \frac{m_1}{16}D^2 = \frac{3.70}{16}200^2 \text{ kg} \cdot \text{mm}^2 = 9250 \text{ kg} \cdot \text{mm}^2$$

$$I = \frac{\pi d^4}{64} = \frac{3.14}{64}15^2 \text{ mm}^4 = 2480 \text{ mm}^4, \quad J_p = \frac{m_1 D^2}{8} = 18\,500 \text{ kg} \cdot \text{mm}^2$$

$$\alpha_{11} = \frac{l_1 l_2^2 + l_2^3}{3EI} = \frac{470 \cdot 66^2 + 66^3}{3 \cdot 2.1 \cdot 10^5 \cdot 2480} \text{ mm/N} = 1.494 \cdot 10^{-3} \text{ mm/N}$$

$$\gamma_{11} = \frac{2l_1 l_2 + 3l_2^2}{6EI} = \frac{2 \cdot 470 \cdot 66 + 3 \cdot 66^2}{6 \cdot 2.1 \cdot 10^5 \cdot 2480} \text{ N}^{-1} = 2.404 \cdot 10^{-5} \text{ N}^{-1}$$

$$\beta_{11} = \frac{l_1 + 3l_2}{3EI} = \frac{470 + 3 \cdot 66}{3 \cdot 2.1 \cdot 10^5 \cdot 2480} \text{ mm}^{-1} \text{ N}^{-1} = 4.275 \cdot 10^{-7} \text{ mm}^{-1} \cdot \text{N}^{-1}$$

Insertion of these numerical values into (5.33) provides the inverse function of the natural frequencies as a function of the speed. The associated values for Ω are calculated with several numerical values for ω in the range from 0 to 5000 s^{-1} and the curves are determined, see Fig. 5.22. The critical speed for the same direction of rotation ($\Omega/\omega = 1$) is $n_k = 6127 \cdot \text{rpm} \, (\hat{=} f_W = 102.1 \text{ Hz})$.

S5.6 The continuous model is obtained by evenly "spreading" the four masses over the beam length. The beam then has a constant mass distribution $\varrho A = m/l$. Table 5.7, Case 2 provides the eigenvalues for this continuous model, in which the beam is hinged at one end and free at the other:

$$\lambda_1 = 0; \qquad \lambda_2^2 = 15.4; \qquad \lambda_3^2 = 50.0. \tag{5.103}$$

Note when comparing the natural circular frequencies to (5.94) that the flap in Fig. 5.13 has the length $L = 4l$ and the mass $4m$. The natural circular frequencies for this model of the continuum beam are therefore:

$$\omega_i^{(K)} = \lambda_i^2 \sqrt{\frac{EI}{\varrho A L^4}} = \lambda_i^2 \sqrt{\frac{EI}{\varrho A 4^4 l^4}} = \frac{\lambda_i^2}{16} \sqrt{\frac{EI}{m l^3}}. \tag{5.104}$$

The natural frequencies of the free-falling flap in Fig. 5.14 are asymptotic on the right margin ($c \to 0$) as early as at $\bar{c} = 100$. The values

$$\omega_1 \sqrt{\frac{ml^3}{EI}} \to 0; \qquad \omega_2 \sqrt{\frac{ml^3}{EI}} \approx 0.89; \qquad \omega_3 \sqrt{\frac{ml^3}{EI}} \approx 2.8 \tag{5.105}$$

can be obtained for the four-mass system. (5.104) with (5.103) provides the following for the continuum

$$\omega_1^* \sqrt{\frac{ml^3}{EI}} = 0; \qquad \omega_2^* \sqrt{\frac{ml^3}{EI}} \approx 0.96; \qquad \omega_3^* \sqrt{\frac{ml^3}{EI}} \approx 3.1. \tag{5.106}$$

The natural circular frequencies of the continuum are above those of the discrete system because the inertia of the continuum beam is smaller due to the fact that the masses are "spread" "inwards". If one compares the natural frequency of the rigid flap ($EI \to \infty$) to the result in Fig. 5.14, by multiplying both sides of this equation by the same factors, a dimensionless expression is obtained from the exact value $\omega^2 = 9c/(30m)$, which makes the comparison to Fig. 5.14 easier:

$$\frac{ml^3 \omega^2}{EI} = \frac{18}{30} \frac{cl^3}{2EI} = \frac{0.6}{\bar{c}}; \qquad \omega \sqrt{\frac{ml^3}{EI}} = \frac{0.775}{\sqrt{\bar{c}}}. \tag{5.107}$$

This curve corresponds to the asymptote for $i = 1$ in Fig. 5.14.

Chapter 6
Linear Oscillators with Multiple Degrees of Freedom

6.1 Introduction

Many machines and their sub-assemblies can be reduced to a linear calculation model with a finite number of degrees of freedom. One can develop such a calculation model in two ways – by modeling as a multibody system and/or as a FEM model. A calculation model that consists of discrete springs (tension, compression, torsional or bending spring) and individual rigid bodies (characterized by mass, positioning of the center of gravity, moments of inertia and centrifugal moments) is called a multibody system. Initially continuous calculation models with distributed elasticity and mass such as beams, plates, disks, solid bodies or shells can also be reduced to calculation models with a finite number of degrees of freedom using the finite element method (FEM).

Calculation models with few degrees of freedom are often sufficient for the dynamic modeling of many machines. However, calculation models with $n > 10^6$ degrees of freedom are sometimes used as well. The calculation effort always increases with the number of degrees of freedom; the accuracy, however, may not. The overall accuracy depends on the accuracy of the input data and on whether the correct parameters are captured. Often one can describe the real behavior sufficiently accurately using a model with few degrees of freedom if one neglects all inessential parameters. Calculation models with several (a finite number of) degrees of freedom can be used to treat

- Translational oscillations (e. g. of coupled vehicles)
- Torsional vibrations (e. g. of shafts and drive systems)
- Bending vibrations (e. g. of machine frames, beams, frames, plates)
- Vibrations of elastically coupled bodies (e. g. foundation blocks, vehicle formations, machine tools)
- Oscillations of bar and plate structures

and arbitrarily coupled models of any geometrical structure.

All linear oscillation phenomena are basically of the same physical nature. They are described uniformly in mathematics by linear differential equations with constant coefficients. Examples of such calculation models are shown in Fig. 6.1. Many of the calculation models discussed in previous sections (see, for example, Sects. 3.2.2, 3.3, 4.2.2, 4.3, and 5.2.5) can be classified as linear oscillators with n degrees of freedom. The vibration system with n degrees of freedom lets one study the behavior of planar and three-dimensional drive and support systems of machines of any structure.

The linear calculation model allows the study of general laws under one aspect, e. g. free vibrations after impact processes, forced vibrations under periodic excitation or for any desired force-time function. This makes critical speeds, time functions and extreme values of the forces, moments, deformations, stresses, etc. of interest calculable as functions of the machine parameters.

The *matrix notation* allows the uniform and transparent treatment of the calculation models regardless of their structure and number of degrees of freedom, which first of all makes the examination of systems with many degrees of freedom easier. Despite the high degree of abstraction, specific machines can be described well while their dynamic examination can be efficiently performed using commercial software.

The matrix notation allows the elegant formulation of fundamental ideas that are typical in the treatment of oscillators with multiple degrees of freedom. This includes building a matrix of the overall system from those of substructures, reducing the degrees of freedom, stating the eigenvalue problem, justifying the modal analysis and a formal description of the computer-compatible generation and solution of all equations.

To illustrate the procedure, this section only uses models with few degrees of freedom. Students should be able to calculate the examples given with a fair amount of effort and time. They should know, however, that these examples are only meant to introduce them to the general approach.

The fundamentals of the advanced treatment of oscillators with n degrees of freedom go back to the 19th century. Generalized coordinates were introduced by J. L. LAGRANGE (1736–1813). They also provide the basis for his equations of motion of the second kind published in 1811. The theory of frame structures that was developed in works by J. C. MAXWELL (1831–1879), CASTIGLIANO (1847–1884), and O. MOHR (1835–1918) marks another point of departure. The discretization of the continuum models was accomplished by generalizing the method published by W. RITZ (1878–1909) in 1905.

Computer technology gave rise in the 1950s for developing the matrix methods to which R. ZURMÜHL (1904–1966), S. FALK, J. ARGYRIS, and E. PESTEL (1914–1988) made major contributions. Other stimuli for designing computer-compatible formalisms originated in multibody dynamics in which the interest has boomed since the 1960s when aerospace engineering and robotics posed novel demands. In the 1960s and 1970s, program systems were developed based on models of multibody systems (MBS) and finite elements (FEM), which are used by the industry. The combination of MBS and FEM programs with commercial CAD applications have been widely used since the 1980s.

6.1 Introduction

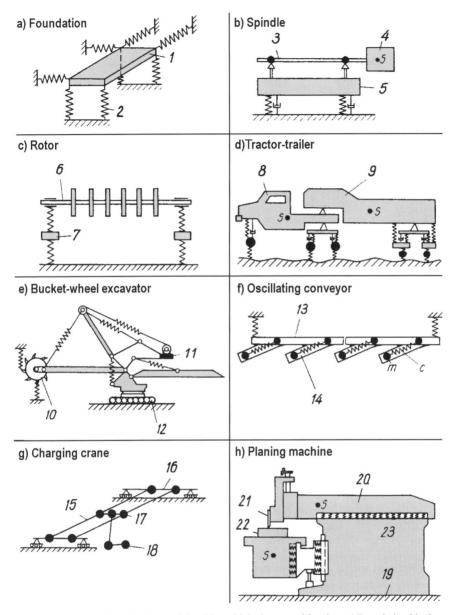

Fig. 6.1 Examples of calculation models with multiple degrees of freedom; *1* Foundation block; *2* Elastic support; *3* Shaft; *4* Body of revolution; *5* Housing; *6* Turbine shaft; *7* Bearing; *8* Driver's cab; *9* Car body; *10* Impeller; *11* Counterweight; *12* Caterpillar; *13* conveying trough; *14* Electromagnetic oscillation exciter; *15* Crane bridge; *16* Chassis; *17* Trolley; *18* Charging tongs; *19* Frame; *20* Carriage; *21* Tool; *22* Workpiece; *23* Elastic and damping layers

6.2 Equations of Motion

6.2.1 Mass, Spring, and Compliance matrix

The motion of an oscillator with n degrees of freedom can be uniquely described by indicating the displacement over time of discrete points or angles of rotation about given axes. Generalized coordinates $q_1, q_2, \ldots, q_n$ that can represent displacement or angles are used to describe the deformation of the system. The totality of these coordinates is arranged in a position vector (synonym: coordinate vector) $\boldsymbol{q}$:

$$\boldsymbol{q} = \begin{bmatrix} q_1 \\ q_2 \\ \vdots \\ q_n \end{bmatrix}, \quad \text{or} \quad \boldsymbol{q}^\mathrm{T} = [q_1, q_2, \ldots, q_n]. \tag{6.1}$$

It is also useful to introduce generalized forces $Q_1, Q_2, \ldots, Q_n$, which act at the discrete points $1, 2, \ldots, n$ in the direction of the generalized coordinates $q_1, q_2, \ldots, q_n$. Generalized forces may be individual forces or moments, see also Sect. 2.4.1.1. They are arranged in the *force vector*

$$\boldsymbol{g}^\mathrm{T} = [Q_1, Q_2, \ldots, Q_n]. \tag{6.2}$$

For the static case, the following linear relations exist between the generalized forces and generalized coordinates for the models of machines under consideration here, in which vibrations about the stable equilibrium position $\boldsymbol{q}^\mathrm{T} = [0, 0, \ldots, 0]$ are of interest:

$$Q_l = \sum_{k=1}^{n} c_{lk} q_k \quad \text{or} \quad \boldsymbol{g} = \boldsymbol{C}\boldsymbol{q}. \tag{6.3}$$

Expressed alternatively:

$$q_l = \sum_{k=1}^{n} d_{lk} Q_k \quad \text{or} \quad \boldsymbol{q} = \boldsymbol{D}\boldsymbol{g}. \tag{6.4}$$

The generalized spring coefficients that characterize the respective system c_{lk} are captured by the matrix $\boldsymbol{C}$, the generalized influence coefficients d_{lk} are contained in the matrix $\boldsymbol{D}$.

$$\boldsymbol{C} = \begin{bmatrix} c_{11} & c_{12} & \cdots & c_{1n} \\ c_{21} & c_{22} & \cdots & c_{2n} \\ \vdots & & & \vdots \\ c_{n1} & c_{n2} & \cdots & c_{nn} \end{bmatrix} \quad (6.5) \quad \text{and} \quad \boldsymbol{D} = \begin{bmatrix} d_{11} & d_{12} & \cdots & d_{1n} \\ d_{21} & d_{22} & \cdots & d_{2n} \\ \vdots & & & \vdots \\ d_{n1} & d_{n2} & \cdots & d_{nn} \end{bmatrix} \quad (6.6)$$

Matrix $\boldsymbol{C}$ is called the **stiffness matrix** (or **spring matrix**), matrix $\boldsymbol{D}$ is called **compliance matrix** (or flexibility matrix). Both matrices are symmetrical, as can

6.2 Equations of Motion

be derived from the theorem by *Maxwell-Betti* for elastic mechanical systems. This means that $c_{lk} = c_{kl}$ and $d_{lk} = d_{kl}$. The specific selection of the matrices C and D is dependent on whether the spring or influence coefficients can be determined more easily in practical problems. Typically, one uses spring constants for torsional oscillators and influence coefficients for bending oscillators.

The deformation work stored in an elastic system depends on the coordinates or on the forces. Linear elastic behavior produces equations of the form

$$W_{\text{pot}} = \frac{1}{2}q^T C q \quad (6.7) \qquad\qquad W_{\text{pot}} = \frac{1}{2}g^T D g \quad (6.8)$$

The deformation work W_{pot} is always positive for any motions and loads. After its determination, one can calculate the elements of the matrices C and D from the second partial derivatives:

$$c_{lk} = c_{kl} = \frac{\partial^2 W_{\text{pot}}}{\partial q_l \partial q_k} \quad (6.9) \qquad\qquad d_{lk} = d_{kl} = \frac{\partial^2 W_{\text{pot}}}{\partial Q_l \partial Q_k} \quad (6.10)$$

For machines where a static calculation is common, such as frames and steel support structures of conveyors, agricultural machines, etc., the influence coefficients d_{lk} can often be calculated more conveniently than c_{lk} using existing software programs.

The d_{lk} can also be determined based on (6.4) from a static deformation measurement. If the deformation q of the system can be measured for a given load (e. g. by photogrammetry), the d_{lk} result from a system of linear equations.

Table 6.1 lists the equations for calculating W_{pot} and d_{lk} for systems that consist of rods and beams. The following applies due to (6.3) and (6.4):

$$g = Cq = CDg; \qquad CD = E. \quad (6.11)$$

E is the identity matrix. The spring matrix is the stiffness matrix of the compliance matrix and vice versa:

$$C = D^{-1}; \qquad D = C^{-1}. \quad (6.12)$$

Based on *d'Alembert's principle*, one can establish a connection between the acceleration vector $\ddot{q}$ and the kinetic forces g that act from the masses onto the elastic system during motion. If motions in which linear or quadratic velocity members occur (e. g. Coriolis forces, gyroscopic moments) are excluded, the following applies to the **inertia forces**, see (2.80) and (2.92):

$$g = -M\ddot{q}. \quad (6.13)$$

$$M = \begin{bmatrix} m_{11} & m_{12} & \cdots & m_{1n} \\ m_{21} & m_{22} & \cdots & m_{2n} \\ \vdots & & & \vdots \\ m_{n1} & m_{n2} & \cdots & m_{nn} \end{bmatrix} \quad (6.14)$$

is the **mass matrix** of the system, which quantifies its inertia properties. The elements of the mass matrix are masses, moments of inertia, centrifugal moments, static moments, or sums of such quantities, which capture the inertia of discrete elements, see the examples in Table 6.2. The elements of the mass matrix can be determined by two different approaches. The first consists of formulating the kinetic energy W_{kin} of a system in general and to form partial derivatives from it. Then

$$m_{kl} = m_{lk} = \frac{\partial^2 W_{\text{kin}}}{\partial \dot{q}_l \partial \dot{q}_k}. \tag{6.15}$$

The second approach starts from equations of motion established with any valid method. After sorting based on the coordinate vector, the mass and stiffness or compliance matrices can be read from them. The kinetic energy is generally:

$$W_{\text{kin}} = \frac{1}{2} \sum_{l=1}^{n} \sum_{k=1}^{n} m_{lk} \dot{q}_l \dot{q}_k \quad \text{or} \quad W_{\text{kin}} = \frac{1}{2} \dot{\mathbf{q}}^{\text{T}} \mathbf{M} \dot{\mathbf{q}}. \tag{6.16}$$

Table 6.1 Equations for calculating the deformation work and influence coefficients d_{lk} of rods and beams

Load type	Deformation work W_{pot}	Influence coefficients d_{lk}
Tensile and compressive forces $F_{Nj}(s_j)$	$W_{\text{pot}} = \sum_{j=1}^{J} \int_{l_j} \frac{F_{Nj}^2}{2E_j A_j} ds_j$	$d_{lk} = \sum_{j=1}^{J} \int_{l_j} \frac{\partial F_{Nj}}{\partial Q_l} \frac{\partial F_{Nj}}{\partial Q_k} \frac{ds_j}{E_j A_j}$
Shear forces $F_{Qj}(s_j)$	$W_{\text{pot}} = \sum_{j=1}^{J} \int_{l_j} \frac{F_{Qj}^2}{2G_j \varkappa_j A_j} ds_j$	$d_{lk} = \sum_{j=1}^{J} \int_{l_j} \frac{\partial F_{Qj}}{\partial Q_l} \frac{\partial F_{Qj}}{\partial Q_k} \frac{ds_j}{G_j \varkappa_j A_j}$
Bending moments $M_j(s_j)$	$W_{\text{pot}} = \sum_{j=1}^{J} \int_{l_j} \frac{M_j^2}{2E_j I_j} ds_j$	$d_{lk} = \sum_{j=1}^{J} \int_{l_j} \frac{\partial M_j}{\partial Q_l} \frac{\partial M_j}{\partial Q_k} \frac{ds_j}{E_j I_j}$
Torsional moments $M_{tj}(s_j)$	$W_{\text{pot}} = \sum_{j=1}^{J} \int_{l_j} \frac{M_{tj}^2}{2G_j I_{tj}} ds_j$	$d_{lk} = \sum_{j=1}^{J} \int_{l_j} \frac{\partial M_{tj}}{\partial Q_l} \frac{\partial M_{tj}}{\partial Q_k} \frac{ds_j}{G_j I_{tj}}$

Where:
l_j Rod length
ds_j Length element of the rod axis
A_j Cross-sectional area
$\varkappa_j$ Shear coefficient for the cross section
j Index of the rod
$Q_i; Q_k$ Forces or moments in directions q_i and q_k
J Number of rods
E_j Modulus of elasticity
G_j Shear modulus
I_j Area moment of inertia with respect to a principal bending axis
I_{tj} Torsional moment of inertia

6.2 Equations of Motion

The elastic restoring forces are at equilibrium with the inertia forces at any time. This is the condition from which the differential equations result that govern the motions and forces of a system. One can find from (6.4), (6.3) and (6.13):

$$Cq = -M\ddot{q} \quad \text{or} \quad q = -DM\ddot{q}. \tag{6.17}$$

The differential equations of the free vibrations follow in three possible forms:

$$M\ddot{q} + Cq = 0 \tag{6.18}$$

$$DM\ddot{q} + q = 0 \tag{6.19}$$

$$\ddot{q} + (DM)^{-1}q = 0 \quad \text{or} \quad \ddot{q} + M^{-1}Cq = 0. \tag{6.20}$$

The vibration process can be described by the time functions of the motion parameters (relative or absolute deflection q, velocity $\dot{q}$, acceleration $\ddot{q}$) or load parameters (force vector g and derivatives $\dot{g}, \ddot{g}$). Likewise, since forces and deformations are mutually dependent variables, differential equations for the forces $g(t)$ can be derived. (6.3) yields the following in combination with (6.20):

$$\ddot{g} = C\ddot{q} = -C(M^{-1}Cq) = -CM^{-1}g. \tag{6.21}$$

This results in the other forms corresponding to (6.18) to (6.20):

$$\ddot{g} + CM^{-1}g = 0, \quad D\ddot{g} + M^{-1}g = 0, \quad MD\ddot{g} + g = 0. \tag{6.22}$$

To summarize, the elastic properties of a linear vibration system are characterized by the matrices C or D and the inertia properties are characterized by the mass matrix M. The first task when analyzing a vibration system is to determine the elements of the matrices C or D and M from the specifications of the real machine. Table 6.2 contains some examples.

Table 6.2 Examples of the mass and stiffness matrices of a calculation model

Case	System with parameters and coordinates	q	Stiffness matrix C	Mass matrix M
1	Vehicle model (relative coordinates)	ξ_1 ξ_2 ξ_3 ξ_4	$\begin{bmatrix} c_1 & 0 & 0 & 0 \\ 0 & c_2 & 0 & 0 \\ 0 & 0 & c_3 & 0 \\ 0 & 0 & 0 & c_4 \end{bmatrix}$	$\begin{bmatrix} m_1 + m_3 s_2^2 + m & & & \text{symmetrical} \\ m_3 s_1 s_2 - m & m_2 + m_3 s_1^2 + m & & \\ m_3 s_2^2 + m & m_3 s_1 s_2 - m & m_3 s_2^2 + m & \\ m_3 s_1 s_2 - m & m_3 s_1^2 + m & m_3 s_1 s_2 - m & m_3 s_1^2 + m \end{bmatrix}$ $m = J_s/(l_1 - l_2)^2, \; s_1 = l_1/(l_1 + l_2), \; s_2 = l_2/(l_1 - l_2)$
2	Vehicle model (absolute coordinates)	ξ_1 ξ_2 φ_3 φ_4	$\begin{bmatrix} c_1 + c_3 & 0 & -c_3 & -c_3 s_1 \\ 0 & c_2 + c_4 & -c_4 & +c_4 s_2 \\ -c_3 & -c_4 & c_3 + c_4 & +c_3 s_1 - c_4 s_2 \\ -s_1 c_3 & c_4 s_2 & c_3 s_1 - c_4 s_2 & c_3 s_1^2 + c_4 s_2^2 \end{bmatrix}$	$\begin{bmatrix} m_1 & 0 & 0 & 0 \\ 0 & m_2 & 0 & 0 \\ 0 & 0 & m_3 & 0 \\ 0 & 0 & 0 & m \end{bmatrix}$ $m = J_s/(l_1 + l_2)^2$

The following procedure is recommended for obtaining the matrix elements:

1. Schematic of the system in deformed state
2. Entry of the selected generalized coordinates (including their positive directions)
3. Derivation of equations for the kinetic and deformation energy, if applicable taking into account any constraints
4. Differentiation according to (6.9) or (6.10) and (6.15) to obtain the matrix elements

The use of *d'Alembert's* principle (establishment of equilibrium conditions) with subsequent comparison of the coefficients in the equations of the form (6.18), (6.19) or (6.20) is in general a suitable method as well. It only is more tedious for coupled systems than the energy-based method that is closely related to *Lagrange's equations* of the second kind.

Substructures are mechanical subsystems, the mechanical behavior of which is described by known matrices and that are suited for building more complex structures. The rationale is the modular principle found elsewhere in engineering. The dynamic behavior of an overall system that consists of sub-assemblies that are put together only during final assembly can be explained from the behavior of the individual sub-assemblies.

Substructures can be such elementary parts as translational, torsional, or bending springs, rigid bodies or point masses. Any combination of those, such as spring-mass systems of a specific geometrical structure, three-dimensional support systems, FEM models of plates, disks, or shells are suited as substructures. The matrices of substructures are obtained using the methods described in Sect. 6.2.1. Table 6.3 presents some of them.

Table 6.3 System matrices of substructures of the examples discussed

Case No.	System schematic	$q^{(r)}$	Stiffness matrix **C**	Mass matrix **M**
1	$m_1\ c_1\ m_2\ c_2\ m_3$; $q_1\ q_2\ q_3$	$\begin{bmatrix} q_1 \\ q_2 \\ q_3 \end{bmatrix}$	$\mathbf{C} = \begin{bmatrix} c_1 & -c_1 & 0 \\ -c_1 & c_1+c_2 & -c_2 \\ 0 & -c_2 & c_2 \end{bmatrix}$	$\mathbf{M} = \begin{bmatrix} m_1 & 0 & 0 \\ 0 & m_2 & 0 \\ 0 & 0 & m_3 \end{bmatrix}$
2	$2m$, $2m$, $2m$, EI = const.	$\begin{bmatrix} q_1 \\ q_2 \\ q_3 \\ q_4 \end{bmatrix}$	$\mathbf{C} = \dfrac{48 EI}{97\, l^3} \begin{bmatrix} 26 & -59 & 9 & -12 \\ -59 & 160 & -54 & 72 \\ 9 & -54 & 74 & -131 \\ -12 & 72 & -131 & 304 \end{bmatrix}$	$\mathbf{M} = m \begin{bmatrix} 1 & 0 & 0 & 0 \\ 0 & 2 & 0 & 0 \\ 0 & 0 & 5 & 0 \\ 0 & 0 & 0 & 2 \end{bmatrix}$
3	l_1, l_2, S, q_1, m_3, J_S, q_2	$\begin{bmatrix} q_1 \\ q_2 \end{bmatrix}$	$\mathbf{C} = 0$ rigid body	$\mathbf{M} = \dfrac{1}{(l_1+l_2)^2} \begin{bmatrix} J_S + m_3 l_2^2 & -J_S + m_3 l_1 l_2 \\ -J_S + m_3 l_1 l_2 & J_S + m_3 l_1^2 \end{bmatrix}$
4	m_1, J_1, x_1, EI, v_2, m_2, J_2, x_2	$\begin{bmatrix} v_1 \\ x_1 \\ v_2 \\ x_2 \end{bmatrix}$	$\mathbf{C} = \dfrac{2EI}{l^3} \begin{bmatrix} 6 & 3l & -6 & 3l \\ 3l & 2l^2 & -3l & l^2 \\ -6 & -3l & 6 & -3l \\ 3l & l^2 & -3l & 2l^2 \end{bmatrix}$	$\mathbf{M} = \begin{bmatrix} m_1 & 0 & 0 & 0 \\ 0 & J_1 & 0 & 0 \\ 0 & 0 & m_2 & 0 \\ 0 & 0 & 0 & J_2 \end{bmatrix}$

6.2 Equations of Motion

An overall system is obtained by expressing the relationships between the local coordinates of the total of R substructures and the global coordinates of the overall system using transformation matrices $\boldsymbol{T}_r$ ($r = 1, 2, \ldots, R$). The matrices $\boldsymbol{T}_r$ are rectangular matrices with nr rows and n columns. Such a matrix $\boldsymbol{T}_r$ describes the coincidence (the geometrical compatibility), i.e. the constraints that exist between local and global coordinates, or, in other words, it indicates where the substructures are connected.

It is assumed that the mass matrix $\boldsymbol{M}_r$ and the stiffness matrix $\boldsymbol{C}_r$ of the rth substructure with the coordinate vector

$$\boldsymbol{q}^{(r)} = \left[q_1^{(r)}, q_2^{(r)}, \ldots, q_{nr}^{(r)}\right]^{\mathrm{T}} \tag{6.23}$$

are known. The connection between the global coordinate vector $\boldsymbol{q}$ of the overall system known from (6.1) and the local coordinate vector of the rth substructure can generally be brought into the following form:

$$\boldsymbol{q}^{(r)} = \boldsymbol{T}_r \boldsymbol{q}, \qquad r = 1, \ldots, R. \tag{6.24}$$

The resulting matrices of the equations of motion of the overall system then are as follows:

$$\boldsymbol{C} = \sum_{r=1}^{R} \boldsymbol{T}_r^{\mathrm{T}} \boldsymbol{C}_r \boldsymbol{T}_r, \qquad \boldsymbol{M} = \sum_{r=1}^{R} \boldsymbol{T}_r^{\mathrm{T}} \boldsymbol{M}_r \boldsymbol{T}_r. \tag{6.25}$$

They can thus be found by performing the following steps:

1. Division of the overall system into R subsystems, the matrices $\boldsymbol{C}_r$ and $\boldsymbol{M}_r$ of which are known.
2. Determination of the global coordinates $\boldsymbol{q}$
3. Expression of the relationships between the local coordinates $\boldsymbol{q}^{(r)}$ of the substructures and the global coordinates $\boldsymbol{q}$ using the transformation matrices $\boldsymbol{T}_r$.
4. Determination of the global matrices of the overall system according to (6.25)

6.2.2 Examples

6.2.2.1 Frame/Load Parameter Method

The matrices $\boldsymbol{D}$, $\boldsymbol{C}$, and $\boldsymbol{M}$ are to be established as an example for the system outlined in Fig. 6.2 consisting of point masses and massless beams.

Given parameters are the bending stiffness EI, the length l, the masses m_1 and m_2 and the coordinates q_1 and q_2 that are to be used.

The relationships between forces and displacements can be expressed in the form (6.4). The influence coefficients can be calculated according to Table 6.1. First, the forces $\boldsymbol{g}^{\mathrm{T}} = (Q_1, Q_2)$ are assumed in the direction of the coordinates $\boldsymbol{q}^{\mathrm{T}} = (q_1, q_2)$. If s_1 and s_2 are the position coordinates for describing the bending

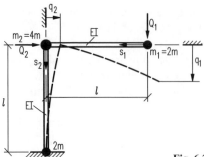

Fig. 6.2 Calculation model of a frame

moments (see Table 6.1), they are as follows:

$$M_1 = Q_1 s_1; \qquad M_2 = Q_1 l + Q_2 s_2. \tag{6.26}$$

The deformation energy (potential energy) for the bending load results from Table 6.1

$$W_{\text{pot}} = \int_0^l \frac{M_1^2}{2EI} \mathrm{d}s_1 + \int_0^l \frac{M_2^2}{2EI} \mathrm{d}s_2$$

$$= \frac{1}{2EI} \left[\int_0^l Q_1^2 s_1^2 \mathrm{d}s_1 + \int_0^l (Q_1 l + Q_2 s_2)^2 \mathrm{d}s_2 \right]. \tag{6.27}$$

Integration yields the potential energy in the form of (6.8)

$$W_{\text{pot}} = \frac{1}{2} \frac{4l^3}{3EI} Q_1^2 + \frac{l^3}{2EI} Q_1 Q_2 + \frac{1}{2} \frac{l^3}{3EI} Q_2^2. \tag{6.28}$$

Forming the first derivative produces the following in accordance with the *Castigliano's theorem* as in (6.4):

$$\frac{\partial W_{\text{pot}}}{\partial Q_1} = q_1 = \frac{4l^3}{3EI} Q_1 + \frac{l^3}{2EI} Q_2 = d_{11} Q_1 + d_{12} Q_2 \tag{6.29}$$

$$\frac{\partial W_{\text{pot}}}{\partial Q_2} = q_2 = \frac{l^3}{2EI} Q_1 + \frac{l^3}{3EI} Q_2 = d_{21} Q_1 + d_{22} Q_2. \tag{6.30}$$

According to (6.10), the influence coefficients then result from the second derivatives:

$$d_{11} = \frac{4l^3}{3EI}; \qquad d_{12} = d_{21} = \frac{l^3}{2EI}; \qquad d_{22} = \frac{l^3}{3EI}. \tag{6.31}$$

This yields matrix D, and after brief calculation, its inverse matrix C, see (6.12):

6.2 Equations of Motion

$$D = \begin{bmatrix} \frac{4l^3}{3EI} & \frac{l^3}{2EI} \\ \frac{l^3}{2EI} & \frac{l^3}{3EI} \end{bmatrix} = \frac{l^3}{6EI}\begin{bmatrix} 8 & 3 \\ 3 & 2 \end{bmatrix}; \quad D^{-1} = C = \frac{6EI}{7l^3}\begin{bmatrix} 2 & -3 \\ -3 & 8 \end{bmatrix}. \quad (6.32)$$

One can check by numerical calculation that $D \cdot C = E$. The kinetic energy is

$$W_{\text{kin}} = \frac{1}{2}m_1\dot{q}_1^2 + \frac{1}{2}m_1\dot{q}_2^2 + \frac{1}{2}m_2\dot{q}_2^2 = \frac{1}{2}m_1\dot{q}_1^2 + \frac{1}{2}(m_1+m_2)\dot{q}_2^2. \quad (6.33)$$

Note that the horizontal motion of the mass has to be considered in addition to its vertical motion m_1 (Fig. 6.2). It follows as a result of forming the partial derivatives according to (6.15):

$$m_{11} = m_1; \quad m_{12} = m_{21} = 0; \quad m_{22} = m_1 + m_2.$$

The mass matrix for $m_1 = 2m$ and $m_2 = 4m$ is therefore:

$$M = \begin{bmatrix} m_{11} & m_{12} \\ m_{21} & m_{22} \end{bmatrix} = \begin{bmatrix} m_1 & 0 \\ 0 & m_1+m_2 \end{bmatrix} = m\begin{bmatrix} 2 & 0 \\ 0 & 6 \end{bmatrix}. \quad (6.34)$$

The stiffness matrix C can also be obtained directly using the beam element discussed in 6.2.2.2, see (6.225).

6.2.2.2 Beam Element/Deformation Method

A beam element with interfaces 1 and 2 onto which F_{L1}, F_{L2}, F_1, F_2, M_1 and M_2 act forms the starting point. The deformations caused by this in the local coordinate system are depicted in Fig. 6.3a.

The equilibrium of forces and moments provides three equations:

$$F_{L1} + F_{L2} = 0, \quad F_1 + F_2 = 0, \quad M_1 + M_2 + F_2 l = 0. \quad (6.35)$$

Due to the linear elastic behavior of the beam, the following applies to the relationships between load and deformation parameters that follow from the bending line:

$$F_{L2} = \frac{EA}{l}(u_2 - u_1) \quad (6.36)$$

$$\chi_2 = \chi_1 + \frac{F_2 l^2}{2EI} + \frac{M_2 l}{EI}, \quad v_2 = v_1 + \chi_1 l + \frac{F_2 l^3}{3EI} + \frac{M_2 l^2}{2EI}. \quad (6.37)$$

These six linear equations can be solved for the load parameters. One finds:

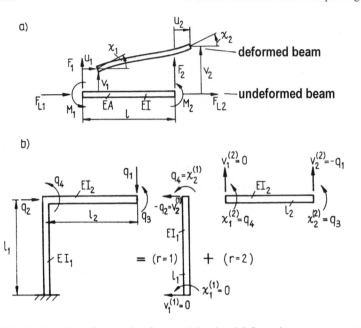

Fig. 6.3 Modeling of supporting frames; **a)** Load and deformation parameters at the beam element, **b)** Division of a frame into two beam elements (substructures $r = 1, 2$)

$$f = \begin{bmatrix} F_{L1} \\ F_1 \\ M_1 \\ F_{L2} \\ F_2 \\ M_2 \end{bmatrix} = \frac{EI}{l^3} \begin{bmatrix} Al^2/I & 0 & 0 & -Al^2/I & 0 & 0 \\ 0 & 12 & 6l & 0 & -12 & 6l \\ 0 & 6l & 4l^2 & 0 & -6l & 2l^2 \\ -Al^2/I & 0 & 0 & Al^2/I & 0 & 0 \\ 0 & -12 & -6l & 0 & 12 & -6l \\ 0 & 6l & 2l^2 & 0 & -6l & 4l^2 \end{bmatrix} \begin{bmatrix} u_1 \\ v_1 \\ \chi_1 \\ u_2 \\ v_2 \\ \chi_2 \end{bmatrix} = Cq.$$

(6.38)

Check if (6.36) is satisfied or solve (6.37) for F_2 and M_2. If one neglects the axial deformation, which is justifiable in most cases, (6.38) shrinks to the following because of $u_1 = u_2 = 0$:

$$f = \begin{bmatrix} F_1 \\ M_1 \\ F_2 \\ M_2 \end{bmatrix} = \frac{2EI}{l^3} \begin{bmatrix} 6 & 3l & -6 & 3l \\ 3l & 2l^2 & -3l & l^2 \\ -6 & -3l & 6 & -3l \\ 3l & l^2 & -3l & 2l^2 \end{bmatrix} \begin{bmatrix} v_1 \\ \chi_1 \\ v_2 \\ \chi_2 \end{bmatrix} = Cq. \qquad (6.39)$$

This stiffness matrix is included in Table 6.3, Case 4.

The stiffness matrix of the frame shown in Fig. 6.3b can be obtained from those of two beam elements with different lengths l_r and bending stiffnesses EI_r. The frame consists of $R = 2$ substructures (beams), see Fig. 6.3c.

The relationships between the local coordinates $q^{(r)}$ and the global coordinates $q^T = (q_1, q_2, q_3, q_4)$ can be established taking into account the boundary conditions and the conditions at the interface between the two beam elements. The

6.2 Equations of Motion

geometrical relations evident from Fig. 6.3b and c are:

$$r = 1: \quad v_1^{(1)} = 0; \quad \chi_1^{(1)} = 0; \quad v_2^{(1)} = -q_2: \quad \chi_2^{(1)} = q_4$$
$$r = 2: \quad v_1^{(2)} = 0; \quad \chi_1^{(2)} = q_4; \quad v_2^{(2)} = -q_1: \quad \chi_2^{(2)} = q_3. \quad (6.40)$$

They are represented by the following matrix equations:

$$q^{(1)} = \begin{bmatrix} v_1^{(1)} \\ \chi_1^{(1)} \\ v_2^{(1)} \\ \chi_2^{(1)} \end{bmatrix} = \begin{bmatrix} 0 & 0 & 0 & 0 \\ 0 & 0 & 0 & 0 \\ 0 & -1 & 0 & 0 \\ 0 & 0 & 0 & 1 \end{bmatrix} \begin{bmatrix} q_1 \\ q_2 \\ q_3 \\ q_4 \end{bmatrix} = T_1 q \quad (6.41)$$

$$q^{(2)} = \begin{bmatrix} v_1^{(2)} \\ \chi_1^{(2)} \\ v_2^{(2)} \\ \chi_2^{(2)} \end{bmatrix} = \begin{bmatrix} 0 & 0 & 0 & 0 \\ 0 & 0 & 0 & 1 \\ -1 & 0 & 0 & 0 \\ 0 & 0 & 1 & 0 \end{bmatrix} \begin{bmatrix} q_1 \\ q_2 \\ q_3 \\ q_4 \end{bmatrix} = T_2 q. \quad (6.42)$$

Unlike the model in 6.2.2.1, the angles of rotation q_4 at the corner and q_3 at the free end were included in the model. Note for the stiffness matrix of the beam elements as given by (6.39) that EI_r and l_r may differ for $r = 1$ and $r = 2$. Using the transformation matrices from (6.41) and (6.42), the stiffness matrix of the overall system can be calculated according to (6.25) ($\beta_r = 2EI_r/l_r^3$, $r = 1, 2$):

$$C = T_1^T C_1 T_1 + T_2^T C_2 T_2$$

$$C = \frac{2EI_1}{l_1^3}\begin{bmatrix} 0 & 0 & 0 & 0 \\ 0 & 6 & 0 & 3l_1 \\ 0 & 0 & 0 & 0 \\ 0 & 3l_1 & 0 & 2l_1^2 \end{bmatrix} + \frac{2EI_2}{l_2^3}\begin{bmatrix} 6 & 0 & 3l_2 & 3l_2 \\ 0 & 0 & 0 & 0 \\ 3l_2 & 0 & 2l_2^2 & l_2^2 \\ 3l_2 & 0 & l_2^2 & 2l_2^2 \end{bmatrix} \quad (6.43)$$

$$C = \begin{bmatrix} 6\beta_2 & 0 & 3\beta_2 l_2 & 3\beta_2 l_2 \\ 0 & 6\beta_1 & 0 & 3\beta_1 l_1 \\ 3\beta_2 l_2 & 0 & 2\beta_2 l_2^2 & \beta_2 l_2^2 \\ 3\beta_2 l_2 & 3\beta_1 l_1 & \beta_2 l_2^2 & 2\beta_1 l_1^2 + 2\beta_2 l_2^2 \end{bmatrix}.$$

6.2.2.3 Vehicle/Energy Method

The example in Table 6.2, Cases 1 and 2, represents the simplest model of a vehicle in which the wheel masses are taken into account. The motion of the center of gravity is considered in y direction only. When selecting fixed coordinates $q_2^T = (\xi_1, \xi_2, y_S, \varphi)$ (Case 2) twice the kinetic energy is

$$2W_{\text{kin}} = m_1 \dot{\xi}_1^2 + m_2 \dot{\xi}_2^2 + m_3 \dot{y}_S^2 + J_S \dot{\varphi}^2 = \dot{q}_2^T M_2 \dot{q}_2. \quad (6.44)$$

The potential energy can be expressed in a simpler way using just the relative coordinates $q_1^T = (\xi_1, \xi_2, \xi_3, \xi_4)$ (Case 1):

$$2W_{\text{pot}} = c_1 \xi_1^2 + c_2 \xi_2^2 + c_3 \xi_3^2 + c_4 \xi_4^2 = q_1^T C_1 q_1. \quad (6.45)$$

The following constraints apply between these two coordinates, which can be derived from geometrical considerations (prerequisite $\varphi \ll 1$):

$$\varphi = \frac{\xi_1 + \xi_3 - \xi_2 - \xi_4}{l_1 + l_2}; \quad y_S = \frac{l_1(\xi_2 + \xi_4) + l_2(\xi_1 + \xi_3)}{l_1 + l_2} \quad (6.46)$$

$$\xi_3 = y_S + l_1 \varphi - \xi_1; \quad \xi_4 = y_S - l_2 \varphi - \xi_2.$$

Insertion of φ and y_S from (6.46) in (6.44) provides

$$2W_{\text{kin}} = m_1 \dot{\xi}_1^2 + m_2 \dot{\xi}_2^2 + m_3 \left[\frac{l_1(\dot{\xi}_2 + \dot{\xi}_4) + l_2(\dot{\xi}_1 + \dot{\xi}_3)}{l_1 + l_2} \right]^2$$

$$+ \frac{J_S}{(l_1 + l_2)^2} (\dot{\xi}_1 + \dot{\xi}_3 - \dot{\xi}_2 - \dot{\xi}_4)^2 = \dot{q}_1^T M_1 \dot{q}_1. \quad (6.47)$$

Insertion of ξ_3 and ξ_4 from (6.46) in (6.45) provides the potential energy as a function of the absolute coordinates:

$$2W_{\text{pot}} = c_1 \xi_1^2 + c_2 \xi_2^2 + c_3(y_S + l_1 \varphi - \xi_1)^2 + c_4(y_S - l_2 \varphi - \xi_2)^2 = q_2^T C_2 q_2. \quad (6.48)$$

The matrices C_1 and M_1 for the coordinates q_1, see Table 6.2, Case 1, follow from (6.45) and (6.47) as partial derivatives according to (6.9) and (6.15). Likewise, the matrices C_2 and M_2 result from (6.48) and (6.44) for the absolute coordinates q_2, see Table 6.2, Case 2. If one expresses the relationships between the two coordinate systems using the transformation matrix T, the following is valid:

$$T = \begin{bmatrix} 1 & 0 & 0 & 0 \\ 0 & 1 & 0 & 0 \\ -1 & 0 & 1 & l_1 \\ 0 & -1 & 1 & l_2 \end{bmatrix}, \quad (6.49)$$

$$q_1 = T q_2, \quad C_2 = T^T C_1 T, \quad M_2 = T^T M_1 T,$$

which one can verify.

6.2.2.4 Support Structure Consisting of Substructures

The support structure consisting of a frame with constant bending stiffness and an elastically supported vibration system, see Fig. 6.4, is viewed as an example of the method described in Sect. 6.2.1 that is suitable for establishing the system matrices.

6.2 Equations of Motion

When dividing into substructures, which represents the first step of this method, one resorts to the matrices in Table 6.3. $R = 5$ substructures occur.

The frame can be composed of two of the boom-shaped support structures that correspond to Case 2 in Table 6.3. The rigid body on top, which represents the third substructure ($r = 3$), exactly matches Case 3 in Table 6.3. The spring-mass systems that connect the rigid body with the frame are considered to be the substructures $r = 4$ and $r = 5$. They represent a special case of Case 1 in Table 6.3 if they are set to $m_1 = m_3 = 0$ and the designations of masses and springs are changed accordingly.

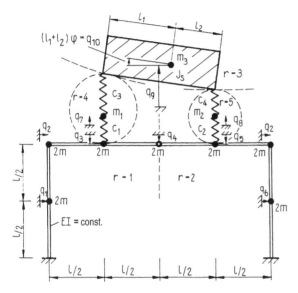

Fig. 6.4 Overall system of a support structure, divided into substructures $r = 1, \ldots, 5$ with global coordinates q_1 and q_{10}

The global coordinates are shown in Fig. 6.4. Now the relationships between the local coordinates that correspond to those in Table 6.3 and these global coordinates have to be expressed. The substructure $r = 1$ represents the left half of the frame, and the following constraints (coincidence) apply:

$$q_1^{(1)} = q_4, \quad q_2^{(1)} = q_3, \quad q_3^{(1)} = q_2, \quad q_4^{(1)} = q_1. \quad (6.50)$$

In view of the $n = 10$ elements of the vector of the global coordinates, these constraints can also be expressed by the following 4×10 transformation matrix, see (6.24):

$$T_1 = \begin{bmatrix} 0 & 0 & 0 & 1 & 0 & 0 & 0 & 0 & 0 & 0 \\ 0 & 0 & 1 & 0 & 0 & 0 & 0 & 0 & 0 & 0 \\ 0 & 1 & 0 & 0 & 0 & 0 & 0 & 0 & 0 & 0 \\ 1 & 0 & 0 & 0 & 0 & 0 & 0 & 0 & 0 & 0 \end{bmatrix}. \quad (6.51)$$

The second substructure consists of the same boom, but mirrored about the vertical axis so that the right half of the frame results. The relationships between the coordinates then are for $r = 2$:

$$q_1^{(2)} = q_4, \quad q_2^{(2)} = q_5, \quad q_3^{(2)} = -q_2, \quad q_4^{(2)} = -q_6. \tag{6.52}$$

The transformation matrix T_2 differs from T_1 but the mass matrices are the same $M_1 = M_2 = M$, and the stiffness matrices $C_1 = C_2 = C$ coincide as well, see Case 2 in Table 6.3. The transformation matrix according to (6.52) is:

$$T_2 = \begin{bmatrix} 0 & 0 & 0 & 1 & 0 & 0 & 0 & 0 & 0 & 0 \\ 0 & 0 & 0 & 0 & 1 & 0 & 0 & 0 & 0 & 0 \\ 0 & -1 & 0 & 0 & 0 & 0 & 0 & 0 & 0 & 0 \\ 0 & 0 & 0 & 0 & 0 & -1 & 0 & 0 & 0 & 0 \end{bmatrix}. \tag{6.53}$$

The rigid body (Case 3 in Table 6.3) is coupled at both ends with the other substructures so that another transformation is required. Using the shortcuts $s_1 = l_1/(l_1+l_2)$ and $s_2 = l_2/(l_1+l_2)$, it looks as follows:

$$q_1^{(3)} = q_9 + s_1 q_{10}, \quad q_2^{(3)} = q_9 - s_2 q_{10}. \tag{6.54}$$

The associated transformation matrix is:

$$T_3 = \begin{bmatrix} 0 & 0 & 0 & 0 & 0 & 0 & 0 & 0 & 1 & s_1 \\ 0 & 0 & 0 & 0 & 0 & 0 & 0 & 0 & 1 & -s_2 \end{bmatrix}. \tag{6.55}$$

Furthermore, $C_3 = 0$ and $M_3 = M$, see the mass matrix of Case 3 in Table 6.3.

Substructure $r = 4$ is the left spring-mass system between the rigid body and the frame, see Fig. 6.4. The constraints are:

$$q_1^{(4)} = -q_3, \quad q_2^{(4)} = q_7, \quad q_3^{(4)} = q_9 + s_1 q_{10}, \tag{6.56}$$

and therefore:

$$T_4 = \begin{bmatrix} 0 & 0 & -1 & 0 & 0 & 0 & 0 & 0 & 0 & 0 \\ 0 & 0 & 0 & 0 & 0 & 0 & 1 & 0 & 0 & 0 \\ 0 & 0 & 0 & 0 & 0 & 0 & 0 & 0 & 1 & s_1 \end{bmatrix}. \tag{6.57}$$

The system matrices result from Table 6.3, Case 1, and should be given the designations used for the parameters in Fig. 6.4. Thus:

$$C_4 = \begin{bmatrix} c_1 & -c_1 & 0 \\ -c_1 & c_1 + c_3 & -c_3 \\ 0 & -c_3 & c_3 \end{bmatrix}, \quad M_4 = \begin{bmatrix} 0 & 0 & 0 \\ 0 & m_1 & 0 \\ 0 & 0 & 0 \end{bmatrix}. \tag{6.58}$$

The same substructure (Table 6.3, Case 1) is on the right side, but for this ($r = 5$) the interface conditions are

$$q_1^{(5)} = -q_5, \quad q_2^{(5)} = q_8, \quad q_3^{(5)} = q_9 - s_2 q_{10} \tag{6.59}$$

6.2 Equations of Motion

and thus the transformation matrix becomes

$$T_5 = \begin{bmatrix} 0 & 0 & 0 & 0 & -1 & 0 & 0 & 0 & 0 \\ 0 & 0 & 0 & 0 & 0 & 0 & 1 & 0 & 0 \\ 0 & 0 & 0 & 0 & 0 & 0 & 0 & 1 & -s_2 \end{bmatrix}. \tag{6.60}$$

The matrices of this substructure match those for $r = 4$ if the following substitutions are made: $c_1 \hat{=} c_2$, $c_3 \hat{=} c_4$, $m_1 \hat{=} m_2$. The global stiffness and mass matrices result according to (6.25) using the matrices provided.

The reader should calculate at least one summand to illustrate the relationships and to comprehend the development of the system matrices. The following are the system matrices of the overall system:

$$\frac{C}{c^*} = \begin{bmatrix} 304 & -131 & 72 & -12 & 0 & 0 & 0 & 0 & 0 & 0 \\ -131 & 148 & -54 & 0 & 54 & -131 & 0 & 0 & 0 & 0 \\ 72 & -54 & 160+\bar{c}_1 & -59 & 0 & 0 & \bar{c}_1 & 0 & 0 & 0 \\ -12 & 0 & -59 & 52 & -59 & 12 & 0 & 0 & 0 & 0 \\ 0 & 54 & 0 & -59 & 160+\bar{c}_2 & -72 & 0 & \bar{c}_2 & 0 & 0 \\ 0 & -131 & 0 & 12 & -72 & 304 & 0 & 0 & 0 & 0 \\ 0 & 0 & \bar{c}_1 & 0 & 0 & 0 & \bar{c}_1+\bar{c}_3 & 0 & -\bar{c}_3 & -\bar{c}_3 s_1 \\ 0 & 0 & 0 & 0 & \bar{c}_2 & 0 & 0 & \bar{c}_2+\bar{c}_4 & -\bar{c}_4 & \bar{c}_4 s_2 \\ 0 & 0 & 0 & 0 & 0 & 0 & -\bar{c}_3 & -\bar{c}_4 & \bar{c}_3+\bar{c}_4 & c_{9\,10} \\ 0 & 0 & 0 & 0 & 0 & 0 & -\bar{c}_3 s_1 & \bar{c}_4 s_2 & c_{10\,9} & c_{10\,10} \end{bmatrix}$$

where

$$c_{9\,10} = \bar{c}_3 s_1 - \bar{c}_4 s_2 = c_{10\,9}, \qquad c_{10\,10} = \bar{c}_3 s_1^2 + \bar{c}_4 s_2^2 \tag{6.61}$$

$$c^* = \frac{48EI}{97l^3}, \quad \bar{c}_k = c_k/c^* \quad \text{for} \quad k = 1, 2, 3, 4. \tag{6.62}$$

and

$$M = \text{diag}[2m, 10m, 2m, 2m, 2m, 2m, m_1, m_2, m_3, J_S/(l_1+l_2)^2]. \tag{6.63}$$

One should look at the stiffness matrix closely and try to explain the structure and origin of each element. One will find quadratic submatrices that partially overlap. The coupling points can be explained physically if one looks at the coordinates in Fig. 6.4 for comparison. In addition to the symmetry of the stiffness matrix that has to be satisfied here as well, one can check where the elements with the $\bar{c}_k$ come from and if their signs are plausible.

The matrix elements in the upper left corner come from the frame whereas the lower right ones come from the connected spring-mass system. A banded structure of the stiffness matrix is a striking feature. The fact that zeros occur in the lower left and upper right corners of the stiffness matrix is typical of such oscillators. This is the result of the sequential numbering of the coordinates used in the calculation model.

Considerable calculation advantages exist for large systems with $n > 10$ if one utilizes the banded structure of the matrices, e. g. when solving the eigenvalue problem. It is, therefore, recommended to introduce the coordinates from the start in such a way that the bandwidth becomes minimal. There are algorithms for automatic bandwidth minimization of matrices that one can use in the case of arbitrary initial coordinate numbering (and the resulting high bandwidth).

6.2.3 Problems P6.1 to P6.3

P6.1 Matrix Elements

For the example in Table 6.2, Case 1, calculate the elements m_{23} and m_{24} of M.

P6.2 Substructure Matrices

Establish the system matrices of the vehicle model that is given in Table 6.2, Case 1, using the substructure matrices specified in Table 6.3.

P6.3 Mass and Stiffness Matrices for a Hoist

Calculate the elements of the mass matrix M and the stiffness matrix C for the coordinate vector $q^T = [x_1, x_2, r\varphi_M/i]$ for the calculation model of an overhead crane (Fig. 6.5). Check if the stiffness matrix is singular and give a physical interpretation.

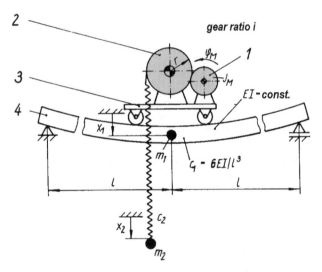

Fig. 6.5 Model of an overhead crane for calculating the dynamic loads when lifting and lowering the load; *1* Motor, *2* Cable drum, *3* Trolley, *4* Crane girder

Given:
Reduced moment of inertia of the hoist J_M
Reduced mass of the crane m_1 (referred to the position of the trolley)
Mass of the hoisted load m_2, gear ratio i, spring constant of the crane girder c_1 (referred to the position of the trolley), axial spring constant of the cable c_2, cable drum radius r

Note: Introducing the parameter $r\varphi_M/i$ as a generalized coordinate q_3 has the advantage that all components of the vector q (and thus the elements of C and M) are equal in dimension.

6.2.4 Solutions S6.1 to S6.3

S6.1 The first partial derivative of the kinetic energy first provides the following, see (6.47):

$$\frac{\partial W_{\text{kin}}}{\partial \dot{\xi}_2} = m_2 \dot{\xi}_2 + m_3 \frac{l_1(\dot{\xi}_2 + \dot{\xi}_4) + l_2(\dot{\xi}_1 + \dot{\xi}_3)}{l_1 + l_2} \frac{l_1}{l_1 + l_2} \qquad (6.64)$$
$$- \frac{J_S}{(l_1 + l_2)^2}(\dot{\xi}_1 + \dot{\xi}_3 - \dot{\xi}_2 - \dot{\xi}_4).$$

After rearranging and using the abbreviations

$$s_1 = l_1/(l_1 + l_2); \qquad s_2 = l_2/(l_1 + l_2); \qquad m = J_S/(l_1 + l_2)^2 \qquad (6.65)$$

one obtains

$$\frac{\partial W_{\text{kin}}}{\partial \dot{\xi}_2} = (m_3 s_1 s_2 - m)\dot{\xi}_1 + (m_2 + m_3 s_1^2 + m)\dot{\xi}_2 \qquad (6.66)$$
$$+ (m_3 s_1 s_2 - m)\dot{\xi}_3 + (m_3 s_1^2 + m)\dot{\xi}_4$$

According to (6.15), the elements of the mass matrix result:

$$\underline{m_{32} = m_{23}} = \frac{\partial^2 W_{\text{kin}}}{\partial \dot{\xi}_2 \partial \dot{\xi}_3} = \underline{m_3 s_1 s_2 - m} \qquad (6.67)$$
$$\underline{m_{42} = m_{24}} = \frac{\partial^2 W_{\text{kin}}}{\partial \dot{\xi}_2 \partial \dot{\xi}_4} = \underline{m_3 s_1^2 + m}$$

S6.2 The overall system can be viewed as a combination of substructures of the rigid body ($r = 1$, see Table 6.3, Case 3) and the two spring-mass systems ($r = 2$ and $r = 3$), the matrices of which are obtained as a special case of Case 1 in Table 6.3. The ξ_k listed in Table 6.2, Case 1, are used as global coordinates and denoted as $\xi_k = q_k$ to adjust them to the naming convention used in Sect. 6.2.1.

The couplings of the substructures are subject to the following constraints between the local coordinates $q^{(r)}$ and the global coordinates $q^{\text{T}} = [\xi_1, \xi_2, \xi_3, \xi_4] = [q_1, q_2, q_3, q_4]$: substructure $r = 1$:

$$q_1^{(1)} = q_1 + q_3, \qquad q_2^{(1)} = q_2 + q_4. \qquad (6.68)$$

These two equations can also be represented in accordance with (6.24) using the transformation matrix T_1, which one can check by multiplying:

$$q^{(1)} = \begin{bmatrix} q_1^{(1)} \\ q_2^{(1)} \end{bmatrix} = \begin{bmatrix} 1 & 0 & 1 & 0 \\ 0 & 1 & 0 & 1 \end{bmatrix} \begin{bmatrix} q_1 \\ q_2 \\ q_3 \\ q_4 \end{bmatrix}, \qquad T_1 = \begin{bmatrix} 1 & 0 & 1 & 0 \\ 0 & 1 & 0 & 1 \end{bmatrix}. \qquad (6.69)$$

The constraint equations $q_1^{(2)} = 0$, $q_2^{(2)} = q_1$ and $q_3^{(2)} = q_1 + q_3$ apply to the spring-mass system located at the bottom left, the substructure $r = 2$. Accordingly,

$$q^{(2)} = \begin{bmatrix} q_1^{(2)} \\ q_2^{(2)} \\ q_3^{(2)} \end{bmatrix}, \qquad T_2 = \begin{bmatrix} 0 & 0 & 0 & 0 \\ 1 & 0 & 0 & 0 \\ 1 & 0 & 1 & 0 \end{bmatrix}. \qquad (6.70)$$

For the substructure $r = 3$, the spring-mass system on the right, the conditions are $q_1^{(3)} = 0$, $q_2^{(3)} = q_2$, $q_3^{(3)} = q_2 + q_4$. The corresponding matrix is:

$$T_3 = \begin{bmatrix} 0 & 0 & 0 & 0 \\ 0 & 1 & 0 & 0 \\ 0 & 1 & 0 & 1 \end{bmatrix}. \tag{6.71}$$

Designations that are in line with the overall system must be introduced for the mass parameters of the first substructure, see Case 3 in Table 6.3 and Case 1 in Table 6.2. Equivalent are therefore $J_S/(l_1+l_2)^2 = m$, $l_1/(l_1+l_2) = s_1$, $l_2/(l_1+l_2) = s_2$. The mass matrix of the first substructure takes the form:

$$M_1 = \begin{bmatrix} m_3 s_2^2 + m & m_3 s_1 s_2 - m \\ m_3 s_1 s_2 - m & m_3 s_1^2 + m \end{bmatrix}. \tag{6.72}$$

Since the rigid body does not contain any springs, $C_1 = 0$. For the substructure $r = 2$, the following notations for the mass and spring parameters with respect to the overall system are valid, see Case 1 in Table 6.3 and the figure in Table 6.2, Case 1: $m_1 = 0$, $m_2 \hat{=} m_1$, $m_3 = 0$, $c_1 \hat{=} c_1$, $c_2 \hat{=} c_3$. The matrices, therefore, take the following forms:

$$C_2 = \begin{bmatrix} c_1 & -c_1 & 0 \\ -c_1 & c_1+c_3 & -c_3 \\ 0 & -c_3 & c_3 \end{bmatrix}, \quad M_2 = \begin{bmatrix} 0 & 0 & 0 \\ 0 & m_1 & 0 \\ 0 & 0 & 0 \end{bmatrix}. \tag{6.73}$$

Likewise, the relationships $m_1 = 0$, $m_2 \hat{=} m_2$, $m_3 = 0$, $c_1 \hat{=} c_2$, $c_2 \hat{=} c_4$ result in the following matrices for the substructure $r = 3$:

$$C_3 = \begin{bmatrix} c_2 & -c_2 & 0 \\ -c_2 & c_2+c_4 & -c_4 \\ 0 & -c_4 & c_4 \end{bmatrix}, \quad M_3 = \begin{bmatrix} 0 & 0 & 0 \\ 0 & m_2 & 0 \\ 0 & 0 & 0 \end{bmatrix}. \tag{6.74}$$

If one performs the multiplication and summation according to (6.25) with all these matrices, one obtains the stiffness and mass matrices specified in Table 6.2, Case 1.

S6.3 The translational energy of the crane and load and the rotational energy of the motor yield the overall kinetic energy

$$2W_{\text{kin}} = m_1 \dot{x}_1^2 + m_2 \dot{x}_2^2 + J_M \dot{\varphi}_M^2 = m_1 \dot{x}_1^2 + m_2 \dot{x}_2^2 + \frac{i^2 J_M}{r^2} \left(\frac{r \dot{\varphi}_M}{i} \right)^2. \tag{6.75}$$

The potential energy with respect to the static equilibrium position of crane and load corresponds to the deformation energy of the crane bridge and the cable. The cable is stretched by the length $(x_2 - x_1 + r\varphi_M/i)$. One can determine the appropriate signs by setting the respective other coordinates to zero. Therefore:

$$2W_{\text{pot}} = c_1 x_1^2 + c_2 (x_2 - x_1 + r\varphi_M/i)^2$$

The first partial derivatives of the energies are

$$\frac{\partial W_{\text{kin}}}{\partial \dot{x}_1} = m_1 \dot{x}_1; \qquad \frac{\partial W_{\text{pot}}}{\partial x_1} = c_1 x_1 - c_2 (x_2 - x_1 + r\varphi_M/i)$$
$$\frac{\partial W_{\text{kin}}}{\partial \dot{x}_2} = m_2 \dot{x}_2; \qquad \frac{\partial W_{\text{pot}}}{\partial x_2} = c_2 (x_2 - x_1 + r\varphi_M/i) \tag{6.76}$$
$$\frac{\partial W_{\text{kin}}}{\partial \left(\frac{r\dot{\varphi}_M}{i} \right)} = \frac{i^2 J_M}{r^2} \left(\frac{r\dot{\varphi}_M}{i} \right); \qquad \frac{\partial W_{\text{pot}}}{\partial \left(\frac{r\varphi_M}{i} \right)} = c_2 (x_2 - x_1 + r\varphi_M/i).$$

The matrices with respect to the specified vector q, the elements of which were selected with equal dimensions, result from (6.9) and (6.15):

$$M = \begin{bmatrix} m_1 & 0 & 0 \\ 0 & m_2 & 0 \\ 0 & 0 & i^2 J_M/r^2 \end{bmatrix}; \quad C = \begin{bmatrix} c_1+c_2 & -c_2 & -c_2 \\ -c_2 & c_2 & c_2 \\ -c_2 & c_2 & c_2 \end{bmatrix}. \tag{6.77}$$

As the resulting determinant is det $C = 0$, this system of equations is singular. The mechanical reason for this is the unconstrained motion that can take place in the system without the occurrence of elastic restoring forces. In this case, the hoisted mass m_2 can move freely together with the motor (J_M) (rigid-body motion). The consequence of this singularity is, as has been mentioned when discussing unconstrained torsional oscillators, that the first natural frequency becomes zero, see Sect. 4.2.

6.3 Free Undamped Vibrations

6.3.1 Natural Frequencies, Mode Shapes, Eigenforces

The equations of motion (6.18) to (6.22) discussed so far have symmetrical mass, stiffness, and compliance matrices. One can demonstrate that the sum of the kinetic and potential energies remains constant when such systems move, i. e. that the energy introduced by the initial conditions is conserved. Unlike damped or excited systems, in which the total energy changes during the motion, these are called *conservative* systems.

If a system is deflected from its equilibrium position (that was assumed for our considerations in Sect. 6.2 at $q = o$) and left to itself, it will perform so-called free vibrations. The free vibrations are a superposition of the various natural vibrations of the system, which themselves are characterized by its natural frequencies and mode shapes. Before discussing free vibrations, there should be a more detailed discussion of *natural vibrations*.

From mathematical point of view, natural vibrations are the general solutions of the homogeneous equations of motion. First, consider (6.18) to (6.20). The vector $\boldsymbol{v} = (v_1, v_2, \ldots, v_n)^T$ in

$$q = v \exp(j\omega t), \quad \ddot{q} = -\omega^2 v \exp(j\omega t) \tag{6.78}$$

contains the amplitudes of the harmonic motions of all coordinates $q = (q_1, q_2, \ldots, q_n)^T$ with an as yet unknown natural circular frequency ω. After a few brief conversions,

$$(C - \omega^2 M)v = o, \tag{6.79}$$

$$(DM - \frac{1}{\omega^2} E)v = o, \tag{6.80}$$

$$(M^{-1}C - \omega^2 E)v = o, \tag{6.81}$$

are obtained after dividing these equations by $\exp(j\omega t)$. These are homogeneous systems of linear equations in the unknown variables $v_1, v_2, \ldots, v_n$. All three forms express the same physical phenomenon. In mathematics, it is called a *general eigenvalue problem* if the form of (6.79) occurs, that is, if two different matrices appear. A *special* eigenvalue problem is what the form

$$(\boldsymbol{A} - \lambda \boldsymbol{E})\boldsymbol{v} = \boldsymbol{0}. \tag{6.82}$$

is called in the mathematical literature. It is useful for the further calculations to use dimensionless quantities. If one relates to an arbitrarily specifiable reference circular frequency ω^*, one can use the dimensionless quantity λ and the dimensionless matrix $\boldsymbol{A}$ for calculation:

$$\lambda = \frac{\omega^{*2}}{\omega^2}, \qquad \boldsymbol{A} = \omega^{*2} \boldsymbol{D} \boldsymbol{M}, \qquad \omega^2 = \frac{\omega^{*2}}{\lambda} \tag{6.83}$$

This follows from (6.80) when multiplying by the factor ω^{*2}. (6.81) can be transformed in the same way, however, another definition, $\lambda = \omega^2/\omega^{*2}$ and $\boldsymbol{A} = \boldsymbol{M}^{-1}\boldsymbol{C}/\omega^{*2}$, has to be used.

(6.82) only has a nonzero solution if its determinant equals zero. This condition yields

$$\det(\boldsymbol{A} - \lambda \boldsymbol{E}) = \lambda^n + a_{n-1}\lambda^{n-1} + \ldots + a_1\lambda + a_0 = 0. \tag{6.84}$$

This so-called *characteristic equation* is satisfied for the n roots $\lambda_1, \lambda_2, \ldots, \lambda_n$, the so-called eigenvalues. Since the mass matrix is positive definite and the stiffness matrix is positive definite or positive semidefinite, all eigenvalues in this eigenvalue problem are real and positive or zero. A vibration system with n degrees of freedom has n natural frequencies $f_i = \omega_i/2\pi$, which can be calculated from the eigenvalues λ_i, see (6.83). Natural frequencies that are zero and repeated frequencies are counted to reflect their multiplicity, as a rule.

The stiffness matrix in elastic vibration systems that can move without the occurrence of restoring forces is singular, i.e. $\det \boldsymbol{C} = 0$. In such systems, one or more of the lowest natural frequencies are identical to zero. The associated mode shapes then are forms of motion of the rigid-body system. The first mode shape in unconstrained torsional oscillators, for example, is an unconstrained rotation because the first natural frequency is zero, see, for example, (4.12), (4.27), and (4.132) in Chap. 4. In the case of unconstrained bending oscillators that can move in one plane, two natural frequencies are zero if the beam can move freely in that plane, see Table 5.7. The first six natural frequencies are zero for a flying object that can move freely in space. One therefore has to distinguish between the first natural frequency (that may be zero) and the **fundamental frequency** of a vibration system which is the lowest natural frequency that is different from zero.

It shall be assumed here that all λ_i are known. If they are inserted one after the other into (6.82), n different homogeneous systems of linear equations are obtained:

$$(\boldsymbol{A} - \lambda_i \boldsymbol{E})\boldsymbol{v}_i = \boldsymbol{0}, \qquad i = 1, 2, \ldots, n \tag{6.85}$$

The unknown variables v_{ki} of this system of equations (k is the number of the coordinate, i is the number of the eigenvalue) can only be determined up to a scaling

6.3 Free Undamped Vibrations

factor to be specified since the right side of the system of equations equals zero. It is specified by a normalization condition, e. g.

$$\mu_i = \boldsymbol{v}_i^{\mathrm{T}} \boldsymbol{M} \boldsymbol{v}_i = 1 \quad \text{or} \quad \gamma_i = \boldsymbol{v}_i^{\mathrm{T}} \boldsymbol{C} \boldsymbol{v}_i = 1 \quad \text{or} \quad (6.86)$$

$$\sum_{k=1}^{n} v_{ki}^2 = 1 \quad \text{or} \quad \max_k(|v_{ki}|) = 1 \quad \text{or} \quad v_{ki\,\max} = 1 \quad \text{for } i = 1, 2, \ldots, n.$$

This is normally done automatically by calculation programs.

What remains from (6.85) are $(n-1)$ linear equations for calculating the amplitude ratios v_{ki}. The totality of amplitude ratios that are characteristic for the eigenvalue λ_i are arranged in the *eigenvector*

$$\boldsymbol{v}_i = [v_{1i}, v_{2i}, \ldots, v_{ni}]^{\mathrm{T}}. \tag{6.87}$$

An eigenvector provides an illustrative description of a mode shape, also simply called *mode*). If one combines all eigenvectors, one obtains the so-called **modal matrix**

$$\boldsymbol{V} = [\boldsymbol{v}_1, \boldsymbol{v}_2, \ldots, \boldsymbol{v}_n] = \begin{bmatrix} v_{11} & v_{12} & \cdots & v_{1n} \\ v_{21} & v_{22} & \cdots & v_{2n} \\ \vdots & \vdots & \ddots & \vdots \\ v_{n1} & v_{n2} & \cdots & v_{nn} \end{bmatrix} = [v_{ki}]. \tag{6.88}$$

As in (6.80), which follows from (6.19), one can obtain the following form of a special eigenvalue problem from (6.22) using a similar assumption as in (6.78)

$$\boldsymbol{g} = \boldsymbol{w} \exp(\mathrm{j}\omega t), \qquad \ddot{\boldsymbol{g}} = -\omega^2 \boldsymbol{w} \exp(\mathrm{j}\omega t): \tag{6.89}$$

$$(\boldsymbol{M}\boldsymbol{D} - \frac{1}{\omega^2}\boldsymbol{E})\boldsymbol{w} = \boldsymbol{o}. \tag{6.90}$$

From (6.82) with (6.83) follows:

$$(\boldsymbol{A}^{\mathrm{T}} - \lambda \boldsymbol{E})\boldsymbol{w} = \boldsymbol{o}. \tag{6.91}$$

This (for physical reasons alone) provides the same eigenvalues λ_i, but the eigenvectors $\boldsymbol{w}_i$ have a different meaning: they represent the *eigenforces*. In the mathematical literature, the vectors $\boldsymbol{v}_i$, which physically represent the *mode shapes*, are called right eigenvectors and $\boldsymbol{w}_i$ left eigenvectors with respect to (6.82), whereas $\boldsymbol{w}_i$ are the right eigenvectors and $\boldsymbol{v}_i$ the left eigenvectors of the eigenvalue problem (6.91).

Because of (6.3) and (6.13), the relationships

$$\boldsymbol{w}_i = \omega_i^2 \boldsymbol{M} \boldsymbol{v}_i = \boldsymbol{C} \boldsymbol{v}_i \tag{6.92}$$

exist between the mode shapes and the eigenforces. One can picture the **eigenforces** as if the forces w_{ki} were applied to the coordinates q_k, see Fig. 6.8. These forces can be viewed as the amplitudes of inertia forces that occur as a result of the vibration at the natural circular frequency ω_i at the ith mode shape. Similarly justified is the notion that these inertia forces act like applied forces and that the mode shape is a consequence thereof.

Since two natural circular frequencies ($\omega_i = \pm\omega^*/\sqrt{\lambda_i}$) formally correspond to each real eigenvalue λ_i according to (6.83), the solution according to (6.78) for the ith natural vibration has two complex conjugate summands:

$$\boldsymbol{q}_i(t) = \frac{1}{2}\boldsymbol{v}_i\left[(a_i - \mathrm{j}b_i)\exp(\mathrm{j}\omega_i t) + (a_i + \mathrm{j}b_i)\exp(-\mathrm{j}\omega_i t)\right] \\ = \boldsymbol{v}_i(a_i\cos\omega_i t + b_i\sin\omega_i t) = \boldsymbol{v}_i\hat{p}_i\sin(\omega_i t + \beta_i) \tag{6.93}$$

Since all $\boldsymbol{q}_i(t)$ are real, Euler's formula provides the specified representation with trigonometric functions. The ith natural vibration thus is harmonic with the natural circular frequency ω_i and with an amplitude distribution that is expressed by the mode shape $\boldsymbol{v}_i$.

Further considerations will show that this is a major finding. One can also say that the *natural vibration* represents a state of motion of the oscillator that is maintained without energy supply.

The complete solution of the equations of motion (6.18) to (6.20) arises from the superposition of n natural vibrations where the $2n$ constants (a_i, b_i, or $\hat{p}_i$, β_i) follow from the initial conditions, see 6.3.3:

$$\boldsymbol{q}(t) = \sum_{i=1}^{n} \boldsymbol{q}_i(t). \tag{6.94}$$

6.3.2 Orthogonality and Modal Coordinates

An important mathematical relationship that exists among the eigenvectors will now be derived. It follows from (6.85) for the ith mode shape that

$$\boldsymbol{A}\boldsymbol{v}_i = \lambda_i \boldsymbol{v}_i. \tag{6.95}$$

Similarly, it follows from (6.91) for the amplitude vector of the load parameters

$$\boldsymbol{A}^\mathrm{T}\boldsymbol{w}_k = \lambda_k \boldsymbol{w}_k. \tag{6.96}$$

If one multiplies (6.95) by $\boldsymbol{w}_k^\mathrm{T}$ and (6.96) by $\boldsymbol{v}_i^\mathrm{T}$ from the left, the result is

$$\boldsymbol{w}_k^\mathrm{T}\boldsymbol{A}\boldsymbol{v}_i = \lambda_i \boldsymbol{w}_k^\mathrm{T}\boldsymbol{v}_i \tag{6.97}$$

$$\boldsymbol{v}_i^\mathrm{T}\boldsymbol{A}^\mathrm{T}\boldsymbol{w}_k = \lambda_k \boldsymbol{v}_i^\mathrm{T}\boldsymbol{w}_k = \lambda_k \boldsymbol{w}_k^\mathrm{T}\boldsymbol{v}_i. \tag{6.98}$$

One can exchange the factors in (6.97) when using the transposed matrix $\boldsymbol{A}^\mathrm{T}$ instead of $\boldsymbol{A}$, resulting in the following

6.3 Free Undamped Vibrations

$$v_i^T A^T w_k = \lambda_i w_k^T v_i = \lambda_i v_i^T w_k. \tag{6.99}$$

If one forms the difference between (6.98) and (6.99), it follows that

$$(\lambda_k - \lambda_i) v_i^T w_k = (\lambda_k - \lambda_i) w_k^T v_i = 0. \tag{6.100}$$

The following is therefore obtained for different eigenvalues ($\lambda_i \neq \lambda_k$)

$$v_i^T w_k = w_k^T v_i = 0 \tag{6.101}$$

and using (6.92) one obtains the generalized orthogonality relations for the eigenvectors (because of $M^T = M$):

$$i \neq k: \quad v_i^T C v_k = 0 \tag{6.102}$$
$$i \neq k: \quad v_i^T M v_k = 0 \tag{6.103}$$
$$i = k: \quad v_i^T C v_i = \gamma_i \tag{6.104}$$
$$i = k: \quad v_i^T M v_i = \mu_i \tag{6.105}$$
$$i \neq k: \quad w_i^T D w_k = \omega_i^2 v_i^T M D M v_k \omega_k^2 = 0 \tag{6.106}$$
$$i = k: \quad w_i^T D w_i = \omega_i^4 v_i^T M D M v_i = \gamma_i \neq 0. \tag{6.107}$$

This defines the modal spring constants γ_i and the modal masses μ_i. They represent a reduction of the inertia and elasticity of the overall system to the ith mode shape.

(6.101) expresses the fact that the forces w_k of the kth mode shape do not perform any work with respect to the ith mode shape with the amplitudes v_i. In other words, the mode shapes do not influence each other because there is no exchange of energy between them. (6.102) to (6.107) represent generalized orthogonality relations, which have the same physical meaning as (6.101).

An important conclusion can be drawn from (6.93). If $\hat{p}_i \sin(\omega_i t + \beta_i)$ is considered as a coordinate $p_i(t)$, one can use this coordinate instead of the previous generalized coordinates $q_i(t)$. The connection between these **modal coordinates**, *normal* or *principal coordinates* that are arranged in the vector $p^T = [p_1, p_2, \ldots, p_n]$ and the generalized coordinates q_k is expressed using the modal matrix V from (6.88) (modal transformation):

$$q_k = \sum_{i=1}^n v_{ki} p_i; \quad q = V p \quad \text{or} \quad p = V^{-1} q. \tag{6.108}$$

If the modal coordinates are used, the potential energy is, see (6.7):

$$W_{\text{pot}} = \frac{1}{2} q^T C q = \frac{1}{2} (Vp)^T C V p = \frac{1}{2} p^T V^T C V p \tag{6.109}$$

and the kinetic energy is, see (6.16):

$$W_{\text{kin}} = \frac{1}{2} \dot{q}^T M \dot{q} = \frac{1}{2} (V\dot{p})^T M V \dot{p} = \frac{1}{2} \dot{p}^T V^T M V \dot{p}. \tag{6.110}$$

Note when multiplying the matrices $V^T CV$ that V^T and V contain the eigenvectors. Therefore,

$$V^T CV = \begin{bmatrix} v_1^T \\ v_2^T \\ \vdots \\ v_n^T \end{bmatrix} [Cv_1, Cv_2, \ldots, Cv_n] = \begin{bmatrix} v_1^T Cv_1 & v_1^T Cv_2 & \cdots & v_1^T Cv_n \\ \vdots & \vdots & & \vdots \\ v_n^T Cv_1 & v_n^T Cv_2 & \cdots & v_n^T Cv_n \end{bmatrix} \quad (6.111)$$

Due to the special properties of the modal matrix V, the relations simplify when observing (6.102) to (6.105). All off-diagonal elements equal zero. The remaining diagonal elements are the modal spring constants γ_i and the modal masses μ_i. Therefore the following applies, see (6.102) to (6.105):

$$V^T CV = \text{diag}(\gamma_i); \qquad V^T MV = \text{diag}(\mu_i). \quad (6.112)$$

The modal stiffnesses γ_i and modal masses μ_i depend on the normalization, see (6.86). The potential and kinetic energies can be expressed by the squares of the principal coordinate or its velocities, see (6.109) and (6.110):

$$W_{\text{pot}} = \frac{1}{2} \sum_{i=1}^{n} \gamma_i p_i^2; \qquad W_{\text{kin}} = \frac{1}{2} \sum_{i=1}^{n} \mu_i \dot{p}_i^2. \quad (6.113)$$

The equations of motion for the free vibrations of a vibration system therefore have a simple form in terms of the modal coordinates. Due to (6.18) with $q = Vp$, see (6.108):

$$MV\ddot{p} + CVp = o \quad (6.114)$$

and after multiplying this equation by V^T

$$V^T MV\ddot{p} + V^T CVp = o \quad (6.115)$$

or simply because of (6.112), n equations of motion of "modal oscillators" result for free vibrations:

$$\mu_i \ddot{p}_i + \gamma_i p_i = 0, \qquad i = 1, 2, \ldots, n. \quad (6.116)$$

They are equivalent to the coupled equations (6.18), see also (6.19) and (6.20). The n natural circular frequencies result from the modal spring constants γ_i and the modal masses μ_i from (6.116), (6.104), (6.107), and (6.105):

$$\omega_i^2 = \frac{\gamma_i}{\mu_i} = \frac{v_i^T Cv_i}{v_i^T Mv_i} = \frac{v_i^T Mv_i}{v_i^T MDMv_i} \qquad i = 1, 2, \ldots, n. \quad (6.117)$$

One usually sorts the natural circular frequencies by magnitude ($\omega_{i+1} > \omega_i$). This relationship makes an interesting connection between each natural circular frequency ω_i and the associated mode shape v_i, see also the Rayleigh quotient in (6.181) and (6.184).

6.3.3 Initial Conditions, Initial Energy, Estimates

A vibration system performs free vibrations if it is supplied with energy at the beginning of the motion, if it is deflected from its equilibrium position and left to itself. Dynamic loads of machines as a result of free vibrations are of practical interest primarily after impact-like excitations (e. g. sudden braking of a motion, hitting against an obstacle, clutching processes) or after sudden loading or unloading, e. g. dropping a load from a crane, tearing a tensioning element, or breaking of a component.

In the case of impulse loads, the initial conditions can frequently be calculated from the principle of conservation of linear momentum and/or the principle of conservation of angular momentum. Vibrations about the equilibrium position $q = o$ occur. A new (changed) equilibrium position results from sudden loading/unloading that the system reaches by transient oscillation. Note that the equations of motion must be established for the changed system because it performs vibrations about the position $q = o$.

The time functions of the deformations and load parameters that are to be calculated for oscillatory machines are obtained as the solution of differential equations of the type of (6.18) to (6.22), taking into account the initial conditions. Which one of these equations is used depends on the respective practical problem or on the person solving it. Equations of the form (6.18) to (6.20), which yield the time functions of the motions, are commonly used. If load parameters are of interest, these are subsequently calculated from (6.3) or (6.21).

The initial conditions define the state of the vibration system at the time $t = 0$, i. e. they specify the initial deflections and velocities:

$$t = 0: \quad q(0) = q_0; \quad \dot{q}(0) = u_0. \tag{6.118}$$

The deflection from the static equilibrium position is associated with the transfer of potential energy and the imparting of an initial velocity corresponds to a sudden transfer of kinetic energy.

The physical fact that the initial conditions also express the transfer of the energy

$$W_0 = W_{\text{pot }0} + W_{\text{kin }0} = \frac{1}{2}q_0^T C q_0 + \frac{1}{2}u_0^T M u_0 = \sum_{i=1}^{n} W_{i0} \tag{6.119}$$

onto the vibration system is important in this respect. This energy is distributed between the individual mode shapes (W_{i0}) and "is dissipated" in the free vibrations that are excited. Extreme deformations, velocities and loads can be estimated if one bears in mind that at most this total energy W_0 is concentrated in the respective element, see (4.95) to (4.99) in Sect. 4.2.6 as well as (6.128) and (6.129). The initial values of the modal coordinates can be calculated from the initial conditions (6.118) of the position coordinates. This does not have to be done using the equations resulting from (6.108)

$$p(0) = p_0 = V^{-1} q_0, \qquad \dot{p}(0) = \dot{p}_0 = V^{-1} u_0 \tag{6.120}$$

for which the inverse matrix of the modal matrix would have to be formed. The relationships

$$V^T M q = V^T M V p = \text{diag}(\mu_i) p,$$
$$V^T C q = V^T C V p = \text{diag}(\gamma_i) p \tag{6.121}$$

result in equations that require fewer operations:

$$p_0 = \text{diag}(1/\gamma_i) V^T C q_0 = \text{diag}(1/\mu_i) V^T M q_0, \tag{6.122}$$

$$\dot{p}_0 = \text{diag}(1/\gamma_i) V^T C u_0 = \text{diag}(1/\mu_i) V^T M u_0. \tag{6.123}$$

The energy distribution of the initial state can be expressed using the initial values of the modal coordinates and amounts to, see (6.113):

$$W_0 = \sum_{i=1}^{n} W_{i0} = \frac{1}{2} \sum_{i=1}^{n} (\gamma_i p_{i0}^2 + \mu_i \dot{p}_{i0}^2)$$
$$= \frac{1}{2} \sum_{i=1}^{n} \left[\frac{(v_i^T C q_0)^2}{\gamma_i} + \frac{(v_i^T M u_0)^2}{\mu_i} \right]. \tag{6.124}$$

The total initial energy W_0 is distributed between the mode shapes, and the share that each mode shape gets does not change during the subsequent natural vibration. Over time, energy is exchanged between the individual masses and springs of the oscillator but each mode shape behaves like an isolated individual oscillator, see (6.116).

The ith summand in (6.124) captures the energy of the ith natural vibration. The initial conditions (6.118), which are transformed according to (6.120), (6.122), and (6.123), take the following form for the ith principal coordinate:

$$t = 0: \quad p_i(0) = p_{i0}, \quad \dot{p}_i(0) = \dot{p}_{i0}. \tag{6.125}$$

The solution of each of the equations (6.116) under these initial conditions is:

$$p_i = p_{i0} \cos \omega_i t + \frac{\dot{p}_{i0}}{\omega_i} \sin \omega_i t = \hat{p}_i \sin(\omega_i t + \beta_i). \tag{6.126}$$

One can consider the motion of the vibration system as the sum of the motions of n individual oscillators. If the system only vibrates at the natural circular frequency ω_i, it assumes the amplitude ratios of the ith eigenvector v_i, and one can say that its motion is uniquely characterized by stating this one principal coordinate p_i.

The position coordinates q can be calculated from the principal coordinates p with the help of (6.108) and (6.126):

$$q(t) = \sum_{i=1}^{n} \frac{v_i v_i^T M}{\mu_i} \left[q_0 \cos \omega_i t + \frac{u_0}{\omega_i} \sin \omega_i t \right] \tag{6.127}$$

6.3 Free Undamped Vibrations

This result is also known from (6.93) and (6.94). The inertia forces of the generalized coordinates thus result from (6.13). They represent a superposition of the eigenforces that change harmonically with the natural frequencies. The motions, bearing responses, or internal forces of interest (e. g. transverse force, axial force, bending moment in beams) that act at specific machine positions can be calculated from the known $q(t)$ or $g(t)$, respectively, taking into account the geometrical relations.

A generally different approach to calculating $q(t)$ for free vibrations is the numerical integration of the differential equations (6.18) with the initial conditions (6.118), e. g. using the *Runge-Kutta* method. It can be even simpler than the method described here. The calculation of the natural frequencies and mode shapes is avoided. The disadvantage of such numerical methods is, however, that essential physical relations become less evident and the analysis of parameter influences is obscured.

One can calculate limits for extreme motion or load parameters at arbitrary positions k inside of a vibration system without integrating the differential equations. If one assumes that the total energy W_0 of the initial state is concentrated there as kinetic energy, the following inequalities result as estimates for the maximum velocity of a single mass m_k or angular velocity of a rotating mass J_k, respectively

$$v_{k\,\max} < \sqrt{2W_0/m_k}; \qquad \Omega_{k\,\max} < \sqrt{2W_0/J_k}. \tag{6.128}$$

The maximum load of any spring can only be that the entire energy is stored in it as potential energy. The inequalities below follow for the maximum force in the translational spring c_k or the maximum moment in the torsional spring c_{Tk}:

$$\begin{aligned} c_k \Delta x_{k\,\max} = F_{k\,\max} < \sqrt{2W_0 c_k} \\ c_{Tk} \Delta \varphi_{k\,\max} = M_{k\,\max} < \sqrt{2W_0 c_{Tk}}. \end{aligned} \tag{6.129}$$

Other estimates result from using the mode shape-related energy portions from (6.190).

6.3.4 Examples

6.3.4.1 Modal Analysis of Machines

It is recommended to start the dynamic analysis of a machine (or sub-assembly) with a modal analysis, in which, first of all, the mode shapes are determined in addition to the natural circular frequencies. Regardless of the absolute magnitude of the excitations and damping, knowledge of the essential natural frequencies and mode shapes is helpful in evaluating the dynamic behavior.

There is software available today with which all eigenvalues and eigenvectors can be determined in a short time so that the engineer does not have to care about the algorithms used for this purpose. The computational effort required to determine the eigenvalues and eigenvectors of a $(n \times n)$ matrix is proportional to n^3 for all methods. Commercial software applications usually display the natural frequencies and mode shapes immediately based on the input data without showing the matrices to the user of the software.

Figures 6.6 and 6.7 give an impression of how the results of a modal analysis are represented. The software typically lets users view the moving mode shapes like in a movie on the screen. The mode depicted shape of the engine housing basically represents a torsion of the entire hollow body. The bending of the lateral wall of the textile machine frame was remarkable since a higher fundamental frequency was achieved by selective stiffening of the wall after noticing this weak spot.

Certain difficulties may occur when using some calculation programs if repeated or very close natural frequencies occur. This can happen with systems of a branched or intermeshed structure, e. g. foundation blocks or machine platforms (where the vicinity of frequencies may be required for limiting the spectrum when the system is low-tuned), spatially extended structures, pipe systems and branched torsional vibration systems.

The preparations such as the selection of the calculation model and the determination of the parameter values take the greatest effort when solving practical tasks. The usefulness of the results of the calculations primarily depends on how well the calculation model reflects the real machine and how accurate the input data are. The geometrical and mechanical parameters of a machine are often known with only 2 or 3 digits accuracy. The data of the mass parameters can generally be determined more reliably than the spring constants.

In reality, modeling requires simplifications, starting with the geometry, e. g. when it comes to ribs and recesses of complex cross-sectional shapes of a frame. The stiffnesses of the fasteners and contact points such as joints, screwed and glued connections, bearings, etc. are frequently not known. Most parameter values that capture damping are inaccurate, which is why the theoretical modal analysis is often performed without considering the damping. If parameter values are inaccurate, one can work around that by performing a modal analysis for specific parameter ranges and by determining the sensitivities with respect to these parameter values.

In addition to a theoretical modal analysis (e. g., using an FEM program), which is a viable option as early as in the project phase of a design, an **experimental modal analysis** on the actual object is recommended. In an experimental modal analysis, it is assumed that the real object behaves like a linear vibration system and that the corresponding theoretical relationships apply, e. g. the principle of superposition, see Sect. 9.1. Since nonlinearities of various kinds can be present, it is advisable when performing an experimental modal analysis to deform and load the real object in such a frequency range that includes the range of its operating states. If the loads and preloads are too small, phenomena such as backlash influences can dominate that would have no significance at higher loads. If another frequency range is traversed during measurement than during real operation, stiffness and damping characteristics may deviate considerably.

A so-called **MAC matrix** (MAC = Modal Assurance Criterion) is commonly used in structural dynamics for comparing calculated and measured mode shapes.

6.3 Free Undamped Vibrations

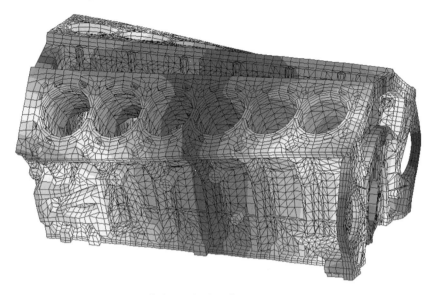

Fig. 6.6 Mode shape of a 6-cylinder engine housing

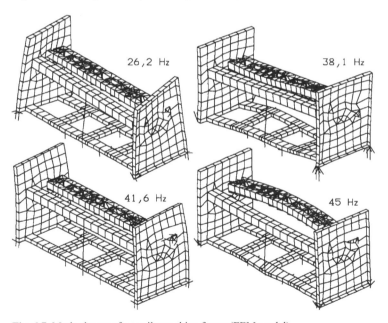

Fig. 6.7 Mode shapes of a textile machine frame (FEM model)

This is a rectangular matrix that has as many columns (j) as mode shapes were measured and as many rows (i) as calculated eigenvectors are available. Each element of this matrix represents the value of a normalized scalar product of a measured eigenvector v_{je} and a calculated eigenvector v_{io} and is used to evaluate the correlation between these two vectors. The matrix element

$$\text{MAC}(i,j) = \frac{(v_{je}^\text{T} v_{io})^2}{(v_{je}^\text{T} v_{je})(v_{io}^\text{T} v_{io})} \qquad (6.130)$$

is a measure of the agreement between the two vectors v_{je} and v_{io}.

If these two vectors are identical (but for a scaling factor), the matrix element equals one, if they are orthogonal to each other, it equals zero. In practice, the matrix elements have values between zero and one because the vectors under review do not meet ideal conditions. Empirically, there is hardly any correlation between the two vectors for a matrix element $\text{MAC}(i,j) < 0.3$ while the calculated and measured vectors match with a high probability for a matrix element $\text{MAC}(i,j) > 0.8$. The MAC matrix is also suitable for differentiating the eigenvectors of natural frequencies that are close together.

One has to select points for comparison in a real-life situation, for in most cases the numbers of elements in the vectors v_{je} and v_{io} do not match. The coordinates of the measuring points are also rarely identical with the coordinates of the calculation model, and they can also not easily be determined with a precision of several digits. In addition to an evaluation using the MAC matrix, there are several other criteria for comparing experimental and calculated values.

The results of an experimental modal analysis can be used to precisely specify the input data for the calculation. It is recommended to collect empirical values from the modal analysis for the product under consideration and to store them in an orderly manner for comparison. There are internal files and lists in the design and development departments of leading industries for some machines and vehicles and their sub-assemblies containing all essential natural frequencies, mode shapes, and modal damping parameters. These are not only used for comparison with competing products but also for the support of diagnostics and rapid fault detection in the event of malfunctions.

Prior to an experimental study, one should also base the proper selection of the measuring points on information about the mode shapes of the machine since many phenomena can be interpreted as their superposition. The time dependent forces, stresses, displacements, velocities or accelerations measured for a machine can be thought of and interpreted as superposition of its natural vibrations.

Table 6.4 contains the natural frequencies of some machines and their sub-assemblies. Note that the lowest natural frequencies of larger objects are around 1 Hz and those of small compact sub-assemblies are several hundred Hertz.

For many objects, only the lower natural frequencies or just a part of the natural frequency spectrum, e. g. the natural frequencies in the vicinity of a given excitation frequency, are of interest. The highest natural frequencies of a *calculation model* only characterize the calculation model, not the real machine. The highest natural frequencies of each real machine are infinitely large.

It often turns out that the actual spring constants are smaller than the calculated ones and that measured natural frequencies are below the calculated ones. The rea-

6.3 Free Undamped Vibrations

Table 6.4 Examples of the lower natural frequencies of machines

Machine	Lower natural frequencies			excitation frequency
	$f_1 = \frac{\omega_1}{2\pi}$ in Hz	$f_2 = \frac{\omega_2}{2\pi}$ in Hz	$f_3 = \frac{\omega_3}{2\pi}$ in Hz	$f = \frac{\Omega}{2\pi}$ in Hz
Bucket wheel excavator SRs 4000	0.46	1.2	1.8	0.8
Machine platform of a 50 MW turbine	1.1	9.1	9.5	50
Luffing and slewing crane DWK 5MPx22m	0.7...0.9	1.2...1.6		
Frame of a grinding machine	3.9	9.2	13	
Household spin drier	4...4.5	9...9.5		
Textile mandrils	6...12	30...90	100...200	160...300
Oscillating conveyor	9.2	15	54	50
Printing machine drive (torsional osc.)	10			
Crankshaft of a marine diesel engine (torsional osc.)	60	140	160	
Shaft of a 200 MW turbogenerator (bending oscillations)	19	23	27	50
Hobbing machine drive	40	70	> 160	
Centrifugal compressor 4 VRZ	60...70	150...180		
Gas turbine blade	160	350	590	
Crankshaft of a motorbike	360	850		
Warped sheet-metal blade of an axial fan	375	760	1300	

son for this is that often not all elastic elements are included in the calculation (see Problem P6.6) and that some joints loosen up during operation so that the occurring backlash actually results in smaller average stiffness values. A match of measured and pre-calculated natural frequencies of approximately under 5 % can rarely be achieved. One can consider a match within 10 % successful for first model calculations. The desired accuracy for real objects is only achieved through step-by-step model improvements.

One can in general expect for various specimen of machines of the same design that their vibration behavior will hardly differ. The observed deviations from the calculated results are often due to the scattering of parameter values of the machine examined. There are machines in which the stiffness of fasteners is mainly determined by the assembly conditions in that the assembler determines the preload, and thus the natural frequencies, with his wrench. Such uncertainties should be avoided in the design. One should ensure that the assumed values in the calculation are reliably put into practice. This requirement implies designing "with keeping calculations in mind" and complying with the measures and conditions the designer stipulates for manufacture.

The following considerations should be made to check the numerical results:

- Check whether only exactly symmetrical and exactly antisymmetrical mode shapes are obtained for systems with a symmetrical geometric structure and sym-

metrical mass and stiffness distributions (not to be confused with the symmetry of C and M!) from the calculation.
- Simplify the overall system, i.e. divide it into decoupled subsystems with few degrees of freedom and calculate partial natural frequencies for these. Depending on whether the simplification makes the system stiffer or "softer" or more or less inert, one can judge whether the estimated natural frequency must be higher or lower than that of the overall system.

6.3.4.2 Impact on a Frame

The calculation model of Table 6.3, Case 2, is considered as a representative of a machine frame. The masses are distributed after dividing the continuum into four sections. Determine the free vibrations that will occur after the end point of the system initially at rest suddenly receives the initial velocity $\dot{q}_1(0) = u_{10}$ due to an impulse.

Given are the coordinate vector q and the elements of the matrices M and D of the calculation model, see Table 6.3, Case 2:

$$q = \begin{bmatrix} q_1 \\ q_2 \\ q_3 \\ q_4 \end{bmatrix}; \quad M = m \begin{bmatrix} 1 & 0 & 0 & 0 \\ 0 & 2 & 0 & 0 \\ 0 & 0 & 5 & 0 \\ 0 & 0 & 0 & 2 \end{bmatrix}; \quad D = \frac{l^3}{48EI} \begin{bmatrix} 64 & 29 & 24 & 6 \\ 29 & 14 & 12 & 3 \\ 24 & 12 & 16 & 5 \\ 6 & 3 & 5 & 2 \end{bmatrix}. \quad (6.131)$$

With the initial condition mentioned above, the initial vectors for $t = 0$ are:

$$q(0) = \begin{bmatrix} q_1(0) \\ q_2(0) \\ q_3(0) \\ q_4(0) \end{bmatrix} = \begin{bmatrix} 0 \\ 0 \\ 0 \\ 0 \end{bmatrix} = q_0; \quad \dot{q}(0) = \begin{bmatrix} \dot{q}_1(0) \\ \dot{q}_2(0) \\ \dot{q}_3(0) \\ \dot{q}_4(0) \end{bmatrix} = \begin{bmatrix} u_{10} \\ 0 \\ 0 \\ 0 \end{bmatrix} = u_0. \quad (6.132)$$

The matrix $A = \omega^{*2} DM$ according to (6.83) results from a matrix multiplication

$$DM = \frac{l^3}{48EI} \begin{bmatrix} 64 & 29 & 24 & 6 \\ 29 & 14 & 12 & 3 \\ 24 & 12 & 16 & 5 \\ 6 & 3 & 5 & 2 \end{bmatrix} m \begin{bmatrix} 1 & 0 & 0 & 0 \\ 0 & 2 & 0 & 0 \\ 0 & 0 & 5 & 0 \\ 0 & 0 & 0 & 2 \end{bmatrix} = \frac{ml^3}{48EI} \begin{bmatrix} 64 & 58 & 120 & 12 \\ 29 & 28 & 60 & 6 \\ 24 & 24 & 80 & 10 \\ 6 & 6 & 25 & 4 \end{bmatrix}.$$

$$(6.133)$$

As a result, the reference circular frequency and the dimensionless eigenvalue are introduced as

$$\omega^{*2} = \frac{48EI}{ml^3}, \quad \lambda = \frac{\omega^{*2}}{\omega^2} = \frac{48EI}{\omega^2 ml^3}. \quad (6.134)$$

For this particular case, it follows from (6.84):

$$\det(A - \lambda E) = \lambda^4 - 176\lambda^3 + 3480\lambda^2 - 5456\lambda + 1940 = 0.$$

6.3 Free Undamped Vibrations

The expansion of the determinant provides this characteristic equation of 4th degree for calculating the eigenvalues. The roots of this equation are (rounded off to 5 significant digits):

$$\lambda_1 = 153.57; \quad \lambda_2 = 20.751; \quad \lambda_3 = 1.1489; \quad \lambda_4 = 0.52990. \tag{6.135}$$

This provides the sought after natural circular frequencies from (6.83):

$$\begin{aligned} \omega_1 &= 0.0807\omega^*; & \omega_2 &= 0.2195\omega^*; \\ \omega_3 &= 0.9326\omega^*; & \omega_4 &= 1.374\omega^*. \end{aligned} \tag{6.136}$$

A maximum of 4 significant digits can be stated in the following calculations due to inevitable round-off errors.

The eigenvector v_1 associated with the eigenvalue λ_1 results from the system of equations (6.85),

$$(\boldsymbol{A} - \lambda_1 \boldsymbol{E})\boldsymbol{v}_1 = 0, \tag{6.137}$$

which takes the following form when expanded:

$$\begin{aligned} (64 - 153.57)v_{11} + & 58v_{21} + & 120v_{31} + & 12v_{41} = 0 \\ 29v_{11} + (28 - 153.57)v_{21} + & 60v_{31} + & 6v_{41} = 0 \\ 24v_{11} + & 24v_{21} + (80 - 153.57)v_{31} + & 10v_{41} = 0 \\ 6v_{11} + & 6v_{21} + & 25v_{31} + (4 - 153.57)v_{41} = 0. \end{aligned} \tag{6.138}$$

No absolute magnitudes of the amplitudes v_{k1} can be calculated since these are natural vibrations. Using the normalization $v_{11} = 1$ according to (6.86), 3 equations are obtained for the 3 unknown variables v_{21}, v_{31} and v_{41} because one of these four equations does not have to be considered. If, for example, the first three equations are used, the following applies:

$$\begin{aligned} 58v_{21} + & 120v_{31} + 12v_{41} &= 89.57 \\ -125.57v_{21} + & 60v_{31} + 6v_{41} &= -29 \\ 24v_{21} - & 73.57v_{31} + 10v_{41} &= -24 \end{aligned} \tag{6.139}$$

One arrives at the solution

$$v_{21} = 0.4774; \quad v_{31} = 0.5014; \quad v_{41} = 0.1431. \tag{6.140}$$

In this way, one obtains the vectors v_2, v_3, and v_4 of the other mode shapes by solving one of the systems of linear equations with 3 unknown variables, see Fig. 6.8.

Associated with the mode shapes according to (6.92) are eigenforce shapes. For the first mode shape, one obtains an eigenforce shape of $\boldsymbol{w}_1 = \omega_1^2 \boldsymbol{M} \boldsymbol{v}_1$:

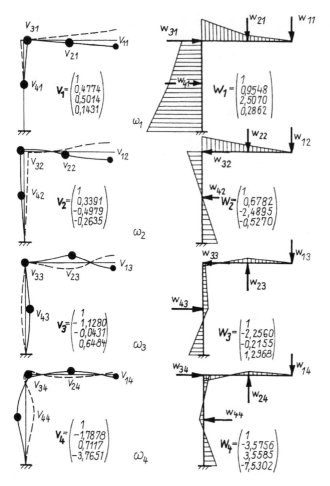

Fig. 6.8 Mode shapes v_i, eigenforce vectors w_i and the resulting deformation and moment curves

$$\boldsymbol{w}_1 = \begin{bmatrix} w_{11} \\ w_{21} \\ w_{31} \\ w_{41} \end{bmatrix} = m\omega_1^2 \begin{bmatrix} 1 & 0 & 0 & 0 \\ 0 & 2 & 0 & 0 \\ 0 & 0 & 5 & 0 \\ 0 & 0 & 0 & 2 \end{bmatrix} \begin{bmatrix} 1 \\ 0.4774 \\ 0.5014 \\ 0.1431 \end{bmatrix} = m\omega_1^2 \begin{bmatrix} 1 \\ 0.9548 \\ 2.5070 \\ 0.2862 \end{bmatrix}. \quad (6.141)$$

The "bending eigenmoments" caused by the eigenforces for the 4 natural frequencies are shown on the right side of Fig. 6.8. The modal matrix in its complete form is:

$$\boldsymbol{V} = \begin{bmatrix} 1 & 1 & 1 & 1 \\ 0.4774 & 0.3391 & -1.1280 & -1.7878 \\ 0.5014 & -0.4979 & -0.0431 & 0.7117 \\ 0.1431 & -0.2635 & 0.6484 & -3.7651 \end{bmatrix}. \quad (6.142)$$

6.3 Free Undamped Vibrations

According to (6.112), one obtains the modal spring constants γ_i and modal masses μ_i ($c^* = 48EI/l^3$):

$$\gamma_1 = 0.01793c^*, \quad \gamma_2 = 0.1257c^*, \quad \gamma_3 = 3.825c^*, \quad \gamma_4 = 72.31c^*$$
$$\mu_1 = 2.754m, \quad \mu_2 = 2.609m, \quad \mu_3 = 4.398m, \quad \mu_4 = 38.18m.$$

According to (6.117) and in accordance with (6.136), it follows that, see (6.134):

$$\omega_1 = 0.08069\omega^*, \quad \omega_2 = 0.2195\omega^*, \quad \omega_3 = 0.9326\omega^*, \quad \omega_4 = 1.374\omega^*. \quad (6.143)$$

The initial conditions in terms of the principal coordinates are derived from (6.122), (6.123), and (6.132):

$$\mathbf{p}_0 = \mathbf{0}, \quad \dot{\mathbf{p}}_0 = \begin{bmatrix} \dot{p}_{10} \\ \dot{p}_{20} \\ \dot{p}_{30} \\ \dot{p}_{40} \end{bmatrix} = \begin{bmatrix} 0.3631 \\ 0.3834 \\ 0.2274 \\ 0.0261 \end{bmatrix} u_{10}. \quad (6.144)$$

Looking at the numerical values from (6.143), (6.144), and the following relation, one can observe the initial energy distribution among the four mode shapes, see (6.124):

$$W_0 = W_{\text{kin }0} = \frac{1}{2}m u_{10}^2 = \frac{1}{2}(\mu_1 \dot{p}_{10}^2 + \mu_2 \dot{p}_{20}^2 + \mu_3 \dot{p}_{30}^2 + \mu_4 \dot{p}_{40}^2)$$
$$= \frac{m}{2} u_{10}^2 (0.3631 + 0.3834 + 0.2274 + 0.0261) \quad (6.145)$$

The energy is thus primarily transferred to the first three mode shapes.

Using the initial values from (6.144), the principal coordinates p_i can be calculated according to (6.126), the coordinates q_k according to (6.127) and the inertia forces from (6.13). The following is valid for $q^* = u_{10}/\omega^*$:

$$p_1(t) = 4.500 q^* \sin\omega_1 t, \quad p_2(t) = 1.747 q^* \sin\omega_2 t,$$
$$p_3(t) = 0.244 q^* \sin\omega_3 t, \quad p_4(t) = 0.019 q^* \sin\omega_4 t. \quad (6.146)$$

The results for the position coordinates and inertia forces are shown in Fig. 6.9.

The upper part of Fig. 6.9a shows the individual components of the motion $q_1(t)/q^*$ and the resulting actual motion. The dotted sinusoidal line corresponds to component $\hat{p}_1 v_{11} \sin\omega_1 t$. Furthermore, the upper part of the figure shows the components of the natural vibrations at the natural circular frequencies ω_2 and ω_3 as solid curves. The amplitude $0.019 q^* \hat{p}_4 v_{14}$ of the highest natural vibration is so small that it is no longer visible in this figure. The resulting motion is nonperiodical because the ratio of the natural frequencies is not rational.

One can interpret the time functions of $q_2(t)$, $q_3(t)$ and $q_4(t)$ analogously. The individual components were not plotted for clarity. There is no portion of the third and fourth natural vibrations visible in $q_3(t)$, unlike in q_1, q_2 and q_4. This is because the amplitudes $\hat{p}_3 v_{3i}$ and $\hat{p}_4 v_{3i}$ are considerably smaller than the other portions.

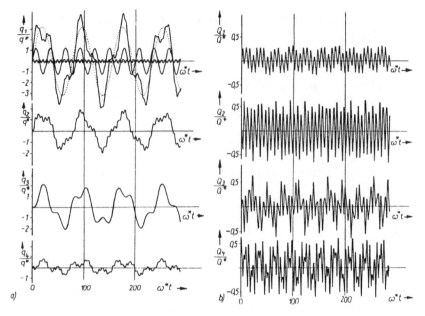

Fig. 6.9 Free vibrations after an impulse to the frame; **a)** Time functions of the coordinates $q_k(\omega^* t)$ where $q^* = u_{10}/\omega^*$, **b)** Time curves of the forces $Q_k(\omega^* t)$ where $Q^* = m u_{10} \omega^*$

The counterpart to Fig. 6.9a is Fig. 6.9b, showing the time functions of the forces arranged in the vector $\boldsymbol{g}(t)$ that result from (6.13). The components Q_1, Q_2, Q_3, and Q_4 represent the inertia forces at the 4 points under review. Like the displacements $\boldsymbol{q}(t)$, these forces also result from a superposition of the 4 natural vibrations. The higher natural vibrations have a greater effect here since the forces are proportional to the accelerations into which the squared circular frequencies ω_i enter and where $\omega_4^2 > \omega_3^2 > \omega_2^2 > \omega_1^2$.

All forces are zero at time $t = 0$ since the coordinates are zero as well, see the initial conditions (6.144). The slope of the forces has a finite value, which can be checked using the derivative of (6.11). $\dot{\boldsymbol{g}}(t) = \boldsymbol{C}\dot{\boldsymbol{q}}(t)$ and therefore $\dot{\boldsymbol{g}}(0) = \boldsymbol{C}\dot{\boldsymbol{q}}(0)$. One can use the known forces to calculate the time function of the moments at any point. The bending stress at this point could then be calculated from the bending moment acting there if the cross-sectional parameters are known, and thus a failure analysis can be conducted. The support moment at the fixed end, for example, is the sum of all products of the inertia forces with their respective leverages, that is, according to Table 6.3, Case 2:

$$M_\text{B} = Q_1 l + Q_2 \frac{l}{2} + Q_3 l + Q_4 \frac{l}{2} = -m\ddot{q}_1 l - 2m\ddot{q}_2 \frac{l}{2} - 5m\ddot{q}_3 l - 2m\ddot{q}_4 \frac{l}{2}$$

$$M_\text{B}(t) = -ml(1,\ 1,\ 5,\ 1)\ddot{\boldsymbol{q}}(t) = -ml(1,\ 1,\ 5,\ 1)\boldsymbol{V}\ddot{\boldsymbol{p}}(t)$$

(6.147)

6.3 Free Undamped Vibrations   393

In general it is not possible to state simple relations between the maximum dynamic loads (e. g. as a result of an impact) and the static loads.

6.3.4.3 Natural Vibrations of a Support Structure

The support structure shown in Fig. 6.4 is used as an example. Its system matrices are known from (6.62) and (6.63). In particular, the following numerical values are assumed:

$$c^* = 6.228 \cdot 10^8 \text{ N/m}, \bar{c}_1 = \bar{c}_2 = 1, \bar{c}_3 = \bar{c}_4 = 2, s_1 = s_2 = 0.5, l = 4 \text{ m}$$
$$m = 2000 \text{ kg}, m_1 = m_2 = 50 \text{ kg}, m_3 = 500 \text{ kg}, J_S/l^2 = 20 \text{ kg}$$

This is a fictitious frame structure with a relatively small elastically supported rigid body. Verify the matrix elements and, using an eigenvalue program, the natural frequencies for comparison purposes.

One finds the following modal matrix, which is stated with an accuracy of 10^{-3} to illustrate numerical effects (the numerical accuracy of common calculation programs is often better than 10^{-7}):

$$\boldsymbol{V} = (\ \boldsymbol{v}_1, \quad \boldsymbol{v}_2, \quad \boldsymbol{v}_3, \quad \boldsymbol{v}_4, \quad \boldsymbol{v}_5, \quad \boldsymbol{v}_6, \quad \boldsymbol{v}_7, \quad \boldsymbol{v}_8, \quad \boldsymbol{v}_9, \quad \boldsymbol{v}_{10}\ )$$

$$\boldsymbol{V} = \begin{bmatrix} -0.059 & 0.399 & -0.004 & 0.001 & -0.318 & -0.610 & -0.009 & -0.996 & 1 & -0.003 \\ 0.0 & 1 & 0.0 & 0.0 & 0.0 & 0.0 & 0.0 & 0.0 & -0.186 & 0.0 \\ 0.410 & 0.163 & 0.034 & -0.004 & 0.766 & 1 & 0.005 & -0.488 & 0.414 & 0.0 \\ 1 & 0.0 & 0.092 & 0.0 & 0.0 & -0.886 & -0.002 & 0.281 & 0.0 & 0.0 \\ 0.410 & -0.163 & 0.034 & 0.004 & -0.766 & 1 & 0.005 & -0.490 & -0.412 & 0.0 \\ 0.059 & 0.399 & 0.004 & 0.001 & -0.318 & 0.610 & 0.009 & 1 & 1 & -0.003 \\ -0.727 & -0.185 & 0.684 & 0.401 & 0.172 & -0.734 & 1.000 & -0.456 & -0.393 & -0.501 \\ -0.727 & 0.185 & 0.684 & 0.401 & -0.172 & -0.734 & 1 & -0.452 & 0.395 & 0.501 \\ -0.862 & 0.0 & 1 & 0.0 & 0.0 & 0.172 & -0.139 & 0.047 & 0.0 & 0.0 \\ 0.0 & -0.380 & 0.0 & 1 & 1 & 0.0 & 0.0 & 0.004 & 1 & 1 \end{bmatrix}$$

(6.148)

The first eight mode shapes and the associated natural frequencies are shown in Fig. 6.10.

The two highest natural frequencies are $f_9 = 118.5$ Hz and $f_{10} = 125.2$ Hz. One can check them using (6.117). The interpretation and physical explanation of the calculation results is strongly recommended to each user of software programs to avoid the risk of "blind trust in computers", which may result in the acceptance of results that do not make any sense.

The modal matrix in (6.148) shows the entry 0.0 in several places. These numbers were written wherever the results were smaller than 10^{-4}. Most of them are exact zeros for physical reasons because $v_{2i} = v_{10i} \equiv 0$ for symmetrical mode shapes and $v_{4i} = v_{9i} \equiv 0$ for antisymmetrical mode shapes. As one can see from the modal matrix and from Fig. 6.10, mode shapes of the orders $i = 1, 3, 6, 7$ and 8 are symmetrical and those of orders $i = 2, 4, 5, 9$ and 10 are antisymmetrical. This is why the following conditions have to be satisfied:

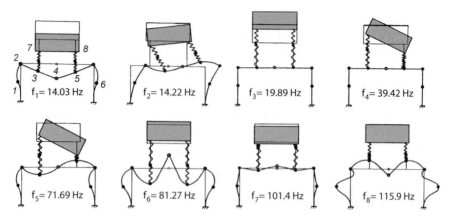

Fig. 6.10 The first eight natural frequencies and mode shapes of the support structure of Fig. 6.4 (with 10 degrees of freedom)

$$v_{1i} = -v_{6i}, \quad v_{3i} = +v_{5i}, \quad v_{7i} = +v_{8i} \quad \text{for} \quad i = 1, 3, 6, 7, 8 \quad (6.149)$$
$$v_{1i} = +v_{6i}, \quad v_{3i} = -v_{5i}, \quad v_{7i} = -v_{8i} \quad \text{for} \quad i = 2, 4, 5, 9, 10. \quad (6.150)$$

One can see that these conditions are exactly satisfied up to three digits for mode shapes v_1 to v_7 while in the vectors v_8 and v_9 deviations apparently occur as early as in the third digit. This has numerical reasons, and quite often the highest calculated mode shapes are the least accurate. In this case it turned out that the round-off errors reached an order of magnitude of 10^{-3}, making it difficult to decide whether the values given as 0.501 for v_{10} are exactly $1/2$ and whether 0.003 is an exact zero.

Next, some estimates for the natural frequencies shall be made. One can view the frame on the one hand and the spring-mass system resting on it on the other hand as subsystems. The latter forms a system with 4 degrees of freedom if one considers the relatively heavy frame to be immobile. In view of the relatively small masses m_1 and m_2 it can be considered as an oscillator with two degrees of freedom that can only perform a pure vertical oscillation and a pure pitching vibration because it is symmetrical. The natural frequencies for these subsystems are

$$f_\mathrm{H} = \frac{1}{2\pi}\sqrt{\frac{2c}{m_3}} \quad \text{und} \quad f_\mathrm{N} = \frac{1}{2\pi}\sqrt{\frac{cl^2}{4J_\mathrm{S}}} \quad (6.151)$$

for the effective spring constant of springs in series $c = c_1 c_3/(c_1 + c_3)$ and $2/3 c^* = 4.152 \cdot 10^6$ N/m. Using the numerical values given above, (6.149) yields the frequencies $f_\mathrm{H} = 20.51$ Hz and $f_\mathrm{N} = 36.26$ Hz, a good approximation of f_3 and f_4. The first two mode shapes are mainly determined by the frame deformations. The vertical oscillation approximately corresponds to the 3rd mode shape. Since the support structure became stiffer and lost mass due to the simplification, the estimated frequency is $f_\mathrm{H} > f_3$.

6.3.5 Problems P6.4 to P6.6

P6.4 Transient Loading Cases of an Overhead Crane

State the equations of motion and initial conditions for the following loading cases of the overhead crane shown in Fig. 6.5:
  a) Load m_2 falls with the initial velocity u into the cable of the resting crane (gripper drops into the holding cables)
  b) Load drops suddenly (breakage of the load handling device)
  c) Sudden motor shutdown during lifting (u_h = const., crane at rest)
  d) Load m_2 is suddenly caught by a rigid obstacle during lifting.

What are the equations for calculating the cable force and the bending moment in the center of the crane girder?

Note: The coordinate origin of x_1 is the static equilibrium position of the unloaded crane, that of x_2 is the end of the unloaded cable.

P6.5 Natural Frequencies and Mode Shapes of a Frame

For the model shown in Fig. 6.2, calculate the natural frequencies and mode shapes and check whether the orthogonality relations are satisfied. Determine the system matrices as a special case of the example given in Sect. 6.3.4.2.

P6.6 Influence of an Elastic Bearing on the Natural Frequencies of a Torsional Oscillator

The measured fundamental frequency of a drive deviated significantly from the one that was calculated using the calculation model according to Fig. 4.3a (torsionally elastic shaft and two rotating masses). The observed difference could not be explained by parametric uncertainties alone. The assumed cause of the deviations was the coupling of torsional and transverse vibrations due to bearing elasticity.

Determine the natural frequencies for the calculation model shown in Fig. 6.11 of a torsionally elastic drive with a transmission step and horizontal bearing elasticity c of the gear 2 and compare them to the ones obtained for $c \to \infty$ (rigid bearing).

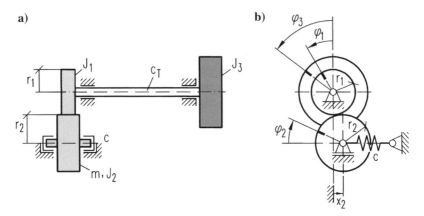

Fig. 6.11 Calculation model of a torsional oscillator with an elastic bearing

Given:
Pitch radii of the gears $\quad r_1 = 0.175$ m; $r_2 = 0.25$ m
Moments of inertia of the gears $J_1 = 0.5$ kg $\cdot$ m^2; $J_2 = 1, 6$ kg $\cdot$ m^2;
$\qquad\qquad\qquad\qquad\qquad J_3 = 0.75$ kg $\cdot$ m^2
Gear mass 2 $\qquad\qquad\qquad m = 52$ kg
Torsional stiffness of the drive $\;\; c_T = 5.6 \cdot 10^4$ N $\cdot$ m
Bearing stiffness $\qquad\qquad\;\; c = 4.48 \cdot 10^6$ N/m
(4.27) as known equation for the fundamental frequency of a torsional oscillator

Find:
1. Constraint between angles of rotation φ_k and bearing displacement x_2
2. Mass and stiffness matrix for the coordinate vector $q^T = (\varphi_1,\ \varphi_2,\ \varphi_3)$
3. Natural frequencies for the torsional oscillator with an elastic bearing

6.3.6 Solutions S6.4 to S6.6

S6.4 The equations of motion taking into account the weight of the load, the motor torque M_M and the matrices from Solution S6.3 are:

$$m_1\ddot{x}_1 + c_1 x_1 + c_2\left(x_1 - x_2 - \frac{r}{i}\varphi_M\right) = 0$$

$$m_2\ddot{x}_2 - c_2\left(x_1 - x_2 - \frac{r}{i}\varphi_M\right) = m_2 g \qquad (6.152)$$

$$\frac{i}{r}J_M\ddot{\varphi}_M - c_2\left(x_1 - x_2 - \frac{r}{i}\varphi_M\right) = iM_M/r$$

The initial conditions, if the time is measured starting from the beginning of the process of interest, are:
$t = 0$:

a) $x_1 = 0;\qquad\quad \dot{x}_1 = 0;\qquad\quad x_2 = 0;\qquad\qquad\qquad\quad \dot{x}_2 = u;$
$\quad\;\varphi_M = 0;\qquad\quad \dot{\varphi}_M = 0$

b) $x_1 = m_2 g/c_1;\; \dot{x}_1 = 0;\qquad\;\; x_2 = m_2 g\left(\dfrac{1}{c_1} + \dfrac{1}{c_2}\right);\; \dot{x}_2 = 0;$
$\quad\;\varphi_M = 0;\qquad\quad \dot{\varphi}_M = 0$

c) $x_1 = m_2 g/c_1;\; \dot{x}_1 = 0;\qquad\;\; x_2 = m_2 g\left(\dfrac{1}{c_1} + \dfrac{1}{c_2}\right);\; \dot{x}_2 = -u_h;$ $\qquad(6.153)$
$\quad\;\varphi_M = 0;\qquad\quad \dot{\varphi}_M = 0$

d) $x_1 = m_2 g/c_1;\; \dot{x}_1 = 0;\qquad\;\; x_2 = m_2 g\left(\dfrac{1}{c_1} + \dfrac{1}{c_2}\right);\; \dot{x}_2 = 0;$
$\quad\;\varphi_M = 0;\qquad\quad \dot{\varphi}_M = u_h i/r$

$c_2 = 0$ is to be set in case b), and 3 systems with one degree of freedom each remain. For all cases, the cable force

$$\underline{F_s = -c_2\left(x_1 - x_2 - \frac{r}{i}\varphi_M\right) = m_2(g - \ddot{x}_2)} \qquad (6.154)$$

and the bending moment

$$\underline{M_b = (m_1 g + c_1 x_1)\frac{l}{2} = (F_s + m_1 g - m_1\ddot{x}_1)\frac{l}{2}} \qquad (6.155)$$

result from the equilibrium conditions.

6.3 Free Undamped Vibrations

S6.5 Deleting the second and fourth rows and columns turns the compliance matrix from (6.131) into the one that applies here because the coordinates q_2 and q_4 do not apply, see (6.32):

$$D = \frac{l^3}{48EI}\begin{bmatrix} 64 & 24 \\ 24 & 16 \end{bmatrix} = \frac{l^3}{6EI}\begin{bmatrix} 8 & 3 \\ 3 & 2 \end{bmatrix},$$

$$C = D^{-1} = \frac{6EI}{7l^3}\begin{bmatrix} 2 & -3 \\ -3 & 8 \end{bmatrix}. \tag{6.156}$$

The kinetic energy $W_{\text{kin}} = \frac{1}{2}\left[2m\dot{q}_1^2 + (4m+2m)\dot{q}_2^2\right]$ yields the mass matrix according to (6.15), see also (6.34):

$$M = \begin{bmatrix} 2m & 0 \\ 0 & 6m \end{bmatrix} = m\begin{bmatrix} 2 & 0 \\ 0 & 6 \end{bmatrix}. \tag{6.157}$$

Using the reference circular frequency of (6.134), one finds:

$$DM = \frac{ml^3}{48EI}\begin{bmatrix} 64 & 24 \\ 24 & 16 \end{bmatrix}\begin{bmatrix} 2 & 0 \\ 0 & 6 \end{bmatrix} = \frac{1}{\omega^{*2}}\begin{bmatrix} 128 & 144 \\ 48 & 96 \end{bmatrix} = \frac{1}{\omega^{*2}} \cdot A \tag{6.158}$$

The eigenvalues $\lambda = \omega^{*2}/\omega^2$ are derived according to (6.84) from

$$|A - \lambda E| = \begin{vmatrix} 128-\lambda & 144 \\ 44 & 96-\lambda \end{vmatrix} = \lambda^2 - 224\lambda + 5376 = 0. \tag{6.159}$$

The solution of this quadratic equation is

$$\underline{\lambda_1 = 196.664}, \qquad \underline{\omega_1^2 = 0.24407\frac{EI}{ml^3}},$$

$$\underline{\lambda_2 = 27.336}, \qquad \underline{\omega_2^2 = 1.75593\frac{EI}{ml^3}}. \tag{6.160}$$

Compare these values to the ones of the model with four degrees of freedom in (6.136). The less accurate mass distribution has caused a decline in frequency! Each of the two eigenvectors results from a system of linear equations according to (6.85):

$$(128-\lambda_i)v_{1i} + 144v_{2i} = 0, \qquad 48v_{1i} + (96-\lambda_i)v_{2i} = 0 \qquad i = 1,2. \tag{6.161}$$

The normalization $v_{1i} = 1$ using the known values of λ_1 and λ_2 results in: $\underline{v_{21} = 0.47683}$, $\underline{v_{22} = -0.69906}$. All elements of the modal matrix V are now known. The following orthogonality relations of (6.102) to (6.105) are obtained:

$$V^T C V = \text{diag}(\gamma_i) = \begin{bmatrix} \gamma_1 & 0 \\ 0 & \gamma_2 \end{bmatrix} = \begin{bmatrix} 0.82110 & 0.0 \\ 0.0 & 8.6604 \end{bmatrix}\frac{EI}{l^3}, \tag{6.162}$$

$$V^T M V = \text{diag}(\mu_i) = \begin{bmatrix} \mu_1 & 0 \\ 0 & \mu_2 \end{bmatrix} = \begin{bmatrix} 3.3642 & 0.0 \\ 0.0 & 4.9321 \end{bmatrix} m. \tag{6.163}$$

They are satisfied within the constraint of limited calculation accuracy.

S6.6 For describing the individual bodies of the calculation model, the angles of rotation φ_1, φ_2 and φ_3 of the three gears as well as the horizontal displacement x_2 of gear 2 are introduced, see Fig. 6.11b. If one stipulates for the origins of the position coordinates that $\varphi_2(\varphi_1 = 0, x_2 = 0) = 0$, the constraint for the condition $|x_2| \ll r_2$ can be stated as

$$\underline{r_1\varphi_1 \approx x_2 + r_2\varphi_2}. \tag{6.164}$$

It can be "read" from Fig. 6.11b if one imagines an infinitely thin gear rack placed between gear 1 and gear 2.

If the vector of the generalized coordinates is defined according to

$$\boldsymbol{q} = [\varphi_1,\ \varphi_2,\ \varphi_3]^T = [q_1,\ q_2,\ q_3]^T, \tag{6.165}$$

the kinetic and potential energy can be stated as follows:

$$W_{\text{kin}} = \frac{1}{2}\left[J_1\dot{\varphi}_1^2 + m\dot{x}_2^2 + J_2\dot{\varphi}_2^2 + J_3\dot{\varphi}_3^2\right] \tag{6.166}$$

$$= \frac{1}{2}\left[J_1\dot{q}_1^2 + m(r_1\dot{q}_1 - r_2\dot{q}_2)^2 + J_2\dot{q}_2^2 + J_3\dot{q}_3^2\right] \tag{6.167}$$

$$W_{\text{pot}} = \frac{1}{2}\left[c_T(\varphi_3 - \varphi_1)^2 + cx_2^2\right] = \frac{1}{2}\left[c_T(q_3 - q_1)^2 + c(r_1q_1 - r_2q_2)^2\right] \tag{6.168}$$

The mass and stiffness matrices follow from the relations (6.15) and (6.9):

$$\boldsymbol{M} = \begin{bmatrix} J_1 + mr_1^2 & -mr_1r_2 & 0 \\ -mr_1r_2 & J_2 + mr_2^2 & 0 \\ 0 & 0 & J_3 \end{bmatrix} = J_3 \cdot \begin{bmatrix} (J_1 + mr_1^2)/J_3 & -mr_1r_2/J_3 & 0 \\ -mr_1r_2/J_3 & (J_2 + mr_2^2)/J_3 & 0 \\ 0 & 0 & 1 \end{bmatrix}$$

$$= J_3 \cdot \overline{\boldsymbol{M}} \tag{6.169}$$

$$\boldsymbol{C} = \begin{bmatrix} c_T + cr_1^2 & -cr_1r_2 & -c_T \\ -cr_1r_2 & cr_2^2 & 0 \\ -c_T & 0 & c_T \end{bmatrix} = c_T \cdot \begin{bmatrix} (c_T + cr_1^2)/c_T & -cr_1r_2/c_T & -1 \\ -cr_1r_2/c_T & cr_2^2/c_T & 0 \\ -1 & 0 & 1 \end{bmatrix}$$

$$= c_T \cdot \overline{\boldsymbol{C}}. \tag{6.170}$$

The stiffness matrix $\boldsymbol{C}$ is singular, i.e., $\det \boldsymbol{C} = 0$ because the system considered here is unconstrained with respect to rotation. To determine the natural frequencies, the determinant of the coefficient matrix of the homogeneous system of equations

$$(\boldsymbol{C} - \omega^2 \boldsymbol{M})\boldsymbol{v} = (c_T \overline{\boldsymbol{C}} - \omega^2 J_3 \overline{\boldsymbol{M}})\boldsymbol{v} = c_T(\overline{\boldsymbol{C}} - \lambda \overline{\boldsymbol{M}})\boldsymbol{v} = \boldsymbol{o} \tag{6.171}$$

has to vanish according to (6.84). After dividing by c_T and using the abbreviation

$$\lambda = \frac{J_3 \omega^2}{c_T}; \qquad \left(\omega^{*2} = \frac{c_T}{J_3}\right) \tag{6.172}$$

as well as the given parameter values, this provides the following condition that is similar to (6.84)

$$\det(\overline{\boldsymbol{C}} - \lambda \overline{\boldsymbol{M}}) = \begin{vmatrix} 3.45 - 2.79 \cdot \lambda & -3.5 + 3.033 \cdot \lambda & -1 \\ -3.5 + 3.033 \cdot \lambda & 5 - 6.467 \cdot \lambda & 0 \\ -1 & 0 & 1 - \lambda \end{vmatrix}$$

$$= -\lambda \cdot (13.56 - 23.867\,56 \cdot \lambda + 8.840\,889 \cdot \lambda^2) \stackrel{!}{=} 0. \tag{6.173}$$

It follows for the eigenvalues:

$$\lambda_1 = 0; \qquad \lambda_2 = 0.812\,919; \qquad \lambda_3 = 1.886\,7. \tag{6.174}$$

Because of (6.172) and for $\omega = 2\pi f$, the resulting natural frequencies are:

$$f_1 = 0 \text{ Hz}; \quad f_2 = \frac{\omega^*}{2\pi}\sqrt{\lambda_2} = 39.21 \text{ Hz}; \quad f_3 = \frac{\omega^*}{2\pi}\sqrt{\lambda_3} = 59.73 \text{ Hz}. \tag{6.175}$$

If the spring and mass of the bearing are not considered, a vibration system according to Fig. 4.3 is obtained, and its natural circular frequencies, taking into account the reduced moment of inertia $J_{\text{red}} = J_1 + J_2(r_1/r_2)^2$ according to (4.27), are

$$\omega_1^2 = 0; \qquad \omega_2^2 = \frac{c_T}{J_3} + \frac{c_T}{J_1 + J_2\left(\dfrac{r_1}{r_2}\right)^2}. \tag{6.176}$$

For comparison, one can also express this using (6.172) in the form

$$\lambda_1 = 0; \qquad \lambda_2 = 1 + \frac{J_3}{J_1 + J_2\left(\dfrac{r_1}{r_2}\right)^2} \tag{6.177}$$

which yields $\lambda_2 = 1.5841$ using the given parameter values. The fundamental frequency that is to be compared to f_2 in (6.175) for rigid support amounts to

$$f_{(2)} = 54.74 \text{ Hz}. \tag{6.178}$$

It is much too high for a "pure torsional oscillator". The bearing elasticity is the reason for a reduction of the "natural torsional frequency" by about 15.5 Hz!

Conclusion: The mode shapes in drive systems are not always pure torsional vibrations. There can also be coupled vibrations. It may be required for explaining or interpreting these vibrations to include bending and/or bearing compliance. These should always be considered if the bearing stiffness values are not very large. Similar effects may occur in drive systems that are initially viewed as torsional oscillators if meshing spur gears are overhung. The bending compliance of the shafts can also result in a transverse motion of the gears within the plane.

6.4 Structure and Parameter Changes

6.4.1 Rayleigh Quotient

The **Rayleigh quotient** provides a method for estimating the lowest natural frequency. Lord RAYLEIGH offered the following consideration in 1878: If a system with a circular frequency ω and a mode shape $\boldsymbol{v}$ vibrates harmonically ($\boldsymbol{q} = \boldsymbol{v}\sin\omega t$), the potential and kinetic energies change as follows according to (6.7) and (6.16):

$$W_{\text{pot}} = \frac{1}{2}\boldsymbol{v}^{\text{T}}\boldsymbol{C}\boldsymbol{v}\sin^2\omega t, \qquad W_{\text{kin}} = \frac{\omega^2}{2}\boldsymbol{v}^{\text{T}}\boldsymbol{M}\boldsymbol{v}\cos^2\omega t. \tag{6.179}$$

The stiffness matrix $\boldsymbol{C}$ must not be singular, i. e. $\det \boldsymbol{C} \neq 0$.

The potential energy has its maximum at $\omega t = \pi/2$. In this case, $W_{\text{kin}} = 0$. However, when passing through the static equilibrium position ($\omega t = \pi$), the potential energy $W_{\text{pot}} = 0$ and the kinetic energy has its maximum value. The following applies to conservative systems:

$$W_{\text{kin}} + W_{\text{pot}} = W_{\text{kin max}} + 0 = \frac{\omega^2}{2} \boldsymbol{v}^{\text{T}} \boldsymbol{M} \boldsymbol{v} = 0 + W_{\text{pot max}} = \frac{1}{2} \boldsymbol{v}^{\text{T}} \boldsymbol{C} \boldsymbol{v}. \quad (6.180)$$

One can use this relationship to estimate ω_1 if one knows an approximation $\boldsymbol{v}$ for the eigenvector $\boldsymbol{v}_1$. The first natural circular frequency can be approximately calculated using an estimated (not exactly known) amplitude distribution of the first mode shape. It can be shown that the natural circular frequency ω_R calculated in this way is always higher than the exact value ω_1. The estimate resulting from the energy balance of (6.180) is called the Rayleigh quotient:

$$\omega_1^2 < \omega_R^2 = \frac{\boldsymbol{v}^{\text{T}} \boldsymbol{C} \boldsymbol{v}}{\boldsymbol{v}^{\text{T}} \boldsymbol{M} \boldsymbol{v}}. \quad (6.181)$$

While this is only an inequality, it does allow multiple estimates. The best (i.e. closest to $\boldsymbol{v}_1$) of all approximations to the fundamental mode shape $\boldsymbol{v}$ is the one that provides the smallest ω_R value.

Another estimate for ω_1^2 is provided by the quotient named after R. GRAMMEL (1889–1964), which can be based on a similar reasoning as the Rayleigh quotient, see (6.117):

$$\omega_1^2 < \omega_G^2 = \frac{\boldsymbol{v}^{\text{T}} \boldsymbol{M} \boldsymbol{v}}{\boldsymbol{v}^{\text{T}} \boldsymbol{M} \boldsymbol{D} \boldsymbol{M} \boldsymbol{v}} = \frac{\boldsymbol{v}^{\text{T}} (\boldsymbol{M} \boldsymbol{v})}{(\boldsymbol{M} \boldsymbol{v})^{\text{T}} \boldsymbol{D} (\boldsymbol{M} \boldsymbol{v})}. \quad (6.182)$$

It is preferable if $\boldsymbol{D}$ is known instead of $\boldsymbol{C}$.

Both the Rayleigh quotient and the Grammel quotient provide an **upper limit** for the lowest natural frequency, i.e., the actual natural frequency is always lower.

6.4.2 Sensitivity of Natural Frequencies and Mode Shapes

Sometimes in design practice the question arises what measures can be taken to influence the natural frequencies and mode shapes of a machine. Of interest is the effect of structural and parameter changes, and what changes in spring, mass, or geometrical parameters have to be made to achieve a specific spectral or modal behavior.

Calculating the influence of parameter changes plays a major role in the identification of calculation models. Closely related is the question of the sensitivity of the results to potential parameter changes.

The variable parameters of a model, e.g. the masses m_k, spring constants c_k, lengths l_k, e.t.c. are uniformly denoted as x_k below. Parameter changes $\Delta x_k = x_k - x_{k0}$ that represent the difference between the original (x_{k0}) and the new parameter values x_k result in changes of the matrices

$$\begin{aligned} \Delta \boldsymbol{C} &= \boldsymbol{C}(x_k) - \boldsymbol{C}(x_{k0}) = \boldsymbol{C} - \boldsymbol{C}_0 \\ \Delta \boldsymbol{M} &= \boldsymbol{M}(x_k) - \boldsymbol{M}(x_{k0}) = \boldsymbol{M} - \boldsymbol{M}_0. \end{aligned} \quad (6.183)$$

If the mode shapes of the original systems are denoted as $\boldsymbol{v}_{i0}$, the following applies because of (6.117) for changed parameter values with initially unknown $\Delta \omega_i$ and

6.4 Structure and Parameter Changes

$\Delta \boldsymbol{v}_i$:

$$\omega_i^2 = \omega_{i0}^2 + \Delta\omega_i^2 = \frac{(\boldsymbol{v}_{i0} + \Delta\boldsymbol{v}_i)^{\mathrm{T}}(\boldsymbol{C}_0 + \Delta\boldsymbol{C})(\boldsymbol{v}_{i0} + \Delta\boldsymbol{v}_i)}{(\boldsymbol{v}_{i0} + \Delta\boldsymbol{v}_i)^{\mathrm{T}}(\boldsymbol{M}_0 + \Delta\boldsymbol{M})(\boldsymbol{v}_{i0} + \Delta\boldsymbol{v}_i)}. \tag{6.184}$$

If one can assume that the mode shape changes insignificantly as a result of the parameter change so that $\Delta\boldsymbol{v}_i \approx \boldsymbol{0}$ and that the parameter change is small ($\|\Delta\boldsymbol{C}\| \ll \|\boldsymbol{C}_0\|$, $\|\Delta\boldsymbol{M}\| \ll \|\boldsymbol{M}_0\|$), (6.184) in first approximation results in

$$\Delta(\omega_i^2) \approx \omega_{i0}^2 \left(\frac{\boldsymbol{v}_{i0}^{\mathrm{T}} \Delta\boldsymbol{C} \boldsymbol{v}_{i0}}{\boldsymbol{v}_{i0}^{\mathrm{T}} \boldsymbol{C}_0 \boldsymbol{v}_{i0}} - \frac{\boldsymbol{v}_{i0}^{\mathrm{T}} \Delta\boldsymbol{M} \boldsymbol{v}_{i0}}{\boldsymbol{v}_{i0}^{\mathrm{T}} \boldsymbol{M}_0 \boldsymbol{v}_{i0}} \right). \tag{6.185}$$

If the parameter changes only affect the masses and springs (or rotating masses and torsional springs, respectively), the variable mass and spring parameters m_k^* and c_l^* enable series expansions with the parameter-independent matrices $\overline{\boldsymbol{C}}_l$ and $\overline{\boldsymbol{M}}_k$, see (6.183) and the example in 6.4.5.2:

$$\begin{aligned} \boldsymbol{C} &= \boldsymbol{C}_u + \sum_l \boldsymbol{C}_l = \boldsymbol{C}_u + \sum_l c_l^* \overline{\boldsymbol{C}}_l \\ \boldsymbol{M} &= \boldsymbol{M}_u + \sum_k \boldsymbol{M}_k = \boldsymbol{M}_u + \sum_k m_k^* \overline{\boldsymbol{M}}_k \end{aligned} \tag{6.186}$$

The changes of the matrices then can be written as

$$\begin{aligned} \Delta\boldsymbol{C} &= \sum_l (\boldsymbol{C}_l - \boldsymbol{C}_{l0}) = \sum_l \frac{\Delta c_l^*}{c_{l0}^*} \boldsymbol{C}_{l0} \\ \Delta\boldsymbol{M} &= \sum_k (\boldsymbol{M}_k - \boldsymbol{M}_{k0}) = \sum_k \frac{\Delta m_k^*}{m_{k0}^*} \boldsymbol{M}_{k0} \end{aligned} \tag{6.187}$$

The asterisk on the parameters is to point to the fact that general mass and spring parameters are meant (e. g. masses, rotating masses, translational, bending, or torsional springs), which can still be multiplied by factors like, for example, in (6.227). The elements of the matrices with a bar on top indicate the place where the respective parameter occurs in the structural matrices, whereas the matrices $\boldsymbol{C}_u$ and $\boldsymbol{M}_u$ are the parameter-independent portions of the structural matrices.

From (6.185) follows the simpler expression

$$\Delta(\omega_i^2) \approx \omega_{i0}^2 \left(\sum_l \gamma_{il} \frac{\Delta c_l^*}{c_l^*} - \sum_k \mu_{ik} \frac{\Delta m_k^*}{m_k^*} \right), \tag{6.188}$$

if one introduces the dimensionless sensitivity coefficients, which can be calculated with the parameter matrices $\boldsymbol{M}_{k0}$ and $\boldsymbol{C}_{l0}$ of the original system and its mode shapes $\boldsymbol{v}_{i0}$ and are independent of the normalization of the mode shapes:

$$\gamma_{il} = \frac{\boldsymbol{v}_{i0}^{\mathrm{T}} \boldsymbol{C}_{l0} \boldsymbol{v}_{i0}}{\boldsymbol{v}_{i0}^{\mathrm{T}} \boldsymbol{C}_0 \boldsymbol{v}_{i0}}; \qquad \mu_{ik} = \frac{\boldsymbol{v}_{i0}^{\mathrm{T}} \boldsymbol{M}_{k0} \boldsymbol{v}_{i0}}{\boldsymbol{v}_{i0}^{\mathrm{T}} \boldsymbol{M}_0 \boldsymbol{v}_{i0}}. \tag{6.189}$$

The μ_{ik} express the ratio of the kinetic energy in the kth mass parameter to the total kinetic energy of the ith mode shape. The γ_{il} represent the ratio of the potential energy in the lth spring parameter to the total potential energy during a vibration at the ith mode shape.

The energy distribution follows from (6.124) in conjunction with (6.189). The energy content of the parameters m_k and c_l, which changes over time with the ith natural frequency, reaches the following maximum values:

$$(W_{\text{kin}})_{ik} = \frac{1}{2}\mu_{ik}\frac{\left(\boldsymbol{v}_i^{\text{T}}\boldsymbol{M}\boldsymbol{u}_0\right)^2}{\mu_i}; \qquad (W_{\text{pot}})_{il} = \frac{1}{2}\gamma_{il}\frac{\left(\boldsymbol{v}_i^{\text{T}}\boldsymbol{C}\boldsymbol{q}_0\right)^2}{\gamma_i}. \qquad (6.190)$$

This correlation can be used like (6.128) and (6.129) for estimates that relate to the ith natural vibration.

Because of $\omega_{i0} = 2\pi f_{i0}$ and $\Delta(\omega_{i0}^2) \approx 2\omega_{i0}\Delta\omega_i$, (6.188) also results in the following approximation for the relative change of the ith natural frequency for small parameter changes:

$$\frac{\Delta f_i}{f_{i0}} \approx \frac{1}{2}\left(\sum_l \gamma_{il}\frac{\Delta c_l^*}{c_l^*} - \sum_k \mu_{ik}\frac{\Delta m_k^*}{m_k^*}\right); \qquad i = 1, 2, \ldots, n. \qquad (6.191)$$

The change of the ith natural frequency for relatively small parameter changes (Δc_l^* and/or Δm_k^*) thus depends on the sensitivity coefficients γ_{il} and μ_{ik}.

As can be deduced from (6.185) to (6.188), each natural frequency typically changes differently in the event of a parameter change Δx_k. The sensitivity coefficients express *quantitatively* what an experienced practitioner can only predict *qualitatively* based on physical concepts. For example, a change of a mass in the vicinity of a node has little influence on the natural frequency, and the sensitivity coefficient of a mass becomes large when this mass is located in the antinode of a mode shape.

The eigenvalue problem (6.79) provides the basis for finding out how parameter changes affect the **mode shapes**. Using the modal matrix $\boldsymbol{V}_0 = (\boldsymbol{v}_{10}, \boldsymbol{v}_{20}, \ldots, \boldsymbol{v}_{n0})$ known from (6.88) and the abbreviation $\boldsymbol{\lambda}_0 = \text{diag}(\omega_{i0}^2)$, the following can be written for the totality of all eigenvalue problems of the original (index 0) and the modified system (with parameter changes $\Delta\boldsymbol{C}$ and $\Delta\boldsymbol{M}$):

$$\begin{aligned}\boldsymbol{C}_0\boldsymbol{V}_0 - \boldsymbol{M}_0\boldsymbol{V}_0\boldsymbol{\lambda}_0 &= \boldsymbol{o} \\ (\boldsymbol{C}_0 + \Delta\boldsymbol{C})(\boldsymbol{V}_0 + \Delta\boldsymbol{V}) - (\boldsymbol{M}_0 + \Delta\boldsymbol{M})(\boldsymbol{V}_0 + \Delta\boldsymbol{V})(\boldsymbol{\lambda}_0 + \Delta\boldsymbol{\lambda}) &= \boldsymbol{o}.\end{aligned} \qquad (6.192)$$

Since the equation $\boldsymbol{C}_0\boldsymbol{V}_0 - \boldsymbol{M}_0\boldsymbol{V}_0\boldsymbol{\lambda}_0 = \boldsymbol{o}$ is satisfied for the unchanged system, (6.192) simplifies, and if one neglects all terms involving second-order changes, the following linear approximation remains:

$$\boldsymbol{C}_0\Delta\boldsymbol{V} + \Delta\boldsymbol{C}\boldsymbol{V}_0 - \boldsymbol{M}_0\boldsymbol{V}_0\Delta\boldsymbol{\lambda} - \boldsymbol{M}_0\Delta\boldsymbol{V}\boldsymbol{\lambda}_0 - \Delta\boldsymbol{M}\boldsymbol{V}_0\boldsymbol{\lambda}_0 = \boldsymbol{o}. \qquad (6.193)$$

If this equation is multiplied from the left by $\boldsymbol{V}_0^{\text{T}}$ and the change in mode shapes is expressed by the initially unknown elements of the sensitivity matrix $\boldsymbol{K} = ((k_{ij}))$

6.4 Structure and Parameter Changes

and the known mode shapes in the form of

$$\Delta V = V_0 K^T \quad \text{or} \quad \Delta v_i = \sum_{j=1}^{n} k_{ij} v_{j0}, \quad i \ne j, \tag{6.194}$$

taking into account (6.112), it follows that

$$\operatorname{diag}(\gamma_i) K^T + V_0^T \Delta C V_0 - \operatorname{diag}(\mu_i)\Delta\lambda - \operatorname{diag}(\mu_i) K^T \lambda_0 - V_0^T \Delta M V_0 \lambda_0 = 0. \tag{6.195}$$

If these matrix equations are written out in detail, the structure of each element becomes clearer. If one requires that the iith element of the system of equations is satisfied, one gets a relation already known from (6.185), from which the change in the natural frequency can be calculated:

$$v_{i0}^T \Delta C v_{i0} - \mu_i \Delta\omega_i^2 - v_{i0}^T \Delta M v_{i0} \omega_{i0}^2 = 0. \tag{6.196}$$

The ijth element of the system of equations (6.195) results in

$$\gamma_i k_{ji} + v_{i0}^T \Delta C v_{j0} - \mu_j k_{ji} \omega_{j0}^2 - v_{i0}^T \Delta M v_{j0} \omega_{j0}^2 = 0. \tag{6.197}$$

The sensitivity coefficients k_{ji} and k_{ij} of the mode shapes can be calculated from this. If one inserts them together with $\gamma_j = \mu_i \omega_{j0}^2$ into (6.194), one gets the sought after linear approximation for calculating the mode shape changes ($i = 1, 2, \ldots, n$):

$$\Delta v_i = \sum_{j=1}^{n} \frac{v_{j0}^T (\Delta C - \omega_{i0}^2 \Delta M) v_{i0}}{\gamma_i - \gamma_j} v_{j0}, \quad i \ne j. \tag{6.198}$$

The normalization $\mu_i = \mu_j = 1$ is a condition in (6.198). The modified mode shape therefore is $v_i \approx v_{i0} + \Delta v_i$. The change of mode shapes has only little influence on the natural frequencies for small parameter changes so that the simple equations (6.185) suffice if only changes of natural frequencies are of interest.

One can derive conditions from (6.185) and (6.198) as to when the natural frequencies or mode shapes do not change despite parameter changes. If, for example, the spring matrix or the mass matrix are changed proportionately in all elements ($\Delta C \sim C$ or $\Delta M \sim M$), all natural frequencies change but not the mode shapes. For example, all natural frequencies increase by a factor of $\sqrt{2}$ if all spring constants are increased by a factor of 2.

There are also parameter changes where only the mode shapes change but not the natural frequencies, see Fig. 6.12. An engineer must keep these facts in mind to prevent him or her from premature conclusions about the correctness of the calculation model based on the matching of measured and calculated frequencies or shapes, which would be a gross mistake. When identifying a calculation model, i.e. when determining the parameter values, both the natural frequencies and mode shapes and the static deformation behavior have to be taken into account.

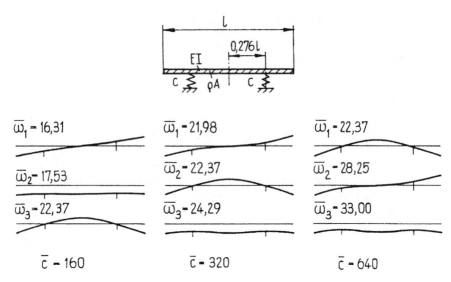

Fig. 6.12 Influence of the bearing spring constant c on the natural frequencies and mode shapes of a beam ($\bar{c} = cl^3/EI$, $\lambda_i^2 = \bar{\omega}_i = \omega_i\sqrt{\varrho Al^4/EI}$)

The dependence of the lowest 7 natural frequencies on the bearing spring constant is shown in Fig. 6.13a.

All natural frequencies increase with an increase in the bearing stiffness c as required by theory, however, to a different extent. The linear relationship expressed in (6.191) applies to small parameter changes only. The curves show that the natural frequencies of various orders i change differently.

Why do the lower three natural frequencies start at zero? This phenomenon corresponds to the "free – free" case in Table 5.7, since the model considered here has three degrees of freedom of a rigid-body for $cl^3/EI = 0$. These curves are strongly nonlinear at the outset since the influence of the bending stiffness increases with the increase of this parameter.

The following rules can be derived from the equations specified with respect to the influence of parameter changes on natural frequencies, all of which are relevant for practical design tasks:

1. When increasing a mass at any point of an oscillator, all n natural frequencies are generally decreased – and increased if the mass is reduced, see Fig. 5.17.
2. When increasing the stiffness at any point of an oscillator, all n natural frequencies are generally increased – and decreased if the stiffness is reduced, see Fig. 6.13.
3. If a mass is located in a node of a mode shape, its change does not affect the associated natural frequency.
4. Mass changes have the greatest influence on the associated natural frequency where the amplitudes of a mode shape are high (antinode).

6.4 Structure and Parameter Changes

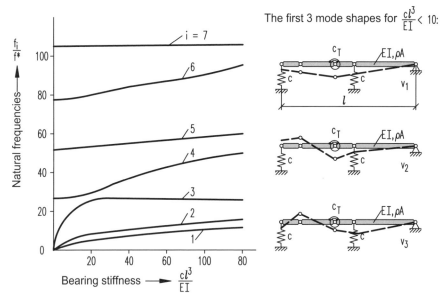

Fig. 6.13 Influence of the bearing stiffness on the natural frequencies of the transverse vibration; **a)** Lowest seven natural frequencies, **b)** Lowest three mode shapes in the range of the characteristic parameter $cl^3/EI < 10$

5. Stiffness changes are most effective at springs with a large deformation work; their effect thus differs for the various mode shapes.
6. Additional fastenings (supports, fixations, bearings) increase all natural frequencies, reduced fastening (additional hinges, reduced degree of static indeterminacy) decreases the natural frequencies, see Fig. 6.15.
7. After long periods of operation, the natural frequencies mostly drop because various fastenings loosen (cracks, backlash).
8. High natural frequencies are achieved if only tensile and compressive forces are transmitted as in truss structures.
9. Mass and stiffness changes should be opposing to shift natural frequencies effectively.
10. Natural frequencies can be strongly increased (or decreased) by shortening (or lengthening) beams or trusses.
11. The natural frequencies in rotors rise and fall in pairs with speed, see Figs. 5.6, 5.12, and 5.22.

6.4.3 Reduction of Degrees of Freedom

Engineers used to be "cautious" when establishing the equations of motion and took into account as many coordinates as required and as few as possible in a model to reduce the calculation effort. The selection of the structure of a minimal model was often based on years of experience. Since computer methods were developed (e. g. FEM) with which a large number of degrees of freedom is obtained formally and rapidly, methods are required for systematic simplification of the calculation models because too many degrees of freedom of a model are often superfluous, that is, physically useless.

The method of static condensation by GUYAN (1965) described below provides a way of reducing the number of degrees of freedom and thus calculation time and storage space. It is suitable, for example, when a complex model is available for static calculation since dynamic analyses can usually be performed with fewer degrees of freedom. This method can be used when assembling substructures to condense the matrices of individual substructures to the coordinates of the interconnection points before assembly of the global stiffness matrix.

The coordinate vector q of a large system is divided into $n1$ **external coordinates (master degrees of freedom)** q_1 that remain in the calculation and $n2$ **internal coordinates (slave degrees of freedom)** q_2 that are removed: $q^T = (q_1^T, q_2^T)$. For the degrees of freedom, it is $n = n1 + n2$. It is recommended to eliminate such coordinates that describe the displacement of relatively small masses or the angles of rotation of relatively small moments of inertia.

The following consideration initially neglects the inertia forces so that the following static relation remains of the equation of motion $M\ddot{q} + Cq = f$ and is partitioned as specified:

$$Cq = \begin{bmatrix} C_{11} & C_{12} \\ C_{21} & C_{22} \end{bmatrix} \begin{bmatrix} q_1 \\ q_2 \end{bmatrix} = \begin{bmatrix} f_1 \\ f_2 \end{bmatrix} = f. \tag{6.199}$$

C_{11} is a $(n1 \times n1)$ matrix, C_{22} a $(n2 \times n2)$ matrix, and $C_{12}^T = C_{21}$ are rectangular matrices. The relation

$$q_2 = -C_{22}^{-1} C_{21} q_1 + C_{22}^{-1} f_2 = -Sq_1 + C_{22}^{-1} f_2. \tag{6.200}$$

follows from the lower row. It is useful to introduce the $(n2 \times n1)$ rectangular matrix as an abbreviation

$$S = C_{22}^{-1} C_{21} = (C_{12} C_{22}^{-1})^T. \tag{6.201}$$

It follows from the upper row of (6.199):

$$f_1 = C_{11} q_1 + C_{12} q_2 = C_{11} q_1 - C_{12} S q_1 + S^T f_2. \tag{6.202}$$

If one defines the reduced force vector for the condensed system with $n1$ degrees of freedom

$$f_{\text{red}} = f_1 - C_{12} C_{22}^{-1} f_2 = f_1 - S^T f_2. \tag{6.203}$$

and the reduced stiffness matrix

$$C_{\text{red}} = C_{11} - C_{12} C_{22}^{-1} C_{21} = C_{11} - S^T C_{21}, \tag{6.204}$$

6.4 Structure and Parameter Changes

one can write $C_{\text{red}} q_1 = f_{\text{red}}$ instead of (6.202).

The mass matrix is partitioned like the spring matrix. It is required that its virtual work on the original system should be of the same magnitude as on the condensed system to approximately consider the influence of the inertia forces:

$$\delta q^T M \ddot{q} = [\delta q_1^T, \delta q_2^T] \begin{bmatrix} M_{11} & M_{12} \\ M_{21} & M_{22} \end{bmatrix} \begin{bmatrix} \ddot{q}_1 \\ \ddot{q}_2 \end{bmatrix} = \delta q_1^T M_{\text{red}} \ddot{q}_1. \quad (6.205)$$

If one eliminates the internal coordinates (slave degrees of freedom) for $f_2 = 0$ according to (6.200), this becomes

$$\delta q_1^T M_{\text{red}} \ddot{q}_1 = \delta q_1^T [E, -S^T] \begin{bmatrix} M_{11} & M_{12} \\ M_{21} & M_{22} \end{bmatrix} \begin{bmatrix} E \\ -S \end{bmatrix} \ddot{q}_1. \quad (6.206)$$

If one takes into consideration the identity $M_{12}^T = M_{21}$, the equation for the reduced mass matrix ($n1 \times n1$) follows as a result of a comparison of coefficients:

$$M_{\text{red}} = M_{11} - M_{21} S - S^T M_{21} + S^T M_{22} S. \quad (6.207)$$

The natural frequencies that result from the reduced eigenvalue problem

$$(C_{\text{red}} - \omega^2 M_{\text{red}}) v_{\text{red}} = o \quad (6.208)$$

provide an approximation for the $n1$ lowest natural frequencies. The natural frequencies of orders $i > n1$ no longer exist; they were neglected with the slave degrees of freedom. The reduced system has a lower inertia but the same stiffness, and therefore all its $n1$ natural frequencies are somewhat higher than those of the same orders of the original system, see also Solution S6.9.

The eigenvectors, which are comparable to those of the original system, result due to (6.200) as follows from the ones of the condensed system:

$$\overline{v}_i = \begin{bmatrix} v_{\text{red}}^{(i)} \\ -C_{22}^{-1} C_{21} v_{\text{red}}^{(i)} \end{bmatrix} = \begin{bmatrix} E \\ -S \end{bmatrix} v_{\text{red}}^{(i)}, \quad i = 1, 2, \ldots, n1. \quad (6.209)$$

6.4.4 Influence of Constraints on Natural Frequencies and Mode Shapes

An effective measure for increasing the natural frequencies of a vibration system are struts in the form of rods, which are virtually undeformable in the longitudinal direction and can be mathematically described as constraints between the coordinates. A general method for determining the influence of such constraints on the natural frequencies is of interest.

The formalism for determining the influence of rigid supports and struts corresponds to the one used for reducing the number of degrees of freedom.

The existing constraints can be expressed by r linear relations between the coordinates in the form

$$q = T \cdot q_1. \tag{6.210}$$

The matrix T is a rectangular matrix with n rows and $(n - r)$ columns. Its elements t_{jk} result from the respective constraints.

As a result of the imposed constraints, a new system emerges from the original one that only has $(n - r)$ degrees of freedom so that as many coordinates suffice to describe its behavior. The remaining coordinates are arranged in the vector q_1.

The relations for the kinetic and potential energies of the original system are converted and expressed in terms of the new coordinates using (6.210), see (6.16):

$$W_{\text{pot}} = \frac{1}{2} q^{\text{T}} \cdot C \cdot q = \frac{1}{2} (T \cdot q_1)^{\text{T}} \cdot C \cdot T \cdot q_1 = \frac{1}{2} q_1^{\text{T}} \cdot C_1 \cdot q_1, \tag{6.211}$$

$$W_{\text{kin}} = \frac{1}{2} \dot{q}^{\text{T}} \cdot M \cdot \dot{q} = \frac{1}{2} (T \cdot \dot{q}_1)^{\text{T}} \cdot M \cdot T \cdot \dot{q}_1 = \frac{1}{2} \dot{q}_1^{\text{T}} \cdot M_1 \cdot \dot{q}_1. \tag{6.212}$$

A comparison of coefficients using the relation

$$(T \cdot q_1)^{\text{T}} = q_1^{\text{T}} \cdot T^{\text{T}} \tag{6.213}$$

provides the matrices for the equations of motion:

$$M_1 \ddot{q}_1 + C_1 q_1 = o \tag{6.214}$$

where

$$C_1 = T^{\text{T}} \cdot C \cdot T; \qquad M_1 = T^{\text{T}} \cdot M \cdot T. \tag{6.215}$$

The natural frequencies and mode shapes of the system stiffened with rigid fasteners result from the solution of the eigenvalue problem

$$(C_1 - \omega^2 M_1) v = o. \tag{6.216}$$

As a result of the additional constraints, all natural frequencies usually increase. Those frequencies, the mode shapes of which are most obstructed, are most affected. Those frequencies, the mode shapes of which are not affected as a result of the support, remain unchanged.

The frame for which the mass and stiffness matrices are given in Table 6.3 (Case 2) is used as an example.

A hinged column according to Fig. 6.14b obstructs the motion of the mass m_2 in the direction of that support, but it allows the motion perpendicular to it. The horizontal displacement of the mass m_2 corresponds to coordinate q_3 and its vertical displacement is q_2, see Fig. 6.14a and b. As a result of the attached column, q_2 and q_3 can no longer change independently. The following constraint applies:

6.4 Structure and Parameter Changes

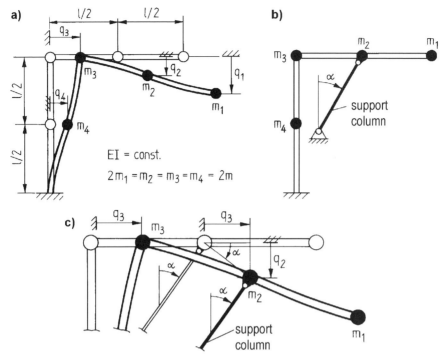

Fig. 6.14 Machine frame; **a)** System schematic with parameters and coordinates, **b)** Identification of the rigid support, **c)** Explanation of the constraint (6.217)

$$q_2 = q_3 \cdot \tan \alpha. \qquad (6.217)$$

It was assumed that the change of angle α as a result of the displacements q_2 and q_3 is negligibly small.

Only three $(n - r = 3)$ coordinates are included in the vector $\boldsymbol{q}_1^\mathrm{T} = [q_1, q_3, q_4]$. In matrix notation, (6.217) according to (6.210) takes the form

$$\boldsymbol{q} = \begin{bmatrix} q_1 \\ q_2 \\ q_3 \\ q_4 \end{bmatrix} = \begin{bmatrix} 1 & 0 & 0 \\ 0 & \tan \alpha & 0 \\ 0 & 1 & 0 \\ 0 & 0 & 1 \end{bmatrix} \cdot \begin{bmatrix} q_1 \\ q_3 \\ q_4 \end{bmatrix} = \boldsymbol{T} \cdot \boldsymbol{q}_1. \qquad (6.218)$$

The transformation matrix $\boldsymbol{T}$ is obtained from (6.218) by a comparison of coefficients. According to (6.215) and using the stiffness matrix given in the problem statement, see Table 6.3, Case 2, the following are obtained:

$$C_1 = \frac{48EI}{97l^3} \cdot \begin{bmatrix} 1 & 0 & 0 & 0 \\ 0 & \tan\alpha & 1 & 0 \\ 0 & 0 & 0 & 1 \end{bmatrix} \cdot \begin{bmatrix} 26 & -59 & 9 & -12 \\ -59 & 160 & -54 & 72 \\ 9 & -54 & 74 & -131 \\ -12 & 72 & -131 & 304 \end{bmatrix} \cdot \begin{bmatrix} 1 & 0 & 0 \\ 0 & \tan\alpha & 0 \\ 0 & 1 & 0 \\ 0 & 0 & 1 \end{bmatrix}$$

(6.219)

$$C_1 = \frac{48EI}{97l^3} \cdot \begin{bmatrix} 26 & -59\tan\alpha + 9 & -12 \\ -59\tan\alpha + 9 & 160\tan^2\alpha - 108\tan\alpha + 74 & 72\tan\alpha - 131 \\ -12 & 72\tan\alpha - 131 & 304 \end{bmatrix}.$$

The mass matrix is determined in a similar way with M from Table 6.3, Case 2:

$$M_1 = m \cdot \begin{bmatrix} 1 & 0 & 0 & 0 \\ 0 & \tan\alpha & 1 & 0 \\ 0 & 0 & 0 & 1 \end{bmatrix} \cdot \begin{bmatrix} 1 & 0 & 0 & 0 \\ 0 & 2 & 0 & 0 \\ 0 & 0 & 5 & 0 \\ 0 & 0 & 0 & 2 \end{bmatrix} \cdot \begin{bmatrix} 1 & 0 & 0 \\ 0 & \tan\alpha & 0 \\ 0 & 1 & 0 \\ 0 & 0 & 1 \end{bmatrix}$$

$$M_1 = m \cdot \begin{bmatrix} 1 & 0 & 0 \\ 0 & 2\tan^2\alpha + 5 & 0 \\ 0 & 0 & 2 \end{bmatrix}.$$

(6.220)

The natural circular frequencies for the unchanged system with the matrices given in the problem statement are known, see (6.143) and Fig. 6.15. Using the matrices according to (6.219) and (6.220), the natural circular frequencies for the system with the support column result from the condition

$$\det\left(C_1 - \omega^2 M_1\right) = 0,$$

(6.221)

see (6.27) and (6.216).

The numerical results obtained using an eigenvalue program that satisfy (6.221) are plotted in Fig. 6.15. The three curves show how much and how differently the natural circular frequencies change as a result of the support tilted by the angle α.

Since the remaining system has one degree of freedom less than the original system, there are only these three natural circular frequencies, and $\omega_4 \to \infty$. All natural circular frequencies are higher than those of the original system due to the support, see the general statements on structural and parameter changes in Sect. 6.4.

The minimum of curve ω_1/ω^* corresponds to the first natural circular frequency of the original system, and this minimum value only occurs if the support does not obstruct the original first mode shape, see Fig. 6.8.

(6.217) provides the angle α_i in the direction of which the mass m_2 vibrates in the original system at the ith order. If the support is perpendicular to this direction of oscillation, it does not influence this shape, i.e. the respective natural circular frequency remains unchanged. Table 6.5 lists the calculated support angles α_i that do not interfere with the original mode shapes at the support site. The curves of the natural circular frequencies in Fig. 6.23 show the extreme values for this angle α_i of the magnitude of the natural circular frequencies of the original system. The

6.4 Structure and Parameter Changes

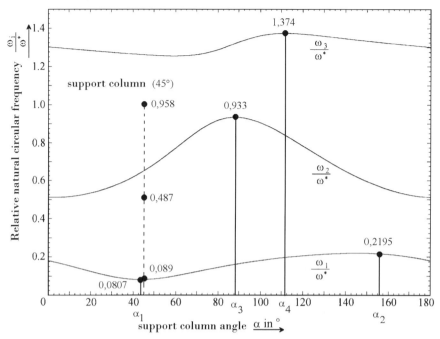

Fig. 6.15 Dependence of the natural frequencies on the support column angle α

fact that these are extreme values can be explained by the uniqueness of the support column angles α_i (in the relevant range of $0° \ldots 180°$) for which the ith mode shape is not obstructed.

Table 6.5 Results of the calculation

Order	Components of the original eigenvectors		Support column tilt	Support column angle
i	v_{2i}	v_{3i}	$\tan \alpha_i = v_{2i}/v_{3i}$	α_i in degrees
1	0.4774	0.5014	0.9521	43.6
2	0.3391	−0.4979	−0.6810	155.7
3	−1.1280	−0.0431	26.1717	87.8
4	−1.7878	0.7117	−2.5120	111.7

6.4.5 Examples of the Reduction of Degrees of Freedom

6.4.5.1 Simple Frame (from Four to Two)

The stiffness matrix that was established in 6.2.2.2 for the frame with four degrees of freedom is to be simplified by no longer taking into account as coordinates the angles of rotation that used to be taken into account by the coordinates q_3 and q_4, see Fig. 6.3.

The vector of the external coordinates (master degrees of freedom) q_1 contains the remaining generalized coordinates. The partitioning of the coordinate vectors

$$q = \begin{bmatrix} q_1 \\ q_2 \end{bmatrix}; \qquad q_1 = \begin{bmatrix} q_1 \\ q_2 \end{bmatrix}; \qquad q_2 = \begin{bmatrix} q_3 \\ q_4 \end{bmatrix} \qquad (6.222)$$

requires a partitioning of the stiffness matrix (6.43) into partial matrices, see (6.199). $l_1 = l_2 = l$ and $\beta_1 = \beta_2 = 2EI/l^3$ are assumed. Of particular interest here are C_{22} and its inverse matrix, which has to be calculated by matrix inversion. One finds

$$C_{11} = \frac{2EI}{l^3}\begin{bmatrix} 6 & 0 \\ 0 & 6 \end{bmatrix}, \qquad C_{12} = \frac{2EI}{l^3}\begin{bmatrix} 3l & 3l \\ 0 & 3l \end{bmatrix} = C_{21}^T,$$

$$C_{22} = \frac{2EI}{l^3}\begin{bmatrix} 2l^2 & l^2 \\ l^2 & 4l^2 \end{bmatrix}, \qquad C_{22}^{-1} = \frac{l}{14EI}\begin{bmatrix} 4 & -1 \\ -1 & 2 \end{bmatrix}. \qquad (6.223)$$

One can use this to calculate S, see (6.201):

$$S = C_{22}^{-1}C_{21} = \frac{l}{14EI}\begin{bmatrix} 4 & -1 \\ -1 & 2 \end{bmatrix}\frac{2EI}{l^3}\begin{bmatrix} 3l & 0 \\ 3l & 3l \end{bmatrix} = \frac{3}{7l}\begin{bmatrix} 3 & -1 \\ 1 & 2 \end{bmatrix}. \qquad (6.224)$$

The reduced stiffness matrix is obtained from (6.223) and (6.224) using (6.204):

$$C_{\text{red}} = C_{11} - S^T C_{21}$$

$$C_{\text{red}} = \frac{2EI}{l^3}\begin{bmatrix} 6 & 0 \\ 0 & 6 \end{bmatrix} - \frac{3}{7l}\begin{bmatrix} 3 & 1 \\ -1 & 2 \end{bmatrix}\frac{2EI}{l^3}\begin{bmatrix} 3l & 0 \\ 3l & 3l \end{bmatrix} \qquad (6.225)$$

$$C_{\text{red}} = \frac{6EI}{7l^3}\begin{bmatrix} 2 & -3 \\ -3 & 8 \end{bmatrix}$$

This is the stiffness matrix of the model with two degrees of freedom known from 6.2.2.1 and determined there in a different way (by forming the inverse matrix of D), see (6.32).

6.4.5.2 Textile Mandril (on Sensitivity)

Figure 6.16a shows the calculation model of a textile mandril, see also Fig. 6.32. The package (mass parameter: m_2, J_2) sits on the stepped shaft. The shaft is supported by a housing, which corresponds to an elastically supported rigid body characterized by the mass parameters m_1 and J_1. Of interest is the influence of the mass parameters on the natural frequencies and mode shapes. It is to be calculated in particular how an increase in the mass m_2 by 20 % affects the first two natural frequencies and the four mode shapes.

6.4 Structure and Parameter Changes

The mass matrix of this system can be represented according to (6.186) as the sum

$$M = m_1^* \overline{M}_1 + m_2^* \overline{M}_2 + m_3^* \overline{M}_3 + m_4^* \overline{M}_4 = M_1 + M_2 + M_3 + M_4$$

$$M = \begin{bmatrix} m_2^* + m_4^* & m_2^* - m_4^* & 0 & 0 \\ m_2^* - m_4^* & m_2^* + m_4^* & 0 & 0 \\ 0 & 0 & m_1 a^2 + m_3^* & m_1 ab - m_3^* \\ 0 & 0 & m_1 ab - m_3^* & m_1 b^2 + m_3^* \end{bmatrix} \quad (6.226)$$

where all the mass parameters have the dimension of a mass. The following is valid with respect to the original system:

$$m_1^* = m_1 = 0.5 \text{ kg}, \qquad m_2^* = \frac{m_2}{4} = 0.25 \text{ kg}$$

$$m_3^* = \frac{J_1}{l_1^2} = 0.0486 \text{ kg}, \qquad m_4^* = \frac{J_2}{l_4^2} = 0.3516 \text{ kg}. \quad (6.227)$$

The associated matrices M_k are ($a = l_5/l_1$, $b = l_6/l_1$):

$$M_1 = \begin{bmatrix} 0 & 0 & 0 & 0 \\ 0 & 0 & 0 & 0 \\ 0 & 0 & a^2 & ab \\ 0 & 0 & ab & b^2 \end{bmatrix} m_1^* \qquad M_2 = \begin{bmatrix} 1 & 1 & 0 & 0 \\ 1 & 1 & 0 & 0 \\ 0 & 0 & 0 & 0 \\ 0 & 0 & 0 & 0 \end{bmatrix} m_2^*$$

$$M_3 = \begin{bmatrix} 0 & 0 & 0 & 0 \\ 0 & 0 & 0 & 0 \\ 0 & 0 & 1 & -1 \\ 0 & 0 & -1 & 1 \end{bmatrix} m_3^* \qquad M_4 = \begin{bmatrix} 1 & -1 & 0 & 0 \\ -1 & 1 & 0 & 0 \\ 0 & 0 & 0 & 0 \\ 0 & 0 & 0 & 0 \end{bmatrix} m_4^*. \quad (6.228)$$

Using the numerical values of the real system, the following applies for the mass matrix

$$M_0 = \begin{bmatrix} 0.6016 & -0.1016 & 0 & 0 \\ -0.1016 & 0.6016 & 0 & 0 \\ 0 & 0 & 0.1354 & 0.0729 \\ 0 & 0 & 0.0729 & 0.2188 \end{bmatrix} \text{ kg} \quad (6.229)$$

and for the compliance matrix

$$D_0 = \begin{bmatrix} 0.54220 & 0.12090 & 0.02722 & -0.16993 \\ 0.12090 & 0.03486 & 0.01355 & -0.03394 \\ 0.02722 & 0.01355 & 0.01019 & 0 \\ -0.16993 & -0.03394 & 0 & 0.10194 \end{bmatrix} \text{ mm/N}. \quad (6.230)$$

With the normalization $v_{i0}^T M_0 v_{i0} = 1 \text{ kg}$ according to (6.86), the modal matrix results from the solution of the eigenvalue problem (6.80):

$$\mathbf{V}_0 = \begin{bmatrix} 1.2834 & 0.1967 & -0.1574 & -0.0180 \\ 0.2856 & 0.4466 & 1.1043 & -0.4586 \\ 0.0625 & 0.5778 & 0.9121 & 2.7983 \\ -0.4150 & 1.7639 & -1.0555 & -1.0827 \end{bmatrix} \quad (6.231)$$

$$= \begin{bmatrix} \mathbf{v}_{10} , & \mathbf{v}_{20} , & \mathbf{v}_{30} , & \mathbf{v}_{40} \end{bmatrix}$$

The mode shapes $\mathbf{v}_{i0}$ are shown in Fig. 6.16.

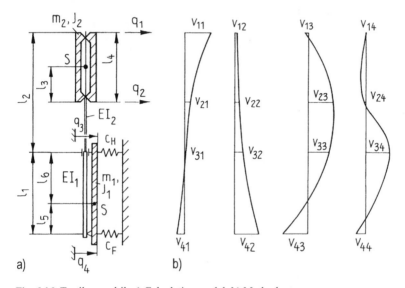

Fig. 6.16 Textile mandril; **a)** Calculation model, **b)** Mode shapes $\mathbf{v}_{i0}$

The sensitivity coefficients μ_{ik} were calculated according to (6.189) to examine the impact of changes in the mass parameters on the natural frequencies. They are given in Table 6.6 together with the natural circular frequencies ω_{i0} and the modal masses μ_i that result from (6.112).

Table 6.6 Sensitivity coefficients μ_{ik} and natural frequencies of the calculation model of a textile mandril

i	Sensitivity coefficients				Natural frequencies
	μ_{i1}	μ_{i2}	μ_{i3}	μ_{i4}	f_{i0} in Hz
1	0.0234	0.6155	0.0111	0.3501	8.777
2	0.8063	0.1035	0.0684	0.0219	44.28
3	0.0278	0.2242	0.1882	0.5598	75.42
4	0.1428	0.0568	0.7319	0.0682	473.3

6.4 Structure and Parameter Changes

The calculation of one of the sensitivity coefficients shall be performed as an example here, see v_{20} in (6.231) and M_3 in (6.228):

$$\mu_{23} = \frac{v_{20}^T M_3 v_{20}}{v_{20}^T M_0 v_{20}} \tag{6.232}$$

$$\mu_{23} = [0.1967;\ 0.4466;\ 0.5778;\ 1.7639] \begin{bmatrix} 0 & 0 & 0 & 0 \\ 0 & 0 & 0 & 0 \\ 0 & 0 & m_3^* & -m_3^* \\ 0 & 0 & -m_3^* & m_3^* \end{bmatrix} \begin{bmatrix} 0.1967 \\ 0.4466 \\ 0.5778 \\ 1.7639 \end{bmatrix}$$

$$= 1.407 \cdot m_3^* = 0.068\,38.$$

One can see from the numerical values in Table 6.6 that there are major differences. For example, one immediately notices that the second natural frequency ($i = 2$) is strongly influenced by the mass $m_1 (\mu_{21} = 0.6155)$ and less by the moment of inertia $J_2 (\mu_{24} = 0.0219)$. A change of m_1 has relatively little effect on the first and third natural frequencies, see Fig. 6.16.

If the mass m_2 is increased from originally $m_2 = 1$ kg by only $\Delta m_2 = 0.2$ kg, the first two natural frequencies change in accordance with the approximation of (6.191):

$$\frac{\Delta f_1}{f_{10}} \approx -\frac{\mu_{12}}{2} \frac{\Delta m_2^*}{m_2^*} = -0.5 \cdot 0.6155 \cdot 0.2 = -0.061\,55$$

$$\Rightarrow f_1 \approx f_{10} \cdot \left(1 + \frac{\Delta f_1}{f_{10}}\right) \approx 8.777\,\text{Hz} \cdot 0.93845 = 8.237\,\text{Hz}$$

$$\frac{\Delta f_2}{f_{20}} \approx -\frac{\mu_{22}}{2} \frac{\Delta m_2^*}{m_2^*} = -0.5 \cdot 0.1035 \cdot 0.2 = -0.010\,35 \tag{6.233}$$

$$\Rightarrow f_2 \approx f_{20} \cdot \left(1 + \frac{\Delta f_2}{f_{20}}\right) \approx 44.28\,\text{Hz} \cdot 0.989\,65 = 43.82\,\text{Hz}$$

The exact values that were obtained by solving the eigenvalue problem with the changed parameter values are $f_1 = 8.28$ Hz and $f_2 = 43.8$ Hz. The approximation is more accurate than 1 %, that means it is usable.

The next step is to calculate the change of the mode shapes if this change in mass $\Delta m_2 = 0.2$ kg is performed. The linear approximation (6.198) results specifically for this case with $\Delta C = 0$ and $\Delta M = \Delta m_2 \cdot \overline{M}_2$, see (6.187) and (6.226):

$$\Delta v_i = -\omega_{i0}^2 \Delta m_2^* \sum_{j=1}^{4} \frac{v_{j0}^T \overline{M}_2 v_{i0}}{\gamma_i - \gamma_j} v_{j0}, \quad i \neq j \tag{6.234}$$

This equation shall be written out in detailed form for $i = 1$ and $i = 2$ to illustrate this expression:

$$\Delta v_1 = -\omega_{10}^2 \Delta m_2^* \left[0 + \frac{v_{20}^T \overline{M}_2 v_{10}}{\gamma_1 - \gamma_2} v_{20} + \frac{v_{30}^T \overline{M}_2 v_{10}}{\gamma_1 - \gamma_3} v_{30} + \frac{v_{40}^T \overline{M}_2 v_{10}}{\gamma_1 - \gamma_4} v_{40} \right]$$

$$\Delta v_2 = -\omega_{20}^2 \Delta m_2^* \left[\frac{v_{10}^T \overline{M}_2 v_{20}}{\gamma_2 - \gamma_1} v_{10} + 0 + \frac{v_{30}^T \overline{M}_2 v_{20}}{\gamma_2 - \gamma_3} v_{30} + \frac{v_{40}^T \overline{M}_2 v_{20}}{\gamma_2 - \gamma_4} v_{40} \right]$$

(6.235)

If one inserts the numerical values for this example that are known for the modal matrix from (6.231) for the natural circular frequencies and modal masses from Table 6.6, $\gamma_i = \mu_i \omega_{i0}^2$ results in the following for the eigenvector changes:

$$\Delta v_1 = (\quad 0 \quad + 135.8 v_{20} + 67.08 v_{30} - 0.8457 v_{40}\,) \omega_{10}^2 \cdot k$$
$$\Delta v_2 = (-135.8 v_{10} + \quad 0 \quad + 41.41 v_{30} - 0.3457 v_{40}\,) \omega_{20}^2 \cdot k$$
$$\Delta v_3 = (-67.08 v_{10} - 41.41 v_{20} + \quad 0 \quad - 0.5235 v_{40}\,) \omega_{30}^2 \cdot k$$
$$\Delta v_4 = (\;0.8457 v_{10} + 0.3457 v_{20} + 0.5235 v_{30} + \quad 0 \quad\,) \omega_{40}^2 \cdot k$$

(6.236)

with $k = \Delta m_2^* \cdot 10^{-7} \text{ kg}^{-1} \text{s}^2$

If $\Delta m_2 = 0.2$ kg, $\Delta m_2^* = 0.05$ kg, and after inserting all numerical values, one obtains the following changes of the eigenvectors:

$$\Delta v_1 = \begin{bmatrix} 0.00024 \\ 0.00205 \\ 0.00209 \\ 0.00258 \end{bmatrix}, \quad \Delta v_2 = \begin{bmatrix} -0.0699 \\ 0.0027 \\ 0.0109 \\ 0.0050 \end{bmatrix},$$

$$\Delta v_3 = \begin{bmatrix} -0.1058 \\ -0.0420 \\ -0.0332 \\ -0.0501 \end{bmatrix}, \quad \Delta v_4 = \begin{bmatrix} 0.0474 \\ 0.0432 \\ 0.0324 \\ -0.0127 \end{bmatrix}.$$

(6.237)

One can see that all changes are small in comparison to the elements of the modal matrix (6.182) and the linear approximation is justified. If one calculates the new eigenvectors $v_i = v_{i0} + \Delta v_i$ of the changed system, one obtains the same values up to three significant digits as in the exact solution of the eigenvalue problem, see also Fig. 6.16.

6.4.5.3 Support Structure (Reduction from Ten to Five)

The support structure with 10 degrees of freedom shown in Fig. 6.4, of which the system matrices are known by (6.62) and (6.63) and the numerical values are known from Sect. 6.3.4.3, is to be reduced to a system with five degrees of freedom. $q_2^T = [q_1, q_4, q_6, q_7, q_8]$ are treated as slave degrees of freedom and $q_1^T = [q_2, q_3, q_5, q_9, q_{10}]$ as master degrees of freedom.

6.4 Structure and Parameter Changes

Based on the mode shapes known from Fig. 6.10 in this example, it should be taken into account that the mass of the rigid body on top partially oscillates in opposite direction to the frame in the lower mode shapes. Small as it may be, this body should not be neglected because its motion would quasi-statically follow the motion of the frame at the spring coupling points.

The elements of the mass and spring matrices must be rearranged in accordance with the partitioning into master and slave degrees of freedom. Since the mass matrix is a diagonal matrix, $M_{12} = M_{21} = 0$. The partitioned mass matrix consists of the two submatrices

$$M_{11} = \mathrm{diag}(10m,\ 2m,\ 2m,\ m_3,\ J_S/l^2) = m^*\mathrm{diag}(2000,\ 400,\ 400,\ 50,\ 2),$$
$$M_{22} = \mathrm{diag}(2m,\ 2m,\ 2m,\ m_1,\ m_2) = m^*\mathrm{diag}(400,\ 400,\ 400,\ 5,\ 5).$$
(6.238)

It is useful to introduce a reference mass $m^* = 10$ kg and a reference spring constant $c^* = 6.228 \cdot 10^6$ N/m for the numerical calculations below.

The partitioned stiffness matrix C_{11} results from the elements at the points of intersection, i.e. from the 2nd, 3rd, 5th, 9th, and 10th rows and columns of the stiffness matrix, see (6.62):

$$C_{11} = c^* \begin{bmatrix} 148 & -54 & 54 & 0 & 0 \\ -54 & 161 & 0 & 0 & 0 \\ 54 & 0 & 161 & 0 & 0 \\ 0 & 0 & 0 & 4 & 0 \\ 0 & 0 & 0 & 0 & 1 \end{bmatrix}, \quad C_{22} = c^* \begin{bmatrix} 304 & -12 & 0 & 0 & 0 \\ -12 & 52 & 12 & 0 & 0 \\ 0 & 12 & 304 & 0 & 0 \\ 0 & 0 & 0 & 3 & 0 \\ 0 & 0 & 0 & 0 & 3 \end{bmatrix}$$

$$C_{12} = c^* \begin{bmatrix} -131 & 0 & -131 & 0 & 0 \\ 72 & -59 & 0 & 1 & 0 \\ 0 & -59 & -72 & 0 & 1 \\ 0 & 0 & 0 & -2 & -2 \\ 0 & 0 & 0 & -1 & 1 \end{bmatrix}.$$
(6.239)

After forming the inverse matrix of C_{22}, one can calculate the matrix S that occurs repeatedly in the subsequent calculations, see (6.201):

$$S = C_{22}^{-1} C_{21} = \begin{bmatrix} -0.4309 & 0.1934 & -0.0434 & 0 & 0 \\ 0 & -1.1000 & -1.1000 & 0 & 0 \\ -0.4309 & 0.0434 & -0.1934 & 0 & 0 \\ 0 & 0.3333 & 0 & -0.6667 & -0.3333 \\ 0 & 0 & 0.3333 & -0.6667 & 0.3333 \end{bmatrix}.$$
(6.240)

The reduced matrices result from (6.204) and (6.207):

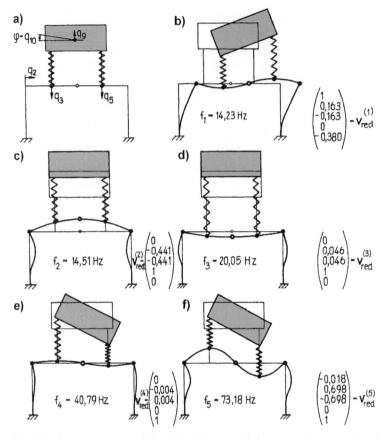

Fig. 6.17 Support structure of Fig. 6.4; **a)** System with five master degrees of freedom (external coordinates) $q^{\mathrm{T}} = (q_2, q_3, q_5, q_9, q_{10})$, **b)** Mode shapes, eigenvectors $v_{\mathrm{red}}^{(r)}$ and natural frequencies f_i of the system reduced to $n1 = 5$ degrees of freedom

$$C_{\mathrm{red}} = c^* \begin{bmatrix} 0.351 & -2.297 & -2.297 & 0 & 0 \\ -2.297 & 8.184 & -6.177 & 0.667 & 0.333 \\ 2.297 & -6.177 & 8.184 & 0.667 & -0.333 \\ 0 & 0.667 & 0.667 & 1.333 & 0 \\ 0 & 0.333 & -0.333 & 0 & 0.333 \end{bmatrix}$$

$$M_{\mathrm{red}} = m^* \begin{bmatrix} 2149 & -40.82 & 40.82 & 0 & 0 \\ -40.82 & 900.3 & 477.3 & -1.111 & -0.556 \\ 40.82 & 477.3 & 900.3 & -1.111 & 0.556 \\ 0 & -1.111 & -1.111 & 54.44 & 0 \\ 0 & -0.556 & 0.556 & 0 & 3.111 \end{bmatrix}. \quad (6.241)$$

Only three to four digits are given here, although more digits were used in the subsequent calculations. The eigenvalue problem was solved with these reduced matrices

6.4 Structure and Parameter Changes

according to (6.208). The values specified in Fig. 6.17 resulted for the natural frequencies and the reduced vectors $v_{\text{red}}^{(i)}$.

Compare these results with those of the original system in Fig. 6.10 and try to explain common features and differences. The comparison confirms the general statement explained in Sect. 6.4.3 that the natural frequencies of the reduced system are always higher than those of the original one.

The mode shapes occurring in the reduced system correspond to the lowest ones of the original system because the slave degrees of freedom (internal coordinates) were selected purposefully. The first mode shape $v_{\text{red}}^{(1)}$, however, is antisymmetrical as opposed to the symmetrical mode shape v_1. Therefore, one has to compare $v_{\text{red}}^{(1)}$ with the mode shape v_2 and $v_{\text{red}}^{(2)}$ with v_1! Such a rearrangement in the order of the mode shapes should also be observed when comparing the associated natural frequencies.

If one compares all components in the mode shapes directly with the modal matrix (6.148) using (6.209), one finds that the qualitative match is quite good, as can be seen from the two Figs. 6.10 and 6.17. The quantitative differences in the mode shapes are larger than the those in the natural frequencies (a frequent occurrence in practice).

6.4.6 Problems P6.7 to P6.9

P6.7 Load during Impact

Extremely large loads occur when a moving machine hits against something. The typical method of calculation shall be demonstrated using the example of a falling flap in continuation of the Example from Sect. 5.2.5.2.

The flap is modeled as a bending oscillator with 4 degrees of freedom and the point of impact as a massless spring, see Fig. 6.18.

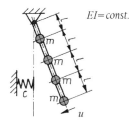

Fig. 6.18 Beam before the impact

Given:
Mass of each point mass $m = 4$ kg
Distance between point masses $l = 0.2$ m
Velocity of the outer point mass upon impact $u = 10$ m/s
Spring constant of the support column $c = 10^6$ N/m
Compliance and mass matrices from (5.62) and (5.63)

The bending stiffness EI of the beam model is given in the form of the nondimensional parameter $\bar{c} = 2EI/cl^3$. It is assumed that the flap remains in contact with the support spring after the impact, i. e. that the two are not separated.

Find:
1. In general form: the initial conditions upon impact
2. For the specific values $\bar{c} = 1/3$; $\bar{c} = 1$ and $\bar{c} \to \infty$: the horizontal forces in joint and support column; bending moment at the support point

P6.8 Increase in Natural Frequencies by a Rigid Strut

A frame is stiffened by a strut, see Fig. 6.19. Determine how the four natural frequencies of the originally frame without stiffening change.

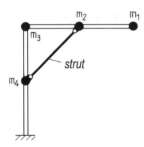

Fig. 6.19 Frame with a rigid strut

Given:
Mass matrix M and stiffness matrix C of the original frame, see Table 6.3, Case 2.

Find:
1. Constraints and transformation matrix T for the case that a rigid strut is hinged at both ends between the masses m_2 and m_4, see Fig. 6.19.
2. Matrices M_1 and C_1 of the frame with strut
3. Comparison of the natural circular frequencies of the original frame with those of the modified frame.

P6.9 Reduction from Four to Two Degrees of Freedom

The mass and stiffness matrices of the calculation model with 4 degrees of freedom shown in Table 6.3, Case 2, are given. Reduce it to a model with 2 degrees of freedom by making the coordinates q_2 and q_4 slave degrees of freedom (internal coordinates). Determine the spring and mass matrices of the reduced system and compare the two natural frequencies and mode shapes with the first two of the original system.

6.4.7 Solutions S6.7 to S6.9

S6.7 Figure 6.20 shows the bending oscillator after the impact in the deformed state. The absolute displacements can be summarized in the coordinate vector

$$q = [q_1,\ q_2,\ q_3,\ q_4]^{\mathrm{T}}. \tag{6.242}$$

6.4 Structure and Parameter Changes

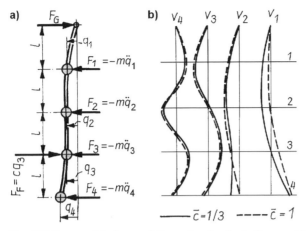

Fig. 6.20 Impact of the beam; **a)** Forces in a free-body diagram, **b)** Mode shapes according to (6.245) and (6.247)

The equation of motion of the system with 4 degrees of freedom according to (6.19) is:

$$\boldsymbol{D} \cdot \boldsymbol{M} \cdot \ddot{\boldsymbol{q}} + \boldsymbol{q} = \boldsymbol{0} \tag{6.243}$$

The beam rotates like a rigid body before the impact. The velocities depend linearly on the distance to the center of rotation so that, at the time of impact ($t = 0$), the initial conditions for the coordinate vector according to (6.118) are:

$$\boldsymbol{q}(0) = \boldsymbol{q}_0 = [0,\ 0,\ 0,\ 0]^\mathrm{T}; \qquad \dot{\boldsymbol{q}}(0) = \boldsymbol{u}_0 = (u/4) \cdot [1,\ 2,\ 3,\ 4]^\mathrm{T}. \tag{6.244}$$

The eigenvalues and eigenvectors were determined using a calculation program and are known from Sect. 5.2.5.2.

The eigenvectors are summarized according to (6.88) in the modal matrix $\boldsymbol{V}$. It has the following form for $\bar{c} = 1/3$ when normalizing according to $\boldsymbol{v}_i^\mathrm{T} \cdot \boldsymbol{v}_i = 1$, see (6.86):

$$\boldsymbol{V} = \begin{bmatrix} 0.3164 & 0.5623 & 0.6285 & -0.4345 \\ 0.3412 & 0.6216 & -0.2808 & 0.6468 \\ -0.0792 & 0.3595 & -0.7027 & -0.6088 \\ -0.8816 & 0.4101 & 0.1800 & 0.1491 \end{bmatrix} = [\boldsymbol{v}_1,\ \boldsymbol{v}_2,\ \boldsymbol{v}_3,\ \boldsymbol{v}_4]. \tag{6.245}$$

The following natural frequencies were determined using a calculation program, see also Fig. 5.14:

$$f_1 = 23.2\ \mathrm{Hz}, \qquad f_2 = 38.8\ \mathrm{Hz}, \qquad f_3 = 100.6\ \mathrm{Hz}, \qquad f_4 = 176.5\ \mathrm{Hz}. \tag{6.246}$$

Check these values using (6.117) with $\boldsymbol{D}$ from (5.62), $\boldsymbol{M}$ from (5.63), and the numerical values from Sect. 5.2.5.2.

For $\bar{c} = 1$, one obtains $f_1 = 35.8\ \mathrm{Hz}$, $f_2 = 50.1\ \mathrm{Hz}$, $f_3 = 155.5\ \mathrm{Hz}$, and $f_4 = 298.9\ \mathrm{Hz}$.

$$\boldsymbol{V} = \begin{bmatrix} 0.0873 & 0.5746 & 0.6646 & -0.4696 \\ 0.0254 & 0.7164 & -0.1400 & 0.6830 \\ -0.3096 & 0.3918 & -0.6787 & -0.5386 \\ -0.9465 & -0.0560 & 0.2795 & 0.1512 \end{bmatrix} = [\boldsymbol{v}_1,\ \boldsymbol{v}_2,\ \boldsymbol{v}_3,\ \boldsymbol{v}_4]. \tag{6.247}$$

The use of modal coordinates is useful for the calculation of the free vibrations. According to (6.112), and using the matrices from (6.245), (6.247) and (5.63), the modal masses result from $\mathrm{diag}(\mu_i) = \boldsymbol{V}^\mathrm{T} \cdot \boldsymbol{M} \cdot \boldsymbol{V}$ and are $\mu_i = m$. The initial deflections equal $\boldsymbol{p}_0 = \boldsymbol{V}^{-1} \cdot \boldsymbol{q}_0 = \boldsymbol{o}$ for all $\bar{c}$ values, see (6.244).

The initial velocities of the modal coordinates are determined using (6.123):

$$\dot{\boldsymbol{p}}_0 = \mathrm{diag}(1/\mu_i)\boldsymbol{V}^\mathrm{T} \cdot \boldsymbol{M} \cdot \boldsymbol{u}_0 = (\dot{p}_{10};\ \dot{p}_{20};\ \dot{p}_{30};\ \dot{p}_{40})^\mathrm{T}. \tag{6.248}$$

The values of $\dot{\boldsymbol{p}}_0$ depend on the relative bending stiffness $\bar{c}$:

$$\begin{aligned}
\bar{c} &= 1/3:\ \dot{\boldsymbol{p}}_0 = u \cdot (-0.6913;\ 1.1311;\ -0.3303;\ -0.0927)^\mathrm{T} \\
\bar{c} &= 1\ \ \ :\ \dot{\boldsymbol{p}}_0 = u \cdot (-1.1442;\ 0.7397;\ -0.1334;\ -0.0287)^\mathrm{T} \\
\bar{c} &\to \infty:\ \dot{\boldsymbol{p}}_0 = u \cdot (\ 1.3693;\ \ \ \ 0\ \ \ \ ;\ \ \ \ 0\ \ \ ;\ \ \ \ 0\ \ )^\mathrm{T}
\end{aligned} \tag{6.249}$$

From this one can determine the energy distribution using (6.113). The kinetic energy (the rotational energy of the rotating flap) is distributed over the four mode shapes, see (6.124) and (6.145):

$$\begin{aligned}
W_\mathrm{kin} &= \frac{1}{2}J_G\dot{\varphi}^2 = 15ml^2 \cdot \left(\frac{u}{4l}\right)^2 = \frac{1}{2}\sum_{i=1}^{4} W_{i0} = \frac{1}{2}\sum_{i=1}^{4} \frac{(\boldsymbol{v}_i^\mathrm{T}\boldsymbol{M}\boldsymbol{u}_0)^2}{\mu_i} \\
&= \frac{1}{2}\sum_{i=1}^{4} \mu_i \dot{p}_{i0}^2 = 0.9375 m u^2.
\end{aligned} \tag{6.250}$$

The following applies for the summands W_{i0}:

$$\begin{aligned}
\bar{c} &= 1/3:\ W_\mathrm{kin} = (0.2389 + 0.6397 + 0.0546 + 0.0043)mu^2 \\
\bar{c} &= 1\ \ \ :\ W_\mathrm{kin} = (0.6546 + 0.2736 + 0.0089 + 0.0004)mu^2 \\
\bar{c} &\to \infty:\ W_\mathrm{kin} = (0.9375 + 0\ \ \ \ \ \ + 0\ \ \ \ \ \ + 0\ \ \ \ \ \ )mu^2.
\end{aligned} \tag{6.251}$$

The stiffer the beam, the more energy is received by the first mode shape.

Each principal coordinate oscillates harmonically according to (6.126) at its natural frequency:

$$p_i = \frac{\dot{p}_{i0}}{\omega_i}\sin\omega_i t;\quad i = 1, 2, 3, 4. \tag{6.252}$$

The real coordinates and their accelerations are obtained according to (6.108) or (6.127):

$$q_k = \sum_{i=1}^{4} v_{ki}\frac{\dot{p}_{i0}}{\omega_i}\sin\omega_i t;\quad \ddot{q}_k = -\sum_{i=1}^{4} v_{ki}\omega_i \dot{p}_{i0}\sin\omega_i t,\quad k = 1, 2, 3, 4. \tag{6.253}$$

The horizontal forces in the support spring and in the revolute joint result from the equilibrium conditions and amount to, see Fig. 6.20:

$$\begin{aligned}
F_\mathrm{F} &= \frac{1}{3}(F_1 + 2F_2 + 3F_3 + 4F_4) = -\frac{m}{3}(\ddot{q}_1 + 2\ddot{q}_2 + 3\ddot{q}_3 + 4\ddot{q}_4) = cq_3 \\
F_\mathrm{G} &= \frac{1}{3}(2F_1 + F_2 - F_4) = -\frac{m}{3}(2\ddot{q}_1 + \ddot{q}_2 - \ddot{q}_4) \\
&= F_1 + F_2 + F_3 + F_4 - F_\mathrm{F}.
\end{aligned} \tag{6.254}$$

Insertion of the accelerations from (6.253) provides these forces in the form:

6.4 Structure and Parameter Changes

$$F_{\rm F} = \frac{m}{3} \sum_{i=1}^{4} (v_{1i} + 2v_{2i} + 3v_{3i} + 4v_{4i}) \omega_i \dot{p}_{i0} \sin \omega_i t = \sum_{i=1}^{4} F_{{\rm F}i} \sin \omega_i t \quad (6.255)$$

$$F_{\rm G} = \frac{m}{3} \sum_{i=1}^{4} (2v_{1i} + v_{2i} - v_{4i}) \omega_i \dot{p}_{i0} \sin \omega_i t = \sum_{i=1}^{4} F_{{\rm G}i} \sin \omega_i t. \quad (6.256)$$

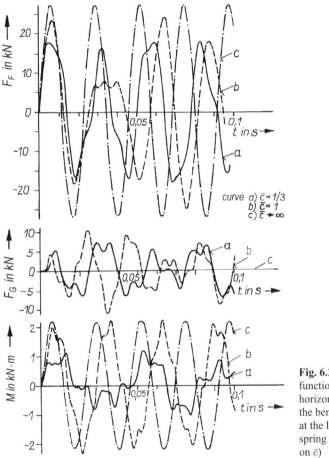

Fig. 6.21 Time functions of the horizontal forces and the bending moment at the location of the spring (dependence on $\bar{c}$)

Likewise, the bending moment at the spring in the beam is the product of the inertia force and the distance, see Figs. 6.20 and (6.253):

$$\underline{\underline{M}} = F_4 l = -m\ddot{q}_4 l = ml \sum_{i=4}^{4} v_{4i}\omega_i \dot{p}_{i0} \sin \omega_i t = \sum_{i=1}^{4} M_i \sin \omega_i t. \quad (6.257)$$

The numerical values of the amplitudes of the forces ($F_{{\rm F}i}$; $F_{{\rm G}i}$) and the moment (M_i) as defined by (6.255) to (6.257) can be calculated using the data from the problem statement and from (6.245) for each natural vibration.

One can see from Fig. 6.21 that the load parameters are primarily determined by the second natural vibration when the bending stiffness EI is small (i. e. small $\bar{c}$ value) and also that the influence of the higher natural vibrations is considerable. The larger the bending stiffness EI becomes in relation to the spring constant c, the larger becomes the influence of the fundamental vibration. For a perfectly rigid beam, an oscillator with only one degree of freedom results, which performs a harmonic motion at its natural frequency.

An interesting fact about these results is that the maximum value of the load parameters that can occur in the course of the vibration increases with increasing bending stiffness EI. Had one used a rigid lever with the support spring for this flap as a calculation model, a slightly too high maximum value (27.39 kN > 25.88 kN) for the bearing force F_F and a slightly too low maximum value (2.191 kN · m < 2.412 kN · m) for the moment would have resulted.

In the case of a rigid lever ($\bar{c} \to \infty$), there is no joint force at all at the center of rotation G because the spring force acts exactly at the center of percussion of this rigid body, see P2.10. Only when taking into account the bending stiffness of the flap with its finite magnitude, dynamic forces also occur in the joint G after the impact, namely as a result of the excited bending vibrations.

Conclusion: The dynamic loads in a component after an impact are proportional to the initial velocity and depend on the stiffness distribution within the impacted vibration system. The natural frequencies increase with increasing stiffness of an element. No general statement can be made regarding the influence of the stiffness on the magnitude of the dynamic loads.

S6.8 The strut between the masses m_2 and m_4 (see Fig. 6.19) forces these masses into matching displacement components in the direction of the strut. The mass m_2 can still move within the plane, and m_4 can still shift horizontally. The constraint is found if one considers the geometrical conditions at small deformations, see Fig. 6.22.

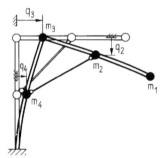

Fig. 6.22 Frame with strut

Since the triangle that connects the three masses remains rigid when the translational deformations are negligible, the following is valid for small angles ($q_k/l \ll 1$):

$$\begin{array}{ll} q_1 = q_1 \\ q_2 = q_3 - q_4 \\ q_3 = q_3 \\ q_4 = q_4 \end{array} \quad \text{or} \quad T = \begin{bmatrix} 1 & 0 & 0 \\ 0 & 1 & -1 \\ 0 & 1 & 0 \\ 0 & 0 & 1 \end{bmatrix}. \quad (6.258)$$

The coordinate q_2 can be eliminated in this way. Applying a consideration as in (6.218), one obtains the transformation matrix for the coordinate vector $\boldsymbol{q}^T = (q_1, q_3, q_4)$.

The mass and spring matrices are then obtained from (6.215) using the matrices given in Table 6.3, Case 2, as in (6.219) and (6.220):

6.4 Structure and Parameter Changes

$$M_1 = m \cdot \begin{bmatrix} 1 & 0 & 0 \\ 0 & 7 & -2 \\ 0 & -2 & 4 \end{bmatrix}; \quad C_1 = \frac{48EI}{97l^3} \cdot \begin{bmatrix} 26 & -50 & 47 \\ -50 & 126 & -165 \\ 47 & -165 & 320 \end{bmatrix} \quad (6.259)$$

The natural circular frequencies of the system stiffened by a strut according to Fig. 6.22 result from (6.221) with $\omega^* = \sqrt{48EI/(ml^3)}$:

$$\omega_1 = 0.088\,84\omega^*; \quad \omega_2 = 0.486\,89\omega^*; \quad \omega_3 = 0.958\,27\omega^*. \quad (6.260)$$

They are shown as points for comparison at $\alpha = 45°$ in Fig. 6.15 and are higher than the ones of the original system, as expected. The support with respect to the fixed reference system increases the second and third natural frequencies more than this strut, however, the first natural frequency is increased more by this strut (if one compares at the angle of $45°$).

S6.9 The method requires that the coordinates be sorted by master degrees of freedom q_1 and slave degrees of freedom q_2 so that the coordinate vector is brought into the new form $q^T = (q_1, q_3, q_2, q_4)$. Therefore, rows two and three and columns two and three in the matrices must be switched. The matrices associated with the rearranged coordinate vector are thus ($c^* = 48EI/l^3$, $m^* = m$):

$$C = c^* \begin{bmatrix} 26 & 9 & -59 & -12 \\ 9 & 74 & -54 & -131 \\ -59 & -54 & 160 & 72 \\ -12 & -131 & 72 & 304 \end{bmatrix}, \quad M = m^* \begin{bmatrix} 1 & 0 & 0 & 0 \\ 0 & 5 & 0 & 0 \\ 0 & 0 & 2 & 0 \\ 0 & 0 & 0 & 2 \end{bmatrix}. \quad (6.261)$$

The submatrices according to the partitioning as in (6.199) or (6.206) are

$$C_{11} = c^* \begin{bmatrix} 26 & 9 \\ 9 & 74 \end{bmatrix}, \quad C_{12} = c^* \begin{bmatrix} -59 & -12 \\ -54 & -131 \end{bmatrix}, \quad C_{22} = c^* \begin{bmatrix} 160 & 72 \\ 72 & 304 \end{bmatrix}$$

$$M_{11} = m^* \begin{bmatrix} 1 & 0 \\ 0 & 5 \end{bmatrix}, \quad M_{12} = 0, \quad M_{22} = m^* \begin{bmatrix} 2 & 0 \\ 0 & 2 \end{bmatrix}. \quad (6.262)$$

If (6.204) is applied, the reduced matrices are obtained after the following numerical calculation:

$$C_{\text{red}} = c^* \left\{ \begin{bmatrix} 26 & 9 \\ 9 & 74 \end{bmatrix} - \begin{bmatrix} -59 & -12 \\ -54 & -131 \end{bmatrix} \begin{bmatrix} 160 & 72 \\ 72 & 304 \end{bmatrix}^{-1} \begin{bmatrix} -59 & -54 \\ -12 & -131 \end{bmatrix} \right\}$$

$$C_{\text{red}} = c^* \begin{bmatrix} 3.464\,28 & -5.196\,43 \\ -5.196\,43 & 13.857\,10 \end{bmatrix}. \quad (6.263)$$

In the same way, it follows from (6.207)

$$M_{\text{red}} = m \begin{bmatrix} 1.314\,41 & 0.084\,18 \\ 0.084\,18 & 5.360\,33 \end{bmatrix}. \quad (6.264)$$

The following squares of the natural circular frequencies are obtained from the eigenvalue problem (6.208) of the reduced system:

$$\omega_1^2 = 0.312\,97 \frac{EI}{ml^3}, \quad \omega_2^2 = 2.3346 \frac{EI}{ml^3}. \quad (6.265)$$

They are slightly higher than those of the original system, see (6.136). The associated eigenvectors are (normalized with $v_{1i} = 1$):

$$v_{\text{red}}^{(1)} = \begin{bmatrix} 1 \\ 0.501\,55 \end{bmatrix}, \qquad v_{\text{red}}^{(2)} = \begin{bmatrix} 1 \\ -0.489\,29 \end{bmatrix}. \tag{6.266}$$

Calculation back to the original position coordinates according to (6.209) results in the complete eigenvectors, which are as follows after normalization for the coordinate vector $q^{\text{T}} = (q_1, q_3, q_2, q_4)$:

$$\bar{v}_1 = \begin{bmatrix} 1 \\ 0.501\,55 \\ 0.473\,46 \\ 0.143\,47 \end{bmatrix}, \qquad \bar{v}_2 = \begin{bmatrix} 1 \\ -0.489\,29 \\ 0.314\,22 \\ -0.245\,79 \end{bmatrix}. \tag{6.267}$$

Note the switch in the order of q_2 and q_3 as compared to (6.142). If one compares the mode shapes obtained for the reduced system with the values of the original system known from (6.142), one will see that they match with sufficient accuracy.

6.5 Forced Undamped Vibrations

6.5.1 General Solution

The basic assumption when treating forced vibrations is that the excitation forces or the motion excitations that act on a mechanical system explicitly depend on time. Such a modeling does not allow **any reaction** of the oscillator on the exciter and assumes an infinitely large energy reservoir of the exciter. Both are never true, strictly speaking, but it is a useful approximation that is frequently justified. The object treated is cut out of its mechanical environment by prescribing a time-dependent mechanical quantity, i. e., the **model boundary** is defined in this way.

The equations of motion for forced vibrations can be established, for example, by means of Lagrange's equations or from the equilibrium conditions in conjunction with d'Alembert's principle. They take the following form in matrix notation, see (6.18) and (6.19):

$$M\ddot{q} + Cq = f(t) \qquad \text{or} \qquad DM\ddot{q} + q = Df(t). \tag{6.268}$$

On the right is the vector of the excitation forces, $f^{\text{T}} = [F_1, F_2, \ldots, F_n]$, the components F_k of which are generalized forces or moments that act in the direction of the generalized coordinates q_k. The time function of interest $q(t)$ could be obtained by numerical integration from (6.268). However, essential physical and mechanical information regarding the spectral and modal behavior of the system is lost when using such methods.

The solution that uses the principal coordinates known from 6.3.2 is considered here. If the transformation according to (6.108) is performed, a system of n decoupled differential equations, as in (6.116), is obtained:

$$\mu_i \ddot{p}_i + \gamma_i p_i = h_i(t), \qquad i = 1, 2, \ldots, n. \tag{6.269}$$

6.5 Forced Undamped Vibrations

The modal masses μ_i and the modal spring constants γ_i are known from (6.104) and (6.105), see also (6.112). The quantities h_i are the excitation forces reduced to the ith mode shape and are called modal forces. The result following from the original load parameters is:

$$\boldsymbol{h}(t) = \boldsymbol{V}^\mathrm{T} \boldsymbol{f}(t) \quad \text{or} \quad h_i(t) = \sum_{k=1}^{n} v_{ki} F_k(t). \tag{6.270}$$

Each modal force h_i represents the work that all load parameters F_k perform in the direction of the deformation that corresponds to mode shape $\boldsymbol{v}_i$. Each of the equations in (6.269) formally corresponds to the equation of motion of a single-degree-of-freedom oscillator that is called a modal oscillator. When using the natural circular frequencies ω_i, see (6.117), the solution, taking into account the initial conditions of (6.125), is as follows:

$$p_i = p_{i0} \cos \omega_i t + \frac{\dot{p}_{i0}}{\omega_i} \sin \omega_i t + \frac{1}{\mu_i \omega_i} \int_0^t h_i(t') \sin \omega_i (t - t') \mathrm{d}t'. \tag{6.271}$$

The integral expression in (6.271) is **Duhamel's integral**. The special case of free vibrations results for $\boldsymbol{h}(t) = \boldsymbol{0}$. The velocities result when using the general rule for differentiating an integral:

$$\dot{p}_i = -\omega_i p_{i0} \sin \omega_i t + \dot{p}_{i0} \cos \omega_i t + \frac{1}{\mu_i} \int_0^t h_i(t') \cos \omega_i (t - t') \mathrm{d}t'. \tag{6.272}$$

The forced vibrations of a system can be calculated for any time function of the force $h_i(t)$ by solving Duhamel's integral. If only the steady-state solution is of interest, such as for harmonic or periodic excitation, one can leave out any terms determined by the initial conditions.

6.5.2 Harmonic Excitation (resonance, absorption)

Harmonic excitations occur frequently in mechanical engineering, e. g. when inertia forces are generated by rotating unbalances. The solutions to the equations of motion for harmonic excitation are of further interest because each periodic excitation can be viewed as a superposition of harmonic excitations. The focus shall now be on the case where n harmonically varying load parameters act on the vibration system at the circular frequency of the excitation Ω, the phase angles β_k of which can be different.

$$F_k = \hat{F}_k \sin(\Omega t + \beta_k), \quad k = 1, 2, \ldots, n. \tag{6.273}$$

The amplitudes of the generalized excitation forces are arranged in the vector $\hat{\boldsymbol{f}}^T = [\hat{F}_1, \hat{F}_2, \ldots, \hat{F}_n]$. According to (6.270), the modal forces are

$$h_i = \sum_{k=1}^{n} v_{ki} \hat{F}_k \sin(\Omega t + \beta_k) = \hat{h}_i \cdot \sin(\Omega t + \alpha_i), \qquad i = 1, 2, \ldots, n \quad (6.274)$$

with the amplitudes

$$\hat{h}_i = \sqrt{\left(\sum_{k=1}^{n} v_{ki} \hat{F}_k \cos \beta_k\right)^2 + \left(\sum_{k=1}^{n} v_{ki} \hat{F}_k \sin \beta_k\right)^2}. \qquad (6.275)$$

The phase angles result from

$$\sin \alpha_i = \frac{\sum_{k=1}^{n} v_{ki} \hat{F}_k \sin \beta_k}{\hat{h}_i}, \qquad \cos \alpha_i = \frac{\sum_{k=1}^{n} v_{ki} \hat{F}_k \cos \beta_k}{\hat{h}_i}. \qquad (6.276)$$

If one selects the initial conditions $\boldsymbol{p}_0 = \boldsymbol{o}$ and $\dot{\boldsymbol{p}}_0 = \boldsymbol{o}$, the deformation of the principal coordinates results from the integral (6.271) using $\gamma_i = \mu_i \omega_i^2$ and for $\alpha_i = 0$:

$$\begin{aligned} p_i &= \frac{\hat{h}_i}{\mu_i \omega_i} \int_0^t \sin \Omega t' \cdot \sin \omega_i (t - t') \mathrm{d}t' \\ &= \frac{\hat{h}_i}{\gamma_i (1 - \Omega^2/\omega_i^2)} \left(\sin \Omega t - \frac{\Omega}{\omega_i} \sin \omega_i t \right). \end{aligned} \qquad (6.277)$$

These motions are composed of vibrations both at the circular frequency of the excitation Ω and at the natural circular frequency ω_i. The virtually omnipresent influence of damping causes the portion that is governed by the natural circular frequency ω_i to decay over time so that it can frequently be neglected, see Fig. 6.23.

Resonance can occur if the excitation frequency $f = \Omega/(2\pi)$ coincides with the natural frequencies ($f_i = \omega_i/(2\pi)$), i. e. when

$$\Omega = \omega_i. \qquad (6.278)$$

One can find out how the motions increase over time in the **resonance case** as a special case of (6.277) by means of a limiting process:

$$p_i = \lim_{\Omega \to \omega_i} \frac{\hat{h}_i \left(\sin \Omega t - \dfrac{\Omega}{\omega_i} \sin \omega_i t \right)}{\gamma_i \left(1 - \dfrac{\Omega^2}{\omega_i^2}\right)} = \frac{\hat{h}_i}{2\gamma_i} (\sin \Omega t - \Omega t \cos \Omega t). \qquad (6.279)$$

6.5 Forced Undamped Vibrations

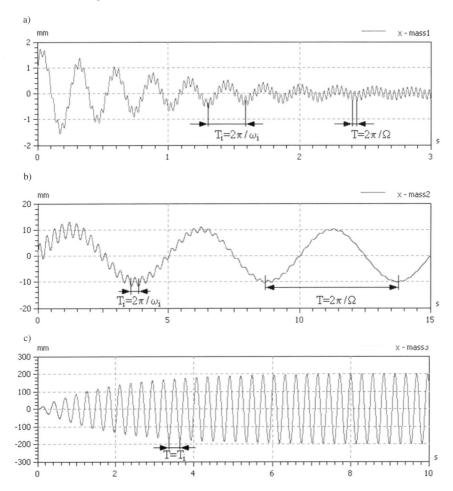

Fig. 6.23 Time functions of forced vibrations for harmonic excitation [34]; **a)** Decay of the natural vibration at $\Omega > \omega_i$, **b)** Decay of the natural vibration at $\Omega < \omega_i$, **c)** Resonance case of the weakly damped oscillator ($\Omega = \omega_i$)

Figure 6.23c illustrates this motion, however, taking into account a weak degree of damping. The linear growth of the amplitudes over time can be observed in $0 \leqq t \leqq 1$ s. Subsequently, one can see the transition into a steady state. If there were no damping, there would be infinite linear growth of the amplitudes with time.

The steady-state solution results for $\eta_i \neq 1$ from

$$\mu_i \ddot{p}_i + \gamma_i p_i = \hat{h}_i \sin(\Omega t + \alpha_i) \tag{6.280}$$

as follows:

$$p_i = \frac{\hat{h}_i \sin(\Omega t + \alpha_i)}{\gamma_i(1 - \Omega^2/\omega_i^2)} = \hat{p}_i \sin(\Omega t + \varphi_i), \qquad i = 1, 2, \ldots, n, \qquad (6.281)$$

that is, for $\alpha_i = 0$, the solution known from (6.277) without the term of natural vibration. Although $\alpha_i = \varphi_i$ applies here, a distinction is made between the angles α_i and φ_i with respect to the damping influence discussed in Sect. 6.6.3.

The motion in the original coordinates is obtained using the transformation (6.108). It follows from

$$\hat{p}_i = \frac{\hat{h}_i}{\gamma_i(1 - \eta_i^2)}; \qquad \eta_i^2 = \frac{\Omega^2 \mu_i}{\gamma_i} = \frac{\Omega^2}{\omega_{i0}^2} \qquad (6.282)$$

that

$$q_k = \sum_{i=1}^{n} v_{ki} p_i = \sum_{i=1}^{n} v_{ki} \hat{p}_i \sin(\Omega t + \varphi_i) = \hat{q}_k \sin(\Omega t + \psi_k). \qquad (6.283)$$

The amplitudes of the coordinates are

$$\hat{q}_k = \sqrt{\left(\sum_{i=1}^{n} v_{ki} \hat{p}_i \cos \varphi_i\right)^2 + \left(\sum_{i=1}^{n} v_{ki} \hat{p}_i \sin \varphi_i\right)^2}. \qquad (6.284)$$

The phase angles result from

$$\cos \psi_k = \frac{\sum\limits_{i=1}^{n} v_{ki} \hat{p}_i \cos \varphi_i}{\hat{q}_k}; \qquad \sin \psi_k = \frac{\sum\limits_{i=1}^{n} v_{ki} \hat{p}_i \sin \varphi_i}{\hat{q}_k}. \qquad (6.285)$$

The solution shows that the system vibrates in the steady state with the excitation frequency for harmonic excitation. The **forced mode shape**, described by the totality of q_k depends on the excitation frequency. There is a major difference between the excitation frequency-independent mode shapes of the forced vibrations and the mode shapes that only depend on the model parameters.

The mode shapes change continuously with the excitation frequency. In the vicinity of a resonance point ($\Omega = \omega_i$), the ith summand in the sum (6.283) dominates the other terms so that the amplitudes of the forced vibrations roughly correspond to the ith mode shape v_i. It is therefore sometimes sufficient to consider the natural vibrations closest to the resonance point for forced vibrations because the amplitudes of the "more distant" ones remain negligibly small.

The mode shapes shown in Fig. 6.24b can be explained by the amplitude responses shown in Fig. 6.24c (not exactly to scale). A frequent practical task is to reconstruct forced mode shapes from measured amplitude responses. Note, therefore, the close correlations between both representations of the same physical phenomenon. If the excitation frequency ($\Omega \to 0$) is very low, the forced mode shape is almost identical with the static deflection curve. "Low" means here that the excita-

6.5 Forced Undamped Vibrations

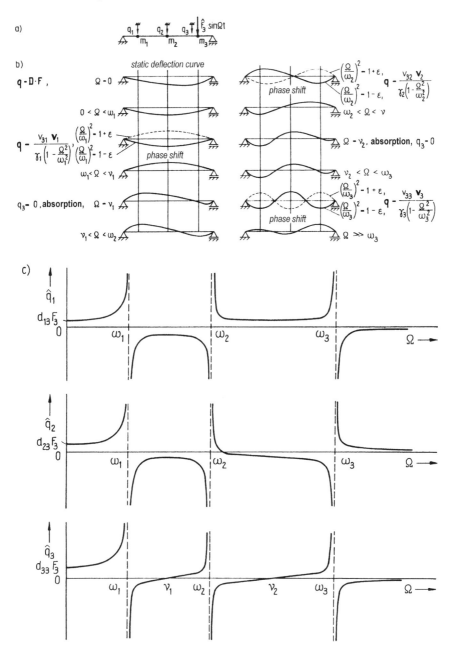

Fig. 6.24 Mode shapes of a beam supported at two points due to a harmonic excitation force $\hat{F}_3 \sin \Omega t$; **a)** Model with three degrees of freedom, **b)** Forced mode shapes, **c)** Amplitude response of the coordinates q_1, q_2, q_3

tion frequency is much smaller than the first natural frequency. Because of the sign change in the denominator of the first summand, the amplitude changes its sign at the resonance point. The resulting **phase jump** is typical for each resonance point, as can be seen for every coordinate in Fig. 6.24c at the points $\Omega = \omega_i$.

Note that a node "slides" from the right bearing to the left into the beam after the phase jump if the frequency is further increased. The so-called absorption frequencies (also called antiresonances) occur when the node of the forced mode shape reaches the application point of the force. One can determine between which natural frequencies an excitation frequency lies from the number and position of the nodes of a forced vibration (e. g. during measurements).

Coincidence of an excitation frequency with a natural frequency does not say anything about the hazardousness of these vibrations. It can happen that $\Omega = \omega_i$ and still no very large amplitudes occur because the modal force is $h_i = 0$. The following applies if the excitation force vector $\boldsymbol{f}$ is orthogonal to the ith mode shape $\boldsymbol{v}_i$

$$h_i = \boldsymbol{v}_i^{\mathrm{T}} \boldsymbol{f} = \boldsymbol{f}^{\mathrm{T}} \boldsymbol{v}_i = \sum_{k=1}^{n} v_{ki} F_k = 0 \tag{6.286}$$

and no amplitudes occur (so-called "pseudo resonance").

It is a special case if an excitation force acts in the vicinity of a node of a mode shape. The amplitudes of a mode shape are zero if the force applies exactly at its node. One should keep in mind that one only has to fear resonances if the excitation forces feed energy into the associated mode shape at the respective resonance frequency.

If only a single harmonic force acts onto the oscillator in the direction of the coordinate $q_s (1 \leq s \leq n)$, the following applies according to (6.274)

$$h_i = v_{si} \hat{F}_s \sin(\Omega t + \beta_s) = \hat{h}_i \sin(\Omega t + \beta_s), \qquad i = 1, 2, \ldots, n \tag{6.287}$$

and the corresponding steady-state solution according to (6.283) is:

$$q_k(t) = \sum_{i=1}^{n} \left[\frac{v_{ki} v_{si}}{\gamma_i (1 - \eta_i^2)} \right] \hat{F}_s \sin(\Omega t + \beta_s) = \hat{q}_k \sin(\Omega t + \beta_s). \tag{6.288}$$

The relation between the amplitudes is

$$\hat{q}_k = D_{sk} \hat{F}_s \tag{6.289}$$

with the elements of the **dynamic compliance matrix** $\boldsymbol{D}(\Omega)$

$$D_{ks}(\Omega) = D_{sk}(\Omega) = \sum_{i=1}^{n} \frac{v_{ki} v_{si}}{\gamma_i (1 - \eta_i^2)} \tag{6.290}$$

between the force application point (q_s) and the coordinate q_k. Evidently $D_{ks} = D_{sk}$. The deformation at the force application point $(k = s)$ is:

6.5 Forced Undamped Vibrations

$$q_s(t) = \sum_{i=1}^{n}\left[\frac{v_{si}^2}{\gamma_i(1-\eta_i^2)}\right]\hat{F}_s\sin(\Omega t+\beta_s) = D_{ss}\hat{F}_s\sin(\Omega t+\beta_s). \quad (6.291)$$

It follows from (6.291) that:

$$D_{ss}(\Omega) = \frac{\hat{q}_s}{\hat{F}_s} = \sum_{i=1}^{n}\frac{v_{si}^2}{\gamma_i(1-\eta_i^2)}. \quad (6.292)$$

The dynamic compliance D_{ss} has n resonance points, as can be seen from (6.292). It follows that there are $n-1$ zeroes between these singular points, that is excitation frequencies where $D_{ss}(\Omega = \nu_j) = 0$. The force application point remains at rest at these excitation frequencies but all other points of the system continue to vibrate, see also Fig. 6.24c. This phenomenon represents vibration absorption or "antiresonance". The corresponding frequencies $f_j = \nu_j/2\pi$ are called **absorption frequencies**.

A system with n degrees of freedom has $n-1$ absorption frequencies with respect to each force application point, see also the detailed discussion in Sect. 4.4. The force application point is immobile during absorption, i.e. one can imagine it to be fixed and consider the existing mode shape as natural mode shapes of the system fixed in q_s direction. The excitation force then appears as a support force. Therefore, the absorption frequencies can be determined as the natural frequencies of a system with $(n-1)$ degrees of freedom using the common eigenvalue programs.

If one puts all summands in (6.292) on a common denominator, its zeroes are the natural circular frequencies. On the other hand, the resulting numerator must be a polynomial of degree $(n-1)$, the roots of which are the absorption frequencies ν_k. If one denotes the product as Π and the remaining numerical value of the polynomials as d_{ss}, one arrives at the following representation:

$$D_{ss}(\Omega) = d_{ss}\frac{\prod_{k=1}^{n-1}(1-\Omega^2/\nu_k^2)}{\prod_{i=1}^{n}(1-\Omega^2/\omega_i^2)}. \quad (6.293)$$

The d_{ss} represent the dynamic compliance at $\Omega = 0$, i.e. the static compliance (principal diagonal elements of the matrix $\mathbf{D}$). (6.293) offers a way to calculate the dynamic compliance from the static compliance if the natural circular frequencies ω_i and the absorption frequencies ν_k of a system are known.

6.5.3 Transient Excitation (Rectangular Impulse)

In addition to harmonic excitation, excitation during a finite excitation time Δt is an important loading case in mechanical engineering. It allows the capture of the loading and unloading that occurs during starting and braking processes. Even **short-**

term loads during clutching processes, when vehicles or machine parts hit against stops, when accepting or ejecting workpieces (gripping, opening, closing) or during processing operations (forming, cutting) often require the calculation of the resulting maximum values of load and motion parameters.

Section 4.3.3 discussed various excitations for torsional oscillators. The results from Fig. 4.33 and Fig. 4.35 can be transferred to modal oscillators. A perfectly rectangular pattern of the modal excitation force is assumed here. In this way, typical correlations can be shown that also occur for other load patterns, e. g. a half-sine force pattern. This also allows consideration of impulse excitation for very short load times $\Delta t \ll T_i$, see Fig. 6.25a.

If a constant single force F_{s0} is applied at point s in the direction of the coordinate q_s, the resulting modal forces according to (6.270) are

$$h_i = v_{si} F_{s0}, \qquad i = 1, 2, \ldots, n. \tag{6.294}$$

If the force F_{s0} suddenly acts onto a system at rest during the time $0 \leq t \leq \Delta t$, the principal coordinates can be found from (6.271):

$$p_i = \frac{h_i}{\mu_i \omega_i} \int_0^t \sin \omega_i (t - t') \mathrm{d}t' = \frac{h_i}{\gamma_i} (1 - \cos \omega_i t). \tag{6.295}$$

The following results for $t > \Delta t$ because $F_S = 0$ and the modal force $h_i = 0$,

$$p_i = \frac{h_i}{\mu_i \omega_i} \int_0^{\Delta t} \sin \omega_i (t - t') \mathrm{d}t' + \int_{\Delta t}^t 0 \mathrm{d}t' = \frac{h_i}{\gamma_i} [\cos \omega_i (t - \Delta t) - \cos \omega_i t] \tag{6.296}$$

and after a few trigonometric transformations

$$p_i = \frac{2 h_i}{\gamma_i} \cdot \sin \frac{\omega_i \Delta t}{2} \sin \omega_i \left(t - \frac{\Delta t}{2} \right) \tag{6.297}$$

is obtained. If one sets $\omega_i = 2\pi/T_i$, one can represent p_i using (6.295) for $0 \leq t/\Delta t \leq 1$ and (6.297) for $t/\Delta t > 1$. Figure 6.25c shows this for various ratios $\Delta t/T_i$. One can see that the maximum value is

$$p_{i\,\mathrm{max}} = \frac{2 h_i}{\gamma_i} \left| \sin \frac{\omega_i \Delta t}{2} \right| = \frac{2 h_i}{\gamma_i} \left| \sin \frac{\pi \Delta t}{T_i} \right|. \tag{6.298}$$

Each principal coordinate thus deforms as a result of a sudden load at most twice as much as for a static load of the same magnitude. The maximum value depends on the ratio of excitation time Δt to period T_i of the respective natural vibration.

The *peak value* of the force generated in the system, the so-called maximum *impulse response* is often used to compare impulse excitations and to simulate laboratory tests. One distinguishes between the initial impulse response that takes into account the highest amplitude during the impulse period ($0 < t < \Delta t$) and the

6.5 Forced Undamped Vibrations

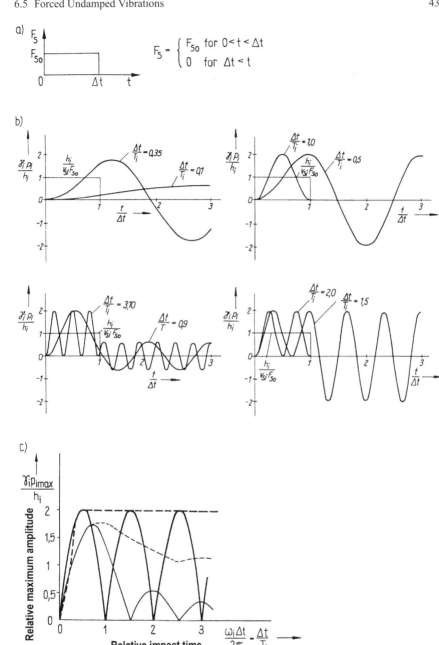

Fig. 6.25 Reaction of a modal oscillator to a force of finite duration $\triangle t$; **a)** Rectangular impulse, **b)** Time functions, **c)** Maximum impulse response (thick lines: rectangular impulse; thin lines: half-sine impulse; dashed: initial maximum value; solid line: residual maximum value)

residual impulse response that expresses the highest amplitude after the excitation time $(t > \Delta t)$.

Figure 6.25c illustrates the impulse response for a rectangular and a half-sine load pattern. The solid curve matches the solution obtained using (6.295) and (6.296) while results from the literature were used for the half-sine impulse.

If the excitation time is an integer multiple of the oscillation period $(\Delta t = nT_i)$, the respective mode shape remains at rest after the force subsides, see (6.297). This extinction of the residual oscillation (rectangular impulse: $\Delta t = T_i + nT_i$, half-sine impulse: $\Delta t = 1,5T_i + nT_i$) is an interesting dynamic effect, see Fig. 6.25c. For a modal hammer, it can result in a mode shape not being excited despite the impulse. The energy supplied in the first half period is removed by the "source of energy" in the second half period.

If the excitation time is small as compared to the period of oscillation, a series expansion for $\Delta t/T_i \ll 1$ results from (6.297) because of $\sin(\pi \Delta t/T_i) \approx \pi \Delta t/T_i$ in the following form

$$p_{i\,\max} = \frac{2\pi h_i \Delta t}{\gamma_i T_i} = \frac{\omega_i}{\gamma_i} h_i \Delta t = \frac{h_i \Delta t}{\sqrt{\gamma_i \mu_i}}. \qquad (6.299)$$

The coordinates q_k are derived from (6.108), (6.294), and (6.297) for $t > \Delta t$:

$$\begin{aligned} q_k &= \sum_{i=1}^{n} v_{ki} \frac{2h_i}{\gamma_i} \sin \frac{\omega_i \Delta t}{2} \sin \omega_i \left(t - \frac{\Delta t}{2}\right) \\ &= 2F_s \sum_{i=1}^{n} \frac{v_{ki} v_{si}}{\gamma_i} \sin \frac{\omega_i \Delta t}{2} \sin \omega_i \left(t - \frac{\Delta t}{2}\right). \end{aligned} \qquad (6.300)$$

They arise as a superposition of all vibrations, and it is not true here, as has been stated for the case discussed in Sect. 4.3.3.2, that the maximum dynamic deformations q_k are twice as large as the static ones. The ratio of static and dynamic deformations has a different value for each coordinate q_k, which is mainly determined by the mode shapes v_i. As a result of the mostly irrational numerical values of ω_i, the maximum values of the principal coordinates are typically not reached at the same time. An estimate for the coordinates for this load case can be given in the form of the following inequality:

$$q_{k\,\max} \leqq 2F_s \sum_{i=1}^{n} \left| \frac{v_{ki} v_{si}}{\gamma_i} \sin \frac{\omega_i \Delta t}{2} \right|. \qquad (6.301)$$

The impact force F_{s0}, the excitation time Δt, the mode shapes v_i and the natural circular frequencies ω_i must be known to evaluate the effect of impact-like loads. Damping prevents the maximum amplitudes from reaching the high value according to (6.301) since the higher mode shapes decay fast and not all peak values add up. For an arbitrary load function $h_i(t)$, the following applies according to (6.271) because of $\sin \omega_i(t - t') = \sin \omega_i t \cos \omega t' - \cos \omega_i t \sin \omega_i t'$:

6.5 Forced Undamped Vibrations

$$p_i = \frac{1}{\mu_i \omega_i} \left\{ \left[\int_0^t h_i(t') \cos \omega_i t' dt' \right] \sin \omega_i t - \left[\int_0^t h_i(t') \sin \omega_i t' dt' \right] \cos \omega_i t \right\}. \tag{6.302}$$

An impulse is kinematically defined as a velocity jump, just like a jerk is defined as an acceleration jump. From a dynamics point of view, the **impact force** is an excitation force that acts for a short time as compared to the period of the excited oscillation. The following consideration does not distinguish between kinematic (displacement, angular) and force excitation since both types of excitation can be transformed into a modal excitation force for each principal vibration. The approximations

$$\omega_i t' \ll 1, \qquad \cos \omega_i t' = 1, \qquad \sin \omega_i t' = 0 \tag{6.303}$$

apply for $(\Delta t \ll T_i)$, and (6.302) with the momentum $I_i = \int_0^t h_i(t') dt'$ approximately yields

$$p_i = \frac{I_i}{\mu_i \omega_i} \sin \omega_i t. \tag{6.304}$$

The amplitude of this residual vibration (also called residual impulse response depends only on the momentum for impacts and not on the time function of the excitation force. Note that the solution according to (6.304) matches the one that occurs for free vibrations with the initial condition $\dot{p}_{i0} = I_i/\mu_i$, see (6.127).

6.5.4 Examples

6.5.4.1 Frame

The first example considered is the frame with 4 degrees of freedom from Table 6.3, Case 2. The sought after parameters are the dynamic compliance $D_{11}(\Omega)$ and the amplitudes of the coordinates as functions of the excitation frequency. The modal matrix V, the diagonal matrices of the modal masses $\text{diag}(\mu_i)$ and the modal spring constants $\text{diag}(\gamma_i)$ are known from 6.3.4.2. Furthermore, $h = V^T f = V^T \cdot [1, 0, 0, 0]^T F_1 = F_1 \cdot [1, 1, 1, 1]^T$ due to the normalization of V where $v_{1i} = 1$ according to (6.142). The static deflections, which one can use for comparison, result from (6.131):

$$q_{\text{st}} = D \cdot f = [64,\ 29,\ 24,\ 6]^T \frac{l^3 F_1}{48 EI}. \tag{6.305}$$

The dynamic compliance at the force application point results from (6.292) for $s = 1$ after introducing a common denominator

$$\frac{48EI}{64l^3}D_{11}(\Omega) = \frac{q_1}{q_{1\,st}} \tag{6.306}$$

$$= \frac{1 - 39.59\left(\dfrac{\Omega}{\omega*}\right)^2 + 77.14\left(\dfrac{\Omega}{\omega*}\right)^4 - 30.31\left(\dfrac{\Omega}{\omega*}\right)^6}{1 - 176\left(\dfrac{\Omega}{\omega*}\right)^2 + 3480\left(\dfrac{\Omega}{\omega*}\right)^4 - 5456\left(\dfrac{\Omega}{\omega*}\right)^6 + 1940\left(\dfrac{\Omega}{\omega*}\right)^8}.$$

The roots of the numerator provide the **absorption frequencies** $f_k = \nu_k/(2\pi)$ from:

$$\nu_1^2 = 0.0266\omega^{*2}, \qquad \nu_2^2 = 0.6710\omega^{*2}, \qquad \nu_3^2 = 1.8478\omega^{*2} \tag{6.307}$$

where $\omega^{*2} = 48EI/ml^3$ according to (6.134). The roots of the denominator are known from (6.136). The dynamic compliance can thus be written in the form of (6.293):

$$D_{11}(\Omega) = \frac{64l^3}{48EI} \frac{\left(1 - \dfrac{\Omega^2}{\nu_1^2}\right)\left(1 - \dfrac{\Omega^2}{\nu_2^2}\right)\left(1 - \dfrac{\Omega^2}{\nu_3^2}\right)}{\left(1 - \dfrac{\Omega^2}{\omega_1^2}\right)\left(1 - \dfrac{\Omega^2}{\omega_2^2}\right)\left(1 - \dfrac{\Omega^2}{\omega_3^2}\right)\left(1 - \dfrac{\Omega^2}{\omega_4^2}\right)}. \tag{6.308}$$

This representation shows both the positions of the natural frequencies and the positions of the absorption frequencies.

The **absorption frequencies** are the natural frequencies of the frame shown in Fig. 6.26a, which is developed from the original frame by providing the right bearing with a rigid support. This bearing ensures that this point cannot move in the vertical direction. If the original frame (not supported at this point), were excited with a harmonic vertical excitation force at the absorption frequency, this point would remain at its equilibrium position and there would be no resonance. The free frame has a node at this point at the absorption frequency.

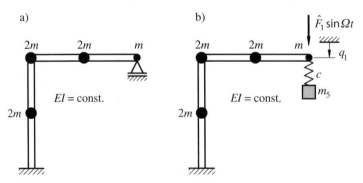

Fig. 6.26 Effect of absorption on the frame; **a)** Frame with additional support, **b)** Frame with vibration absorber

6.5 Forced Undamped Vibrations

To keep the force application point at rest at a given circular frequency of the excitation Ω, a **vibration absorber** in the form of a spring-mass system can be provided as shown in Fig. 6.26b. This additional oscillator causes all natural frequencies of the original frame to shift. The mode shape of the natural frequency that matches the excitation frequency then has a node at the force application point. The condition $\Omega^2 = c/m_5$ must be satisfied for vibration absorption to occur.

This can theoretically be achieved using a large mass (and a stiff spring constant) or with a small mass (and small spring constant). The force application point remains at rest because the dynamic force of the absorber is at any point in time equal in magnitude and opposite in direction to the excitation force. The amplitudes of the absorber mass are therefore inversely proportional to the spring constant (and absorber mass). While damping impairs the perfect absorption effect, it also widens the frequency range in which smaller amplitudes occur.

6.5.4.2 Oscillating Conveyor

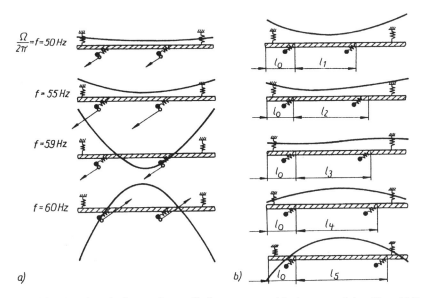

Fig. 6.27 Forced mode shapes of an oscillating conveyor; **a)** In the range of $f = 50 \ldots 60$ Hz (The arrows indicate the directions of motion of the exciter masses and their lengths are proportional to the amplitudes.), **b)** At $f = 50$ Hz and various distances l_i of the vibration exciters

The forced vibrations of an oscillating conveyor shall be analyzed as a second example (Fig. 6.27). The excitation at $f = 50$ Hz is provided by electromagnetic forces that act between the depicted excitation masses and the vibrating trough. Figure 6.27a shows a few calculated mode shapes for the excitation frequency range of $f = 50$ to 60 Hz. One natural frequency is at approximately $f = 59.5$ Hz, which can be seen from the fact that a phase jump occurs between 59 Hz and 60 Hz.

It can be seen from Fig. 6.27 that the forced mode shape in the subcritical range approaches the mode shape of a spring-suspended rigid beam. The amplitudes of the beam increase near resonance. The vibrating trough performs strong bending vibrations so that the conveying process would be impaired because the material being conveyed would accumulate at the nodes. It would therefore be unfavorable to operate the vibrating trough near resonance. The phase jump of the mode shape after passing the resonance frequency, which is typical for all oscillators, is remarkable, see Fig. 6.24.

Since a vibrating conveyor has to be operated at 50 Hz due to the fixed frequency of the electrical power line , the change of the mode shapes when varying the application point of one of the two exciters was investigated. The result of the analysis is shown in Fig. 6.27b. One can see that mounting the exciters at a spacing l_3 allows an even flow of the material to be conveyed while at a spacing of l_1 one can expect accumulation of material in the center of the trough, and at a spacing of l_5 even a counterflow of material in portions of the trough due to a vibration node.

6.5.5 Problems P6.10 to P6.12

P6.10 Equations of Motion in Modal Coordinates

What are the equations of motion in terms of modal coordinates for the system shown in Fig. 6.2 if the two excitation forces F_1 and F_2 act in the directions of the coordinates q_1, q_2? At what magnitudes of F_1 and F_2 is the second mode shape not excited?

P6.11 Excitation by a Force Jump

A vertical force is applied suddenly to the beam end of the system discussed in Problem P6.10, i. e. $F_1 \neq 0$ and $F_2 = 0$. Calculate the time variation of the coordinates q_1 and q_2 over time and estimate the magnitude of the maximum moment at the clamping point by deriving an equation similar to (6.301) for that case.

P6.12 Unbalancing Excitation of a Foundation Block

A rotor with an unbalance $U = me$ rotates at an angular velocity Ω about its vertical axis. The mode shapes of the foundation block are known, see Fig. 3.7 and Sect. 3.2.2.1. Determine the excitation force vector assuming that the centrifugal force remains independent of the vibrations of the foundation block so that the reaction of the foundation vibrations onto it is negligibly small.

Given:
Centrifugal force of the rotor: $F = me\Omega^2$
Horizontal distances of the rotor axis from the vertical axis through the center of gravity: ξ_A, η_A
Distance of the unbalance plane from the horizontal plane through the center of gravity: ζ_A
Reference length l^* (it is introduced so that all generalized coordinates and all load parameters have equal dimensions.)
Coordinate vector, see Fig. 6.28:

6.5 Forced Undamped Vibrations

$$\boldsymbol{q}^{\mathrm{T}} = [q_1,\ q_2,\ q_3,\ q_4,\ q_5,\ q_6] = [x_S,\ y_S,\ z_S,\ l^*\varphi_x,\ l^*\varphi_y,\ l^*\varphi_z]; \quad |\varphi_x|,\ |\varphi_y|,\ |\varphi_z| \ll 1$$

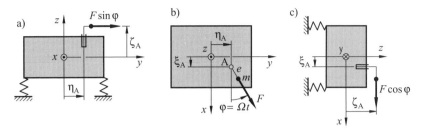

Fig. 6.28 Unbalance excitation on a foundation block

Excitation force vector:

$$\boldsymbol{f}^{\mathrm{T}} = [F_1,\ F_2,\ F_3,\ F_4,\ F_5,\ F_6] = [F_x,\ F_y,\ F_z,\ M_x^S/l^*,\ M_y^S/l^*,\ M_z^S/l^*]$$

Modal matrix:

$$\boldsymbol{V} = [\boldsymbol{v}_1,\ \boldsymbol{v}_2,\ \boldsymbol{v}_3,\ \boldsymbol{v}_4,\ \boldsymbol{v}_5,\ \boldsymbol{v}_6] = \begin{bmatrix} -0.1 & 0.2 & 0 & 0 & 0 & -0.3 \\ 0 & 0 & 0 & 0 & 0.7 & 0 \\ 0 & 0 & 1 & 0 & 0 & 0 \\ 1 & 0 & 0 & 0 & 1 & 0 \\ 0 & 1 & 0 & 0 & 0 & 1 \\ 0 & 0 & 0 & 1 & 0 & 0 \end{bmatrix} \quad (6.309)$$

Find:
1. Excitation forces F_x, F_y, F_z
2. Excitation moments M_x^S, M_y^S, M_z^S
3. Vector $\boldsymbol{h}$ of the modal excitation forces

6.5.6 Solutions S6.10 to S6.12

S6.10 The forces acting in the direction of the coordinates are summarized in the vector $\boldsymbol{f}^{\mathrm{T}} = [F_1,\ F_2]$. The equations of motion (6.268) with the matrices $\boldsymbol{C}$ and $\boldsymbol{M}$ known from (6.32) and (6.34) are therefore:

$$m \begin{bmatrix} 2 & 0 \\ 0 & 6 \end{bmatrix} \begin{bmatrix} \ddot{q}_1 \\ \ddot{q}_2 \end{bmatrix} + \frac{6EI}{7l^3} \begin{bmatrix} 2 & -3 \\ -3 & 8 \end{bmatrix} \begin{bmatrix} q_1 \\ q_2 \end{bmatrix} = \begin{bmatrix} F_1 \\ F_2 \end{bmatrix}. \quad (6.310)$$

The eigenvectors were calculated in S6.6 so that the modal matrix is already known, likewise the modal masses and springs. The decoupled equations of motion in terms of modal coordinates according to (6.269) and (6.270) with the results from S6.5 are:

$$3.364 m \ddot{p}_1 + 0.821 \frac{EI}{l^3} p_1 = F_1 + 0.4768 F_2 = h_1 \quad (6.311)$$

$$4.932 m \ddot{p}_2 + 8.660 \frac{EI}{l^3} p_2 = F_1 - 0.6991 F_2 = h_2. \quad (6.312)$$

If the modal excitation force is $h_2 = 0$, only the first mode shape is excited, i.e. when $F_1 = 0.6991 F_2$.

S6.11 The solution of the equations of motion (6.269) in terms of modal coordinates according to (6.295) is

$$p_1 = \frac{h_1}{\gamma_1}(1 - \cos\omega_1 t), \qquad p_2 = \frac{h_2}{\gamma_2}(1 - \cos\omega_2 t). \tag{6.313}$$

The position coordinates, according to (6.108), have the values

$$q_1 = \left[\frac{v_{11}}{\gamma_1}(1 - \cos\omega_1 t) + \frac{v_{12}}{\gamma_2}(1 - \cos\omega_2 t)\right] F_1$$

$$q_2 = \left[\frac{v_{21}}{\gamma_1}(1 - \cos\omega_1 t) + \frac{v_{22}}{\gamma_2}(1 - \cos\omega_2 t)\right] F_1. \tag{6.314}$$

With the numerical values from S6.5, the results for the related parameters are

$$\frac{EIq_1}{l^3 F_1} = 1.2179(1 - \cos\omega_1 t) + 0.1155(1 - \cos\omega_2 t)$$

$$\frac{EIq_2}{l^3 F_1} = 0.5807(1 - \cos\omega_1 t) - 0.0807(1 - \cos\omega_2 t) \tag{6.315}$$

In the static case, the deformations for $\boldsymbol{q}_{\text{st}} = \boldsymbol{D}\boldsymbol{f}_{\text{st}}$ would be:

$$\boldsymbol{q}_{\text{st}} = \frac{l^3}{6EI}\begin{bmatrix} 8 & 3 \\ 3 & 2 \end{bmatrix}\begin{bmatrix} 1 \\ 0 \end{bmatrix} F_1 = \frac{1}{6}\begin{bmatrix} 8 \\ 3 \end{bmatrix}\frac{F_1 l^3}{EI} = \begin{bmatrix} 1.333 \\ 0.5 \end{bmatrix}\frac{F_1 l^3}{EI}. \tag{6.316}$$

They are obtained in the special case that no vibrations take place: $\cos\omega_1 t$ and $\cos\omega_2 t$ do not occur then and should be set to zero. The maximum dynamic deflections then are within the limits

$$2 \cdot 1.2179 = 2.4358 \leq \left(\frac{EIq_1}{l^3 F_1}\right)_{\max} \leq 2 \cdot (1.2179 + 0.1155) = 2.6668$$

$$2 \cdot 0.5807 = 1.1614 \leq \left(\frac{EIq_2}{l^3 F_1}\right)_{\max} \leq 2 \cdot (0.5807 + 0.0807) = 1.3228. \tag{6.317}$$

The fixed-end moment results from the equilibrium of moments, see Fig. 6.2:

$$M = F_1 l + 2ml\ddot{q}_1 + 6ml\ddot{q}_2. \tag{6.318}$$

The accelerations in nondimensional form are:

$$\frac{m\ddot{q}_1}{F_1} = \frac{mv_{11}}{\gamma_1}\omega_1^2\cos\omega_1 t + \frac{mv_{12}}{\gamma_2}\omega_2^2\cos\omega_2 t$$

$$= 0.2972\cos\omega_1 t + 0.2028\cos\omega_2 t$$

$$\frac{m\ddot{q}_2}{F_1} = \frac{mv_{21}}{\gamma_1}\omega_1^2\cos\omega_1 t + \frac{mv_{22}}{\gamma_2}\omega_2^2\cos\omega_2 t$$

$$= 0.1417\cos\omega_1 t - 0.1417\cos\omega_2 t$$

$$\tag{6.319}$$

The dynamic fixed-end moment is thus

6.5 Forced Undamped Vibrations

$$M = F_1 l \left[1 + (2v_{11} + 6v_{21}) \frac{m\omega_1^2}{\gamma_1} \cos\omega_1 t + (2v_{12} + 6v_{22}) \frac{m\omega_2^2}{\gamma_2} \cos\omega_2 t \right]$$
$$= F_1 l (1 + 1.444 \cos\omega_1 t + 0.444 \cos\omega_2 t) \tag{6.320}$$

The moment is "almost periodic", and its extreme values can be up to:

$$M_{\min} = -0.888 F_1 l, \qquad M_{\max} = 2.888 F_1 l \tag{6.321}$$

As a result of the vibrations, the moment may temporarily act in the opposite direction to the static moment, and the static value is considerably exceeded ($2.888 > 1$).

These typical dynamic effects cannot simply be captured by a "dynamic coefficient". This would have to be told to certain "static calculation advocates" who absurdly hold the opinion that one only needs to multiply the result of a static calculation by a factor to obtain the loads for the dynamic case.

S6.12 The excitation force is rotating, so it has two components in the horizontal direction. The moments result from the product of the load components and the respective leverages, see Fig. 6.28. Note the positive coordinate directions and distances. In all, the excitation vector, with respect to the coordinates mentioned, is

$$\boldsymbol{f}(t) = \begin{bmatrix} F_x \\ F_y \\ F_z \\ M_x^S/l^* \\ M_y^S/l^* \\ M_z^S/l^* \end{bmatrix} = \begin{bmatrix} F\cos\Omega t \\ 0 \\ 0 \\ 0 \\ (\zeta_A/l^*)F\cos\Omega t \\ -(\eta_A/l^*)F\cos\Omega t \end{bmatrix} + \begin{bmatrix} 0 \\ F\sin\Omega t \\ 0 \\ -(\zeta_A/l^*)F\sin\Omega t \\ 0 \\ (\xi_A/l^*)F\sin\Omega t \end{bmatrix} \tag{6.322}$$

Using the given modal matrix $\boldsymbol{V}$, the modal excitation forces from (6.270) are

$$\boldsymbol{h} = \boldsymbol{V}^T \boldsymbol{f}(t) = \begin{bmatrix} h_1 \\ h_2 \\ h_3 \\ h_4 \\ h_5 \\ h_6 \end{bmatrix} = F \left\{ \begin{bmatrix} -0.1 \\ 0.2+\zeta_A/l^* \\ 0 \\ -\eta_A/l^* \\ 0 \\ -0.3+\zeta_A/l^* \end{bmatrix} \cos\Omega t + \begin{bmatrix} -\zeta_A/l^* \\ 0 \\ 0 \\ \xi_A/l^* \\ 0.7-\zeta_A/l^* \\ 0 \end{bmatrix} \sin\Omega t \right\}.$$
$$\tag{6.323}$$

One can see from this that a single unbalance results in an excitation vector with five components (it excites five mode shapes) and that no modal excitation of the vertical mode shape v_3 occurs ($h_3 = 0$). When the excitation frequency coincides with the third natural frequency, no resonance amplitudes have to be feared ("pseudo resonance"). Five resonance peaks can be expected in the amplitude response. The modal forces all have the dimension of a force.

The final results for the real motion and load parameters are independent of the reference length l^* and result in the correct dimensions. This absolute magnitude of the coordinates can only be calculated in conjunction with the mass and stiffness matrices, see Sect. 6.5.

6.6 Damped Vibrations

6.6.1 Determination of Damping

Vibrations of machines are always damped because resisting forces (dissipative forces) counteract the motion so that a loss of mechanical energy occurs. A portion of the mechanical energy is converted into heat (so-called energy dissipation, damping) as stated by the second law of thermodynamics.

It is less a matter of calculating the exact development over time of the vibrations than of the decay behavior of the amplitudes. The exact function of the damping forces within a period of oscillation is generally not of significance. Damping forces are assumed to be proportional to the velocity to simplify the mathematical treatment, even though they are practically never exactly proportional to the velocity. This is also called viscous damping. The great benefits of this approach are

1. that one obtains linear differential equations that can be treated easily,
2. that this approach expresses a loss of mechanical energy during the vibrations (warming), and
3. that one can make do with a few parameter values only, see Sect. 1.4.

Components with defined damping properties are intentionally used in various branches of mechanical engineering, e. g. sleeve springs for textile mandrils (Fig. 6.32c), hydraulic torsional vibration dampers in marine diesel engines (Fig. 4.45), frictional dampers on crankshafts (Fig. 4.42), e. t. c. Rubber springs, rubber couplings, rubber tires, cables, V belts, leaf and disk springs are often used because they have good damping properties. Concrete has in some cases taken the place of cast metal structures for machine tool frames due to its high damping characteristics. If one starts from discrete **damping elements**, the **damping forces** referred to the coordinates q are obtained in general form as

$$\boldsymbol{f}_\mathrm{d} = \boldsymbol{B}\dot{\boldsymbol{q}} \tag{6.324}$$

All damping coefficients can be arranged in a damping matrix $\boldsymbol{B} = [b_{lk}]$. The coefficients b_{lk} result from the damping constants of the individual dampers.

If the damping parameters are not known, it is advantageous to account for damping using a global approximation approach. This is often selected in such a way that it allows the transformation to principal coordinates, which cannot be achieved for arbitrary damping matrices $\boldsymbol{B}$.

It is advisable to calculate forced vibrations only in conjunction with damping values, for otherwise one will not get usable amplitude values in the vicinity of resonance. Since damping constants of real machines are often unavailable, it is common to introduce modal damping values.

It can be shown mathematically that modal damping occurs in a damped vibration system if the damping matrix $\boldsymbol{B}$ satisfies the condition

6.6 Damped Vibrations

$$CM^{-1}B = BM^{-1}C \quad \text{or} \quad B = M \sum_{k=1}^{K} a_k (M^{-1}C)^{k-1}. \quad (6.325)$$

It is most often assumed that the damping matrix only shows a linear dependence on the mass and/or stiffness matrix ($k = 2$):

$$B = a_1 M + a_2 C. \quad (6.326)$$

This special case of the approach according to (6.325), which is called **Rayleigh damping**, is used in many large program systems to account for damping. It has been found in experiments that material damping is better accounted for if it is assumed to be inversely proportional to the circular frequency of the excitation.

6.6.2 Free Damped Vibrations

The equations of motion of free damped vibrations can be written as follows in matrix notation

$$M\ddot{q} + B\dot{q} + Cq = o. \quad (6.327)$$

If the damping matrix matches (6.325) or (6.326), the equations of motion can be decoupled using the eigenvectors of the associated undamped system. This decoupling is performed, as for the undamped system, using the modal matrix of the undamped system V known from (6.88) and results in

$$\ddot{p}_i + 2\delta_i \dot{p}_i + \omega_{0i}^2 p_i = 0, \quad (i = 1, 2, \ldots, n). \quad (6.328)$$

ω_{0i} is the ith natural circular frequency of the *undamped system*. It can be calculated according to (6.117). The **modal decay rate** δ_i is found using (6.326):

$$2D_i \omega_{0i} = 2\delta_i = \frac{v_i^T B v_i}{v_i^T M v_i} = a_1 + \omega_{0i}^2 a_2. \quad (6.329)$$

v_i is the mode shape of the undamped system associated with ω_{i0}. The natural circular frequencies of the damped system are

$$\omega_i = \sqrt{\omega_{0i}^2 - \delta_i^2} = \omega_{0i} \sqrt{1 - D_i^2}, \quad D_i = \frac{\delta_i}{\omega_{0i}} \quad (6.330)$$

It is assumed in the following considerations that the vibration system is not "pathological", which would require special treatment. It is assumed that no unconstrained rigid-body motions are possible (so all natural circular frequencies $\omega_{0i} > 0$) and that no creeping motions occur as a result of overdamping (so all $\delta_i < \omega_{0i}$). If relative coordinates are used, one can always meet the condition mentioned first. If the second condition is violated (e. g. if too many degrees of freedom are taken into account for modeling), not only conjugated complex eigenvalues and mode shapes but also real ones occur.

D_i is the modal damping ratio of the ith principal vibration. It can be estimated using (6.329) if the matrices M, B, C and an approximation for the eigenvector v_i of the undamped system are known.

In vibration systems with little damping ($D_i \ll 1$), the lowest natural circular frequencies of the damped system (ω_i) do not differ much from those of the undamped system (ω_{0i}). The differences can be considerable at higher natural frequencies because the damping ratios D_i are often no longer small. Very high orders of the mode shapes frequently do not form at all since the damping ratios are then supercritical ($D_i > 1$).

In experiments, one can often determine the associated D_i from the free response of a mode shape via the logarithmic decrement, see Sect. 1.4. In this way, the required data and, if required, the elements of a damping matrix B can be determined using (6.326).

Complete decoupling occurs if the damping matrix B transformed with the modal matrix V is a pure diagonal matrix, i.e. in the case of approaches according to (6.325) or in rare special cases only. The matrix $V^T BV$ is generally fully occupied. If the diagonal elements of the matrix $V^T BV$ are considerably larger than the other elements of this matrix, neglecting the off-diagonal elements still represents an approximation for the actual damping behavior of the system.

Analogously to free undamped vibrations, the motions of the modally damped system can be calculated for given initial conditions (6.118) or (6.125), respectively. The solution of (6.325) for the initial conditions of (6.125) is derived in accordance with (1.94):

$$p_i = \frac{e^{-D_i \omega_{0i} t}}{\sqrt{1 - D_i^2}} \left[\sqrt{1 - D_i^2} p_{i0} \cos \omega_i t + \left(\frac{\dot{p}_{i0}}{\omega_{0i}} + D_i p_{i0} \right) \sin \omega_i t \right]. \qquad (6.331)$$

The decay behavior of free vibrations is of practical interest since it is relevant for the number of cycles, e.g. when calculating the endurance limit or structural durability.

It is relevant for periodic impacts whether the excited vibrations decay as a result of damping before the next impact occurs. If there is no cancellation until the next impact, there can be resonance-like stimulations for periodic impacts. Results for typical cases of periodic impacts are contained in [19].

The assumption that the matrix B is symmetrical and real does not apply to some models. Taking into account gyroscopic forces as they occur on rotating shafts due to the gyroscopic effect results in an antisymmetrical matrix B, see Sect. 2.4. However, a purely antisymmetrical matrix B does not cause damping.

Terms that are proportional to the velocity also occur in the equations of motion if the mechanical structures are influenced by controllers such as in magnetic levitation systems, rotors and active absorbers, see [2], [18], [23].

6.6 Damped Vibrations

6.6.3 Harmonic Excitation

The equation of motion for forced damped vibrations is

$$M\ddot{q} + B\dot{q} + Cq = f(t). \tag{6.332}$$

It represents an extension of (6.268) by the damping forces or of (6.327) by the excitation forces **f**(t). In the case of a kinematic excitation, one can transform the applicable equations of motion into the same form.

Only the steady state of **harmonic excitation** shall be analyzed below for which a good match of calculated and measured results has been confirmed for the low mode shapes. A **steady state** occurs if the vibrations excited by the initial conditions have completely decayed. The decay time depends on the decay rates δ_i and amounts to about $t^* > \pi/\delta_{\min}$. The steady-state solution of (6.332) is then described by the particular solution.

The harmonic excitation forces $F_k(t)$, which act onto the vibration system in the direction of the coordinates q_k and were introduced in (6.273), are captured by their amplitudes $\hat{F}_k$ and phase angles β_k or by their cosine and sine terms:

$$\begin{aligned} F_k(t) &= \hat{F}_k \sin(\Omega t + \beta_k) = \hat{F}_k (\sin \beta_k \cos \Omega t + \cos \beta_k \sin \Omega t) \\ &= F_{ak} \cos \Omega t + F_{bk} \sin \Omega t; \qquad k = 1, 2, \ldots, n. \end{aligned} \tag{6.333}$$

They all contain the same circular frequency of the excitation Ω. One can therefore write (6.332) with the force vectors

$$\boldsymbol{f}_a = [F_{a1}, F_{a2}, \ldots, F_{an}]^T; \qquad \boldsymbol{f}_b = [F_{b1}, F_{b2}, \ldots, F_{bn}]^T \tag{6.334}$$

during a harmonic excitation with the circular frequency Ω in the form

$$M\ddot{q} + B\dot{q} + Cq = f(t) = \boldsymbol{f}_a \cos \Omega t + \boldsymbol{f}_b \sin \Omega t. \tag{6.335}$$

The algebra becomes shorter and more elegant if complex numbers are introduced. Nothing changes in the physical relations when formally adding the imaginary equation

$$\mathrm{j}(M\ddot{q}^* + B\dot{q}^* + Cq^*) = \mathrm{j}f^*(t) = \mathrm{j}(\boldsymbol{f}_a \sin \Omega t - \boldsymbol{f}_b \cos \Omega t) \tag{6.336}$$

to (6.335). The complex excitation forces are denoted by a tilde:

$$\tilde{\boldsymbol{f}}(t) = \boldsymbol{f}(t) + \mathrm{j}\boldsymbol{f}^*(t) = \hat{\tilde{\boldsymbol{f}}} \exp(\mathrm{j}\Omega t) = (\boldsymbol{f}_a - \mathrm{j}\boldsymbol{f}_b) \exp(\mathrm{j}\Omega t). \tag{6.337}$$

The real part of the complex force vector $\tilde{\boldsymbol{f}}$ corresponds to the right side of (6.335). Summation of (6.335) and (6.336) together with Euler's relation $\exp(\mathrm{j}\Omega t) = \cos \Omega t + \mathrm{j} \sin \Omega t$ yields the equation of motion for the **complex coordinates** $\tilde{q} = q + \mathrm{j}q^*$ due to **complex excitation forces**:

$$M\ddot{\tilde{q}} + B\dot{\tilde{q}} + C\tilde{q} = \hat{\tilde{f}}\exp(\mathrm{j}\Omega t). \tag{6.338}$$

Material damping is sometimes modeled by a complex damping. A complex equation of motion emerges, see Sect. 1.4:

$$M\ddot{\tilde{q}} + (C + \mathrm{j}B^*)\tilde{q} = \hat{\tilde{f}}\exp(\mathrm{j}\Omega t). \tag{6.339}$$

For linear oscillators, the forced steady-state vibrations are at the excitation frequency so that the solution is sought in real form

$$q = q_\mathrm{a}\cos\Omega t + q_\mathrm{b}\sin\Omega t \tag{6.340}$$

and thus in complex form, as in (6.337), using the expression

$$\tilde{q} = q + \mathrm{j}q^* = \hat{\tilde{q}}\exp(\mathrm{j}\Omega t) = (q_\mathrm{a} - \mathrm{j}q_\mathrm{b})\exp(\mathrm{j}\Omega t). \tag{6.341}$$

Insertion of this expression into (6.338) and (6.339) provides the following system of linear equations for calculating the complex amplitude vector after a comparison of coefficients in $\exp(\mathrm{j}\Omega t)$:

$$[-\Omega^2 M + \mathrm{j}\Omega B + C]\hat{\tilde{q}} = \hat{\tilde{f}} \tag{6.342}$$

or

$$[-\Omega^2 M + C + \mathrm{j}B^*]\hat{\tilde{q}} = \hat{\tilde{f}}, \tag{6.343}$$

respectively. Note that the same type of equation results for both damping models. The damping matrices of viscous damping and complex damping (for excitation at the circular frequency Ω) satisfy the following relation:

$$B = \frac{B^*}{\Omega}. \tag{6.344}$$

The solution of (6.342) is the complex amplitude

$$\hat{\tilde{q}} = [-\Omega^2 M + \mathrm{j}\Omega B + C]^{-1}\hat{\tilde{f}} = H(\mathrm{j}\Omega)\hat{\tilde{f}}. \tag{6.345}$$

It follows due to $\hat{\tilde{q}} = q_a - \mathrm{j}q_b$, see (6.341):

$$q_a = \mathrm{Re}(\hat{\tilde{q}}); \qquad q_b = -\mathrm{Im}(\hat{\tilde{q}}). \tag{6.346}$$

The steady-state solution according to (6.340) is determined as a function of the circular frequency of the excitation Ω.

(6.345) defines the **matrix of the complex frequency responses**:

$$H(\mathrm{j}\Omega) = [-\Omega^2 M + \mathrm{j}\Omega B + C]^{-1} = [H_{kl}(\mathrm{j}\Omega)] = [H_{lk}(\mathrm{j}\Omega)]. \tag{6.347}$$

The matrix of complex frequency responses is also symmetrical for symmetrical matrices C, B and M. Its diagonal elements H_{ll} are called direct frequency responses

6.6 Damped Vibrations

and the off-diagonal elements $H_{lk}(l \neq k)$ are called cross-frequency responses. The significance of the complex frequency response is that it provides insights into the dynamic behavior in the frequency range for the system to be examined. The complex frequency response plays a central role in signal analysis and is often used for comparing calculated and measured results, especially for tasks involving the identification of real systems.

Each individual complex frequency response $H_{lk}(j\Omega)$ characterizes the linear vibration system with respect to its amplitude and phase at point k as a result of a harmonic excitation at point l. For $l = k$, $H_{ll}(j\Omega)$ is the dynamic compliance that was calculated in (6.288) for an undamped system.

An illustrative interpretation can be achieved by dividing the complex frequency response $H_{lk}(j\Omega)$ into the amplitude frequency response $D_{lk}(\Omega) = D_{kl}(\Omega)$ and the phase frequency response $\psi_{lk}(\Omega)$ using the representation

$$H_{lk}(j\Omega) = D_{lk}(\Omega) e^{j\psi_{lk}(\Omega)} = \text{Re}(H_{lk}) + j\text{Im}(H_{lk}). \tag{6.348}$$

Note, for example, Figs. 3.5c, 3.11, 3.14, 4.24, 4.41, 4.46, 6.24, 6.30 and 6.34. With the real (Re) and imaginary (Im) parts of this complex function, the following is valid:

$$D_{lk} = \sqrt{\text{Re}^2(H_{lk}) + \text{Im}^2(H_{lk})} \tag{6.349}$$

$$\sin\psi_{lk} = \frac{\text{Im}(H_{lk})}{D_{lk}}; \quad \cos\psi_{lk} = \frac{\text{Re}(H_{lk})}{D_{lk}}. \tag{6.350}$$

The angle ψ_{lk} and its quadrant can be uniquely determined from the sine and cosine according to (6.350).

In the limiting case $\Omega \to 0$, the **dynamic influencing coefficient** ($D_{lk} = D_{kl}$) matches the static influence coefficient ($d_{lk} = d_{kl}$), see (6.4).

Therefore, the following applies to the complex amplitude of the kth coordinate:

$$\hat{q}_k = q_{ak} - jq_{bk} = \sum_{l=1}^{n} H_{kl}(j\Omega)\hat{f}_l = \sum_{l=1}^{n} D_{kl}(\Omega)\exp[j\psi_{kl}(\Omega)](F_{al} - jF_{bl}) \tag{6.351}$$

and the sine and cosine parts in real form result as

$$q_{ak} = \sum_{l=1}^{n} D_{kl}(F_{al}\cos\psi_{kl} + F_{bl}\sin\psi_{kl}) = \sum_{l=1}^{n} D_{kl}\hat{F}_l \sin(\beta_l + \psi_{kl})$$
$$q_{bk} = \sum_{l=1}^{n} D_{kl}(F_{bl}\cos\psi_{kl} - F_{al}\sin\psi_{kl}) = \sum_{l=1}^{n} D_{kl}\hat{F}_l \cos(\beta_l + \psi_{kl}) \tag{6.352}$$

and finally using (6.340)

$$q_k = q_{ak} \cos \Omega t + q_{bk} \sin \Omega t \qquad (6.353)$$

$$= \sum_{l=1}^{n} D_{kl} \hat{F}_l \sin(\Omega t + \beta_l + \psi_{kl}) = \hat{q}_k \sin(\Omega t + \psi_k)$$

$$\hat{q}_k = \sqrt{q_{ak}^2 + q_{bk}^2}; \qquad \cos \psi_k = \frac{q_{bk}}{\hat{q}_k}; \qquad \sin \psi_k = \frac{q_{ak}}{\hat{q}_k}.$$

Amplitude $\hat{q}_k$ and phase angle ψ_k both depend on the circular frequency of the excitation Ω. Note that the phase angles differ for all coordinates. This means that each coordinate reaches its extreme position at a different point in time and the mode shapes known from the undamped vibrations do not occur.

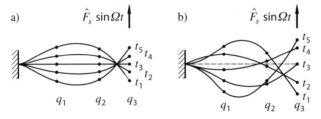

Fig. 6.29 Forced mode shapes of a beam ($s = 3$) at times $t_k = (k - 3)T/4$ for $k = 1$ to 5; **a)** Undamped, **b)** Damped

If only a single excitation force $F_s(t) = \hat{F}_s \sin(\Omega t + \beta_s)$ acts on the oscillator at coordinate $l = s$, one obtains from (6.353) for all coordinates

$$q_k = D_{ks} \hat{F}_s \sin(\Omega t + \beta_s + \psi_{ks}), \quad \hat{q}_k = D_{ks} \hat{F}_s; \quad k = 1, 2, \ldots, n. \qquad (6.354)$$

There is the same formal relationship between the force and displacement amplitudes as in (6.289). The displacement amplitude is proportional to the force amplitude. This is a consequence of the linearity of the vibration system. Note, however, the difference to (6.288). The dynamic compliances do not become zero for the damped oscillator at any excitation frequency, i. e., the amplitudes have **finite values at the resonance points as well**. The forced mode shapes, which result from the totality of the coordinates q_k, have **no fixed nodes** for the damped oscillator, unlike the undamped oscillator. One can recognize that since all phase angles are the same for the undamped oscillator, see (6.288) and Fig. 6.24, but they are of different magnitude for the damped oscillator.

Figure 6.29 illustrates this fact using a simple example. It shows the synchronously changing deflections for an undamped bending oscillator as compared to a damped bending oscillator during half a period of the forced oscillation ($0 < t < \pi/\Omega = T/2$).

The graphic representation of a complex frequency response $H_{lk}(j\Omega)$ provides an planar curve called **locus**, see Fig. 6.30c. It contains important information about the behavior of an oscillator.

6.6 Damped Vibrations

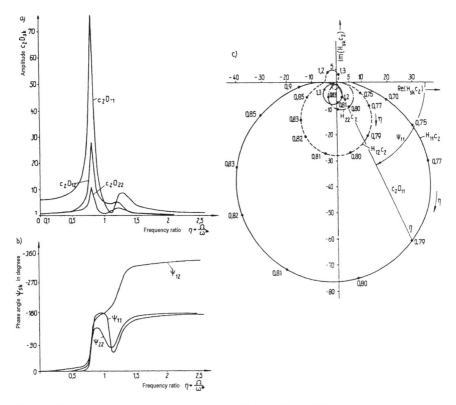

Fig. 6.30 Frequency responses of the two-mass oscillator of Fig. 4.40 for $\mu = 0.2$, $\zeta = 1$, $D = 0.1$; **a)** Amplitude frequency response D_{lk}, **b)** Phase frequency response ψ_{lk}, **c)** Loci of the complex frequency responses H_{lk}

Loci are commonly used in aircraft design, machine design, rotor dynamics and other fields. Loci are obtained by plotting in the complex plane $D_{lk}(\Omega)$ as radius and $\psi_{lk}(\Omega)$ as angle in polar coordinates. It is often sufficient to plot the amplitude frequency response $D_{lk}(\Omega)$. Unlike for the undamped system, it has no singular points at resonance in the damped vibration system. The number of its maxima at most equals n and is often much smaller than n.

Unlike for the undamped system (6.290), the elements of the **dynamic compliance matrix** $D(\Omega)$ for **modal damping** are calculated as

$$D_{sk}(\Omega) = \sqrt{\left(\sum_{i=1}^{n} \frac{2v_{ki}v_{si}D_i\eta_i V_{1i}^2}{\gamma_i}\right)^2 + \left(\sum_{i=1}^{n} \frac{v_{ki}v_{si}(1-\eta_i^2)V_{1i}^2}{\gamma_i}\right)^2} \quad (6.355)$$

with the nondimensionalized amplitude functions ($\eta_i = \Omega/\omega_{i0}$)

$$V_{1i} = \frac{1}{\sqrt{(1-\eta_i^2)^2 + 4D_i^2\eta_i^2}}. \qquad (6.356)$$

If the excitation frequency coincides with the ith natural frequency of the undamped system ($\eta_i = 1$), the approximate maximum value of the amplitude frequency response, the so-called **residuum**, is obtained:

$$D_{sk\,\max} = \frac{|v_{ki} v_{si}|}{2D_i \gamma_i}. \qquad (6.357)$$

6.6.4 Periodic Excitation

Periodic excitation is an important loading case in mechanical engineering practice. All machines that work in cycles (cycle time $T_0 = 2\pi/\Omega$) cause periodic bearing and joint forces in steady-state operation, which excite vibrations both in the subassemblies ("in the interior of the machine") and the frame. Periodic inertia forces occur frequently in textile and packaging machines that include cam mechanisms and linkages while the technological processes in presses, forming and cutting machines cause major periodic forces, see Figs. 2.20, 2.24, 2.32, and 4.50.

(6.332) formally corresponds to the equations of motion for the periodic excitation. It is disputed not only for arbitrary excitation force curves, but even for a periodic excitation (if a Fourier series is on the right side) whether the damping forces are captured in physically correct form above. There are hardly any experimental results for nonharmonic excitation regarding the vibration behavior of real objects or only such results that conflict with the calculated values using the approach (6.324). The damping for excitations that are nonharmonic must be described by nonlinear functions if one wants calculated and experimental results to match.

Instead of (6.333), this calculation does not only involve harmonic, but also **periodic excitation forces**:

$$\begin{aligned} F_k(t) = F_k(t+T_0) &= \sum_{m=1}^{\infty} \hat{F}_{km} \sin(m\Omega t + \beta_{km}) \\ &= \sum_{m=1}^{\infty} (F_{akm} \cos m\Omega t + F_{bkm} \sin m\Omega t), \end{aligned} \qquad (6.358)$$

where the Fourier coefficients F_{akm} and F_{bkm} are known, see for example Figs. 1.25, 2.37 and 3.5. Since the letter k is used for the index of the coordinate, *the order of the harmonic is denoted as* m.

The equations for multiple excitation forces are found by summation of all possible other force application points. Here, the dynamic behavior for a single periodic force that acts in the direction of the coordinate q_s is examined:

6.6 Damped Vibrations

$$F_s(t) = \sum_{m=1}^{\infty} \hat{F}_{sm} \sin(m\Omega t + \beta_{sm}) \tag{6.359}$$

Due to the superposition principle, the result known from (6.354) may be used to obtain the mth harmonic of the periodic response at point q_k:

$$q_k(t) = \sum_{m=1}^{\infty} \hat{q}_{km} \sin(m\Omega t + \psi_{km}) \tag{6.360}$$

The following applies for each mth harmonic as in (6.354)

$$\hat{q}_{km} = D_{ks}(m\Omega) \hat{F}_{sm} \tag{6.361}$$

This product of the dynamic compliance and the amplitude spectrum of the excitation force yields the amplitude spectrum of the coordinate q_k.

The amplitude spectrum of the displacement is clearly changed compared with the excitation spectrum due to the dynamic compliance. This also results in a periodic function but it does not resemble that of the excitation force. It is also different for different speeds. Such calculation or measurement results are sometimes amazing, see the result in Fig. 3.5a and Fig. 3.5b. They can be explained by the described physical relationships.

The maximum amplitudes $|q_k|_{\max}$ in the steady-state operating speed range are frequently of interest. They can be determined by numerical sampling of the function $q_k(t)$ within a kinematic period and plotted over the fundamental circular frequency of the excitation Ω. The resonance curve obtained in this way is not just the sum of the amplitude frequency responses of the individual harmonics, since the phase angles are of importance in a superposition.

The individual coordinates of the vibration system according to (6.360) result in the following for the special case of modal damping:

$$q_k(t) = \sum_{i=1}^{n} \frac{v_{ki}}{\gamma_i} \sum_{m=1}^{\infty} \frac{v_{si}\hat{F}_{sm} \sin(m\Omega t + \varphi_{im})}{\sqrt{(1-m^2\eta_i^2)^2 + 4D_i^2 m^2 \eta_i^2}} \tag{6.362}$$

This sum is often determined by a single summand in the case of resonance, see (6.278),

$$m\eta_i = 1 \quad \text{or} \quad m\Omega = \omega_{0i}, \tag{6.363}$$

(the *resonance of mth order*). The motion is approximately proportional to the ith mode shape:

$$q_k(t) \approx \frac{v_{ki} v_{si} \hat{F}_{sm} \sin(m\Omega t + \alpha_{im})}{2\gamma_i D_i} \sim v_{ki} \sin(\omega_{0i} t + \psi_{km}). \tag{6.364}$$

(6.364) can also be used for the case of higher-order resonance to estimate velocities, accelerations, forces, moments and other mechanical parameters. Note that the phase angles for coordinates q_k differ from each other, see (6.362). The forced mode shapes do not have fixed nodes for damped oscillators, even in the case of resonance, see also Fig. 6.29b.

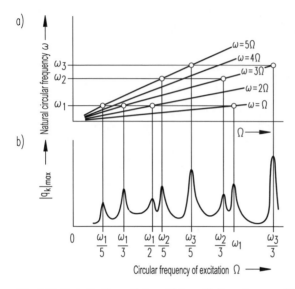

Fig. 6.31 Periodically excited oscillator with three degrees of freedom; **a)** Campbell diagram, **b)** Resonance curve

A typical resonance curve is shown in Fig. 6.31b. It is not identical with an amplitude frequency response, since it represents the **amount of the maximum value** of a coordinate q_k that results from multiple harmonics that do not all reach their maximum at the same time, see (6.362). There are, however, resonance peaks at points that satisfy the condition (6.363), that is, at integral fractions of natural frequencies (i.e. if $\Omega = w_i/m$). Compare in this respect Figs. 3.5, 4.22, 4.23, 4.31, and 4.43.

Three natural frequencies and five harmonics are included in the Campbell diagram, Fig. 6.31a. The intersections of the family of excitation frequencies with the straight lines of the natural frequencies characterize potential higher-order resonance points. A comparison of Fig. 6.31a and Fig. 6.31b indicates that resonance peaks do not emerge at all intersections. In the example at hand, the fourth excitation harmonic was zero so that pseudo-resonance occurs instead at the intersections with the respective straight line.

During the start-up of a machine, many resonances are accordingly far below the first natural frequency. One can reduce these findings to the classic statement that "resonance occurs when the excitation frequency equals the natural frequency" if one considers all excitation frequencies that are contained in the periodic excitation (and all natural frequencies).

The resonance curves of damped systems always remain finite. The resonance curve differs only slightly from that of the undamped system in areas outside resonance, if the damping ratios D_i are small. Damping frequently becomes supercrit-

6.6 Damped Vibrations

ical at higher natural frequencies ($D_i > 1$) so that no resonance peaks form, see e.g. Figure 6.30. Amplitude frequency responses can be calculated for each harmonic of $q_k(t)$ and recorded as functions of Ω as so-called cascade diagram.

6.6.5 Examples

6.6.5.1 Textile Mandril

Textile mandrils work at high speeds and belong to the machine sub-assemblies that cannot be designed or improved without an exact dynamic analysis. Figure 6.32 shows the design drawing of a textile mandril and the corresponding calculation model.

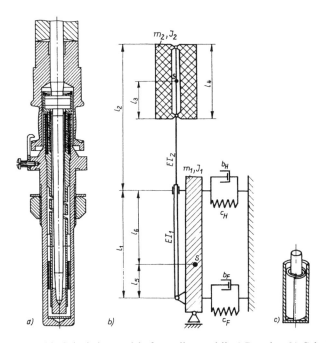

Fig. 6.32 Calculation model of a textile mandril; **a)** Drawing, **b)** Calculation model, **c)** Sleeve spring damper element of the mandril bearing

Note that hydraulic dampers with a damping spiral (sleeve spring) as shown in Fig. 6.32c are used at both bearings. Figure 6.33a shows the calculated amplitude frequency response of the footstep bearing force, Fig. 6.33b that of the displacement of the mandril tip as a result of the unbalance-excited vibrations. The damping constant b_F of the footstep bearing was varied to find out at which values the amplitude maxima to be passed through during start-up and coast-down remain as small as possible.

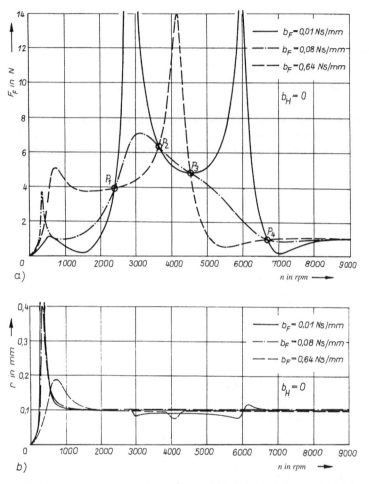

Fig. 6.33 Resonance curves of the textile mandril of Fig. 6.32 as a result of unbalancing excitation; **a)** Bearing force amplitude, **b)** Amplitude of the mandril tip

Figure 6.33 shows how much the resonance amplitudes are determined by the damping of the footstep bearing. Note that there is no proportionality between the amplitudes of the mandril tip and the bearing force. One can therefore not draw any conclusions regarding the bearing forces from the motion of the mandril tip. The three resonance peaks first decline with increasing damping. The family of curves has four damping-independent fixed points P_1 to P_4. The maximum deflection cannot drop below the limit determined by them. The optimum is in the vicinity of $b_F = 0.3 \text{ N} \cdot \text{s/mm}$. If damping is strong, the natural frequencies shift significantly and remarkably the amplitudes rise again. The strong damping makes the bearing

6.6 Damped Vibrations

so inflexible that a degree of freedom of the oscillator is lost and only two instead of three resonance points remain. If the designer were guided by static considerations, he or she could think of limiting the mandril motion by a strong damping. This could worsen the situation since such stiffening would result in a change of the natural frequencies.

6.6.5.2 Belt Drive

Pretensioned belt drives are used to transmit moments. The elasticity of the pretensioning device, together with stiffness differences of the driving and slack strands, causes coupled translational and rotational vibrations.

Under the condition of small vibrations, it should be checked for a pretensioned V-belt drive according to Fig. 6.34 with an elastic tensioning device to what extent the dynamic strand forces influence the pretensioning required for moment transfer as a result of rotational and translational vibrations caused by the residual unbalance $U = me$ of the motor armature. The equations of motion and the equations for calculating the two strand forces and the motor torque shall be specified for the case of a constant speed.

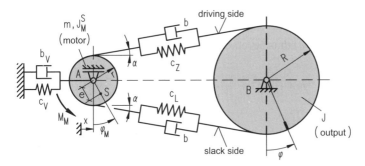

Fig. 6.34 Calculation model of a belt drive with an elastic tensioning device

The belt mass is negligibly small. It is assumed that the change of angle α, as a result of translational vibrations, as well as slippage are negligible. Weak damping can be presupposed so that modal damping is sufficient for calculating the vibration amplitudes. Differences in stiffness between the driving and slack strands result from the nonlinear material behavior.

The deformations of the driving and slack strands depend on three coordinates. The following applies:

$$\Delta l_Z = r\varphi_M - R\varphi - x\cos\alpha; \qquad \Delta l_L = R\varphi - r\varphi_M - x\cos\alpha. \qquad (6.365)$$

The kinetic and potential energy, as well as the virtual work, are formulated as follows using the coordinates defined in Fig. 6.34 ($x = 0$ and $r\varphi_M = R\varphi$ characterizes the pretensioned but vibration-free state):

$$2W_{\text{kin}} = J\dot\varphi^2 + J_M^S \dot\varphi_M^2 + m\left[(\dot x + e\dot\varphi \cos\varphi_M)^2 + e^2 \dot\varphi_M^2 \sin^2\varphi_M\right]$$
$$= m\cdot(\dot x^2 + 2e\dot x\dot\varphi_M \cos\varphi_M) + J\dot\varphi^2 + (J_M^S + me^2)\dot\varphi_M^2 \tag{6.366}$$
$$2W_{\text{pot}} = c_V x^2 + c_Z \cdot (r\varphi_M - x\cos\alpha - R\varphi)^2 + c_L \cdot (R\varphi - r\varphi_M - x\cos\alpha)^2 \tag{6.367}$$

$$\delta W = + M_M \cdot \delta\varphi_M - b_V \dot x \cdot \delta x$$
$$- b\cdot(r\dot\varphi_M - R\dot\varphi - \dot x\cos\alpha)\cdot(r\cdot\delta\varphi_M - R\cdot\delta\varphi - \cos\alpha\cdot\delta x)$$
$$- b\cdot(R\dot\varphi - r\dot\varphi_M - \dot x\cos\alpha)\cdot(R\cdot\delta\varphi - r\cdot\delta\varphi_M - \cos\alpha\cdot\delta x) \tag{6.368}$$

If $\varphi_M(t) = \Omega t$, that is $\dot\varphi_M \equiv \Omega =$ const. is presupposed, it is useful for the subsequent derivations to introduce a new coordinate according to

$$q_2 = R\varphi - r\Omega t \tag{6.369}$$

which describes the rotational vibrations of the output that are superimposed on the rigid rotation. Observing $\delta t = 0$ makes $\delta\varphi_M = \dot\varphi_M \delta t = 0$ valid as well so that the following applies after formally renaming x into q_1:

$$x = q_1; \qquad \dot x = \dot q_1; \qquad \ddot x = \ddot q_1; \qquad \delta x = \delta q_1 \tag{6.370}$$
$$\varphi = \frac{r}{R}\Omega t + q_2/R; \quad \dot\varphi = \frac{r}{R}\Omega + \dot q_2/R; \quad \ddot\varphi = \ddot q_2/R; \quad \delta\varphi = \delta q_2/R. \tag{6.371}$$

The functions (6.366) to (6.368) thus become:

$$2W_{\text{kin}} = m\dot q_1^2 + \frac{J}{R^2}\cdot(r\Omega + \dot q_2)^2 + \frac{(J_M + me^2)^2}{r^2}\cdot(r\Omega)^2$$
$$+ 2m\frac{e}{r}\dot q_1 r\Omega \cos\Omega t \tag{6.372}$$
$$2W_{\text{pot}} = c_V q_1^2 + c_Z \cdot (-q_1 \cos\alpha - q_2)^2 + c_L \cdot (q_2 - q_1 \cos\alpha)^2 \tag{6.373}$$
$$\delta W = -\left(b_V + 2b\cos^2\alpha\right)\dot q_1 \delta q_1 - 2b\dot q_2 \delta q_2. \tag{6.374}$$

If one introduces the coordinate vector

$$\boldsymbol{q} = [q_1,\ q_2]^T = [x,\ R\varphi - r\Omega t]^T, \tag{6.375}$$

the system of the equations of motion results from Lagrange's equations of the second kind as follows:

$$\boldsymbol{M}\cdot\ddot{\boldsymbol{q}} + \boldsymbol{B}\cdot\dot{\boldsymbol{q}} + \boldsymbol{C}\cdot\boldsymbol{q} = \boldsymbol{f}(t) \tag{6.376}$$

where

$$\boldsymbol{M} = \begin{bmatrix} m & 0 \\ 0 & J/R^2 \end{bmatrix} \tag{6.377}$$

is the mass matrix,

$$\boldsymbol{C} = \begin{bmatrix} c_V + (c_L + c_Z)\cos^2\alpha & -(c_L - c_Z)\cos\alpha \\ -(c_L - c_Z)\cos\alpha & c_L + c_Z \end{bmatrix} \tag{6.378}$$

6.6 Damped Vibrations

is the stiffness matrix,

$$\boldsymbol{B} = \begin{bmatrix} b_V + 2b\cos^2\alpha & 0 \\ 0 & 2b \end{bmatrix} \quad (6.379)$$

is the damping matrix, and

$$\boldsymbol{f} = \begin{bmatrix} U\Omega^2 \sin\Omega t; & 0 \end{bmatrix}^T. \quad (6.380)$$

is the right-hand side vector ($U = me$).

One would also have obtained the system of the equations of motion (6.376) if the equilibrium conditions had been established at both wheels in the free-body diagram, taking into account d'Alembert's principle, see Fig. 6.35. The reader is encouraged to verify this way of establishing the equations of motion for himself/herself.

The forces in the slack and driving strands result both from the pretensioning and from the deformations caused by the vibrations (see (6.365)) and their variations over time:

$$F_L = \frac{F_v}{2\cos\alpha} + c_L\underbrace{(R\varphi - r\Omega t - q_1\cos\alpha)}_{=q_2} + b\underbrace{(R\dot\varphi - r\Omega - \dot q_1\cos\alpha)}_{=\dot q_2} \quad (6.381)$$

$$F_Z = \frac{F_v}{2\cos\alpha} + c_Z\underbrace{(r\Omega t - R\varphi - q_1\cos\alpha)}_{=-q_2} + b\underbrace{(r\Omega - R\dot\varphi - \dot q_1\cos\alpha)}_{=-\dot q_2}. \quad (6.382)$$

The motor torque required for generating the predefined motion $\varphi_M(t) = \Omega t$ is derived from the equilibrium of moments at the motor armature:

$$M_M = (F_Z - F_L)r + U\cos\Omega t \cdot \ddot q_1. \quad (6.383)$$

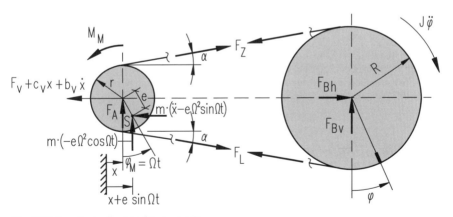

Fig. 6.35 Free-body diagram for $\dot\varphi_M \equiv \Omega$

6.6.6 Problems P6.13 to P6.16

P6.13 Decay Process

Calculate the decay process after a velocity jump u_{10} at coordinate q_1 for the model of a machine frame shown in Table 6.3, Case 2. The Rayleigh damping is used as the damping model. According to (6.326), the following is assumed with $\omega^{*2} = 48EI/ml^3$: $B = 0.008\omega^* M + 0.08C/\omega^*$, i.e., $a_1 = 0.008\omega^*$, $a_2 = 0.08/\omega^*$.

The eigenvalue problem of the undamped vibration system was solved in Sect. 6.3.4.2 so that the natural circular frequencies are known from (6.136). (6.144) specifies the vectors of the initial values in principal coordinates. Calculate the modal damping ratios D_i and derive the equations for calculating the coordinates $q_k(t)$ and inertia forces $Q_k(t)$. Evaluate these as compared to the undamped vibrations (Fig. 6.9).

P6.14 Evaluation of a Locus

Figure 6.36 shows the simplified calculation model of the frame of a milling machine and its two lowest mode shapes. The surface quality during milling depends on the relative motion between tool and workpiece at point A.

The locus for the coordinate q_6 was determined for an excitation F_6 in the frequency interval of interest, see Fig. 6.36. Determine the natural frequencies from the locus and comment on the amplitude frequency response based on the mode shapes shown.

P6.15 Complex Frequency Response

Examine the calculation model shown in Fig. 4.40 for the harmonic excitation given there.

Given:
$\bar{\mu} = J_1/J_2 = 0.2;\ \bar{\gamma} = c_{T1}/c_{T2} = 0.2;\ D = b_T/(2J_1\omega^*) = 0.1;\ \omega^{*2} = c_{T2}/J_2$

Find:
1. Matrices M, C, B and excitation vector f
2. Complex frequency response $H_{22}(j\Omega)$

P6.16 Determination of Natural Frequencies from the Amplitude Frequency Response

The measured (or calculated) amplitude frequency response of a periodically excited system typically includes several resonance peaks at the excitation frequencies f_{0i}. How can one decide which mth-order resonance and which natural frequencies are involved?

Analyze an example in which resonance peaks occur in the range from 12 Hz to 20 Hz at the frequencies 13 Hz, 13.5 Hz, 15.6 Hz, 18 Hz and 19.5 Hz if one could estimate that there will be natural frequencies in this system in the range from 40 Hz to 90 Hz.

What is the magnitude of the natural frequencies f_i that can be identified from these results? What resonance orders are there?

6.6.7 Solutions S6.13 to S6.16

S6.13 If one inserts the damping model into (6.329) one obtains the decay rates using (6.104) and (6.105):

$$\delta_i = 0.004\left(1 + 10\frac{\omega_{i0}^2}{\omega^{*2}}\right)\omega^*. \tag{6.384}$$

Since ω_{i0} are the circular frequencies of the undamped system, using the specified values one finds $\delta_1 = 0.004\,26\omega^*$, $\delta_2 = 0.005\,93\omega^*$, $\delta_3 = 0.038\,82\omega^*$, $\delta_4 = 0.079\,49\omega^*$. The modal damping ratios result from (6.330):

6.6 Damped Vibrations

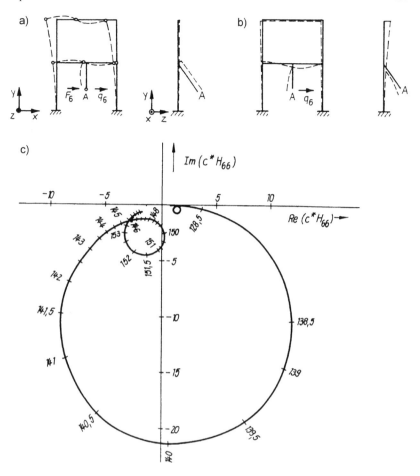

Fig. 6.36 Milling machine frame; a) and b) Mode shapes v_1 and v_2, c) Locus for the coordinate q_6

$$\underline{\underline{D_1 = 0.0528}}, \quad \underline{\underline{D_2 = 0.0270}}, \quad \underline{\underline{D_3 = 0.0416}}, \quad \underline{\underline{D_4 = 0.0579}}. \quad (6.385)$$

Starting from the initial conditions in terms of principal coordinates, (6.331) can be used to express the motions in principal coordinates:

$$p_i(t) = \frac{\dot{p}_{i0}}{\omega_i} e^{-\delta_i t} \cdot \sin \omega_i t, \quad i = 1, 2, 3, 4. \quad (6.386)$$

ω_i according to (6.330) are the natural circular frequencies of the damped system. One finds, see (6.143):

$$\omega_1 = 0.0805\omega^*, \quad \omega_2 = 0.2194\omega^*, \quad \omega_3 = 0.9321\omega^*, \quad \omega_4 = 1.3714\omega^*. \quad (6.387)$$

The initial values p_{i0} are known from (6.144). Based on the amplitudes, which deviate only slightly from the ones according to (6.146) due to weak damping, one can see that the first and second natural frequencies dominate in the response signal. Since the decay rates δ_3

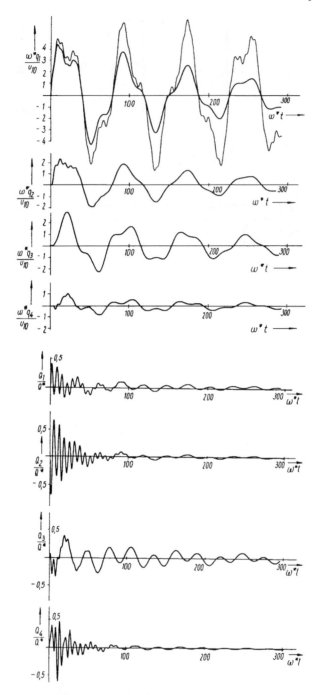

Fig. 6.37 Free damped vibrations of a frame (model of Table 6.3, Case 2); **a)** Coordinates $q_k(t)$, **b)** Inertia forces $Q_k(t)$ (reference value $Q^* = mu_{10}\omega^*$)

6.6 Damped Vibrations

and δ_4 are considerably larger than δ_1 and δ_2, the vibrations with the 3rd and 4th natural frequencies decay very fast. This can also be seen clearly for all coordinates $q_1, \ldots, q_4$ in Fig. 6.37. The position coordinates result from (6.108) and the inertia forces with (6.13):

$$q_k(t) = \sum_{i=1}^{4} v_{ki} p_i(t), \qquad Q_k(t) = -m_{kk} \ddot{q}_k(t) \tag{6.388}$$

Figure 6.37a still shows the undamped motion known from Fig. 6.9a for q_1. One can see that the decay rates δ_i are more telling than the damping ratios D_i for the free response. It is also interesting that the extreme values of the deformations and forces, which occur shortly after an impulse excitation, are only slightly reduced by the damping. It is therefore often permissible to calculate these peak values using an undamped calculation model, compare Fig. 6.9 to Fig. 6.37.

S6.14 The natural frequencies of the machine frame can be found in the locus since the amplitudes take relative extreme values there and the phases change relatively fast. This is how one finds the natural frequencies $f_2 = 140$ Hz and $f_3 = 152$ Hz. The residuum at f_1 is small because the force application point vibrates in opposite phase to the upper frame portion during a fundamental oscillation and moves only slightly.

The residuum at $\Omega = \omega_2$ is that large because the second natural vibration shows strong deflections at point A (Fig. 6.36b). If the cutter engagement frequency were in the vicinity of the first natural frequency, this would be much less hazardous than an excitation in the vicinity of the second natural frequency.

S6.15 The amplitude frequency response $D_{22}(\Omega)$ for this model has already been calculated in Sect. 4.4.3. (4.191) and Fig. 4.41 show it as a nondimensionalized amplitude function $V = \hat{\varphi}_2/\varphi_{\text{st}} = \hat{\varphi}_2 c_2/\hat{M}$. The matrices $\boldsymbol{M}, \boldsymbol{B}, \boldsymbol{C}$ and the excitation vector $\boldsymbol{f}$ are derived from a comparison of coefficients from the equations of motion (4.190):

$$\boldsymbol{q} = \begin{bmatrix} \varphi_1 \\ \varphi_2 \end{bmatrix}, \quad \boldsymbol{M} = \begin{bmatrix} J_1 & 0 \\ 0 & J_2 \end{bmatrix}, \quad \boldsymbol{B} = \begin{bmatrix} b_T & -b_T \\ -b_T & b_T \end{bmatrix},$$

$$\boldsymbol{C} = \begin{bmatrix} c_{T1} & -c_{T1} \\ -c_{T1} & c_{T1}+c_{T2} \end{bmatrix}, \quad \boldsymbol{f} = \begin{bmatrix} 0 \\ \hat{M} \end{bmatrix} \cos \Omega t \tag{6.389}$$

The principal determinant is:

$$\Delta(j\Omega) = \begin{vmatrix} -\Omega^2 J_1 + j\Omega b_T + c_{T1} & -j\Omega b_T - c_{T2} \\ -j\Omega b_T - c_{T2} & -\Omega^2 J_2 + j\Omega b_T + c_{T1} + c_{T2} \end{vmatrix} \tag{6.390}$$

$$\Delta(j\Omega) = \Omega^4 J_1 J_2 - j\Omega^3 b_T (J_1+J_2) - \Omega^2[J_2 c_{T1}+J_1(c_{T1}+c_{T2})]+j\Omega b_T c_{T2}+c_{T1}c_{T2}$$

Since the excitation moment is applied at mass 2, $\hat{F}_s = \hat{F}_2 = \hat{M}$. The complex frequency responses H_{22} are calculated using the determinant

$$\Delta_{22}(j\Omega) = \begin{vmatrix} -\Omega^2 J_1 + j\Omega b_T + c_{T1} & 0 \\ -j\Omega b - c_1 & 1 \end{vmatrix} = -\Omega^2 J_1 + j\Omega b_T + c_{T1} \tag{6.391}$$

If one introduces the dimensionless characteristic parameters according to the problem statement, and in addition $\xi^2 = \overline{\gamma}/\overline{\mu}$ and $\eta = \Omega/\omega^*$, it follows that

$$\Delta(j\eta) = \frac{c_{T1}^2}{\gamma^2} \left\{ \eta^4 \overline{\mu} - j2\eta^3 D \overline{\mu}(1+\overline{\mu}) - \eta^2 \left[\overline{\gamma}+\overline{\mu}(1+\overline{\gamma})\right]+j2\eta D\overline{\mu}+\overline{\gamma} \right\} \tag{6.392}$$

$$\Delta_{22}(j\eta) = \frac{c_{T1}}{\xi^2} \left(-\eta^2+j2\eta D+\xi^2\right).$$

Using the abbreviations

$$a_1 = \xi^2 - \eta^2, \qquad a_2 = \eta^4\bar{\mu} - \eta^2[\bar{\gamma} + \bar{\mu}(1+\bar{\gamma})] + \bar{\gamma}, \qquad (6.393)$$
$$a_3 = 2\eta D, \qquad a_4 = 2\eta D\bar{\mu}\left[\eta^2(1+\bar{\mu}) - 1\right]$$

one finds

$$H_{22} = \frac{\Delta_{22}}{\Delta} = \frac{\bar{\gamma}\bar{\mu}}{c_{T1}}\frac{a_1 + ja_3}{a_2 - ja_4} = \frac{\bar{\mu}}{c_{T2}}\left[\frac{a_1 a_2 - a_3 a_4}{a_2^2 + a_4^2} + j\frac{a_3 a_2 + a_1 a_4}{a_2^2 + a_4^2}\right]. \qquad (6.394)$$

The curves for the amplitude and phase frequency response shown in Fig. 6.30a and b and the locus in Fig. 6.30c correspond to the solution for the numerical values from the problem statement. One can recognize the two resonance frequency ratios $\eta_1 = 0.8$; $\eta_2 = 1.25$ and the absorption frequency (antiresonance) at $\eta = 1.02$.

S6.16 One can use the following heuristic approach: The natural frequencies at resonance according to (6.363) are always an integral multiple of the fundamental excitation frequency f_0, i. e. the measured resonance frequencies are the quotients of the natural frequencies f_i and small integers m, that is $f_{0R} = f_i/m$. To find these natural frequencies, one has to multiply the resonance frequencies f_{0R} with the resonance orders $(m = 1, 2, 3, \ldots)$ to see whether there are matching numerical values. For the example at hand, one obtains the sequences of numbers specified in the table below:

i/m	1	2	3	4	5	6	7
1	13	26	39	52	65	78	91
2	13.5	27	40.5	54	67.5	81	94.5
3	15.6	31.2	46.8	62.4	78	93.6	109.2
4	18	36	54	72	90	108	126
5	19.5	39	58.5	78	97.5	107	126.5

In this way, one finds two numerical values that result from different resonance frequencies. These are the natural frequencies $f_1 = 54$ Hz and $f_2 = 78$ Hz, i. e. the five resonance peaks can be explained by two natural frequencies and the resonance orders $m = 3$ to 6: $13 = 78/6$; $13.5 = 54/4$; $15.6 = 78/5$; $18 = 54/3$; $19.5 = 78/4$.

Chapter 7
Simple Nonlinear and Self-Excited Oscillators

7.1 Introduction

The equations of motion regarding free vibrations for nonlinear and self-excited oscillators do not differ from each other. However, the equations of motion of self-excited oscillators contain, from a physical point of view, a "stimulating" expression so that free vibrations do not always subside like it is the case with "common" nonlinear oscillators. While the equations of motion of self-excited oscillators do not explicitly depend on time, they do for forced nonlinear vibrations. Time-dependent excitation may occur, for example, in the form of motion excitation force excitation when starting, braking, during transitional processes or even during steady-state processes (periodic excitation).

H. POINCARÉ (1854–1912) already knew around 1900 that deterministic nonlinear equations do not necessarily have regular solutions. In the past decades, the term deterministic *chaos* was coined to express that nonlinear systems may also exhibit an irregular dynamic behavior that cannot be predicted for specific points in time [5]. It a characteristic of chaotic systems that minor causes can have major effects, i.e. that the dynamic behavior is highly sensitive to parameter changes (including initial conditions). Whoever wishes to attain a deeper understanding of the field of nonlinear vibrations should refer to the more detailed introductions in [2], [6], and [24]. Examples of nonlinear vibrations in drives are treated in [26].

The equation of motion for *forced vibrations* that describes the translational motion of a mass m along a straight path $x(t)$ results from the equilibrium of the inertia force, the nonlinear damping force $F_D(x, \dot{x})$, the nonlinear restoring force $F(x)$, and the excitation force $F^{(e)}(t)$:

$$m\ddot{x} + F_D(x, \dot{x}) + F(x) = F^{(e)}(t). \tag{7.1}$$

Table 7.1 shows examples of nonlinear restoring forces $F(x)$, see also the special case (4.32). If the restoring force increases faster than in a linear spring, this is called a *progressive* (or superlinear) *spring characteristic*. If it stays below a straight line, it is called a *degressive* (or sublinear). The nonlinearity is determined by geometry

Table 7.1 Examples of nonlinear restoring forces

Case	System	Characteristic	Restoring function $F(x)$	
1	Stepped springs		$F(x) = \begin{cases} c_2 x_1 + (c_1+c_2)x & \text{für } x < -x_1 \\ c_1 x & \text{für } -x_1 \leq x \leq x_1 \\ -c_2 x_1 + (c_1+c_2)x & \text{für } x_1 \leq x \end{cases}$	(1)
2	Oscillator with spring preload		$F(x) = cx + F_V \, \text{sign}(x)$	(2)
3	Spring-mass system F_V = preload force		$F(x) = \dfrac{2 F_V x}{\sqrt{l^2+x^2}} + 2cx\left(1 - \dfrac{l}{\sqrt{l^2+x^2}}\right)$ for $x \ll l$ holds $F(x) = 2F_V \dfrac{x}{l} + cx \dfrac{x^2}{l^2} + \ldots$	(3)

in the three cases shown in Table 7.1 (like in coil springs, see (1.48) and Fig. 1.14), while it is determined by the material in some materials such as rubber or plastics.

The nonlinearity is used, for example, in shock absorbers, overload springs of vehicles, and in couplings, vibrating screens and oscillating conveyors for influencing the resonance curves, see Sect. 7.2.3.1. Many machine elements have nonlinear spring characteristics, e. g. roller bearings, air springs, disk springs, cables, couplings, dampers, and tires. Friction and backlash in gear mechanisms, joints, e. t. c. are also important nonlinear parameters.

In addition to the examples of nonlinear spring forces described here, there are nonlinear damping forces and nonlinearities as a result of inertia forces. Nonlinear expressions as a result of inertia forces develop from nonlinear geometrical constraints or Coriolis forces and gyroscopic moments where products of velocities and angular velocities occur, see (2.155). Examples were discussed as torsional oscillators with a mechanism coupled to them, see (4.219) and Problem P4.4.

In the case of **forced** vibrations of nonlinear oscillators and harmonic excitation in the steady state, the following effects can occur:

- poly-harmonic (periodic) motions (line spectrum);
- vibrations at frequencies kf_0/n with $k = 1, 2, \ldots$ and $n = 1, 2, \ldots$ (k and n very small integer numbers);

7.1 Introduction

- vibrations at integral multiples of the excitation frequency, e.g. $2f_0$, $3f_0$, ... so-called *superharmonics*;
- vibrations at integral ratios of the excitation frequency, e.g. $f_0/2$, $f_0/3$, ... so-called *subharmonics*;
- chaotic motions (continuous spectrum).

It is a major characteristic of nonlinear oscillators that the superposition principle does not apply. Typical nonlinear effects are that:

- the period of undamped oscillations depends on the initial conditions (initial energy), that is, on the oscillation amplitude;
- the time functions of undamped oscillations are not harmonic but periodic, e.g. the free vibration of mechanisms, see Sect. 2.4.4.
- in the case of harmonic excitation, the amplitude of the motion (steady-state response) is not proportional to the excitation amplitude
- the amplitude and phase of forced steady-state vibrations may change abruptly depending on the excitation frequency, see Fig. 7.1
- under certain conditions energy that is supplied at a frequency f_1 is also transferred into vibrations with another frequency f_2 and that self-synchronization and a rectifying effect may occur, see Sect. 7.2.3.3
- when passing through resonance, the response amplitude at an increasing excitation frequency differs from the one occurring at decreasing excitation frequency, see Figures 7.1 and 7.8
- combined vibrations at the frequencies $mf_1 + nf_2$ and $mf_1 - nf_2$ (m and n very small integer numbers) may occur when the system is simultaneously excited with two different frequencies f_1 and f_2.

Chaotic motions are "permanent irregularly oscillating variations of state variables in deterministic systems with a strong sensitivity to changes of the initial conditions" [24]. While the superharmonics and subharmonics can be characterized by a line spectrum, the spectrum of chaotic motions is continuous (in which individual lines may occur as well). The time functions of chaotic motions cannot be exactly calculated in advance. Terms such as "attractor", "period doubling", "Poincaré chart" and concepts typical of probability theory are used for their characterization, see [2], [24]. The type of vibrations that occur depends on the parameter values in the equation of motion. Such a parameter chart in which the ranges of existence of the various motion processes are mapped is shown in [24] for a Duffing oscillator.

Nonlinear vibrations can typically be calculated by approximation methods only [6], [24]. Important and common methods, some of which are used in the section below, include:

- Numerical integration using software, e.g. SimulationX® [34]
- Asymptotic methods (small parameter method, averaging methods), see Sect. 7.2.3.3
- the Ritz-Galerkin method
- the equivalent linearization method, see Sect. 7.2.2.1
- the Krylov-Bogolyubov-Mitropolsky method.

7.2 Nonlinear Oscillators

7.2.1 Undamped Free Nonlinear Oscillators

Free vibrations develop when the system is initially supplied with (kinetic and/or potential) energy and then left to itself. No excitation is active during the vibration. The equation of motion of an undamped free oscillator for the generalized coordinate q, which can represent a displacement or an angle, is:

$$m\ddot{q} + F(q) = 0. \tag{7.2}$$

The discussion below will be limited to odd functions to which $F(q) = -F(-q)$ applies. The initial conditions describe the state of the system at the beginning of the motion:

$$t = 0: \quad q = q_0; \quad \dot{q} = v_0. \tag{7.3}$$

A transferred energy that does not change during the vibration corresponds to this initial state:

$$W_0 = W_{\text{kin}\,0} + W_{\text{pot}\,0} = W_{\text{kin}} + W_{\text{pot}} \tag{7.4}$$

$$= \frac{1}{2}m(v_0)^2 + \int_0^{q_0} F(q*)dq* = \frac{1}{2}m\dot{q}^2 + \int_0^{q} F(q^*)dq^*. \tag{7.5}$$

This energy balance immediately leads to the velocity as a function of the displacement:

$$\dot{q}(q) = \sqrt{(v_0)^2 - \frac{2}{m}\int_{q_0}^{q} F(q^*)dq^*}. \tag{7.6}$$

The inverse function of $q(t)$ can be determined by another integration:

$$t(q) = \int_{q_0}^{q} \frac{dq^*}{\dot{q}(q^*)}. \tag{7.7}$$

The *period of oscillation* results from the periodicity condition $q(t) = q(t+T)$. For an odd function $F(q)$, the motions in the range $0 < q < \hat{q}$ are symmetrical to those in the range $-\hat{q} < q < 0$. It is therefore sufficient to consider a quarter of an oscillation to calculate the period of oscillation. The following applies to the amplitude $\hat{q}$

$$T = 4\int_0^{\hat{q}} \frac{dq^*}{\dot{q}(q^*)}. \tag{7.8}$$

7.2 Nonlinear Oscillators

The time function $q(t)$ is **periodic but not harmonic** like, for example, the solution (7.16). Therefore, this motion is not characterized by a natural frequency but by the period of oscillation. Nonlinear oscillators with the same period of oscillation may have completely different time functions. The time functions – or the dependence of the instantaneous velocity on the deflection specified in (7.6) – differentiate the vibrations of the nonlinear systems from those of linear systems and from each other. They are called the **proper motions** of the system.

At deflections in the range of $-\hat{q} < q < \hat{q}$ during the linearization of the nonlinear characteristic, one can substitute a *mean spring constant* c_m that depends on the amplitude $\hat{q}$ and results from the condition

$$\int_0^{\hat{q}} [(F(q) - c_m q)q]^2 \, dq = \text{min.!} \tag{7.9}$$

The nonlinear restoring force is thus approximated by $F_c(q) = c_m q$. There are two simple methods to determine this mean spring constant.

When it comes to the displacements, the mean spring constant, in accordance with the above requirement, results from the following integral (*equivalent linearization method*):

$$c_m(\hat{q}) = \frac{5}{\hat{q}^5} \int_0^{\hat{q}} q^3 F(q) \, dq. \tag{7.10}$$

One can also average in the time domain and require, in accordance with (7.9), that the equilibrium of forces (7.2) be satisfied on average with the assumed expression for a harmonic motion $\bar{q} = \hat{q} \cos \omega t$ with the initially unknown circular frequency ω:

$$\int_0^{2\pi/\omega} \bar{q}(t) [F(\bar{q}) - c_m \bar{q}(t)] \, dt = 0. \tag{7.11}$$

This corresponds to the *method of harmonic balance*. It provides the average spring constant as

$$c_m(\hat{q}) = \frac{\omega}{\pi \hat{q}} \int_0^{2\pi/\omega} F(\bar{q}(t)) \cos \omega t \, dt. \tag{7.12}$$

The values that one obtains from (7.10) do not always exactly match those from (7.12) since both are approximations. The mean spring constant c_m can be used to approximate the *period of oscillation T of the natural vibration* of a nonlinear oscillator that can be obtained more easily than (7.35). The following applies:

$$T \approx 2\pi \sqrt{\frac{m}{c_m}}. \tag{7.13}$$

The mean spring constant and the period of oscillation depend on the amplitude for all nonlinear oscillators. If one uses a mean spring constant for the calculation, one obtains a harmonic time function as an approximation. It averages the function in a specific deflection range but neglects the frequently much higher harmonics in the solution.

A typical representative of nonlinear oscillators is the oscillator studied as early as in 1918 by the German engineer GEORG DUFFING (1861–1944) (and named after him) that is governed by the differential equation (7.2):

$$F(q) = cq(1 + \varepsilon q^2). \tag{7.14}$$

The nonlinearity of the restoring force is described using a single parameter ε. This characteristic can be used to describe progressive ($\varepsilon > 0$), linear ($\varepsilon = 0$), and degressive characteristics ($\varepsilon < 0$) and study typical nonlinear effects. Such characteristics exist, for example, for the preloaded spring-mass system (Case 3 in Table 7.1) and the pendulum with a large amplitude. The nonlinear restoring functions of nonlinear springs (due to the geometry or the material) can often be captured using (7.14), see Problem P1.3.

The free vibration follows from (7.6) in the form of the velocity-time function for $v_0 = 0$:

$$\dot{q}(q) = \sqrt{\frac{2}{m} \int_{q_0}^{q} cq^*(1 + \varepsilon q^{*2}) dq^*} = \sqrt{\frac{c}{m}\left[q^2 - q_0^2 + \frac{1}{2}\varepsilon(q^4 - q_0^4)\right]}. \tag{7.15}$$

This function can be plotted in the q-$\dot{q}$ plane, the so-called *phase plane* [6], [24]. Each initial condition results in a specific **phase curve**, the shape of which allows conclusions regarding the state of a nonlinear system. An approximate solution for the time function can be determined using the small parameter method, see [6]:

$$q(t) = q_0 \left[\cos \omega t + \frac{\varepsilon q_0^2}{32}(\cos 3\omega t - \cos \omega t) + \left(\frac{\varepsilon q_0^2}{32}\right)^2 (\cos 5\omega t - \cos \omega t) + \ldots\right]. \tag{7.16}$$

The fundamental circular frequency ω is related to the period of oscillation T and the natural circular frequency ω_0 of the linear oscillator and changes with ε and with the initial deflection q_0:

$$\omega = \frac{2\pi}{T} \approx \sqrt{\frac{c}{m}}\left(1 + \frac{3\varepsilon q_0^2}{8} - \frac{21\varepsilon^2 q_0^4}{256}\right); \qquad \omega^2 \approx \omega_0^2\left(1 + \frac{3\varepsilon q_0^2}{4}\right). \tag{7.17}$$

The mean spring constant also depends as follows on the deflection $\hat{q}$ in accordance with the approximation from (7.12):

$$c_m = c\left(1 + \frac{3\varepsilon \hat{q}^2}{4}\right). \tag{7.18}$$

For comparison: the mean spring constant according to (7.10) is $c_m = c(1+5\varepsilon\hat{q}^2/7)$ and differs little from it. The solution using the Fourier series (7.16) that only contains the odd harmonics up to the 5th order converges fast and is sufficient for many practical cases. The exact solution is a sum of infinitely many odd-order harmonics.

7.2.2 Forced Vibrations with Harmonic Excitation

7.2.2.1 First Harmonic with Nonlinear Stiffness

The equation of motion of an oscillator with linear damping, nonlinear restoring force and harmonic excitation force is the following special case of (7.1):

$$m\ddot{q} + b\dot{q} + F(q) = \hat{F}\cos\Omega t. \qquad (7.19)$$

One can obtain an approximate solution for the amplitude of the first harmonic for steady-state vibrations. The restoring force in the equation of motion is linearized for this purpose so that the calculation is performed using an amplitude-independent mean spring constant c_m instead of (7.19), see (7.10) or (7.12):

$$m\ddot{q} + b\dot{q} + c_m(\hat{q})q = \hat{F}\cos\Omega t. \qquad (7.20)$$

The solution to this now linear equation is known from Sect. 3.2.1.2:

$$q(t) = \hat{q}\cos(\Omega t - \varphi). \qquad (7.21)$$

The amplitude $\hat{q}$ and the phase angle φ result as in (3.10) from the following equations:

$$\hat{q} = \frac{\hat{F}}{\sqrt{[c_m(\hat{q}) - m\Omega^2]^2 + (b\Omega)^2}} \qquad (7.22)$$

$$\tan\varphi = \frac{b\Omega}{\sqrt{[c_m(\hat{q}) - m\Omega^2]^2 + (b\Omega)^2}}. \qquad (7.23)$$

They apply to the first harmonic of the linearized nonlinear oscillator. The amplitude cannot be calculated directly since it is also contained in the denominator of (7.22) in c_m. It follows from a nonlinear equation that can in general only be solved numerically. The inverse function $\Omega^2(\hat{q})$ of the *amplitude frequency response for the first harmonic* is more easily calculated from a quadratic equation for Ω^2 that results from (7.22) after a brief rearrangement:

$$\Omega^4 + \Omega^2\frac{b^2 - 2mc_m(\hat{q})}{m^2} + \frac{c_m^2(\hat{q}) - \left(\frac{\hat{F}}{\hat{q}}\right)^2}{m^2} = 0. \qquad (7.24)$$

The averaged spring constant for the Duffing oscillator is known from (7.18). Insertion of the dimensionless characteristic parameters D (damping ratio), ε^* (nonlinearity) and the nondimensionalized amplitude function V according to

$$2D = \frac{b}{\sqrt{cm}}; \qquad \varepsilon^* = \varepsilon \left(\frac{\hat{F}}{c}\right)^2; \qquad V = \frac{c\hat{q}}{\hat{F}} \qquad (7.25)$$

into (7.24) provides the following quadratic equation for $\eta^2 = (\Omega/\omega_0)^2$:

$$\eta^4 + 2\eta^2 \left[2D^2 - 1 - \frac{3}{4}\varepsilon^* V^2\right] + \left[1 + \frac{3}{4}\varepsilon^* V^2\right]^2 - \left(\frac{1}{V}\right)^2 = 0. \qquad (7.26)$$

It has the following two solutions:

$$\eta^2_{1,2} = 1 + \frac{3\varepsilon^*}{4} V^2 - 2D^2 \pm \sqrt{\frac{1}{V^2} - 4D^2 \left[1 + \frac{3}{4}\varepsilon^* V^2 - D^2\right]}. \qquad (7.27)$$

One obtains one or two values of η for each given value of the nondimensionalized amplitude function V (only the real roots have physical relevance) so that the desired resonance curve $V(\eta, D, \varepsilon^*)$ is obtained as the inverse function of $\eta(D, V, \varepsilon^*)$. Vice versa, one, two or three different amplitudes can be associated with an η value, see Fig. 7.1. This amplitude frequency response for the first harmonic differs significantly from that of a linear oscillator.

The dashed line in the center is the so-called *skeleton line*. It describes the dependence of the natural frequency on the amplitude. The skeleton line is bent to the right for progressive restoring forces because the natural frequency increases with the amplitude in this case. The skeleton line is bent to the left for degressive characteristics. There are unique amplitude values for low (left of D) and for high excitation frequencies (right of B). When the damping is small, there theoretically are three solutions at the "overhanging" curve branches in the central range between ω^* and ω^{**}. It is known from theory [6], [24] that no stable vibrations are possible on the branch of the curve between A and B. If the excitation frequency is increased slowly from low frequencies, e. g. during a start-up process, it moves along the upper branch of the curve until, at point A, the amplitude jumps from a large value to a much smaller value at point B. If, on the other hand, the speed is slowly reduced, e. g. when braking, the amplitude jumps from the lower branch to the higher branch of the curve at the latest at point C.

The maximum of the nondimensionalized amplitude function results from

$$V^2_{\max} = \frac{2(1 - D^2)}{3\varepsilon^*} \left[\sqrt{1 + \frac{3\varepsilon^*}{4(1 - D^2)D^2}} - 1\right] \qquad (7.28)$$

at a frequency ratio

$$\eta = \sqrt{1 + \frac{3}{4}\varepsilon^* V^2_{\max} - D^2}. \qquad (7.29)$$

7.2 Nonlinear Oscillators

The result (7.28) shows: **The amplitude of the vibration is not proportional to the amplitude of the excitation force.** For practical tasks in vibration engineering, the *amplitude frequency response of nonlinear oscillators* is therefore of particular interest because one can obtain a wider resonance range than that of linear oscillators. If one wishes to generate large vibration amplitudes, e. g. in vibrating screens, oscillating conveyors or vibrating compactors, the drive must be configured robust enough so that it provides the required minimum amplitudes.

In the nondimensionalized amplitude function of a linear oscillator (see Fig. 3.4), the resonance peak is very narrow and its peak value is inversely proportional to the damping ratio, see (3.20). Even minor changes in damping or inevitable parameter variations (mass, stiffness, frequency) can cause the amplitude maximum to "slip", because a minor change of the optimum frequency ratio would result in losing the benefits of resonance amplification.

7.2.2.2 Superharmonics and Subharmonics in an Undamped Duffing Oscillator

The following consideration is to show that forced vibration with harmonic excitation is not just harmonic. The equation of motion

$$\ddot{q} + \omega_0^2 q(1 + \varepsilon q^2) = \frac{\hat{F}}{m} \cos \Omega t \tag{7.30}$$

follows from (7.19) with $b = 0$ for the undamped Duffing oscillator with (7.14). One can rearrange (7.30) to become

$$\ddot{q} + \Omega^2 q = (\Omega^2 - \omega_0^2)q - \omega_0^2 \varepsilon q^3 + \frac{\hat{F}}{m} \cos \Omega t \tag{7.31}$$

and seek a solution to the problem in the form $q(t) = q^{(1)}(t) + q^{(2)}(t)$. The solution $q^{(1)} = \hat{q} \cos \Omega t$ from (7.21) is used with $b = 0$ as first approximation, which for $\varepsilon = 0$ is the exact solution and approximately satisfies (7.30). If one inserts this into (7.31), an equation for the second approximation $q^{(2)}$ is obtained:

$$\begin{aligned}\ddot{q}^{(2)} + \Omega^2 q^{(2)} &= (\Omega^2 - \omega_0^2)q^{(1)} - \omega_0^2 \varepsilon q^{(1)3} + \frac{\hat{F}}{m} \cos \Omega t \\ &= \left[(\Omega^2 - \omega_0^2)\hat{q} - \frac{3}{4}\omega_0^2 \varepsilon \hat{q}^3 + \frac{\hat{F}}{m}\right] \cos \Omega t - \frac{1}{4}\omega_0^2 \varepsilon \hat{q}^3 \cos 3\Omega t.\end{aligned} \tag{7.32}$$

The trigonometric identity $\cos^3 \Omega t = 1/4(3 \cos \Omega t + \cos 3\Omega t)$ was used for the derivation. The expression in the square brackets is zero because it is identical with (7.22) for $b = 0$, from which the amplitude $\hat{q} = a_1$ (resonance curve of the first harmonic) has already been calculated. The solution of (7.32) therefore provides a second approximation that satisfies the term in the second line in (7.32):

$$a_3 = \frac{\omega_0^2 \varepsilon \hat{q}^3}{32\Omega^2}; \qquad q(t) = q^{(1)} + q^{(2)} = a_1 \cos\Omega t + a_3 \cos 3\Omega t. \tag{7.33}$$

If one continued this approach, one could obtain other approximations by successive approximation. One would find other higher harmonics with the circular frequencies 5Ω, 7Ω, etc. Since these frequencies are above the excitation frequency ($k\Omega$), they are called *superharmonics*. The harmonic excitation force forces the nonlinear oscillator to oscillate not harmonically, but periodically.

Other vibrations may occur that besides superharmonics also include *subharmonics* (circular frequencies Ω/k). They are searched for using the following approach:

$$q = a_1 \cos \Omega t + a_{1/3} \cos(\Omega t/3). \tag{7.34}$$

Insertion into (7.30) provides a finite trigonometric series. It is obtained after using a few trigonometric identities:

$$\left[(\omega_0^2 - \Omega^2)a_1 + \varepsilon \omega_0^2 \frac{3a_1^3 + 6a_1 a_{1/3}^2 + a_{1/3}^3}{4} - \frac{\hat{F}}{m} \right] \cos \Omega t + \tag{7.35}$$

$$\left[\left(\omega_0^2 - \frac{\Omega^2}{9} \right) a_{1/3} + 3\varepsilon \omega_0^2 \frac{2a_1^2 a_{1/3} + a_1 a_{1/3}^2 + a_{1/3}^3}{4} \right] \cos(\Omega t/3) + \ldots = 0.$$

The other occurring functions $\cos 5\Omega t/3$ and $\cos 7\Omega t/3$ are neglected for averaging purposes. Then (7.30) is approximately satisfied if (7.35) is satisfied. If one sets the expressions in the two pairs of square brackets to zero, two coupled nonlinear algebraic equations are obtained. Their solution is complicated and shall not be detailed here. The second expression in brackets is zero for $a_{1/3} = 0$. This results in the solution for $b = 0$, known from (7.22):

$$a_1 = \hat{q} = \frac{\hat{F}}{c_m(\hat{q}) - m\Omega^2}. \tag{7.36}$$

A general case, in which the *amplitude of the subharmonic* is dominant ($a_{1/3} \gg a_1$), results from the second bracket. It results in a quadratic equation for $a_{1/3}$:

$$(a_{1/3})^2 + a_1 a_{1/3} + 2\hat{q}^2 + \frac{4(9\omega_0^2 - \Omega^2)}{27\varepsilon \omega_0^2} = 0. \tag{7.37}$$

Its solution can be calculated in conjunction with the approximation known from (7.36):

$$a_{1/3} = -\frac{a_1}{2} \left[1 \pm \sqrt{\frac{16(\Omega^2 - 9\omega_0^2)}{27\varepsilon \omega_0^2 \hat{q}^2} - 7} \right]. \tag{7.38}$$

Solutions only exist if the expression under the square root is positive. This is the case if the following condition is satisfied for a progressive characteristic ($\varepsilon > 0$), see also (7.17):

7.2 Nonlinear Oscillators

$$\Omega > 3\omega_0 \sqrt{1 + \frac{21\varepsilon \hat{q}^2}{16}} > 3\omega_0 \sqrt{1 + \frac{3\varepsilon \hat{q}^2}{4}} = 3\omega. \tag{7.39}$$

The inequality sign has to be reversed for $\varepsilon < 0$. The amplitudes of the subharmonics are large when the excitation frequency is over triple the fundamental frequency. The forced vibrations then oscillate at a frequency that is approximately equal to the natural frequency, that is about one third of the excitation frequency.

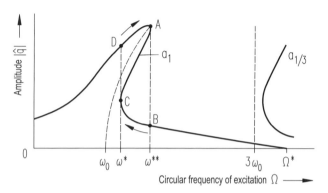

Fig. 7.1 Amplitude frequency response of the first harmonic and the third subharmonic of the Duffing oscillator

Figure 7.1 also shows the skeleton line for the third subharmonic. It turns out that the amplitudes of the subharmonic considerably exceed those of the fundamental harmonic in the area $\Omega > 3\omega_0$. There is also a combination of parameter values at which only the subharmonic occurs ($\hat{q} = 0$). From the first pair of brackets in (7.35), one first finds the relation $\varepsilon \omega_0^2 a_{1/3}^3 = 4\hat{F}/m$ (the displacement amplitude is proportional to the cubic root of the force amplitude). The second pair of brackets leads to the circular frequency of the excitation at which this motion may occur:

$$\Omega^* = 3\omega_0 \sqrt{\frac{1 + 3\varepsilon a_{1/3}^2}{4}}. \tag{7.40}$$

Subharmonic vibrations of the 2nd and 3rd order have often been observed in rotors. It is baffling to observe, for example, slow vibrations at a speed of 5 000 rpm or 3 300 rpm at an operating speed of 10 000 rpm in a rotor. The main cause of subharmonic resonances are nonlinear characteristics of the bearings. One can eliminate these nonlinear effects in most cases by reducing the backlash and increasing the preload.

7.2.2.3 Harmonically Excited Frictional Oscillator with Viscous Damping

Damping is nonlinear in most practical cases, see Sect. 1.4. This section discusses the case where both viscous damping and Coulomb friction occur. This model with a combination of both cases that were explained in Table 1.8 describes the material damping of many materials in good approximation. The equation of motion for a harmonic excitation force according to (1.83) and (1.84) is:

$$m\ddot{q} + b\dot{q} + F_R \text{sign}(\dot{q}) + cq = \hat{F}\sin\Omega t. \tag{7.41}$$

This is not about searching for the analytical solution. It is possible to reach in intervals if one pieces together the solutions of the linear equation. This results in complex expressions in which parameter dependence is hard to detect.

An approximate solution using the equivalent linearization method would only provide the first harmonic but neglect other periodic terms, see the comments in Sect. 7.1.

It is always advisable to introduce dimensionless characteristic parameters before solving such nonlinear differential equations numerically. The original differential equation contains three parameters less after its transformation so that parametric influences can be analyzed more clearly and transparently [5]. In this example, five dimensionless characteristic parameters ("similarity numbers")

$$\xi = \frac{cq}{F_R}; \quad \eta = \frac{\Omega}{\omega_0}; \quad D = \frac{b}{2m\omega_0}; \quad \varkappa = \frac{\hat{F}}{F_R}; \quad \tau = \omega_0 t \tag{7.42}$$

where $\omega_0 = \sqrt{c/m}$ is the circular reference frequency are introduced to eliminate the eight dimensional physical variables x, m, b, F_R, c, $\hat{F}$, Ω and t that appear in (7.41). Derivation with respect to τ is indicated by a prime. The differential equation in (7.41) becomes one for the dimensionless displacement coordinate ξ:

$$\xi'' + 2D\xi' + \text{sign}(\xi') + \xi = \varkappa \sin\eta\tau \tag{7.43}$$

The simulation model shown in Fig. 7.2a was used to calculate the free response and the excitation in resonance ($\eta = 1$). One can see from Fig. 7.2b that the free vibration comes to a rest in a finite time, which would not occur with viscous damping alone. Figure 7.2c reveals another difference in behavior of a nonlinear oscillator as compared to a linear oscillator. The 1.5 times higher excitation force amplitude does not cause a 1.5 times higher, but about a $42/18 = 2.3$ times higher displacement amplitude. This is a typical nonlinear effect.

7.2.3 Examples

7.2.3.1 Oscillating Conveyor with Stepped Springs

Oscillating conveyors with a slider crank drive can reach high amplitudes when operated in the vicinity of resonance. The resonance range is very narrow when using a linear spring. The resonance range of linear oscillators is highly sensitive to variations in the parameter values (load, excitation frequency, spring stiffnesses).

7.2 Nonlinear Oscillators

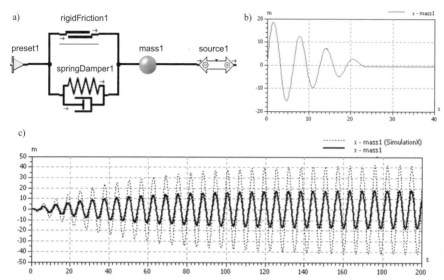

Fig. 7.2 Frictional oscillator with viscous damping; **a)** Model representation in SimulationX®, **b)** Free response, **c)** Displacement-time function for harmonic excitation in resonance (dashed line at $\varkappa = 3$; solid line at $\varkappa = 2$)

The amplitudes would change considerably for small parameter changes and the conveyor would not work reliably. It is therefore advantageous and common for robust operation to use nonlinear spring systems (with stepped springs) to achieve large amplitudes over a wide range of frequency ratios.

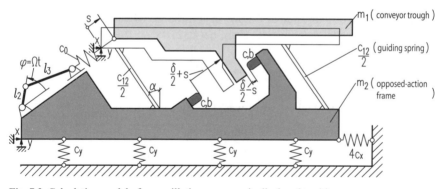

Fig. 7.3 Calculation model of an oscillating conveyor in displaced position

Figure 7.3 shows a lever-guided oscillating conveyor, that is driven by a slider crank ($l_3 \gg l_2$) with an elastic push rod. The calculation model, which actually has 12 degrees of freedom due to the two coupled rigid bodies, is simplified to a model with 3 degrees of freedom using the following assumptions:

- Symmetry with respect to the drawing plane is assumed.
- The tilting motion is neglected (since it is not excited).
- The direction of thrust is perpendicular to the direction of motion.
- The linear spring c_0 is unloaded for $s = l_2 \cos \Omega t$ and $s = 0$ describes the undeformed state of the spring c_{12}.

The equations of motion (for the relative displacement s as well as for the absolute displacements x and y) are obtained when neglecting the damping in the steel and lever springs as compared to the damping in the rubber buffer springs using Lagrange's equations of the second kind, for which the 3 scalar functions W_{kin}, W_{pot} and the virtual work $\delta W^{(e)}$ are needed. If the nonlinear force (due to backlash) in the rubber buffers (the behavior of which is assumed to be linear with respect to stiffness and damping) is designated as $F(s, \dot{s}, \delta, b, c)$, the following is obtained:

$$W_{\text{kin}} = \frac{m_1}{2} \cdot [(\dot{x} + \dot{s} \cos \alpha)^2 + (\dot{y} + \dot{s} \sin \alpha)^2] + \frac{m_2}{2} \cdot (\dot{x}^2 + \dot{y}^2)$$

$$W_{\text{pot}} = \frac{1}{2} \left[4c_x x^2 + 4c_y y^2 + c_{12} s^2 + c_0 \cdot (s - l_2 \cos \Omega t)^2 \right] \quad (7.44)$$

$$\delta W^{(e)} = -F(s, \dot{s}, \delta, b, c) \cdot \delta s$$

The following equations of motion result using Lagrange's formalism:

$$m_1(\ddot{s} + \ddot{x} \cos \alpha + \ddot{y} \sin \alpha) + (c_{12} + c_0)s = c_0 l_2 \cos \Omega t - F(s, \dot{s}, \delta, b, c)$$

$$m_1(\ddot{s} \cos \alpha + \ddot{x}) + m_2 \ddot{x} + 4c_x x = 0 \quad (7.45)$$

$$m_1(\ddot{s} \sin \alpha + \ddot{y}) + m_2 \ddot{y} + 4c_y y = 0.$$

The buffer force can be expressed as follows in accordance with Case 1 in Table 7.1 when taking into account damping, see also Eq. (4.32):

$$F(s, \dot{s}, \delta, b, c) = \frac{1}{2} \left[b\dot{s} + c \cdot \left(s - \frac{\delta}{2} \text{sign}(s) \right) \right] \cdot \left(1 + \text{sign}(|s| - \frac{\delta}{2}) \right) \quad (7.46)$$

The motions of the oscillating conveyor were studied for two variants.

Parameter values that are identical for both variants:

Crank radius	$l_2 = 30$ mm
Angle of inclination of the lever springs	$\alpha = 25$ degrees
Mass of the conveyor trough	$m_1 = 600$ kg
Mass of the opposed-action frame	$m_2 = 2700$ kg
Stiffness of a steel spring package	$c_x = 0.06 \cdot 10^6$ N/m; $c_y = 0.16 \cdot 10^6$ N/m
Axial stiffness of the push rod	$c_0 = 0.2 \cdot 10^6$ N/m
Total damping constant of the buffer springs in s direction	$b = 8000$ N $\cdot$ s/m

7.2 Nonlinear Oscillators

Different parameter values for variant 1 and variant 2

	Variant 1	Variant 2
Backlash between buffer and conveyor trough	$\delta = 6$ mm	$\delta = 0$ mm
Total stiffness of the lever springs in s direction	$c_{12} = 1.5 \cdot 10^6$ N/m	$c_{12} = 0.8 \cdot 10^6$ N/m
Total stiffness of the buffer springs in s direction	$c = 1.5 \cdot 10^6$ N/m	$c = 0$ N/m

The displacement and acceleration functions of all three coordinates were calculated using SimulationX® [34]. Figure 7.4 shows some typical curves of the oscillating conveyor as a function of the input angle φ for about 4.5 revolutions. The influence of the impact after passing through the backlash can be seen when looking at the acceleration curve. The acceleration is larger than the gravitational acceleration, otherwise the material to be conveyed would not be lifted up. One can see that the displacement-time functions are not purely harmonic. This is due to the stepped spring characteristic. The third subharmonic occurs in addition to the components with the excitation frequency and higher harmonics, which represents a typical nonlinear effect.

Figure 7.5 shows the nonlinear buffer force F and the relative displacement of the conveyor trough at the input angular velocity $\Omega = 57.16$ rad/s ($\eta = 3.5$). The buffer force F equals zero for $s < 3 = \delta/2$ mm because in that case there is no contact between the conveyor trough and the buffer spring.

For Fig. 7.6, the resonance curves of the oscillating conveyor were calculated using the "Steady state" module of SimulationX® [34].

The third resonance point for variant 2 has a clearly higher frequency than the third natural frequency of variant 1. The spring characteristic (between mass m_1 and mass m_2) is not linear and progressive due to the backlash. The averaged spring constant increases with decreasing backlash because of the buffer spring, see (7.10) and (7.12). Without a buffer spring, the total spring characteristic is softer than initially. This is why all amplitudes for variant 2 are smaller than for variant 1.

The *resonance curves* in Fig. 7.6 represent the maximum deflections as a function of the circular frequency of the excitation ($\eta^* = \Omega/\omega^*$) relative to $\omega^* = \sqrt{c_y/m_1} = 16.33$ s^{-1}. Two resonance points, which describe extensive movements of the opposed-action frame, can be detected in the lower excitation frequency range. They correspond to those of a linear oscillator and have to be passed through during start-up. In the upper frequency range, the curves are bent to the right since a progressive stiffness was implemented. Since the *resonance curves for both variants* differ significantly only for the coordinate s, those for x and y were drawn only once into Fig. 7.6.

The oscillating conveyor is operated in the range of the frequency ratio $3 < \eta^* < 4$. In this range, the opposed-action frame moves in antiphase to the conveyor unit so that the *dynamic foundation load* is considerably smaller than it would be without this frame.

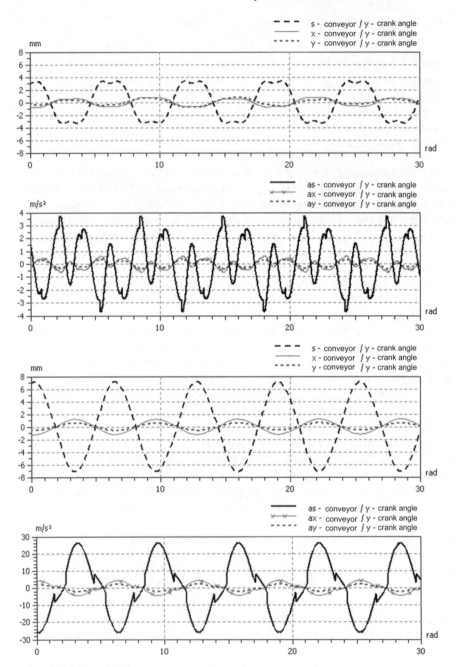

Fig. 7.4 Displacements and accelerations of the conveyor trough (s) and the frame (x, y) for variant 1; upper two figures: $\eta = 1.2$ ($\Omega = 19.6\,\text{rad/s}$); lower two figures: $\eta = 3.5$ ($\Omega = 57.16\,\text{rad/s}$)

7.2 Nonlinear Oscillators

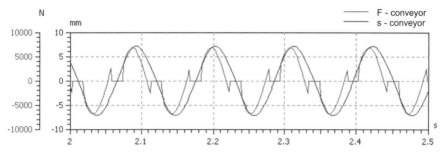

Fig. 7.5 Buffer force F and relative displacement of the conveyor trough at $\eta = 3.5$ ($\Omega = 57.16\,\text{rad/s}$)

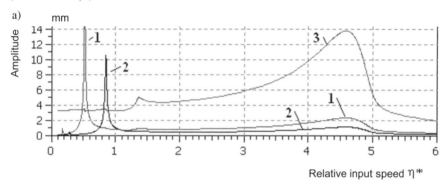

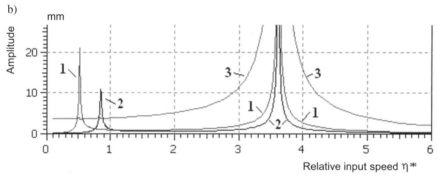

Fig. 7.6 Resonance curves of the oscillating conveyor (result of SimulationX®); **a)** variant 1, **b)** variant 2

7.2.3.2 Starting and Braking of a Processing Machine with a Nonlinear Coupling

Couplings with nonlinear characteristics are intentionally used in drive systems. One of the reasons is that their amplitudes in the resonance range are smaller than those of linear couplings. The solution of the applicable nonlinear equations of motion can only be approximated analytically and requires a great effort. Numerical simulation, that represents the numerical integration of the nonlinear equations of motion, however, is an effective way of studying the dynamic behavior.

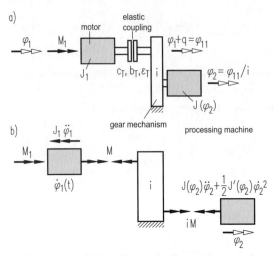

Fig. 7.7 Drive with a nonlinear coupling; **a)** System schematic, **b)** Free-body diagram of the system with the occurring moments

In the drive system outlined in Fig. 7.7a, the motor armature with a moment of inertia J_1 is connected via a nonlinear elastic coupling and a transmission (assumed to be without inertia and rigid) with a transmission ratio $i = \varphi_{11}/\varphi_2$ to a processing machine. The reduced moment of inertia of the processing machine is position dependent, like in a slider-crank mechanism, see (2.277) in Sect. 2.4.7:

$$J(\varphi_2) = J_0(1 + \mu \sin^2 \varphi_2); \qquad \frac{\mathrm{d}J}{\mathrm{d}\varphi_2} = 2J_0\mu \sin \varphi_2 \cos \varphi_2. \qquad (7.47)$$

The restoring moment of the coupling depends on the relative twist q, for which the following is valid

$$q = \varphi_{11} - \varphi_1 = i\varphi_2 - \varphi_1 \qquad \text{or} \qquad \varphi_2 = \frac{\varphi_1 + q}{i} \qquad (7.48)$$

as follows:

$$M = c_T(q + \varepsilon_T q^3) + b_T \dot{q}. \qquad (7.49)$$

Of interest is the dynamic load in the coupling both for the starting and the braking processes. The equations of motion can be established using the equilibrium of moments on both shafts, see Fig. 7.7b:

$$J_1 \ddot{\varphi}_1 - M = M_1 \qquad (7.50)$$

$$J(\varphi_2)\ddot{\varphi}_2 + \frac{1}{2}J'(\varphi_2)\dot{\varphi}_2^2 = -iM. \qquad (7.51)$$

7.2 Nonlinear Oscillators

These are two coupled nonlinear differential equations for the two angles φ_1 and φ_2. Not $M_1(t)$ but $\varphi_1(t)$ is predefined in this problem so that there is only one nonlinear differential equation to solve. It is derived from (7.47) and (7.49) after insertion into (7.51):

$$J_0(1 + \mu \sin^2 \varphi_2)\ddot{\varphi}_2 + \frac{1}{2}J_0\mu \sin 2\varphi_2 \dot{\varphi}_2^2 = -i[c_T(q + \varepsilon_T q^3) + b_T \dot{q}]. \qquad (7.52)$$

If one eliminates the coordinate φ_2 in (7.52) using (7.48), a nonlinear differential equation for the relative angle of rotation q is obtained that one can solve for the angular acceleration:

$$\ddot{q} = -\ddot{\varphi}_1 - \frac{i^2\left[c_T(q + \varepsilon_T q^3) + b_T \dot{q}\right] - \frac{1}{2}J_0\mu \sin\left(2\frac{\varphi_1 + q}{i}\right)\frac{(\dot{\varphi}_1 + \dot{q})^2}{i}}{J_0 + J_0\mu \sin^2\left(\frac{\varphi_1 + q}{i}\right)} \qquad (7.53)$$

One can use the function for the angle $\varphi_1(t)$ to numerically integrate (7.53) for predefined initial conditions, from which $q(t)$ and $\dot{q}(t)$ result. The coupling moment then follows from (7.49) and the angle φ_2 from (7.48). After calculating φ_2, one can calculate the input torque $M_1(t)$ required to produce the motion from (7.50).

For a special drive with the parameter values

Transmission ratio	$i = 2$
Moments of inertia	$J_0 = 2 \text{ kg} \cdot \text{m}^2; J_1 = 0.75 \text{ kg} \cdot \text{m}^2$
Pulsation depth of the moment of inertia	$\mu = 0.1$
Torsional spring constant	$c_T = 5000 \text{ N} \cdot \text{m}$
Torsional damping constant	$b_T = 1 \text{ N} \cdot \text{m} \cdot \text{s}$
Characteristic parameter of nonlinearity	$\varepsilon_T = 1.25$
Angular acceleration (plus for start-up, minus for braking)	$\ddot{\varphi}_1 = \alpha = \pm 0.002 c_T/J_0 = 5/\text{s}^2$

this integration was performed using the MathCad program.

The speed shows a linear change at a constant angular acceleration, i. e. it is assumed that the vibration system does not influence the input motion, see in contrast Sect. 5.2.

Figure 7.8 shows the solutions obtained by numerical integration for an angular velocity that shows a slow linear increase and decrease over time in the range of the frequency ratio $0 < \eta = \Omega/\omega_0 < 3$. η is the ratio of the instantaneous angular velocity $\dot{\varphi}_1 = \Omega(t) = \alpha t$ to the natural circular frequency $\omega_0 = i\sqrt{c_T/J_0}$ of the linear undamped oscillator. The resonance peak occurs at $\eta \approx 1.12$ during start-up, while it occurs at $\eta \approx 0.95$ during braking. This difference is due to the nonlinearity of the coupling, and it can be explained by the "jumping" of the steady-state amplitudes from one branch of the resonance curve to the other as was mentioned when discussing the Duffing oscillator in Sect. 7.2.2.

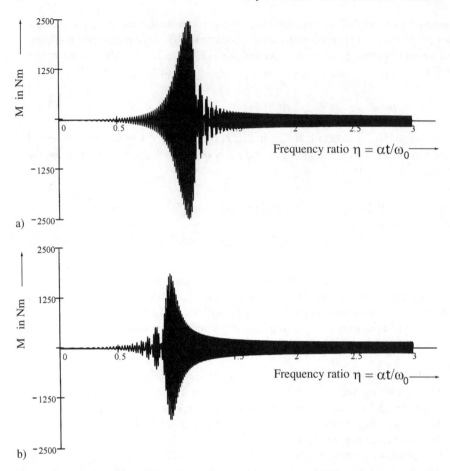

Fig. 7.8 Coupling moment M for **a)** Start-up, **b)** Braking

7.2.3.3 Self-Synchronization of Unbalance Exciters

Dutch physicist HUYGENS observed as early as in the 17th century that pendulum clocks that stood on a common board would swing synchronously after a certain amount of time. At the end of the 1940s, it was noticed that rotary motions of unbalanced rotors placed on the same support system influence each other. One had also observed accidentally that two rotors ran in sync although only one was connected to the power outlet (broken cable). This self-synchronization was brought about by the vibration on the common support.

Here we first look at the elementary process at which the pivot point of a rotor is moved at the circular frequency of the excitation Ω vertically with the amplitude $\hat{y}$:

$$y_A = \hat{y} \sin \Omega t. \tag{7.54}$$

7.2 Nonlinear Oscillators

The equation of motion is that of the physical pendulum with a moving pivot point and results from the equilibrium of moments about point A, see Fig. 7.9:

$$J_A \ddot{\varphi} + m(\ddot{y}_A + g)\xi_S \cos\varphi = -M_R \frac{\dot{\varphi}}{\Omega}. \tag{7.55}$$

A friction moment $M_R \dot{\varphi}/\Omega$ acts in the opposite direction of the instantaneous angular velocity.

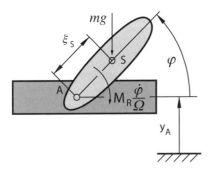

Fig. 7.9 Rotor or physical pendulum, excited vertically

If the acceleration resulting from (7.54) is inserted into (7.55), one obtains

$$J_A \ddot{\varphi} + M_R \frac{\dot{\varphi}}{\Omega} + m(g - \hat{y}\Omega^2 \sin\Omega t)\xi_S \cos\varphi = 0. \tag{7.56}$$

This is a nonlinear differential equation for the angle of rotation φ with variable coefficients that has various solutions. Of interest here shall only be the conditions under which the function

$$\varphi = \Omega t + \varphi^*; \qquad \dot{\varphi} = \Omega; \qquad \ddot{\varphi} = 0 \tag{7.57}$$

can be a solution of (7.56). φ^* is a constant phase angle with respect to the excitation (7.54). Inserting φ from (7.57) into (7.56) results in the relation

$$\xi_S \cos(\Omega t + \varphi^*)m(g - \hat{y}\Omega^2 \sin\Omega t) = -M_R. \tag{7.58}$$

Since the left side of this equation is variable and the right side is constant, it can obviously not be satisfied at all times. Viewed mathematically, (7.57) does not describe an exact solution of (7.56) but can be interpreted as an approximate solution if one only requires that (7.58) be satisfied on average. This is a rationale of the averaging methods, see the comments at the end of Sect. 7.1. If one considers what each of the terms in (7.58) means, one sees that there is a braking moment on the right side and a driving moment (lever arm by force) on the left side. The left side of (7.58) is a periodic function with the cycle time $T_0 = 2\pi/\Omega$ the mean value of which can be found by integrating over the entire period:

$$\frac{m\xi_S}{T_0}\int_0^{T_0}\cos(\Omega t+\varphi^*)(g-\hat{y}\Omega^2\sin\Omega t)\mathrm{d}t = \frac{1}{2}m\xi_S\hat{y}\Omega^2\sin\varphi^*. \tag{7.59}$$

The following expression follows from (7.58) and (7.59) for the phase angle

$$\sin\varphi^* = -\frac{2M_R}{m\xi_S\hat{y}\Omega^2}. \tag{7.60}$$

One can also conclude from this equation that such a phase angle becomes real (i.e. exists) only if the condition

$$\left|\frac{2M_R}{m\xi_S\hat{y}\Omega^2}\right| < 1 \tag{7.61}$$

is satisfied. (7.61), therefore, is a **condition for the self-synchronization of a rotor** on a vibrating support. The larger the unbalance $U = m\xi_S$ and the circular frequency of the excitation Ω, the easier the rotor can be "taken along". The friction moment must not be too large, otherwise this condition is violated as well.

Multiple unbalance exciters that rotate at the same angular velocity and defined phase differences are used for vibration excitation in many vibrating machines. A directed excitation force is frequently generated by two unbalanced rotors, the synchronous motion of which is forced by a pair of gears, see Fig. 7.10a. The disadvantage of such an arrangement is that wear and tear occurs in the toothing and noise develops because the load direction changes. Such an arrangement cannot be used if the vibration exciters are at large distances from each other, which is required for some vibrating machines, in particular for long oscillating conveyors. These shortcomings can be avoided if these drives synchronize themselves, see Fig. 7.10b.

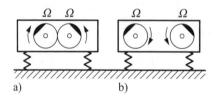

Fig. 7.10 Unbalance exciters on an oscillating body; **a)** Forced synchronization by a gear pair, **b)** Self-synchronization of two independent rotors

Instead of the excitation by constrained gears, a directed vibration excitation by two or more unbalance exciters that are **not** constrained is possible. The dynamic coupling is indirect via the translatory motion of the oscillating body in which the rotational axes are supported. Such excitation systems have been used more and more in past decades for vibration excitation in screening machines, oscillating conveyors, vibrating compactors and other machines, because they are easy to design and can do without mechanical or electrical shafts.

The conditions for stable operating states during **self-synchronization** shall be stated for an example here. The support system considered is an elastically set-up oscillating body that is excited at high excitation frequencies ($\Omega \gg \omega_i$) supercritically (low tuned) so that the inertia forces are dominant and the restoring forces can

7.2 Nonlinear Oscillators

be neglected ($k \to 0$). The rotors have to pass through the low critical speeds to reach their operating state. The axes of rotation of the two equal vibration exciters are parallel to a central principal axis of inertia of the oscillating body and are at the same distance ξ from the center of gravity S of the body, see Fig. 7.11.

It is a planar problem, i. e., the centers of gravity of the unbalanced masses are moving in the same plane. It follows from the theory that in this case synchronization occurs faster and more stably the larger the expression $m\xi^2/J_S$ is, i.e., the distance of the pivot points of the unbalance exciters from the center of gravity S has a strong influence.

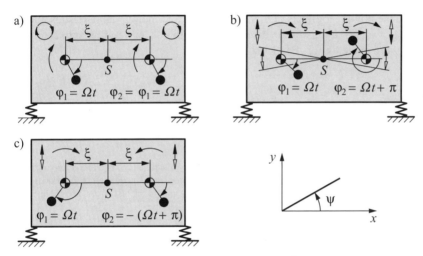

Fig. 7.11 Oscillating body with two unbalance exciters ; **a)** $m\xi^2/J_S > 2$, Circular translation, **b)** $m\xi^2/J_S < 2$, Torsional vibration about S, **c)** Parallel translational vibration for rotation in opposite direction

It follows that in the first case, the rotations with in-phase unbalances and in the second case, the rotations with antiphase unbalances are stable. If the rotors move into the same directions of rotation and the condition

$$m\xi^2 > 2J_S \tag{7.62}$$

is satisfied, a **circular translatory motion** ("circular translation") of the oscillating body occurs, see Fig. 7.11a. The amplitudes are

$$\hat{x} = \hat{y} = \frac{2m_1 e}{m}. \tag{7.63}$$

If the opposite inequality ($m\xi^2 < 2J_S$) is satisfied, **torsional vibrations of the oscillating body** occur (see Fig. 7.11b) with the angular amplitude

$$\hat{\psi} = \frac{2m_1 e \xi}{J_S}. \tag{7.64}$$

If the rotors rotate in opposite direction and $m\xi^2 > J_S$ applies, the opposite rotation with the phase difference π is stable. Straight-line **translatory vibrations** of the oscillating body are caused, see Fig. 7.11c. The amplitudes are

$$\hat{x} = 0; \qquad \hat{y} = \frac{2m_1 e}{m}. \tag{7.65}$$

There are currently more than 300 patents for machines and equipment with vibration exciters that utilize self-synchronization. The principles of self-synchronization cannot be predicted solely based on intuition or experimental trial and error. The dynamic behavior of such machines can be predicted using nonlinear calculation models that take the nonlinear effects of the inertia forces and motor characteristics into account. Major contributions to the development of this theory and its application in industry were made by I. I. BLEKHMAN [2], [31].

7.2.4 Problems P7.1 to P7.2

P7.1 Impact on a Nonlinear Spring

A body with mass m hits a buffer spring at a velocity v_0. Establish the equations for calculating the maximum displacement and the maximum force that is generated in the process. The buffer spring has a nonlinear spring characteristic. Compare the results that are obtained for a progressive, a linear, and a degressive spring characteristic.

Given:
  Impact velocity $v_0 = 4$ m/s
  Mass $m = 400$ kg
  Spring constant $c = 100$ kN/m
  Spring characteristic $F(q) = cq(1 + \varepsilon q^2)$
  progressive: $\varepsilon = 10$ m^{-2}; linear: $\varepsilon = 0$; degressive: $\varepsilon = -5$ m^{-2}

Find: Maximum deformations $q_{\max}$ and maximum forces $F_{\max}$.

P7.2 Mean Spring Constant for a Kinked Spring Characteristic

A practically relevant case of progressive spring characteristics is created when using stepped springs, see also Sect. 7.2.3.1. Only one spring acts for a small deflection, and an additional second spring acts from a specific deflection denoted as q_1.

Given: Spring-mass system according to Table 7.1, Case 1, with restoring force

$$F(q) = \begin{cases} c_1 q & \text{for } 0 < |q| < q_1 \\ (c_1 + c_2)q - c_2 q_1 & \text{for } q_1 < |q| < \hat{q}. \end{cases} \tag{7.66}$$

Find: Mean spring constant c_m for calculating the period of oscillation of the fundamental frequency for $\hat{q} > q_1$.
  1. In general for a kinked characteristic
  2. For a characteristic with backlash
  3. For a characteristic with preload

7.2 Nonlinear Oscillators

7.2.5 Solutions S7.1 and S7.2

S7.1 After the mass has hit the buffer spring, a free vibration starts, the maximum amplitude of which after the first semi-oscillation is of interest. The problem can be solved using an energy balance. The kinetic energy of the impacting mass is

$$W_{kin0} = \frac{1}{2}m(v_0)^2 = 3200 \text{ N} \cdot \text{m}. \tag{7.67}$$

The potential energy that can be stored in the buffer spring is

$$W_{pot\,max} = \int_0^{q_{max}} F(q)\,dq = \int_0^{q_{max}} cq(1+\varepsilon q^2)\,dq = \frac{1}{2}c(q_{max})^2 \left[1 + \frac{\varepsilon}{2}(q_{max})^2\right]. \tag{7.68}$$

$W_{pot\,max} = W_{kin0}$ results from the principle of work and energy. For nonlinear springs, this results in a quadratic equation for $(q_{max})^2$:

$$(q_{max})^4 + \frac{2}{\varepsilon}(q_{max})^2 - \frac{2mv_0^2}{c\varepsilon} = 0. \tag{7.69}$$

The solution is

$$q_{max} = \sqrt{\frac{1}{\varepsilon}\left(\sqrt{1 + \frac{2m\varepsilon v_0^2}{c}} - 1\right)}. \tag{7.70}$$

The maximum force is obtained from (7.14):

$$F_{max} = cq_{max}\left[1 + \varepsilon(q_{max})^2\right]. \tag{7.71}$$

Figure 7.12 shows the results. Using the parameter values given,

$$\underline{q_{max} = 226 \text{ mm}}; \qquad \underline{F_{max} = 34.1 \text{ kN}} \tag{7.72}$$

results for the progressive spring ($\varepsilon = 10 \text{ m}^{-2}$), and

$$\underline{q_{max} = 282 \text{ mm}}; \qquad \underline{F_{max} = 16.9 \text{ kN}} \tag{7.73}$$

for the degressive spring ($\varepsilon^* = -5 \text{ m}^{-2}$). The result for the linear spring ($\varepsilon = 0$) is

$$\underline{q_{max} = 253 \text{ mm}}; \qquad \underline{F_{max} = 25.3 \text{ kN}}. \tag{7.74}$$

Conclusion: A shorter spring deflection (but a larger force) is obtained for a progressive spring as compared to a linear spring, whereas the spring deflection becomes longer but the force becomes smaller for a degressive spring. It is therefore beneficial to use a degressive spring if one wishes to reduce the maximum impact force for an impacting process.

S7.2 $\hat{q} > q_1$ is assumed so that indeed a nonlinear suspension is obtained. It follows from (7.10) that

$$\underline{c_m} = \frac{5}{\hat{q}^5}\int_0^{q_1} c_1 q^4\,dq + \int_{q_1}^{\hat{q}} \left[(c_1 + c_2)q^4 - c_2 q_1 q^3\right]dq$$

$$= \frac{c_1}{4} + \frac{c_2}{4}\left[4 + \left(\frac{q_1}{\hat{q}}\right)^5 - 5\frac{q_1}{\hat{q}}\right]. \tag{7.75}$$

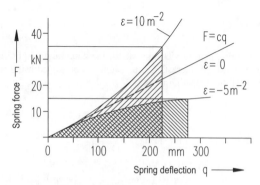

Fig. 7.12 Forces and displacements upon impact

The special case $c_1 = 0$ results in the technologically important case of *oscillators with backlash*, see also Sect. 4.2.1.2. The following results for the *oscillator with preload* (see Table 7.1, Case 2)

$$\underline{\underline{c_m}} = \frac{5}{\hat{q}^5} \int_0^{\hat{q}} (F_v + cq) q^3 \mathrm{d}q = \underline{\underline{c + \frac{5 F_v}{4\hat{q}}}}. \tag{7.76}$$

7.3 Self-Excited Oscillators

7.3.1 General Perspective

During self-excited vibrations, the oscillator itself controls the energy supply from an *external energy source*. Examples of utilized self-excited vibrations include the steam engine, the balance spring of a clock, and the sounds of wind and string instruments. Self-excited vibrations often are nuisances in mechanical engineering, e. g. the chattering of cutting machine tools, the frictional vibrations ("stick-slip") at slow slip velocities, the squeaking of brakes, the screeching of saws, the humming of rolling-mill frames, the whistling of valves when a fluid flows through them, the wobbling of windmill vanes, e. t. c.

Unlike forced and parameter-excited vibrations, the **equations of motion** of self-excited vibrations do not contain a term that explicitly depends on time. They correspond to the equations of motion of free nonlinear oscillators. Self-excited vibrations do not develop as a result of a single deflection from the static equilibrium position as is the case for free vibrations.

Both the formulation of the nonlinear equations of motion (model generation) and their mathematical solution often meet with certain difficulties. In general, these equations cannot be solved analytically. The numerical integration requires parameter values, which are only inaccurately known for the nonlinear terms. The exact time function of self-excited vibrations, however, is hardly of interest, since the goal mostly is to prevent these interfering vibrations altogether.

7.3 Self-Excited Oscillators

The energy source of self-excited oscillators is in general not time-dependent. The energy rations are transmitted for instance by a liquid, a gas or by a body via friction. Each period of oscillation of a self-excited oscillator has an interval of energy supply and an interval of energy decrease. If more energy is supplied (dissipated) than dissipated (supplied), the amplitudes increase (decrease). In the steady state, the energy balance per period is balanced and constant amplitudes develop. One can also interpret the energy balance in a way as if each period had discrete intervals of instability (stimulation) and stability (energy decrease). Often one would merely like to know whether there are any parameter regions of dynamic, temporally limited instability and thus of self-excitation. One can limit oneself to analyzing the linearized equations of motion to determine these because one can tell from the eigenvalues of a linear vibration system whether there is any stimulation.

The linear equations of motion have the form of (6.18). For conservative systems, the matrices M and C are symmetrical, and there are no negative eigenvalues. A linear system with an asymmetrical matrix can have negative eigenvalues. Instable domains of self-excited oscillators exist in those parameter regions in which negative eigenvalues occur. From a mathematical perspective, self-excited vibrations may occur when the linearized system of the equations of motion has an **asymmetrical stiffness matrix**. Such asymmetrical matrices are typical for journal bearings [8], [20] for oscillators with friction and for oscillators in flows, see, for example, (7.95).

7.3.2 Examples

7.3.2.1 Stick-Slip Vibrations

Friction-excited vibrations are mostly undesirable in machinery because they cause noise or wear and tear and reduce the accuracy of an output motion. They include the squeaking of brakes, the chattering of slow guiding motions, clutch grabbing, etc. Stick-slip occurs when for two surfaces in contact, intervals of sticking alternate with those of slipping during one period of oscillation. This is called the **stick-slip effect**.

A frictional oscillator will be studied here, in which a spring-mass system slides on a moving support [6], [24] which in this case represents the energy source. Figure 7.13a shows it as a moving belt but it could also have been a non-moving support and a moving spring-mass system. What is important here is the relative velocity, which in the ideal case is $v = $ const..

It is assumed that the coefficient of static friction μ_0 differs from the coefficient of sliding friction μ ($\mu_0 > \mu$) and that μ is constant, see Fig. 7.13b. This model for the sudden change in friction force from sticking to slipping is the simplest for the more complicated function of the friction force that in reality shows a nonlinear dependence on the velocity, normal force, temperature, time and the material parameters. The dashed friction force characteristic in Fig. 7.13b indicates such a function.

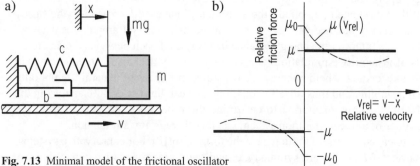

Fig. 7.13 Minimal model of the frictional oscillator
a) Calculation model, **b)** Friction force characteristic

It is assumed that the mass initially sticks to the support that moves at a constant velocity v and that the slipping motion starts at the moment when the mass breaks away from the support. The equation of motion then becomes (without damping)

$$m\ddot{x} + cx = \mu mg; \qquad \dot{x} \leqq v. \tag{7.77}$$

The initial conditions describe the fact that initially the spring force $c\,x_0$ is as large as the maximum adhesive force and the velocity of the mass m is equal to that of the support:

$$t = 0: \qquad x(0) = x_0 = \frac{\mu_0 mg}{c}; \qquad \dot{x}(0) = v_0 = v. \tag{7.78}$$

The solution obtained for the first interval ($0 \leq t \leq t_1$) is $\omega_0^2 = c/m$, see Fig. 7.14a:

$$\begin{aligned} x(t) &= \frac{\mu mg}{c} + (\mu_0 - \mu)\frac{mg}{c}\cos\omega_0 t + \frac{v}{\omega_0}\sin\omega_0 t \\ &= \frac{\mu g}{\omega_0^2} + \hat{s}\cos\omega_0(t - t^*) \\ \dot{x}(t) &= -\omega_0 \hat{s}\sin\omega_0(t - t^*) \end{aligned} \tag{7.79}$$

The following applies in the range $0 < \omega_0 t^* = \arctan\dfrac{v\omega_0/g}{\mu_0 - \mu} < \pi/2$

$$\sin\omega_0 t^* = \frac{v}{\hat{s}\omega_0}; \qquad \cos\omega_0 t^* = \frac{(\mu_0 - \mu)g}{\hat{s}\omega_0^2} \tag{7.80}$$

and the amplitude is

7.3 Self-Excited Oscillators

$$\hat{s} = \sqrt{\left[(\mu_0 - \mu)^2 \frac{mg}{c}\right]^2 + \left(\frac{v}{\omega_0}\right)^2} = \frac{mg}{c}\sqrt{(\mu_0 - \mu)^2 + \left(\frac{v\omega_0}{g}\right)^2}. \quad (7.81)$$

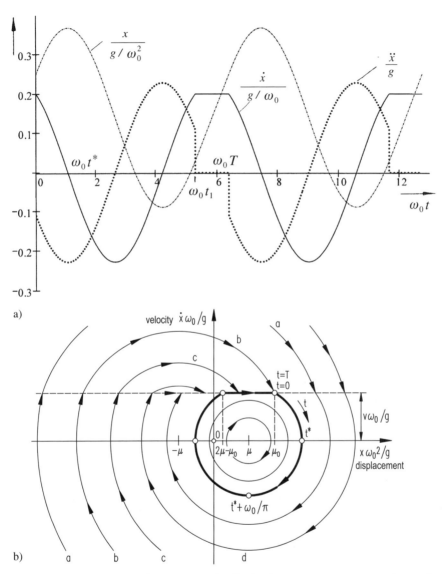

Fig. 7.14 Kinematic variables during a stick-slip vibration ($\mu_0 = 0.25$; $\mu = 0.14$; $\omega_0 v/g = 0.2$);
a) Time functions, dashed/dotted line: displacement $x\omega_0^2/g$, solid line: $\omega_0 \dot{x}/g$; dotted line: $\ddot{x}/g$
b) Phase curves

The velocity of the mass at first drops as a result of the braking friction force before it increases and once again reaches the velocity of the belt at the time t_1, see Fig. 7.14a. The condition

$$\dot{x}(t_1) = v = -\omega_0 \hat{s} \sin \omega_0 (t_1 - t^*) \tag{7.82}$$

results in

$$\omega_0 t_1 = \omega_0 t^* + \arcsin[v/(\omega_0 \hat{s})] + \pi. \tag{7.83}$$

The first interval is completed at the time t_1 and the displacement

$$x(t_1) = \frac{\mu m g}{c} + \hat{s} \cos \omega_0 (t_1 - t^*) = \frac{\mu m g}{c} + \hat{s}\sqrt{1 - \sin^2 \omega_0 (t_1 - t^*)} \tag{7.84}$$

$$= \frac{\mu m g}{c} + \sqrt{\hat{s}^2 - \left(\frac{v}{\omega_0}\right)^2} = \frac{\mu m g}{c} - (\mu_0 - \mu)\frac{mg}{c} = (2\mu - \mu_0)\frac{mg}{c}$$

is reached as follows from (7.79) with (7.81) and (7.82). At this moment, the second interval starts, in which the mass is carried along by the adhesive force. The velocities of mass and belt are constant and equal while the displacement and the spring force are increasing:

$$x(t) = x(t_1) + v(t - t_1) = (2\mu - \mu_0)\frac{mg}{c} + v(t - t_1); \qquad t_1 \leq t \leq T. \tag{7.85}$$

This interval ends when the adhesive force once again reaches its limit. This occurs after a full cycle of motion at the time T, which is why the initial values have to be valid again (period of oscillation T). Inserting the values from (7.78) results in:

$$F_{\max} = cx(T) = (2\mu - \mu_0)mg + cv(T - t_1) = \mu_0 mg. \tag{7.86}$$

In this equation, only the period of oscillation of the steady-state vibration considered is unknown. It is derived from it as follows

$$T = t_1 + 2(\mu_0 - \mu)\frac{mg}{cv} = t_1 + 2(\mu_0 - \mu)\frac{g}{v\omega_0^2}. \tag{7.87}$$

The curves for displacement, velocity and acceleration are shown in Fig. 7.14a.

This calculation was used to demonstrate the stable motion that develops on the so-called limit cycle, the closed curve in the phase plane (Fig. 7.14b). The curves run into this periodic motion for all initial conditions outside of the limit cycle. Studies with various other friction characteristics show that such a limit cycle exists for all descending characteristics (and taking damping into account). The connection of the (interfering) amplitude with the system parameters is quite interesting, see (7.81).

This problem was solved using SimulationX® [34], taking into account additional viscous damping. The calculated phase plot for the model in Fig. 7.13 is shown in Fig. 7.15. One can see that damped vibrations occur for initial conditions that describe a state inside of the limit cycle since all phase curves project spirally

7.3 Self-Excited Oscillators

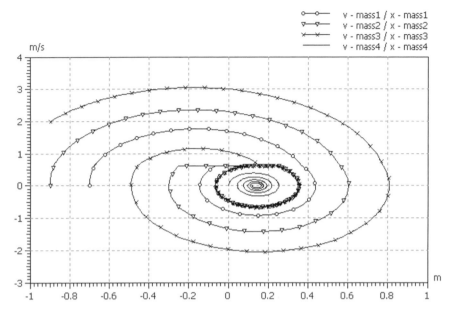

Fig. 7.15 Phase curves for three different initial states of damped frictional oscillators as shown in Fig. 7.13a calculated using SimulationX® ($\mu_0 = 0.25$; $\mu = 0.14$; $D = 0.06$ at the center, otherwise $D = 0.02$)

from there towards a vortex point. The motions that start outside of the limit cycle converge to the same limit cycle that is known from Fig. 7.14.

One can reduce or prevent the stick-slip vibrations by

- Reduction of the normal force F_N acting onto the surfaces with friction
- Increase of damping or friction
- Interference with the time function of the normal force, e. g. additional vibrations
- Lower slope of the descending friction characteristic, e. g. by another material
- Increase of the natural frequency (smaller mass, stiffer spring).

7.3.2.2 Flutter of a Plate with an Incident Flow

The critical speed for the severely simplified calculation model of an elastically supported plate that is exposed to an incident flow parallel to the plate plane is to be determined, see Fig. 7.16. In the horizontal position, static equilibrium exists and the hydro- or aerodynamic forces are zero. If the plate is deflected by an angle, a flow-dependent lifting force emerges, which, according to the laws of hydro- and aerodynamics, depends on a coefficient k, the density ϱ, the flow rate v, the length l, and the angle of the plate deflection φ:

$$F = k\varrho v^2 l \varphi. \tag{7.88}$$

The force application point that is determined by the geometry of the component exposed to an incident flow is at a distance e from the center of gravity S. The force F is assumed to be acting perpendicularly to the plate, neglecting higher-order effects.

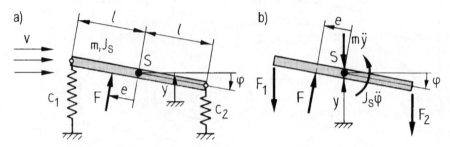

Fig. 7.16 Plate in a flow field; **a)** Parameters, **b)** Coordinates and forces

Figure 7.16a shows the calculation model with two degrees of freedom for the plate exposed to an incident flow in its deflected state. The bearings are taken into account by the spring constants c_1 and c_2 and the inertia properties of the plate by the mass m and the moment of inertia J with respect to the axis through the center of gravity. The static forces (static weight, spring forces) are in equilibrium and are omitted here. The associated free-body diagram is shown in Fig. 7.16b.

The restoring forces of the springs result from the product of the spring constants with the spring deflections, which depend on the deflection from the center of gravity y and the angle of rotation $\varphi \ll 1$ ($\sin \varphi \approx \varphi$, $\cos \varphi \approx 1$):

$$F_1 = c_1 \cdot (y + l\varphi), \qquad F_2 = c_2 \cdot (y - l\varphi). \tag{7.89}$$

The equilibrium conditions are therefore

$$\uparrow: \quad -F_1 - F_2 + F - m\ddot{y} = 0 \tag{7.90}$$

$$\widehat{S}: \quad -F_1 l + F_2 l + Fe - J_S \ddot{\varphi} = 0. \tag{7.91}$$

If one inserts the forces captured by (7.89) into (7.92) and (7.93), one obtains the equations of motion of the system where $q_1 = y$ and $q_2 = l\varphi$ in the form

$$m_1 \ddot{q}_1 + c_{11} q_1 + c_{12} q_2 = 0 \tag{7.92}$$

$$m_2 \ddot{q}_2 + c_{21} q_1 + c_{22} q_2 = 0 \tag{7.93}$$

with the mass and spring parameters

7.3 Self-Excited Oscillators

$$m_1 = m; \qquad m_2 = \frac{J_S}{l^2} = \frac{m}{12}$$
$$c_{11} = c_1 + c_2; \qquad c_{12} = c_1 - c_2 - k\varrho v^2 \qquad (7.94)$$
$$c_{21} = c_1 - c_2; \qquad c_{22} = c_1 + c_2 - k\varrho v^2 \frac{e}{l}.$$

If one looks for a solution of the equations of motion in the form $q_j = v_j e^{\lambda t}$, one obtains two linear equations for the amplitudes v_j and the eigenvalue λ:

$$\begin{aligned}\left(c_{11} + m_1 \lambda^2\right) v_1 + c_{12} v_2 &= 0 \\ c_{21} v_1 + \left(c_{22} + m_2 \lambda^2\right) v_2 &= 0.\end{aligned} \qquad (7.95)$$

After setting the coefficient determinant to zero, the characteristic equation

$$p(\lambda) = \begin{vmatrix} c_{11} + m_1 \lambda^2 & c_{12} \\ c_{21} & c_{22} + m_2 \lambda^2 \end{vmatrix}$$
$$= m_1 m_2 \lambda^4 + (c_{11} m_2 + c_{22} m_1) \lambda^2 + c_{11} c_{22} - c_{12} c_{21} = 0 \qquad (7.96)$$

follows. The roots of this biquadratic equation are the eigenvalues:

$$\lambda_{1,2,3,4} = \pm \sqrt{-\frac{c_{11} m_2 + c_{22} m_1}{2 m_1 m_2} \left[1 \pm \sqrt{1 - 4\frac{(c_{11} c_{22} - c_{12} c_{21}) m_1 m_2}{(c_{11} m_2 + c_{22} m_1)^2}}\right]}.$$
$$(7.97)$$

The plate motion is stable if the real parts of all $\lambda_i \leq 0$. Since, according to (7.97), both signs are in front of the first root, the expression under the outer root must be real and negative. Thus, there are only purely imaginary roots. All other possibilities are excluded because they result in at least one root with a positive real part, which means instability. One stability condition is therefore

$$c_{11} m_2 + c_{22} m_1 > 0. \qquad (7.98)$$

It must also be ensured that the inner root in (7.97) is real and smaller than one, otherwise conjugate complex roots occur in which one of the real parts may be positive. The following relation has to be satisfied:

$$0 < \frac{4 (c_{11} c_{22} - c_{12} c_{21}) m_1 m_2}{(c_{11} m_2 + c_{22} m_1)^2} < 1. \qquad (7.99)$$

7.3.2.3 Chattering of Machine Tools during Cutting Operations

Vibrations at the contact point between workpiece and tool often interfere with the machining precision in machine tools. Self-excited vibrations can occur after exceeding certain cutting parameters (e. g. cutting depth limit, cutting rate, tool parameters). A designer is less interested in the time function of these nonlinear vibrations that are called "chattering", but rather in the question at which parameters chattering starts, for this limit should be set as "high" as possible. The main causes of self-excited vibrations during cutting are the so-called *position coupling* and the *regenerative effect*.

The term "position coupling" is a vibration process at which two or more mode shapes of natural frequencies that are close together are coupled via the cutting operation and a vibration is stimulated. These normal mode shapes influence each other at the contact point of workpiece and tool if they reach specific amplitudes there so that a transmission of energy onto the machine frame may occur during cutting. If one of the respective normal mode shapes is excited, this acts via the cutting force as an excitation of the other "position-coupled" normal mode shape.

During milling and turning, a vibration at a natural frequency of the machine that was excited by a minor interference causes a corrugation of the machined workpiece surface. After one revolution of the workpiece during turning or after passing across the angle $2\pi/n_z$ during milling with n_z cutters, the corrugation produced by the previous cut is cut into again. This causes a variation in the cutting force that parting turn, under specific conditions, can maintain, that is "regenerate" or even stimulate, the previous vibrations. As for forced vibrations (see Sect. 3.3.3), it depends on the phasing of the next excitation and on the damping whether the vibration between workpiece and tool increases (stimulation, instability) or whether it subsides.

The cutting force and thus the intensity of the excitation depend directly on the cutting depth a. It has been observed many times that exceeding a specific cutting depth results in chattering. The following is meant to provide an introduction into the topic by presenting a simple calculation model for recessing.

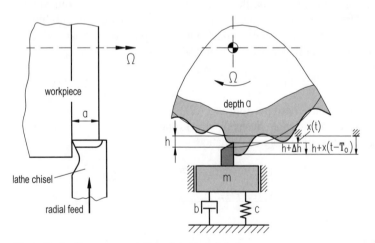

Fig. 7.17 Cutting process; **a)** Side view, **b)** Axial view

$x = 0$ is the static equilibrium position of the oscillator if it is cut vibration-free at a chip thickness h, see Fig. 7.17. As a result of a minor perturbation, a displacement x occurs that results in a change in the chip thickness. If one considers the previous cut

7.3 Self-Excited Oscillators

that was made one period of oscillation (time of a revolution $T_0 = 2\pi/\Omega$) earlier, that is, at the time $t - T_0$, the resulting change in the chip thickness is

$$\Delta h = x(t - T_0) - x(t). \tag{7.100}$$

For the radial cutting force change ΔF, the cutting depth a and the change in the chip thickness Δh can be assumed as a first approximation in the form

$$\Delta F = ak_x(h)\Delta h = ak_x[x(t - T_0) - x(t)]. \tag{7.101}$$

The chip thickness coefficient k_x depends on the material, the cutter geometry, the cutting angle and other cutting parameters. Only the modal parameters of the calculation model shown in Fig. 7.17 are of interest, because it characterizes a normal mode shape of the machine, which results in a radial motion between workpiece and tool. The following equation of motion is obtained:

$$\ddot{x} + 2D\omega_0\dot{x} + \omega_0^2 x - \alpha\omega_0^2[x(t - T_0) - x(t)] = 0; \qquad \alpha = a\frac{k_x}{c}. \tag{7.102}$$

The dynamic properties of the machine are captured by the natural circular frequency ω_0 (in undamped state) and the damping ratio D. The factor α, which contains the cutting parameters, is proportional to the cutting depth. If the solution is assumed to be of the form $x = \hat{x}\exp(\lambda\omega_0 t)$ and it is inserted into (7.102), the resulting characteristic equation is

$$\lambda^2 + 2D\lambda + 1 + \alpha[1 - \exp(-\lambda\omega_0 T_0)] = 0. \tag{7.103}$$

The regenerative effect, that is, the influence of the waviness present from the previous pass, is expressed here by the exponential function.

This is a transcendental equation for λ that can have real or conjugate complex roots. It depends on the sign of the real part if a damped or a stimulated vibration is produced. The stability limit is where the real part is zero, i.e., it can be found using $\lambda = 0 + j\omega/\omega_0 = 0 + j\nu$ and Euler's relation. The limit α_g, from which the cutting depth limit that is of practical interest can be determined, occurs at the stability limit. $\omega = \nu\omega_0$ is the still unknown circular chattering frequency. At the stability limit, therefore, the condition

$$1 - \nu^2 + \alpha_g(1 - \cos\omega T_0) + j(2D\nu + \alpha_g \sin\omega T_0) = 0. \tag{7.104}$$

holds, which follows after inserting the assumed solution form into (7.103). This complex equation is satisfied if its real and imaginary parts are zero. After elimination of the trigonometric functions, its solution provides the limit for the cutting parameter and the associated pass times as functions of $\nu = \omega/\omega_0$:

$$\alpha_g = \frac{(1-\nu^2)^2 + (2D\nu)^2}{2(\nu^2 - 1)}; \qquad T_0 = \frac{2}{\nu\omega_0}\mathrm{arccot}\left(\frac{2D\nu}{1-\nu^2} + k\pi\right) \tag{7.105}$$

$$k = 0, 1, 2, \ldots$$

The minimum of α_g is of particular interest. It is reached at $\nu = \sqrt{1+2D}$ and amounts to

$$\alpha_{g\,min} = 2D(1+D). \qquad (7.106)$$

The obtainable cutting depth therefore depends on the damping ratio. This minimum, however, occurs at specific pass times only. For $\nu = \sqrt{1+2D}$, (7.105) results in a relation for those angular velocities $\Omega = 2\pi/T_0$, below which stable cutting is possible:

$$\omega_0 T_0 \big|_{\alpha_g = \alpha_{g\,min}} = \frac{2}{\sqrt{1+2D}} \left[\left(k + \frac{1}{2}\right)\pi + \arctan\sqrt{1+2D} \right]; \qquad (7.107)$$
$$k = 0, 1, 2, \ldots$$

For example, for the damping ratio $D = 0.08$ this results in $\alpha_{g\,min} = 0.1728$ and $\omega_0 T_0 = 4.444$ for $k = 0$. One can calculate the stability plot from the equations (7.105) by varying ν. If the cutting parameters are known, this will provide a picture of the possible cutting depths as a function of speed.

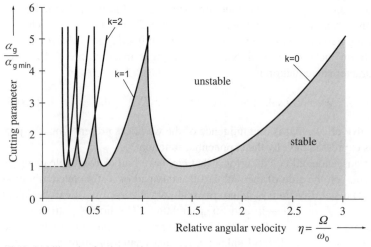

Fig. 7.18 Stability plot for the cutting operation

Figure 7.18 shows the result for a practical example. The curve has the typical shape of a garland-type curve because various local minima line up.

7.3.3 Problems P7.3 and P7.4

P7.3 Instability of a Brake

An elastically anchored mass slides along an planar rough surface at a constant velocity v, see Fig. 7.19. Only the principal stiffness that is oriented in the slanted direction (of the

7.3 Self-Excited Oscillators

actually two-dimensional anchoring) is taken into account by the spring constant c, see Table 1.4. Adhesion is assumed to be excluded, i. e. the velocity of the mass is $|\dot{x}| < v$. The initial angle is in the range of $0 < \beta_0 < \pi/2$.

Given:

Length of the unstretched spring	l_0
Angle of inclination of the unstretched spring	β_0
Gravitational acceleration	g
Mass	m
Spring constant	c
Coefficient of sliding friction	μ

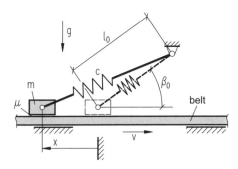

Fig. 7.19 Minimal model for the motion of a brake mass

Find:
1. Free-body diagram of the mass with equations of motion and constraints
2. Spring length $l(x)$ and nonlinear differential equation
3. Linearized equation of motion for $x \ll l_0$
4. Condition for the system to become instable, especially the instability range for the angle β_0 at $\mu = 0.25$.

P7.4 Stability Plot

Establish the stability plot for the flutter vibrations of the plate with an incident flow studied in Sect. 7.3.2.2, from which parameter regions can be detected, at which this vibration becomes stable or instable.

Given: The stability conditions (7.98) and (7.99) from Sect. 7.3.2.2.
Dimensionless characteristic parameters

$$\gamma = \frac{c_1}{c_2}; \qquad \varepsilon = \frac{e}{l}; \qquad \varkappa = \frac{k\varrho v^2}{c_2}. \tag{7.108}$$

Find:
1. Stability conditions, expressed in terms of the dimensionless characteristic parameters
2. Entry of curves for the limits of the stability regions into a diagram $\varkappa = \varkappa(\gamma)$
3. Discussion of parameter influences

7.3.4 Solutions S7.3 and S7.4

S7.3 The friction force does not reverse its direction due to $|\dot{x}| < v$. The following results from the equilibrium of forces, see Fig. 7.20:

$$\text{horizontal:} \quad \mu F_N + c(l - l_0)\cos\beta + m\ddot{x} = 0 \quad (7.109)$$
$$\text{vertical:} \quad F_N + c(l - l_0)\sin\beta - mg = 0 \quad (7.110)$$

The following constraints apply to the angles, the displacement and the spring length:

$$\text{horizontal:} \quad x + l_0\cos\beta_0 = l\cos\beta \quad (7.111)$$
$$\text{vertical:} \quad l_0\sin\beta_0 = l\sin\beta \quad (7.112)$$

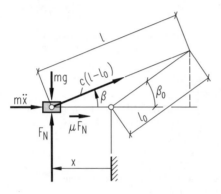

Fig. 7.20 Free-body diagram for the calculation model of the brake

(7.109) to (7.112) provide the four basic equations, from which the four unknown variables x, β, l and F_N are to be calculated from the six parameters (μ, c, g, m, l_0, and β_0).

First, l and β are calculated as functions of x. Squaring and summing of (7.111) and (7.112) after a brief calculation provides

$$l^2 = (x + l_0\cos\beta_0)^2 + (l_0\sin\beta_0)^2 = x^2 + 2xl_0\cos\beta_0 + l_0^2 \quad (7.113)$$

and thus the spring length

$$\underline{l(x) = \sqrt{x^2 + 2xl_0\cos\beta_0 + l_0^2}}. \quad (7.114)$$

Insertion of F_N from (7.110) into (7.109) provides a single equation of motion:

$$m\ddot{x} + \mu[mg - c(l - l_0)\sin\beta] + c(l - l_0)\cos\beta = 0 \quad (7.115)$$

or

$$m\ddot{x} + c(l - l_0)(\cos\beta - \mu\sin\beta) + \mu mg = 0. \quad (7.116)$$

After taking into account (7.111) and (7.112), this becomes

$$\underline{m\ddot{x} + c[1 - l_0/l(x)](x + l_0\cos\beta_0 - \mu l_0\sin\beta_0) + \mu mg = 0.} \quad (7.117)$$

Inserting the expression for $l_0/l(x)$ results in a nonlinear differential equation for the coordinate x from which the vibrations can be calculated. The linearized form of this differential equation suffices for evaluating the stability behavior. Expansion of $l_0/l(x)$ into a Taylor

7.3 Self-Excited Oscillators

series for $x/l_0 \ll 1$ provides

$$\frac{l_0}{l} = \frac{l_0}{\sqrt{x^2 + 2xl_0 \cos\beta_0 + l_0^2}} = \left(1 + \frac{2x}{l_0}\cos\beta_0 + \frac{x^2}{l_0^2}\right)^{-1/2} \tag{7.118}$$

$$= 1 - \frac{x}{l_0}\cos\beta_0 - \frac{1}{2}\left(\frac{x}{l_0}\right)^2 \cdot (1 - 3\cos^2\beta_0) + \ldots$$

It follows from (7.117) and (7.118) when neglecting all second- and higher-order terms

$$m\ddot{x} + cx(\cos\beta_0 - \mu\sin\beta_0)\cos\beta_0 + \mu mg = 0. \tag{7.119}$$

Vibrations are stimulated if the factor of the linear term becomes negative. This means the motion becomes instable at $\cos\beta_0 - \mu\sin\beta_0 < 0$, i.e. at steep angles in the region of self-locking

$$\tan\beta_0 > \frac{1}{\mu}, \tag{7.120}$$

that means initially exponentially increasing displacements occur for angles $\beta_0 > \arctan(1/\mu)$. The function of the snap-through motion could be calculated by solving (7.117). Since (7.117) contains the nonlinearity of $l(x)$, the deflections remain finite. It has to be taken into account after the snap through that the direction of the friction force and thus the sign of the coefficient of sliding friction change. Self-excited vibrations could be calculated when taking into account an additional vertical motion in the model of Fig. 7.19 [4], [5]. A limit angle $\beta_0^* = 75.96$ Grad is found for the coefficient of sliding friction of $\mu = 0.25$.

S7.4 The expression (7.98), taking into account the parameters and characteristic numbers from (7.108) as defined in (7.94), takes the following form:

$$c_{11}m_2 + c_{22}m_1 = \frac{(c_1 + c_2)m}{12} + \left(c_1 + c_2 - \frac{k\varrho v^2 e}{l}\right)m \tag{7.121}$$

$$= c_1 m \frac{\left[1 + \frac{c_1}{c_2} + 12 + 12\frac{c_1}{c_2} - 12\frac{k\varrho v^2 e}{c_2 l}\right]}{12} \tag{7.122}$$

$$= \frac{[13(1 + \gamma) - 12\varkappa\varepsilon]c_1 m}{12}. \tag{7.123}$$

Thus, the first stability condition in dimensionless form is

$$13(1 + \gamma) - 12\varkappa\varepsilon > 0. \tag{7.124}$$

A simple condition results from the inequality (7.99) on the left

$$c_{11}c_{22} - c_{12}c_{21} > 0. \tag{7.125}$$

Using the parameters according to (7.94), this turns into the stability condition

$$4\gamma - (1 + \gamma)\varkappa\varepsilon - (1 - \gamma) \cdot \varkappa > 0. \tag{7.126}$$

The right side of (7.99) provides another stability condition

$$4(c_{11}c_{22} - c_{12}c_{21})m_1 m_2 < (c_{11}m_2 + c_{22}m_1)^2. \tag{7.127}$$

If this condition is transformed, it becomes

$$4\gamma - (1+\gamma)\varkappa\varepsilon - (1-\gamma)\varkappa < 3\left[\frac{13}{12}(1+\gamma) - \varkappa\varepsilon\right]^2. \qquad (7.128)$$

The critical flow rates can be found in dimensionless form from the three stability conditions (7.124), (7.126), and (7.128). If $\varepsilon > 0$ is assumed, it follows from (7.124) that the system is stable if

$$\varkappa < \frac{13(1+\gamma)}{12\varepsilon} \qquad (7.129)$$

is valid. The stability condition results from (7.126)

$$\varkappa \lesseqgtr \frac{4\gamma}{\varepsilon(1+\gamma) - \gamma + 1} \qquad \text{for} \qquad \gamma \lesseqgtr \frac{1+\varepsilon}{1-\varepsilon}, \qquad (7.130)$$

the upper and lower relation operators being associated with each other. It follows from (7.128):

$$\varkappa < \frac{1}{12\varepsilon^2} \cdot \Big[11\varepsilon \cdot (1+\gamma) + 2 \cdot (\gamma - 1) \\ -2 \cdot \sqrt{(\gamma - 1) \cdot [11\varepsilon \cdot (\gamma + 1) - (\gamma + 1) \cdot (12\varepsilon^2 - 1)]} \Big]. \qquad (7.131)$$

One can also plot the curves that result from (7.129) to (7.131) for the equality signs into a γ-$\varkappa$ diagram. Each of these curves splits the domain into a stable and an instable region.

The smaller of the two roots from the quadratic equation that results for $\varkappa$ was used since it provides the lower $\varkappa$ values.

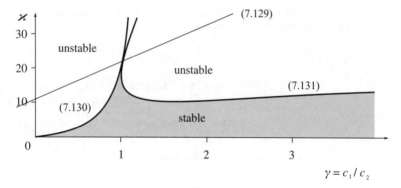

Fig. 7.21 Stability plot for the fluttering of a plate

Figure 7.21 shows the limits of the region in which these inequalities are satisfied for the value $\varepsilon = 0.1$. The critical speed of the flow from which fluttering can occur is the smallest of these values from the three conditions (7.129) to (7.131). This figure shows the different effects of the two springs. Compare for instance the values for $\gamma = 2.0$ with those for $\gamma = 1/2$, for which the same natural frequencies occur without a flow.

Chapter 8
Rules for Dynamically Favorable Designs

The term "dynamically favorable design" is vague and ambiguous. It can be about saving material or energy (power input), preventing excessive wear and tear, increasing service life, reliability, and productivity (speed), improving the working conditions for humans, or utilizing vibrations and impact forces for technological purposes.

From a mechanical point of view, the problem of the synthesis of mechanical structures with respect to dynamic criteria arises when designing a machine. Classical mechanics with its traditional methods provides answers for the question how a mechanical structure behaves at a given excitation. However, when designing a machine, the question is which mechanical structure should be used and what measures should be taken to achieve dynamically favorable behavior in a machine. There is no deterministic algorithm for solving such problems in which a mechanical structure is sought for a specific processing or transport operation (motion or force function at the application point).

There are catalogs, taxonomies, and heuristic methods for finding solutions for some classes of problems. In this section, some rules that are sorted by six aspects and cover a range of practical problems in machine dynamics will be discussed. They can serve as guidelines when searching for dynamically favorable solutions, but they will not suffice to capture the inexhaustible variety of practical problems. One can get an idea of how vast this field is by just remembering that hundreds of patent applications are filed each year in which novel solutions for specific machines are proposed.

Aspect 1: Clarification of the Relevant Physical Phenomena

A machine is usually not created "from scratch". Mostly, an existing design is perfected and improved, or certain shortcomings of such a design are eliminated. Based on experience, the designer has to assign a physical cause to the problem at hand or, in the worst case, start searching based on a working hypothesis.

The dynamic phenomena that cause trouble (or are to be utilized) should, for best results, be sorted by their cause of excitation or be evaluated qualitatively using various models. The engineer has to decide if he or she is dealing with

- forces of the rigid machine (e. g. unbalance excitation, mechanisms, processing forces)
- vibration forces of free, forced, or parameter-excited vibrations
- force parameters from nonlinear or self-excited vibrations,

see the three model stages in Sect. 1.1.1. The borders between these phenomena (which appear to be clearly delimitable from a mathematical point of view) are not as clear-cut from a physical point of view. Strictly speaking, all real processes are not free from reaction and therefore have to be viewed as free (autonomous) vibrations.

Different calculation models are associated with a machine throughout its historical development, which mostly involves an increase in speed. Many cases of damage can be traced back to using a calculation model that is too simple (and to a lack of understanding of what is actually going on physically). During the improvement of a machine, one frequently ventures into domains in which the calculation models used before (and sometimes even sanctioned by regulations) no longer apply.

Many aspects of model generation and mathematical problem formulation discussed for the entire fields of applied mathematics and mechanics naturally apply to machine dynamics as well. Requirements that a model be adequate, simple, and optimal, questions regarding the influence of factors not taken into account, phenomenological and semi-empirical laws, the number of degrees of freedom, the hierarchy of the models and variables are discussed there and will therefore only be referenced here.

Certain design solutions can only be implemented with the new possibilities offered by microelectronics, computers, and new physical effects. Suitable open-loop or closed-loop control, for example, in conjunction with vibration sensors for acquiring measured data, can effect self-adaptive changes of the mass, spring, or damping characteristics of a system during operation and the application of forces that oppose interfering forces. Most recently, many new controllers have been proposed and the respective patent applications filed that will enable an active form of mass and power balancing, vibration isolation, or limit monitoring and diagnostics. It is frequently about early detection of abnormal vibration states and response to abnormal frequencies, amplitudes, or more complex characteristics. Advanced options include comparing measured instantaneous data with stored data for normal or permissible vibration states and, if predetermined limits are exceeded, to output such signals that initiate a change of state using electrical, mechanical, hydraulic and other effects.

Aspect 2: Selection of the Kinematic System

Far-reaching decisions are made already when selecting the operating principle. It is a designer's job to define the topological structure of the drive system and the support structure in such a way that the force flow within the mechanism between motor and application point (input and output) and within the frame between application point and the foundation is as short as possible. The dynamic forces are mostly less determined by the time function of the technological process (e. g. press force) than by the inertia forces from excited natural vibrations of the design, i. e. the chosen structure itself determines major loads and deformations. It is recommended

8 Rules for Dynamically Favorable Designs

to apply the methods and tools of mechanism systematics, design systematics, drive engineering and the use of databases, such as patent references.

Certain vibration aspects have to be taken into account as early as during mechanism synthesis when the structure and dimensions of a mechanism are defined based on geometric requirements at the output link. This includes the requirement to use mechanisms with a minimum number of links. Using computers for synthesis and optimization can make it possible to meet requirements with six-bar linkages, cam mechanisms, and simple geared linkages that used to be realizable only with multi-bar linkages. The fewer links and joints, the smaller typically are the deformations, the impact of backlash, and the noise.

Other rules under this aspect include:

- to prefer force-closed joints to form-closed joints,
- to keep velocities and accelerations as constant as possible (avoid discontinuities),
- to start from the required sequence of motions at the application point and only to comply with the technological minimum requirements,
- to perform the required input motion only as fast as necessary,
- to move intermediate links as slowly as possible,
- to replace the absolute motion of an output link at the application point with relative motions of two mechanism links,
- to move as few bodies nonuniformly as possible,
- to apply closed trajectories rather than reversing motions and use circular trajectories as often as possible,
- to perform as small motions as possible,
- to provide adjustable links to compensate for deviations from the perfect geometrical structure,
- to prefer area contact to line contact and line contact to point contact.

Aspect 3: Influencing Kinetostatic Force Functions

An analysis of the model of the rigid machine (kinetostatic analysis) is the starting point for questions regarding dynamic loading. Kinematics has a major influence on the kinetostatic force functions since the mass distribution effectively determines only the proportionality factors for the accelerations. Reducing the kinetostatic forces is a key issue of dynamic synthesis.

The following rules should be observed for this purpose:

- Transmit the force along the shortest path in space,
- Transmit tensile and compressive forces and no bending and torsional moments, if possible,
- Ensure statically determinate transmission of forces,
- Make the degree of static indeterminacy independent of manufacturing processes, e. g. compensate misalignment,
- Build as small and compact as possible,
- Ensure permanent force closure by preloading,

- Avoid changes in force direction or perform passage of the force through zero as slowly as possible,
- Balance rotating bodies,
- Balance masses in mechanisms by changing the mass parameters or implementing auxiliary mechanisms,
- Generate electromagnetic, hydraulic, or pneumatic compensatory forces,
- Provide compensatory springs and optimize spring parameters,
- Balance the force by an opposing force as closely to the application point as possible,
- Provide mutual compensation of excitation forces, e. g. by optimum crankshaft offset angles,
- Use compensators for storage of kinetic or potential energy,
- Adapt motor and braking torque to the required moment curve.

It has to be noted for measures based on kinetostatics that they often are of a conflicting nature. Masses attached for mass balancing can result in a reduction of the natural frequencies and an increase of the variable portion of the reduced moment of inertia and thus in increased excitation of torsional vibrations. One therefore often has to look beyond the limits of the kinetostatic model since the dynamic problems are complex. If possible, one should estimate the consequences with respect to the elastic model.

Aspect 4: Prevention of Resonances

In addition to the kinetostatic inertia forces that can be calculated from geometrical relations, vibration forces that often exceed the kinetostatic forces can result from natural vibrations. It should be prevented that the time-dependent kinetostatic forces excite natural vibrations in resonance.

The *qualitative* understanding of the occurring physical processes can often be gained by just looking at an oscillator with one degree of freedom. If one identifies a process as principal vibration (modal vibration), many things become explainable. This requires knowledge from the linear theory of vibrations. The spectral and modal properties of a system become more comprehensible if one can determine the excitation and natural frequency spectrum and the modal matrix (normal mode shapes).

The requirement to avoid resonances by placing the natural frequencies far away from the regions of the excitation frequencies is generally known.

Most processing machines follow the rule that they operate without dynamic interference if their first natural frequency is at least 10 to 15 times higher than the fundamental excitation frequency of the drive mechanism. One therefore always tries to get into this region. As is known, one can increase a natural frequency by increasing its modal stiffness, that is, by increasing cross sections or decreasing the distance between supports of beams. One can also reduce the modal mass for the same purpose by applying lightweight design principles or using light metal alloys. Important issues in this respect are the parameter influences and the sensitivity of the natural frequencies and normal mode shapes to parameter changes.

8 Rules for Dynamically Favorable Designs

Rules for this aspect include:

- Reduce major harmonics of excitation, e. g. cutters with uneven tooth spacing,
- Develop cam profiles with a minimum number of harmonics (HS profiles),
- Optimize linkages with respect to the excitation harmonics,
- Detune meshing frequencies with respect to the natural frequencies,
- Pass through the resonance domain fast,
- Tune the impact sequence to the period of oscillation of the natural vibrations,
- Detune chain engagement frequencies and chain rotation frequencies with respect to the natural frequencies,
- Attach auxiliary oscillators and tune them to the principal excitation frequency (vibration absorbers),
- Initiate sensor-controlled auxiliary motions (actuators),
- Reduce resonance amplitudes by materials with strong damping properties (e. g. rubber springs rather than steel springs),
- Perform strength tests if resonance is inevitable,
- Use damping elements (e. g. for torsional vibrations).

Aspect 5: Reduction of Impact Loads

Transient processes such as starting, braking, loading, unloading and impacting play an important role in machines, in addition to periodic processes. The most important characteristic number is the ratio $\nu = \Delta t/T$ of force change time to period of the natural vibration for forces that increase or decrease fast. It is decisive for the magnitude of the excited vibration amplitudes.

If $\nu \leq 0.25$, the system responds to this (kinetostatically – steady) excitation like to a sudden jump, regardless of the time function of the force during the increase time. A system responds to loads of very short duration like to a jump in velocity. It is often important to avoid or minimize such impact loads for practical purposes. The following rules apply:

- Reduce the relative velocity of the impacting masses,
- Make impacting masses as small as possible,
- Keep the stiffness at the impact point as small as possible,
- Couple the impacted mass or subsystem as resiliently as possible to the overall system,
- Reduce the mechanical work at the impact point by influencing the direction of impact,
- Let the impact force act onto the node of the vibration (far away from its antinode),
- Minimize clearance in joints or eliminate it by preload,
- Place a predetermined breaking point near the force application point,
- Absorb the impact force by an opposing force near the impact point (e. g. counterblow for hammers),
- Reduce clearance by classifying components (clearance pairs),

- Tune the application time of the excitation to the period of the principal natural vibrations, e. g. starting, braking, or reversing time.

If the real excitation forces and motions cannot be reduced, one should try to utilize their relations to the normal mode shapes.

Aspect 6: Reduction of Modal Excitation Forces

The key here is to influence the normal mode shapes in space. These are only known exactly after calculating or measuring the vibrations, but they can often be estimated. One should at least know the position of the nodes or the spatial distribution of the amplitudes of the vibrations.

It is often possible to change the normal mode shapes by influencing the static deformation behavior so that the function-related mass distribution can be retained. A simple example is the gripping point of a hand on a hammer that remains immobile during the blow as the center of percussion of the rigid body so that the "support force" on the hand remains zero. In general, the following rules apply to this aspect:

- Influence the normal mode shapes by such structural changes as additional support columns, clamping, adding or omitting bearings and/or joints,
- Change the normal mode shapes in such a way that the work of various excitation forces is mutually reduced or neutralized,
- Let the excitation force act onto the center of percussion of rigid bodies,
- Utilize the fact that symmetrical excitation forces do not excite antisymmetrical normal mode shapes and applying excitation forces antisymmetrically does not excite symmetrical normal mode shapes in a symmetrical system,
- Split the overall system into subsystems so that excitations that act onto one of the subsystems do not affect the others.

To summarize, it has to be noted that, despite the use of computers, a designer's job of solving problems of machine dynamics has not become any easier due to the rapid technological development. A designer has to consider new operating principles and test their applicability. This does not only apply to the "electronic" options mentioned above, but to purely mechanical/physical improvements, e. g. the use of fluid spring elements, the piezo-electric effect, micromechanical innovations, the use of electro-viscous liquids (where a change in viscosity is achieved by applying a voltage), magnetic dampers, shape-memory alloys, etc.

The software that has been developed in recent years for almost all classical problems of machine dynamics make a designer's work easier only up to a certain point. It is not enough to generate output data from input data (if even available to a sufficient extent) using the computer in order to find solutions to vibration problems.

Chapter 9
Relations to System Dynamics and Mechatronics

9.1 Introduction

System dynamics deals with objects that may originate from different disciplines, that is, not only from mechanical and electrical engineering, but from all fields of physics, chemistry, biology, medicine, sociology and economics.

In system dynamics, a system is defined as a set of elements (characterized by their parameter values) that are coupled to each other and may interact with each other (interaction). It is characterized by its topology and several state variables that can change over time and capture its motion. For a system, the relations between input and output variables are described (simulation), and the system can be evaluated based on its dynamic behavior (stability). By its design, the system's behavior over time and with respect to various criteria can be influenced (open-loop and closed-loop control). Influences that act from the outside (input) onto the system are defined at the system boundaries.

"Mechatronics describes the functional and spatial integration of components from the fields of mechanics, electronics and information processing" [31]. This section will only show the relations of the equations of motion of system dynamics with those of mechanics and mechatronics.

The **state** of a system is a fundamental term of system dynamics. It is described by the time functions of N components (the state variables x_n). They are combined in the state vector

$$\boldsymbol{x} = [x_1, x_2, \ldots, x_N]^\mathrm{T} \tag{9.1}$$

which has N elements and defines the **state space**. The motion of a system is described by the variation of its state variables over time which can be associated with the movement along the trajectory of a point in the N"=dimensional state space.

The differential equations of an **autonomous system** do not explicitly contain time and are as follows

$$\boldsymbol{G}(\dot{\boldsymbol{x}}, \boldsymbol{x}) = \boldsymbol{0}. \tag{9.2}$$

(9.2) – the most general form of the so-called state equations – says that changes as a function of time $\dot{x}$ and the instantaneous state x influence each other in the way described by the vector function G. In this abstract form, the equations of system dynamics apply to many physical, biological, economic and social systems. If there are explicit time-dependent influences on a system from the outside, this is called a **heteronomous system**. Its equations of motion have the form

$$\dot{x} = g(x, t). \tag{9.3}$$

A **linear system** satisfies the equation of motion

$$\dot{x} = Ax + b(t). \tag{9.4}$$

A is the $(N \times N)$-**system matrix** and $b(t)$ is an $(N \times 1)$ excitation vector, which captures the time-dependent influence "from the outside". For example, all equations of motion of forced vibrations of linear systems that were discussed in Chap. 6 can be brought into this form.

The state of a mechanical system is generally described uniquely for discrete models in machine dynamics using the generalized coordinates $q(t)$ and their velocities $\dot{q}(t)$ with the following state vector x, see (9.2):

$$x = \left[q^{\mathrm{T}}, \dot{q}^{\mathrm{T}}\right]^{\mathrm{T}} = [q_1, q_2, \ldots, q_n, \dot{q}_1, \dot{q}_2, \ldots, \dot{q}_n]^{\mathrm{T}}. \tag{9.5}$$

The equations of motion (6.331) of linear mechanical vibration systems, that is $M\ddot{q} + B\dot{q} + Cq = f(t)$, can be transformed into the form of (9.4):

$$\dot{x} = \frac{\mathrm{d}}{\mathrm{d}t}\begin{bmatrix} q \\ \dot{q} \end{bmatrix} = \begin{bmatrix} 0 & E \\ -M^{-1}C & -M^{-1}B \end{bmatrix}\begin{bmatrix} q \\ \dot{q} \end{bmatrix} + \begin{bmatrix} 0 \\ M^{-1} \end{bmatrix} f(t). \tag{9.6}$$

The system matrix A and the excitation vector $b(t)$ are thus defined as follows:

$$A = \begin{bmatrix} 0 & E \\ -M^{-1}C & -M^{-1}B \end{bmatrix}; \qquad b(t) = \begin{bmatrix} 0 \\ M^{-1} \end{bmatrix} f(t). \tag{9.7}$$

Note that the matrix A in Eq. (6.83) has a different meaning than in (9.4). Because of (9.5), the number of state variables N in conservative mechanical systems is therefore twice as high as the number of the degrees of freedom n: $N = 2n$.

The solution of the differential equation (9.6) is, see, for example, [2], [31]:

$$x(t) = \exp(At)x_0 + \int_0^t \exp[A(t - t^*)] b(t^*) \mathrm{d}t^*, \tag{9.8}$$

where $x_0 = x(t = 0)$ is the state at time $t = 0$ (the initial conditions). In this compact mathematical form, (9.8) describes both the free vibrations (if $b \equiv o$) and the forced vibrations of linear systems. Specific solutions of (9.8) were discussed in Sects. 6.3, 6.5 and 6.6, see also Duhamel's integral in (6.271).

9.1 Introduction

The methods of system dynamics can be used to study general properties of the solutions mathematically, e. g. regarding stability, open-loop and closed-loop control, and optimization of systems. A more detailed discussion of system theory can be found, for example, in [2], [11], [14], or [24].

An important indicator for the characterization of dynamic behavior of linear dynamic systems is the **principle of superposition**. It states that the "effect of the sum" of individual causes is the same as the "sum of effects" of the individual causes.

If a linear system according to (9.4) is influenced by exactly two "individual causes" one after the other, e. g. the excitation functions $b_1(t)$ and $b_2(t)$, then there exist the solutions $x_1(t)$ and $x_2(t)$ for the two individual causes:

$$\dot{x}_1 = A\,x_1 + b_1(t); \qquad \dot{x}_2 = A\,x_2 + b_2(t). \tag{9.9}$$

The excitation $b_1(t)$ (cause 1) results in a motion x_1 (effect) and cause 2 in an effect x_2. If the "sum of the two causes", i.e. the excitation

$$b(t) = b_1(t) + b_2(t), \tag{9.10}$$

acts onto the system at the same time, the "effect of the sum" (of the excitations) is the solution to the differential equation

$$\dot{x} = A\,x + b_1(t) + b_2(t). \tag{9.11}$$

The "sum of the effects" is $x = x_1 + x_2$, and because of $Ax = A(x_1 + x_2) = Ax_1 + Ax_2$ it is identical with the "effect of the sum" of the causes, since the following applies, according to (9.9):

$$\begin{aligned}\dot{x} = \dot{x}_1 + \dot{x}_2 &= [Ax_1 + b_1(t)] + [Ax_2 + b_2(t)] \\ &= A\,(x_1 + x_2) + [b_1(t) + b_2(t)] = Ax + b(t).\end{aligned} \tag{9.12}$$

The following applies to nonlinear systems, however

$$g\,(x,\,t) = g\,(x_1 + x_2,\,t) \neq g\,(x_1,\,t) + g\,(x_2,\,t)\,.$$

Conclusion: **The principle of superposition is only valid for linear systems.** It has been used, without explicitly pointing it out, for example, when applying the *Fourier series* in the above Sections.

9.2 Closed-Loop Controlled Systems

9.2.1 General Perspective

If electromechanical systems (e. g. machines with an electric drive) are extended by sensors and actuators in such a way that the original system can be influenced via a suitable controller based on the system state, then these are called closed-loop controlled systems. The difference between an open-loop control and a closed-loop control is that the open-loop control just responds to external influences, whereas a closed-loop control influences the parameter to be changed by feedback based on the system response determined. A major characteristic of closed-loop controlled systems, therefore, is the presence of a feedback, see Figure 9.1. Open-loop controlled systems do not have this state-dependent feedback, i. e. $u(t)$ contains only predefined control functions.

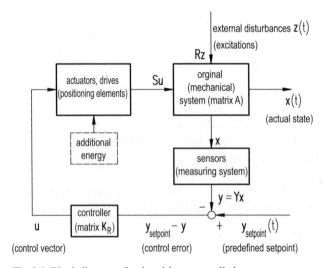

Fig. 9.1 Block diagram of a closed-loop controlled system

If one limits oneself to linear systems, the following equations extended as compared to (9.4) apply, taking into account the feedback:

$$\left. \begin{array}{l} \dot{x} = Ax + Su + Rz(t) \\ u = K_R \cdot (y_{\text{setpoint}}(t) - y) \\ y = Yx \end{array} \right\}. \quad (9.13)$$

They can be summarized in the following form:

$$\dot{x} = (A - SK_RY)x + SK_R y_{\text{setpoint}}(t) + Rz(t). \quad (9.14)$$

9.2 Closed-Loop Controlled Systems

A is the system matrix of the original uncontrolled system known from (9.7), S the control matrix and u the control vector. K_R is the closed-loop control matrix containing the closed-loop control constants and Y the monitoring or measuring matrix, which makes the connection between the measurements combined in the measuring vector y and the state vector x. The disturbance matrix R, together with the disturbance vector $z(t)$, describes the excitation variables that act onto the system from the "outside" in the form of known time functions (e. g. periodic or transient excitations). $y_{\text{setpoint}}(t)$ can be used to specify "desired" time functions of measurable quantities.

A closed-loop controlled system is designed to improve the dynamic system behavior. This mostly means that a specific target function is made the extremum, see e. g. [2]. Such targets may include:

- minimum deviation from a given trajectory
- minimum amplitude of a suspended load during crane motions
- minimum heating-up of a servo motor
- shortest trajectory of a tool moved by a robot arm
- shortest motion time, etc.

It is often attempted to "mitigate" interfering vibrations using closed-loop control. The literature contains a number of methods and procedures on the respective optimum configuration of the closed-loop control system, see [2], [14].

Sensors that capture the current state of the system, at least pointwise within the respective measuring system, are a requirement for any closed-loop control system. Sensors can use different operating principles (inductive, capacitive, piezo-electric, magnetostrictive, optical, ...) to capture and transmit kinematic parameters as well as load parameters of the mechanical system. They represent components that measure external changes or internal state variables within the system, thereby obtaining information that can be processed by the closed-loop control unit. Typical sensors include strain gauges, inductive or capacitive transmitters and fiber-optic, piezo-electric and magnetostrictive sub-assemblies. They measure strains, displacements, velocities, accelerations or other mechanical parameters and convert them into electric signals. Micromechanical sensors with dimensions below the millimeter range have a wide field of applications in machine drives.

Actuators are components that supply mechanical forces, stresses, displacements, or strains to the real object depending on the information received from the controller at suitable points and in predefined directions and thus change the dynamic system properties. The classic electromagnetic actuators that include, for example, servo and stepping motors, use electromagnetic fields to generate the required forces/moments or displacements/angles of rotation. In piezo-actuators, the electrical field causes a mechanical strain in the one-thousandth range that causes mechanical stresses of ca. $300 \, \text{N/mm}^2$ but effects only minor displacements. They are often used in components subject to bending loads and, due to their fast response rates, have proven their worth for rotating shafts that are exposed to bending vibrations. Hydraulic actuators often operate with controllable valves, with which the flow resistance of a liquid is influenced. In magneto- or electro-rheological actuators, the viscosity is changed by magnetic or electrical fields. They can be used to build controllable damping elements that can produce changing forces depending on the actual system state.

Important questions with regard to a closed-loop controlled system involve stability, observability and controllability. The stability and controllability can be influenced by the system parameters that are included in the matrix A and by the actuators (including their positions and directions of action), that is by the matrix K_R. The observability depends on the selection of measured quantities, measuring points and measuring directions and is defined by the measuring matrix Y.

Statements on the stability of linear closed-loop controlled systems can be obtained via the eigenvalues of the matrix $(A - SK_R Y)$. A system is **stable** if its motions do not grow unboundedly after minor disturbances but rather return to their original state.

A system is completely **observable** if all free motions of the system can be captured by the measuring vector y. For example, a normal mode shape of a bending oscillator could be measured at the antinode while a sensor at the node could not register this normal mode shape, i. e. the respective normal mode shape would not be observable there.

A system is completely **controllable** if it can be brought from an arbitrary initial state into an arbitrary end state, i. e. if all free motions can be excited by the control vector u where the control matrix S captures the suitable points and directions. If, for example, an actuator acted at the node of a normal mode shape of a bending oscillator, it could not excite the respective normal mode shape, i. e. this normal mode shape could not be influenced by that actuator and thus not all types of motion could be implemented.

There are mathematical methods and criteria for evaluating the stability, observability and controllability, which are described, for example, in [2], [14] and are recommended for further study.

9.2.2 Example: Influencing Frame Vibrations by a Controller

9.2.2.1 Analytical Interrelations

The simple model of a machine frame with two degrees of freedom according to Fig. 6.2, as known from Sect. 6.2.2.1, shall be considered here, however, extended by two interfering excitation forces $F_1(t)$ and $F_2(t)$ as well as measuring points B and D and a control force F_R in the direction of the coordinate q_1 that is to reduce the excited vibrations, see Fig. 9.2.

The vertical velocity v_{Bv} is measured in B and the difference in strain $\Delta\varepsilon_D$ is measured in D using a strain gauge as the strain has to be determined from both sides to compensate for the influence of the axial force.

The mass and stiffness matrices for $q = [q_1, q_2]^T$ are known from Sect. 6.2.2.1:

9.2 Closed-Loop Controlled Systems

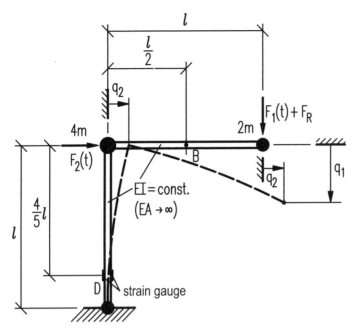

Fig. 9.2 Calculation model of a machine frame

$$M = 2m \cdot \begin{bmatrix} 1 & 0 \\ 0 & 3 \end{bmatrix}, \qquad C = \frac{6EI}{7l^3} \begin{bmatrix} 2 & -3 \\ -3 & 8 \end{bmatrix} \qquad (9.15)$$

$$M^{-1} = \frac{1}{6m} \begin{bmatrix} 3 & 0 \\ 0 & 1 \end{bmatrix}, \qquad M^{-1}C = \frac{3EI}{7ml^3} \begin{bmatrix} 2 & -3 \\ -1 & 8/3 \end{bmatrix}$$

To consider damping, the stiffness-proportional approach $B = a_2 \cdot C$ according to (6.326) is selected, and from the virtual work $\delta W^{(e)} = (F_1(t) + F_R)\delta q_1 + F_2(t)\delta q_2$ the right side results as follows:

$$f = [(F_1(t) + F_R), F_2(t)]^{\mathrm{T}} = [F_1(t), F_2(t)]^{\mathrm{T}} + [F_R, 0]^{\mathrm{T}}. \qquad (9.16)$$

The following equations of motion apply:

$$M\ddot{q} + a_2 C\dot{q} + Cq = f \quad \text{or} \quad \ddot{q} = M^{-1}f - M^{-1}C(a_2\dot{q} + q). \quad (9.17)$$

If, according to (9.5), the state vector

$$x = [x_1, x_2, x_3, x_4]^{\mathrm{T}} = [q^{\mathrm{T}}, \dot{q}^{\mathrm{T}}]^{\mathrm{T}} = [q_1, q_2, \dot{q}_1, \dot{q}_2]^{\mathrm{T}} \qquad (9.18)$$

is selected, the system matrix according to (9.7) with (9.15) and $\omega^{*2} = \dfrac{48EI}{ml^3}$ is obtained:

$$A = \begin{bmatrix} \mathbf{0} & \mathbf{E} \\ -\mathbf{M}^{-1}\mathbf{C} & -a_2\mathbf{M}^{-1}\mathbf{C} \end{bmatrix} = \omega^{*2} \begin{bmatrix} 0 & 0 & \dfrac{1}{\omega^{*2}} & 0 \\ 0 & 0 & 0 & \dfrac{1}{\omega^{*2}} \\ -\dfrac{1}{56} & \dfrac{3}{112} & -\dfrac{a_2}{56} & \dfrac{3a_2}{112} \\ \dfrac{1}{112} & -\dfrac{1}{42} & \dfrac{a_2}{112} & -\dfrac{a_2}{42} \end{bmatrix}. \quad (9.19)$$

The control force F_R of an actuator shown in Fig. 9.2 acts onto the structure so that the control vector contains only one element, i. e. $\boldsymbol{u} = [F_\mathrm{R}]$. If the excitation forces that act as disturbance here are captured in the disturbance vector $\boldsymbol{z}(t) = [F_1(t), F_2(t)]^\mathrm{T}$, one can determine the matrices $\boldsymbol{S}$ and $\boldsymbol{R}$ that occur in (9.13) and (9.14). Taking into account (9.16) as well as the definitions above for $\boldsymbol{u}$ and $\boldsymbol{z}(t)$, from a comparison of coefficients, the following results for the control matrix $\boldsymbol{S}$ and the disturbance matrix $\boldsymbol{R}$ from $\boldsymbol{M}^{-1}\boldsymbol{f}$ in (9.17):

$$\boldsymbol{S} = \frac{1}{2m}\begin{bmatrix} 0 \\ 0 \\ 1 \\ 0 \end{bmatrix}, \quad \boldsymbol{R} = \frac{1}{6m}\begin{bmatrix} 0 & 0 \\ 0 & 0 \\ 3 & 0 \\ 0 & 1 \end{bmatrix}. \quad (9.20)$$

The measuring vector

$$\boldsymbol{y} = [\Delta\varepsilon_\mathrm{D}, v_\mathrm{Bv}]^\mathrm{T} \quad (9.21)$$

contains the difference in strain $\Delta\varepsilon_\mathrm{D}$ and the vertical velocity v_Bv as measurements. It is known from the bending theory that the difference in strain is proportional to the bending moment at D. The bending moment at point D is found from the equilibrium of moments with the leverages shown in Fig. 9.2 and the auxiliary forces Q_1 and Q_2 formally introduced in the direction of the coordinates q_1 and q_2 according to Sect. 6.2.2.1:

$$\Delta\varepsilon_\mathrm{D} \sim M_\mathrm{D} = l\cdot\left[1, \frac{4}{5}\right]\cdot\begin{bmatrix} Q_1 \\ Q_2 \end{bmatrix} = l\cdot\left[1, \frac{4}{5}\right]\boldsymbol{C}\boldsymbol{q} = \frac{6EI}{7l^2}\left[-\frac{2}{5}, \frac{17}{5}\right]\cdot\begin{bmatrix} q_1 \\ q_2 \end{bmatrix}. \quad (9.22)$$

The linear relationship between these forces and the displacement coordinates via the stiffness matrix $\boldsymbol{C}$ was taken into account here, see (6.11) and (9.15).

Using k_ε as the proportionality factor that also considers the bending stiffness EI, one eventually obtains:

$$\Delta\varepsilon_\mathrm{D} = k_\varepsilon\,[-2,\ 17,\ 0,\ 0]\begin{bmatrix} \boldsymbol{q} \\ \dot{\boldsymbol{q}} \end{bmatrix} = \boldsymbol{Y}_\varepsilon\boldsymbol{x}. \quad (9.23)$$

The vertical displacement at B can be determined using the compliance matrix $\boldsymbol{D}$ of the frame model with four degrees of freedom specified in Sect. 6.3.4.2. If the corresponding associations between the parameters from 6.3.4.2 and those of the current problem are used, the following results for the vertical displacement from

9.2 Closed-Loop Controlled Systems

the second row of the matrix D in (6.131) in conjunction with the auxiliary forces occurring here:

$$q_B = \frac{l^3}{48EI} [29, 12] \begin{bmatrix} Q_1 \\ Q_2 \end{bmatrix}. \tag{9.24}$$

Finally, the displacement determined in this way can be expressed using the stiffness matrix known from (9.15) as a function of the coordinates q_1 and q_2, from which the vertical velocity of B follows after differentiation with respect to time:

$$v_{Bv} = \frac{11}{28}\dot{q}_1 + \frac{9}{56}\dot{q}_2 = \begin{bmatrix} 0, & 0, & \frac{11}{28}, & \frac{9}{56} \end{bmatrix} \begin{bmatrix} q \\ \dot{q} \end{bmatrix} = Y_v x. \tag{9.25}$$

The relation between measurements and state variables is thus written as follows:

$$y = \begin{bmatrix} \Delta \varepsilon_D \\ v_{Bv} \end{bmatrix} = \begin{bmatrix} Y_\varepsilon \\ Y_v \end{bmatrix} x = Y x \quad \Rightarrow \quad Y = \begin{bmatrix} -2k_\varepsilon & 17k_\varepsilon & 0 & 0 \\ 0 & 0 & \frac{11}{28} & \frac{9}{56} \end{bmatrix}. \tag{9.26}$$

If one assumes the linear dependence on the measurements known from (9.13) for the control force, one gets for $y_{\text{setpoint}}(t) \equiv 0$:

$$[F_R] = u = -K_R y = -K_R Y x, \qquad K_R = [\varkappa_1, \varkappa_2]. \tag{9.27}$$

The controller matrix K_R contains the two controller constants $\varkappa_1$ and $\varkappa_2$ that eventually have to be selected from a permissible (technically feasible) domain so that the occurring vibrations are as small as possible.

The differential equations of the closed-loop controlled systems are finally found according to (9.14), here in particular with $y_{\text{setpoint}}(t) \equiv 0$ and the matrix that describes the problem at hand

$$A - SK_R Y = \tag{9.28}$$

$$\begin{bmatrix} 0 & 0 & 1 & 0 \\ 0 & 0 & 0 & 1 \\ -\frac{\omega^{*2}}{56} + \frac{k_\varepsilon \varkappa_1}{m} \frac{3\omega^{*2}}{112} & -\frac{17k_\varepsilon \varkappa_1}{2m} & -\left(a_2 \frac{\omega^{*2}}{56} + \frac{11\varkappa_2}{56m}\right) & 3a_2 \frac{\omega^{*2}}{112} - \frac{9\varkappa_2}{112m} \\ \frac{\omega^{*2}}{112} & -\frac{\omega^{*2}}{42} & a_2 \frac{\omega^{*2}}{112} & -a_2 \frac{\omega^{*2}}{42} \end{bmatrix}.$$

These equations are equivalent to those of (9.4) if instead of A the expression $(A - SK_R Y)$ according to (9.28) and

$$b(t) = R z(t) = \frac{1}{6} \begin{bmatrix} 0, & 0, & \frac{3F_1(t)}{m}, & \frac{F_2(t)}{m} \end{bmatrix}^T \tag{9.29}$$

are set. This means that the solution can also be determined according to (9.8), however, it would then also be a function of the controller constants $\varkappa_1$ and $\varkappa_2$.

The real parts of the eigenvalues of $(A - SK_RY)$ provide insight whether the controlled-loop system will behave stably or not.

9.2.2.2 Numerical Example

It will be assumed for simplification that only one harmonically variable excitation or disturbance force acts in the minimal model of a machine frame shown in Fig. 9.2 so that:

$$F_1(t) \equiv 0; \qquad F_2(t) = \hat{F} \sin \Omega t. \tag{9.30}$$

This excites both normal mode shapes, as can be seen from Fig. 9.3. Both normal mode shapes can be influenced using an actuator that generates a control force F_R in the direction of the coordinate q_1 because the free beam end oscillates in the direction of action of F_R for both shapes. The task is to determine the controller force F_R of the actuator (to be exact: the controller constants) from the two measured signals in such a way that the vibration amplitude $q_2(t)$ is reduced.

To obtain a general result, dimensionless characteristic parameters (see (9.34)) and three fundamental parameters are introduced for the calculation model: length l, bending stiffness EI and mass m. The circular reference frequency ω^* that results from these parameters and is known from (6.133) and (9.19) is taken into account. The use of dimensionless characteristic parameters has the advantage that one obtains results for mechanically similar objects. In addition, the numerical results become more precise if there are no major differences in magnitude for the numerical values used in the calculation operations that are performed by the simulation program.

The two natural circular frequencies for the undamped system without a controller are known from (6.160):

$$\begin{aligned}\omega_{01} &= \sqrt{\frac{0.244\,07}{48}} \sqrt{\frac{48EI}{ml^3}} = 0.071\,308\omega^*, \\ \omega_{02} &= \sqrt{\frac{1.755\,93}{48}} \sqrt{\frac{48EI}{ml^3}} = 0.191\,264\omega^*.\end{aligned} \tag{9.31}$$

The normal mode shapes shown in Fig. 9.3 are associated with them.

It is useful for the further analysis to introduce the dimensionless time $\tau = \omega^* t$ as well as a dimensionless state vector $\overline{x}$, which is connected as follows to the state vector x containing dimensionless components used so far:

$$x = \begin{bmatrix} q \\ \dot{q} \end{bmatrix} = l \cdot \mathrm{diag}\,[1, 1, \omega^*, \omega^*] \begin{bmatrix} \overline{x}_1 \\ \overline{x}_2 \\ \overline{x}_3 \\ \overline{x}_4 \end{bmatrix} = T\overline{x}. \tag{9.32}$$

For the differentiation with respect to time, one finds:

9.2 Closed-Loop Controlled Systems

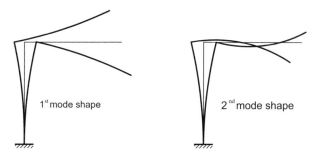

Fig. 9.3 Normal mode shapes of the original system

$$\frac{d}{dt}(\ldots) = (\ldots)^{\cdot} = \frac{d(\ldots)}{d\tau}\omega^* = (\ldots)'\omega^*$$

The differentiation with respect to τ is indicated by a prime.

Using these conversion rules, the now dimensionless equation of motion (9.14) (as in 9.2.2.1, conditional on $y_{\text{setpoint}}(t) \equiv 0$) can be stated:

$$\overline{x}' = \frac{1}{\omega^*}T^{-1}(A - SK_R Y)T\overline{x} + \frac{1}{\omega^*}T^{-1}Rz(t). \tag{9.33}$$

If – as indicated above – the dimensionless characteristic parameters

$$\pi_1 = a_2\omega^*; \quad \pi_2 = \frac{k_\varepsilon \varkappa_1}{m\omega^{*2}}; \quad \pi_3 = \frac{\varkappa_2}{m\omega^*}; \quad \pi_4 = \frac{\hat{F}l^2}{EI}; \quad \eta = \frac{\Omega}{\omega^*} \tag{9.34}$$

are introduced, the matrices known from (9.28) and (9.29) as transformed according to (9.33) can be stated:

$$\frac{1}{\omega^*}T^{-1}(A - SK_R Y)T$$

$$= \frac{1}{112}\begin{bmatrix} 0 & 0 & 112 & 0 \\ 0 & 0 & 0 & 112 \\ 112\pi_2 - 2 & 3 - 952\pi_2 & -2(\pi_1 + 11\pi_3) & 3\pi_1 - 9\pi_3 \\ 1 & -8/3 & \pi_1 & -8\pi_1/3 \end{bmatrix} \tag{9.35}$$

$$\frac{1}{\omega^*}T^{-1}Rz(\tau) = \left[0, 0, 0, \frac{\pi_4}{288}\sin(\eta\tau)\right]^{\mathrm{T}}. \tag{9.36}$$

The following numerical values were assumed for the simulation:

- Damping coefficient: $\pi_1 = 0.5$ (corresponds to modal damping ratios of $D_1 \approx 0.018$ and $D_2 \approx 0.05$ in the uncontrolled system)
- Characteristic numbers associated with the controller constants:

$$\pi_2 = \begin{cases} -0.002; & \text{Variant A} \\ +0.002; & \text{Variant B} \end{cases}, \quad \pi_3 = \begin{cases} 0.75; & \text{Variant A} \\ 10; & \text{Variant B} \end{cases}$$

- Excitation or disturbance force coefficient: $\pi_4 = 0.03$ (corresponds to a force amplitude that causes a static displacement of ca. 1 % of the beam length)
- Frequency ratio: $\eta = 0.07$ for the simulation and $0 < \eta \leq 0.3$ for the amplitude frequency response.

The determination of the eigenvalues of the matrix $(\frac{1}{\omega^*})T^{-1}(A - SK_R Y)T$ from (9.35) according to the form $\overline{x} = \hat{\overline{x}} \cdot \exp(\lambda \tau)$ for solving the system of homogeneous differential equations provides the following result:

- Variant A: $\lambda_1 = -0.026\,779$, $\lambda_{2,3} = -0.029\,487 \pm 0.190\,318\text{j}$, $\lambda_4 = -0.082\,401$
- Variant B: $\lambda_1 = -0.005\,386$, $\lambda_{2,3} = -0.007\,278 \pm 0.165\,422\text{j}$, $\lambda_4 = -1.965\,177$
- uncontrolled, undamped: $\lambda_{1,2} = \pm 0.071\,308\text{j} = \pm(\omega_{01}/\omega^*)\text{j}$,
$\lambda_{3,4} = \pm 0.191\,264\text{j} = \pm(\omega_{02}/\omega^*)\text{j}$.

$\lambda_{2,3} = (-\delta_2 \pm \text{j}\omega_2)/\omega^*$ applies to the complex conjugated eigenvalues if δ_2 is the decay rate and ω_2 is the natural circular frequency of the respective natural vibration.

All real parts are negative for the two variants of the closed-loop controlled system, i.e. there are no increasing natural vibrations, so the system is stable. Creeping processes may occur in addition to subsiding natural vibrations, which can be seen from the purely real eigenvalues. The eigenvalues of the uncontrolled and undamped system correspond to the natural circular frequencies according to (9.31).

The simulation results for the two variants are shown in Fig. 9.4. At time $\tau = 0$, the controller, which determines the actuator force, was added to the vibration system, which was in steady state and near resonance.

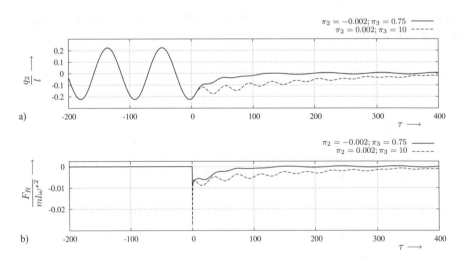

Fig. 9.4 Simulation result for the coordinate q_2 and the force of the actuator;
a) Coordinate $\overline{x}_2 = q_2 l$, **b)** Function of the force $F_R/(ml\omega^{*2})$ caused by the controller of the actuator

9.2 Closed-Loop Controlled Systems

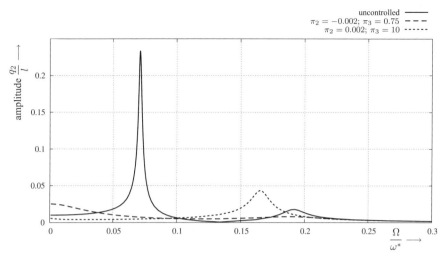

Fig. 9.5 Amplitudes of the uncontrolled and closed-loop controlled systems as functions of the frequency ratio $0 < \eta = \Omega/\omega^* \leq 0.3$

Before the time $\tau = 0$, the frame oscillates in the steady state at the frequency of excitation.

As one can see, a vibration graph similar to a free response forms from $\tau = 0$, initially at a frequency that corresponds to the respective natural circular frequency of $\omega_2 \approx 0.1903\omega^*$ (variant A) or $\omega_2 \approx 0.1654\omega^*$ (variant B) of the closed-loop controlled and damped system. In the course of the decay of the vibrations at this frequency, the frequency of the original and ongoing excitation (or disturbance) $F_2(t)$ can be seen again. This becomes particularly evident in the time function of the controller force (Fig. 9.4b), where the vibrations of variant A decay much faster due to the magnitude of the real parts of the eigenvalues.

The objective of a clear reduction of the forced frame vibrations is reached with both variants after a few natural vibration cycles.

The amplitude frequency responses shown in Fig. 9.5 also show the clear difference between the uncontrolled and the closed-loop controlled system. While variant B still has a resonance point that clearly forms for the complex eigenvalue due to the small decay rate, variant A does not show resonance peaks in the frequency range considered.

Symbols

Latin letters

A	rotational transformation matrix; system matrix
b	damping constant; width
B	damping matrix $B = [b_{kl}]$
c	spring constant
d	diameter
C	stiffness matrix $C = [c_{kl}]$
D	Lehr's damping factor $\equiv$ damping ratio
D	compliance matrix $D = [d_{kl}]$
e	eccentricity of an unbalance mass
F	force vector
F_ξ, F_η, F_ζ	components of the force F in the body-fixed coordinate system
f	frequency
g	gravitational acceleration
g	vector of the generalized load parameters $g = [Q_1, Q_2, \ldots, Q_n]^{\mathrm{T}}$
h_i	ith modal excitation force
h	height of the application point of the excitation force above the base plane
i	transmission ratio; number of the natural frequency
I	area moment of inertia; number of links in a mechanism
j	imaginary unit $\mathrm{j} = \sqrt{-1}$
J_{kl}^{S}	element of the central inertia tensor (with respect to the center of gravity S)
J_{kl}^{O}	element of the inertia tensor (with respect to the reference point O)
k	number (order) of the harmonic in the Fourier series
l	length
m	mass; modulus of a gearing;

m_u	unbalance mass
M	moment
$\boldsymbol{M}$	mass matrix $\boldsymbol{M} = [m_{kl}]$; moment vector
m_{kl}	element of the mass matrix
n	number of degrees of freedom
$\boldsymbol{p}$	parameter vector; vector of the modal coordinates (principal coordinates) $\boldsymbol{p} = [p_1, p_2, \ldots, p_i, \ldots, p_n]^\mathrm{T}$
p_i	ith modal coordinate
P	power
Q_k	kth generalized load parameter (force or moment)
q	generalized coordinate (displacement or angle)
$\boldsymbol{q}$	vector of the generalized coordinates $\boldsymbol{q} = [q_1, q_2, \ldots, q_k, \ldots, q_n]^\mathrm{T}$
q_k	kth generalized coordinate
r, R	radius
S	center of gravity
t	time
T	period of an oscillation ($T = 2\pi/\omega$)
T_0	cycle time ($T_0 = 2\pi/\Omega$) of the drive
T_i	period of the ith natural vibration ($i = 1, 2, \ldots, n$), $T_i = 2\pi/\omega_i$
u	transmission ratio (as an alternative to i), gear ratio
$U(\varphi)$	position function of a mechanism
U	e.g. $U_i = m_i r_i$ unbalance of the mass m_i
V	nondimensionalized amplitude function
$\boldsymbol{V}$	modal matrix $\boldsymbol{V} = [\boldsymbol{v}_1, \boldsymbol{v}_2, \ldots, \boldsymbol{v}_i, \ldots, \boldsymbol{v}_n] = [v_{ki}]$
W	moment of resistance
x, y, z	coordinates in the fixed reference system
$\boldsymbol{Y}_i$	Jacobian matrix of a translation
z	number of teeth of a gear
$\boldsymbol{Z}_i$	Jacobian matrix of a rotation

Greek letters (preferred for dimensionless parameters)

α_k	angle between the ξ axis and axis k
β_k	angle between the η axis and axis k
γ_i	ith modal spring constant
γ_{ik}	sensitivity coefficient (spring parameter)
γ_k	angle between the ζ axis and axis k
δ	coefficient of speed fluctuation; decay rate ($\delta = D\omega_0$); longitudinal backlash

Symbols

Δ	difference, e. g. Δt (time difference)
ξ, η, ζ	coordinates in a body-fixed reference system
Γ_{klp}	Christoffel symbol
$\varkappa$	dimensionless factor (meaning may differ locally for various problems)
Λ	logarithmic damping decrement
λ	crank ratio ($\lambda = l_2/l_3$); eigenvalue ($\lambda = \omega^2/\omega^{*2}$)
μ_0	coefficient of static friction
μ	coefficient of sliding friction
μ_i	ith modal mass
μ_{ik}	sensitivity coefficient (mass parameter)
ν	circular frequency of absorption; Poisson's ratio
η	efficiency; frequency ratio ($\eta = \Omega/\omega_0$); body-fixed coordinate
π_i	ith similarity number (dimensionless parameter)
ϱ	density
σ	stress; normal stress
τ	shear stress
φ	phase angle; loss angle
φ_0	crank angle ($\varphi_0 = \Omega t$)
ψ	phase angle of a coordinate; relative damping
ω	magnitude of angular velocity; circular frequency of the damped oscillator $\omega = \omega_0\sqrt{1-D^2}$
ω_0	circular frequency of the undamped oscillator
ω_i	ith natural circular frequency ($\omega_i = 2\pi f_i$)
$\boldsymbol{\omega}$	vector of angular velocity
Ω	angular velocity of the drive; circular frequency of the excitation ($\Omega = 2\pi f$)

Indices (at most two consecutive)

A	start-up, e. g. t_A (start-up time)
a	axial, e. g. J_a (axial moment of inertia)
an	input, e. g. M_{an} (input torque)
b	reactive, e. g. P_b (reactive power)
B	braking, e. g. φ_B (braking angle)
D	damping, e. g. F_D (damping force)
eff	effective value, e. g. P_{eff} (effective power)
e	excitation, e. g. f_e (excitation frequency)
eig	eigen or natural, e. g. f_{eig} (eigenvalue, natural frequency)

i	number of a body (link) in a mechanism $(i = 1, 2, ..., I)$;
	number of a natural frequency or normal mode shape $(i = 1, 2, ..., n)$
j	number of a step in a sequence of steps $(j = 1, 2, ..., J)$
k	number of a coordinate $(k = 1, 2, ..., n)$;
	number of a principal axis $(k = 1, 2, 3$ or $k = I, II, III)$;
	order of a harmonic $(k = 1, 2, ..., K)$
l	number of a coordinate as an alternative to k $(l = 1, 2, ..., n)$
m	order of a harmonic as an alternative to k $(m = 1, 2, ..., K)$
kin	kinetic, e. g. W_{kin} (kinetic energy)
M	motor, e. g. J_M (moment of inertia of a motor)
m	mean, e. g. M_m (mean moment)
max	maximum
min	minimum
H	horizontal, e. g. horizontal force F_H
N	normal or perpendicular, e. g. normal force F_N
0	initial, e. g. v_0 (initial velocity)
p	polar, e. g. J_p (polar moment of inertia)
red	reduced
R	frictional, e. g. M_R (frictional moment)
s	number of a selected coordinate $(s = 1, 2, ..., n)$
S	center of gravity, e. g. x_S (distance from the center of gravity in x direction);
	backlash, e. g. φ_S (angle of the transmission backlash)
st	static; e. g. M_{st} (static moment)
t	processing(-related), e. g. F_t processing(-related) force
T	torsional, e. g. c_T (torsional spring constant)
v	loss, e. g. P_v power loss
x, y, z	components in the fixed reference system (e. g. F_y is the y component of the force $\boldsymbol{F}$ in the fixed reference system)
zul	permissible, e. g. σ_{zul} (permissible stress)
ξ, η, ζ	components in the body-fixed reference system (e. g. F_ξ is the ξ component of the force $\boldsymbol{F}$ in the body-fixed reference system)
I, II, III	Roman numbers for principal axes

Exponents and superscript symbols

e	applied, e. g. $\boldsymbol{F}^{(e)}$ (applied force)
O	with respect to the fixed origin
$\overline{O}$	with respect to a body-fixed reference point
T	transposed, e. g. transposed vector

Symbols

S	center-of-gravity
z	constraint, e. g. $\boldsymbol{F}^{(z)}$ (constraint force)
*	asterisk as indication of a peculiarity e. g. ω^* (circular reference frequency); imaginary part of a complex number
$\overline{}$	bar on top: for vectors and tensors: refers to a body-fixed system
$\sim$	tilde on top: skew-symmetric matrix of the respective vector; complex number

References

[1] *Biezeno, C. B.; Grammel, R.*: Engineering Dynamics. Vol. 1-4. – London: Blackie&Son, 1954-1956
[2] *Bremer, H.*: Dynamik und Regelung mechanischer Systeme. – Stuttgart: B. G. Teubner Verlag, 1988
[3] *Bremer, H.*: Elastic Multibody Dynamics : A Direct Ritz Approach. – Dordrecht: Springer, 2008
[4] *Dresig, H.*: Schwingungen mechanischer Antriebssysteme. – 2. Aufl. – Berlin; Heidelberg: Springer-Verlag, 2005
[5] *Dresig, H.; Schreiber, U.*: From Simulation Results to the Physical Cause. Result Interpretation for Simulations and Measurements of Vibrations. Course Material. *(In German)* – ITI GmbH, Dresden, 2009
[6] *Fischer, U.; Stephan, W.*: Mechanische Schwingungen. – 3. Aufl. – Leipzig: Fachbuchverlag, 1993
[7] *Gasch, R.; Knothe, K.*: Strukturdynamik. Bd. 1: Diskrete Systeme (1987), Bd. 2: Kontinua und ihre Diskretisierung. – Berlin; Heidelberg: Springer-Verlag, 1989
[8] *Gasch, R.; Nordmann, R.; Pfützner, H.*: Rotordynamik. – 2. Aufl. – Berlin; Heidelberg: Springer-Verlag, 2002
[9] *Hafner, K. E.; Maass, H.*: Torsionsschwingungen in der Verbrennungskraftmaschine. – Wien; New York: Springer-Verlag, 1985
[10] *Hagedorn, P.*: Technische Schwingungslehre, Bd. 2: Lineare Schwingungen kontinuierlicher mechanischer Systeme. – Berlin; Heidelberg: Springer-Verlag, 1989
[11] *Hardtke, H.-J.; Heimann, B.; Sollmann, H.*: Lehr- und Übungsbuch Technische Mechanik II. – Leipzig: Fachbuchverlag, 1997
[12] *Harris, C. M., Crede, C. E.*: Shock and Vibration Handbook. – 2. Aufl. – New York: Mc Graw-Hill Book Company, 1976
[13] *Irretier, H.; Nordmann, R.; Springer, H.*: Schwingungen in rotierenden Maschinen. Tagungsbände 1 bis 8. – Braunschweig; Wiesbaden: Vieweg-Verlag, 1991–2008
[14] *Heimann, B.; Gerth, W.; Popp, K.*: Mechatronik. Komponenten-Methoden-Beispiele. – 2. Aufl. – München; Wien: Fachbuchverlag Leipzig, 2001
[15] *Klein, U.*: Schwingungstechnische Beurteilung von Maschinen und Anlagen. – 3. Aufl. – Düsseldorf: Verlag Stahleisen, 2008
[16] *Klotter, K.*: Technische Schwingungslehre. Schwinger mit mehreren Freiheitsgraden. Bd. 2. – 2. Aufl. – Berlin; Heidelberg: Springer-Verlag, 1960
[17] *Kolerus, J.; Wassermann, J.*: Zustandsüberwachung von Maschinen. – 4. Aufl. – Renningen: expert-Verlag, 2008
[18] *Kortüm, H.; Lugner, P.*: Systemdynamik und Regelung von Fahrzeugen. – Berlin; Heidelberg: Springer-Verlag, 1994

[19] *Korenev, B. G.; Rabinovič, I. M.*: Handbuch Baudynamik. – Berlin: VEB Verlag für Bauwesen, 1980
[20] *Krämer, E.*: Maschinendynamik. – Berlin; Heidelberg: Springer-Verlag, 1984
[21] *Laschet, A.*: Simulation von Antriebssystemen. – Berlin; Heidelberg: Springer-Verlag, 1988
[22] *Linke, H.*: Stirnradverzahnung. – München; Wien: Carl Hanser Verlag, 1996
[23] *Magnus, K.*: Kreisel. Theorie und Anwendungen. – Berlin; Heidelberg: Springer-Verlag, 1971
[24] *Magnus, K.*: Vibrations. – London; Glasgow: Blackie&Son, 1965
[25] *Major, A.*: Berechnung und Planung von Maschinen- und Turbinenfundamenten. – Berlin: VEB Verlag für Bauwesen; – Budapest: Verlag der Ungar. Akad. der Wiss., 1961
[26] *Pfeiffer, F.*: Mechanical System Dynamics. – Berlin: Springer, 2008
[27] *Popp, K.; Schiehlen, W.*: Fahrzeugdynamik. – Stuttgart: B. G. Teubner Verlag, 1993
[28] *Schiehlen, W.*: Advanced Multibody System Dynamics. Simulation and Software Tools. – Dordrecht; Boston; London: Kluwer Academic Publishers, 1993
[29] *Schneider, H.*: Balancing Technology. – Darmstadt: Schenk, 1991
[30] *Szabo, I.*: Technische Mechanik. – Berlin; Heidelberg: Springer-Verlag, 2002
[31] *Hering, E.; Steinhart, H. (Eds.)*: Taschenbuch der Mechatronik. – 2. Aufl. – München; Wien: Fachbuchverlag Leipzig, 2004
[32] *Zurmühl, R.; Falk, S.*: Matrizen 1, Grundlagen. – 7. Aufl. – Berlin; Heidelberg: Springer-Verlag, 1997
[33] GERB Corporate Publication: Vibration Isolation Systems. – 10th ed. – Berlin: 2000 – www.GERB.com
[34] User Manual SimulationX® – Dresden: ITI GmbH, 2008 – www.simulationx.com
[35] VDI Guideline 2149, Transmission dynamics – Rigid body mechanisms. – Part 1 – Berlin: Beuth-Verlag, 2008
[36] VDI Guidelines can be purchased from www.beuth.de

Index

absorber, 200, 203, 283–286, 439
absorption, 203, 438
absorption condition, 295
absorption frequency, 202, 285, 432, 433, 438, 464
acceleration, 8–10, 73–81, 95, 106, 141–143, 180, 315, 361, 479
accuracy, 9, 15, 22, 56, 128, 348, 355, 384, 393, 426
actuator, 179, 509, 515, 520, 522
adhesive force, 492, 494
adjustment calculus, 24, 39, 41
air spring, 466
amplitude, 156, 184–189, 201–210, 430, 448, 453
amplitude frequency response, 451, 452, 455, 460, 464, 471, 472, 522, 523
amplitude ratio, 230, 321, 328
amplitude response, 201, 202, 205, 206, 349, 430, 431
analysis
 –, kinematic, 101, 143
angular acceleration, *see* rotational acceleration, 76, 78, 88–95, 109, 128, 150, 232, 315, 483
angular velocity, 74, 75, 95, 101, 107, 125, 155, 282, 485
antifriction bearing, 35
antinode, 240, 402, 404
antiresonance, 432, 433, 464
antisymmetrical matrix, 446
anvil, 218
approach, 444, 474, 517
approximate solution, 470, 485
approximation, 49, 181, 346, 400, 401, 407, 446, 469, 473
approximation method, 240, 467

articulated shaft, 299
asymptote, 278, 322, 334, 354
asynchronous motor, 59, 61
automotive engineering, 106
averaging, 467, 474, 485
axis of rotation, 20, 23, 81, 89, 94, 106, 156, 316

backlash, 231–234, 251–255, 385–387, 466, 479, 490
balancing, 155–157, 162, 315
 –, complete, 162–164, 169, 207
 –, harmonic, 162, 165, 167
balancing condition, 164, 165, 167, 170, 172, 174
balancing machine, 155, 157–159
balancing mass, 154, 156, 157, 164, 168, 170, 171
ball bearing, 31, 34
banded structure, 371
beam, 318, 327, 359, 365, 440
 –, cylindrical, 344
 –, with distributed mass, 338
bearing, 34
 –, elastic, 335, 353, 395
bearing elasticity, 328, 351, 395
bearing force, 35, 89, 93, 94, 97, 99, 154, 156, 157, 171, 210, 335, 337, 424, 456
 –, dynamic, 156, 336
bearing stiffness, 34, 35, 405
belt conveyor, 340
belt drive, 32, 457
belt-type stacker, 133
bending line, 338
bending moment, 364, 423, 518
bending oscillator, 326–328, 335, 345, 348, 450

533

bending stiffness, 338, 343, 424, 518
bending vibration, 338, 424, 440
Bestehorn sinoid, 268, 271, 272
block foundation, 191
boundary condition, 25, 241, 307, 341
brake, 127, 484, 500, 502
brake squeaking, 491
braking, 258
braking process, 279, 433
braking time, 281, 282
breakage, 289, 395
breakdown moment, 127, 135
breakdown slippage, 59, 135
bucket wheel excavator, 357, 387
buckling force, 340
buffer force, 479, 481
buffer spring, 478, 488, 489

cable, 374, 395, 466
CAD program, 83
calculation model, 5–7, 67, 237, 238, 293, 305, 307, 332, 336, 347, 350, 355, 386, 406
calculation program, 311, 377, 384, 421
calculation result, 264, 268, 288, 393
cam mechanism, 162, 268, 299, 300, 303
cam profile, 272, 299, 509
cam stepping mechanism, 269
Campbell diagram, 261, 265, 454
car engine, 24, 25
cardan angle, 71, 75, 109
carousel, 76
cascade diagram, 455
Castigliano's theorem, 364
causality, 44
cause, 248, 283, 299, 305, 505
center of gravity, 14, 15, 17, 22, 82, 84, 143, 146, 152, 168, 187, 197, 199, 325
center of percussion, 152, 424, 510
center-of-gravity coordinate, 101, 113, 170
center-of-gravity theorem, 98
center-of-mass trajectory, 161, 164, 314
central principal axis, 85, 487
centrifugal compressor, 387
centrifugal force, 325
centrifugal moment, *see* product of inertia, 84
centrifugal pendulum, 294
centrifuge, 315, 324, 328
centroidal distance, 17
chaos, 465
chaotic motion, 467
characteristic, 12, 469
 –, nonlinear, 475, 481
characteristic equation, 235, 240–242, 322, 327, 376, 389, 499

characteristic parameter, 55, 308, 316, 331, 333, 345, 349, 405, 472, 476
 –, dimensionless, 44, 64, 463, 476, 501
chattering, 498
Christoffel symbol, 104, 111
circular cylinder, 94, 100
circular translation, 487
clearance, 34, 35, 100, 234, 509
closed-loop control, 514
clutch engagement, 121
coast-down, 61, 64, 65
coefficient of restitution, 44, 61, 214, 216
coefficient of sliding friction, 491, 503
coefficient of speed fluctuation, 125–127, 134
coil spring, 33, 183, 187, 192, 198, 203
coincidence, 363, 369
combined vibrations, 467
compensator, 303
compensatory mechanism, 163, 168, 171, 172
complex damping, 43, 44, 448
complex equation, 174, 448, 499
compliance
 –, dynamic, 433, 437, 438
compliance matrix, 333, 358, 397, 413
 –, dynamic, 432, 451
compressor, 165, 167, 207
computational effort, 383
computer program, 14, 121, 150, 346, 351, 433
condensed system, 406, 407
connecting rod, 15, 135
conservative system, 375, 399, 491
constrained, 67, 229
constraint, 79, 116, 118, 129, 174, 245, 309, 368, 397, 407, 502
constraint equations, 103
contact point, 42, 90
contact ratio, 30, 305
contact stiffness, 29, 32
continuous spectrum, 467
continuum, 37, 237, 240–242, 307, 334, 338, 340, 342, 354
control, 514, 515
controllability, 516
controller, 446, 506, 515, 519, 520, 522
coordinate
 –, complex, 447
 –, generalized, 100, 109, 358, 442, 461
 –, global, 363, 369
 –, local, 363, 369
 –, modal, 378, 379, 442, 461
coordinate system, 69
 –, body-fixed, 68, 75, 85, 99
 –, co-rotating, 312
coordinate transformation, 68, 70

Index

coordinate vector, 235, 358, 363
Coriolis force, 105
cornering, 76
correlation, 386
Coulomb friction, 43, 475
countermeasure, 247
counterweight, 10, 163
coupled vibration, 399
coupling, 21, 31, 252, 466, 481, 482, 486
coupling force, 150
coupling impact, 233, 252
crane, 15, 21, 226, 251, 372, 374, 387, 395
crank angle, 120, 134, 139, 149, 166, 170, 480
crank orientation, 173
crank press, 135, 140
crank ratio, 56, 134, 137, 170
crank shears, 169
crank-rocker mechanism, 118, 129, 161
crankshaft, 19, 38, 56, 137, 166, 169, 265, 266, 288, 289, 387
creeping motion, 445
criterion, 8, 9, 100, 155, 185, 346, 384
critical speed, 9, 153, 279–283, 311, 323, 337, 340, 348–351, 487
cross product, 68, 69
cross-section, 338
cutting, 498–500
cutting machine, 124, 135
cycle, 44, 49, 122, 129, 135, 139, 219, 260
cycle time, 131, 215, 452, 485

d'Alembert's principle, 362, 426
damage, 8, 10, 234, 506
damper, 51, 283, 286–288
damping, 42, 43, 286, 298, 444, 456
 –, complex, 43, 448
 –, hysteresis, 43
 –, modal, 444
 –, nonlinear, 475
 –, optimum, 290
 –, relative, 48, 50
 –, viscous, 43, 44, 203, 221, 444
damping constant, 50, 263, 290
 –, torsional, 286, 290, 296
damping decrement
 –, logarithmic, 48, 446
damping force, 42, 43, 444
damping matrix, 444, 446, 459
damping model, 42, 43, 50, 448, 460
damping parameter, 47, 48, 50
damping ratio, 44, 48, 50, 52, 198, 446
 –, modal, 446, 460
damping-free point, 291
decay behavior, 446

decay rate, 45, 49, 221, 297, 298, 445, 447, 460, 461, 463, 522, 523, 526
decay time, 447
decoupling, 195, 445
deformation energy, 25, 104, 212, 362, 364, 374
deformation method, 365
deformation work, 115, 359, 360
degree of freedom
 –, master, 425
 –, number, 8
 –, slave, 425
density, 14, 19, 153, 168, 198, 307, 351
derivative, 43, 73–75, 104, 105, 114, 122, 301, 361
 –, partial, 101, 104, 108, 120, 359, 360, 373
design documents, 5, 6, 14, 148
design measure, 278
design solution, 506
design systematics, 507
designer, 158, 302, 505
determinant, 194, 201, 321, 327, 376, 463
development, 311, 371, 386, 444, 488, 506, 510
diesel engine, 197, 279, 289, 291
differential, 107
differential equation, 338, 361, 512
 –, linear, 355
 –, nonlinear, 105, 476, 483, 485, 502
differentiation
 –, implicit, 120
DIN, 30, 35, 61, 155, 179, 180
directional cosine, 22, 24, 85, 86
disk, 14, 37, 100, 229, 260, 273, 312–317, 324–326
dissipation, 444
distance to the center of gravity, 293, 307
disturbance matrix, 515, 518
drawing, 12, 14, 18, 329, 350, 455
dredging shovel, 102
drilling machine, 76
drive motor, 127, 139, 141
drive shaft, 21, 60, 153, 260
drive system, 122, 133, 136, 146, 223, 252, 285, 297, 305, 307, 482, 506
drive train, 224, 248, 249, 266, 283
driving power, 21, 109, 123
dual-mass flywheel, 248, 250, 266, 267
Duffing oscillator, 467, 472, 473, 475, 483
Duffing, Georg, 470
Duhamel's integral, 427, 512
Dunkerley, 346, 352
dwell, 269–272, 299, 300
dyad, 142, 143

dynamic compliance, 432, 433, 437, 438, 449, 451, 453
dynamic coupling, 486
dynamic effect, 436, 443, 476, 505
dynamic load, 231, 255, 424

edge mill, 77, 90
effect of vibration, 179
efficiency, 44, 123, 133
eigenforce, 377, 389
eigenmotion, 125, 194
eigenvalue, *see* natural frequency, 24, 204, 242, 344–346, 376–379, 389, 433, 497
eigenvalue problem, 40, 85, 193, 246, 309, 356, 376, 377, 402, 408, 416
eigenvector, *see* mode shape, 247, 377, 383, 384, 389, 421
electric motor, 123, 141, 155, 348
electromechanical vibration, 59, 514
elevator, 226
empirical, 21, 29, 51–54, 62, 178, 214, 386, 506
energy
 –, kinetic, 82, 86, 103, 115, 127, 144, 148, 360, 367, 383
 –, potential, 115, 368, 383
energy balance, 139, 400, 468, 489, 491
energy distribution, 243, 382, 391, 402, 422
energy requirement, 139, 141
energy source, 6–8, 349, 490, 491
engine, 8, 25, 67, 165, 187, 198, 204, 208, 224, 248, 263
equation
 –, characteristic, 235, 376, 389, 499
 –, transcendental, 65, 167, 282, 308, 499
equation of motion, 24, 43, 64, 116, 121, 188, 261, 296, 447, 468, 485, 512
equilibrium position, 358, 375
equivalent linearization method, 469, 476
equivalent mass, 137, 138
estimate, 198, 215, 257, 309, 346, 347, 381, 394, 400, 402, 436
Euler, 67, 80, 87, 378, 447, 499
Euler's gyroscope equations, 88, 91, 95
evaluation, 49, 155, 177, 179, 180, 238, 240, 312, 386
excitation
 –, harmonic, 46, 284, 313, 427, 447
 –, motion, 261, 305
 –, nonharmonic, 452
 –, periodic, 8, 55, 190, 259, 427, 452
 –, transient, 8, 56, 272, 433
excitation force, 426, 441

–, harmonic, 164, 183, 201, 431, 447, 471, 474, 476
–, modal, 260, 437, 441–443, 510
–, periodic, 142, 188, 452
excitation frequency, 8, 9, 191, 201, 240, 316, 349, 387, 430, 439, 450, 467
excitation moment, 260, 284, 286, 297, 463
excitation spectrum, 58, 149, 453
excitation vector, 443, 512
experimental result, 83, 452
experimentally, 7, 14, 29, 38, 123, 129, 291
extra mass, 15, 20, 21
extreme case, 231, 252, 257
extreme values, 127, 356, 410, 411, 443, 463

fastener, 26, 384, 387
fault detection, 386
feedback, 514
FEM, 3, 178, 311, 348, 355, 362, 385, 406
FFT, 56
fixed coordinate system, 68, 70, 94, 312
flywheel, 127, 129, 132, 134, 136, 138, 140, 141, 155, 241
force, 105
 –, generalized, 100, 104, 115, 358
 –, internal, 106, 141, 224, 383
 –, kinetic, 156
 –, kinetostatic, 105, 146, 507
 –, modal, 427, 432, 434
 –, periodic, 58, 142, 149, 452
force application point, 432, 433, 437, 439, 463
force excitation, 208, 307, 308
force theorem, *see* momentum theorem
force vector, 69, 73, 192, 326, 358, 361, 441, 447
forced mode shape, 440, 454
forced vibration, 259, 294, 305, 314, 432, 473
forging hammer, 211–213
forming, 121
forming force, 135, 139, 140
forming process, 135, 299
foundation, 153, 166, 177, 178, 181, 187, 197, 198
foundation block, 147, 187, 192, 194, 196–198, 204, 212, 440
foundation force, 220, 221
foundation load, 141, 479
foundation mass, 196, 197, 220
foundation modulus, 199, 217, 218
foundation vibration, 218, 440
four-bar linkage, 118, 120, 174
four-cylinder engine, 169, 173, 242, 248
four-stroke engine, 260, 279

Index

Fourier coefficient, 55, 56, 142, 145, 149, 189, 219, 260, 262, 271, 452
Fourier series, 55, 57, 166, 171, 188, 222, 260, 262, 471
fractional order, 260
frame, 96, 147, 363, 388, 411, 420, 437
frame vibration, 167, 516, 523
Fredholm integral equation, 341
free damped vibration, 445, 462
free response, 45, 48
free vibration, 7, 45, 193, 226, 229, 392
free-body diagram, 95, 226, 496, 502
free-response test, 47, 51
frequency equation, 230, 240, 241, 322
frequency ratio, 46, 188–190, 198, 209, 219–221, 262, 270–272, 285, 298, 314, 464, 472, 484, 522, 523
frequency response, 47, 190, 448–450
friction, 42, 51, 104, 491
–, Coulomb, 43, 475
friction coefficient, 121
friction damper, 291
friction damping, 47, 290
friction force, 60, 121, 492
friction moment, 60, 63, 65, 123, 129, 131, 486
friction-excited vibrations, 491
frictional oscillator, 475, 491, 492
fundamental frequency, 181, 240, 282, 346, 376

gap coefficient, 291, 292
gear, 29, 106, 116, 225, 226, 258, 278, 280, 486
gear backlash, 231, 251
gear mechanism, 12, 13, 21, 29, see mechanism, 117, 226, 252, 303
gear meshing frequency, 278, 280, 281
gear ratio, 107, 117, 135, 245, 372
gearing, 30, 305
gearing stiffness, 303, 304
generalized coordinate, 100, 109, 358, 442, 461
generalized force, 73, 89, 100, 104, 358
generalized mass, 103, 115
generalized orthogonality relation, 379
Grammel, 68, 223, 311, 400
gravitational acceleration, 90, 141, 181, 479, 501
grinding machine, 387
grinding spindle, 350
grindstone, 77, 90
guideline, see VDI Guideline, 178, 179
Guyan, 406
gyrobus, 128

gyroscope, 75, 109, 110
gyroscopic effect, 92, 284, 316, 320, 322, 324, 331, 335, 337, 340, 446
gyroscopic moment, 81, 88, 316, 320, 329

half width method, 50, 62
half-sine impulse, 435, 436
hammer, 217
hammer foundation, 214–216
harmonic, 55, 162, 168, 187
–, higher, 162, 203, 324
–, order, 167, 189, 452
–, second, 301
harmonic balancing, 162, 165, 167
harmonic excitation, 183, 427, 447
heavy metal, 153
helical gear, 30, 303
hinged column, 408
historical, 7, 8, 10, 223, 506
hitting, 381
hoist, 372
hoisting gear, 116, 117
housing load, 207
HS profile, 162, 268, 271, 272, 509
Huygens, 484
hypothesis, 9, 214, 505
hysteresis, 43, 47–50, 52

ICD damper, 289
identification, 6, 400, 403, 449
idle speed, 267
image shaft, 227, 228, 257
impact, 63, 215, 217, 253, 333, 388, 419–421, 488
–, periodic, 446
impact force, 222, 436, 437, 489
impact load, 211, 231, 509
impact sequence, 217, 219
implicit differentiation, 120
impulse, 388, 392, 437
impulse load, 381
impulse response, 434, 436
–, residual, 437
indicator diagram, 260
inequality, 346, 383, 400, 436, 475, 487, 503
inertia force, 89, 95, 144, 146, 147, 341, 359
– first-order, 164
– kth-order, 167
– second-order, 164
inertia force balancing, 162
inertia moment, 21, 88, 95, 144, 160, 329
inertial system, 70
inertial tensor, 22

influence coefficient, 318, 319, 326, 337, 358–360, 364, 449
influence function, 340, 341
initial condition, 124, 128, 213, 215, 256, 381, 382, 388, 396, 467, 512
initial energy, 125, 257, 381, 382, 391, 467
inline reciprocating engine, 224
input coordinate, 113, 114
input data, 10, 14, 386
input shaft, 295, 296, 351
input torque, 93, 97, 113, 115, 116, 120, 125, 131–133, 135–137, 139, 224, 276, 305, 483
instability, 298, 299, 340, 497, 500
instable domain, 491
installation site, 181, 183, 187
instantaneous center of rotation, 152
integral equation, 341, 343
integration, 87, 256
 –, numerical, 68, 122, 128, 297, 426
integro-differential equation, 341
interaction, 42, 511
internal combustion engine, 266
internal force, 141, 224, 383
interpretation, 2, 128, 279, 307, 372, 393, 449
inverse matrix, 364, 412
ISO, 153, 155, 179, 180, 311, 312

Jacobian matrix, 102, 103, 108
joint force, 141–144, 148, 151, 152, 424
journal bearing, 35, 349, 491
jump, 30, 217, 220, 272–276, 278, 279, 282, 283, 437, 460, 472, 509

kinematic analysis, 101, 143
kinematic excitation, *see* motion excitation
kinematic parameter, 9, 73, 105, 515
kinematic schematic, 133, 148, 161, 169
kinematics, 68, 69, 76, 77, 80, 101, 107, 121, 506, 507
kinetic energy, 82, 86, 103, 144, 148, 360
kinetic force, 89, 105, 156, 359
kinetic moment, 88, 147
kinetic power, 145, 148
kinetostatic force, 105, 106, 141, 145, 507, 508
kinetostatic moment, 12, 146, 224, 225, 233, 234, 297, 301, 303
kinetostatics, 67, 224, 507
knitting machine, 143
Kronecker delta, 342

Lagrange, 67, 246, 356, 478
Lagrange's equation, 104, 115, 192, 246, 254, 294, 362, 426, 458

large press, 121
large surface mining equipment, 284
length
 –, reduced, 26
lifting, 395
lightweight construction, 10, 153
lightweight design, 223, 303, 508
limit, 346–348, 400, 506
limit cycle, 494
limiting process, 340, 428
linearization, 295, 296, 317, 467, 469, 476
linkage
 –, eight-bar, 148
 –, multi-bar, 507
load, *see* dynamic load, *see* preload, 273, 315
load parameter, 319, 361, 363, 365, 378, 424, 427, 440, 515
loading, 34, 37, 42, 50, 141–152, 181–222, 325, 381
locus, 450, 451, 460, 461, 463, 464
logarithmic decrement, 46, 48, 49, 51, 53, 446
longitudinal lifting mechanism, 302
longitudinal oscillator, 200, 223, 234, 235, 284
longitudinal vibration, 307
loop, 103, 118
loss angle, 46, 48, 50
lower limit, 347, 348

MAC matrix, 384, 386
machine frame, 146, 196, 388, 463
machine setup, 178, 183
machine tool, 196, 223, 497
machine type, 143, 223
mandril, 328
manipulator, 299
marine diesel engine, 143, 197
marine engine, 223, 224
mass
 –, generalized, 103, 115
 –, modal, 236, 379, 380, 391, 414, 416, 427, 437, 508
 –, reduced, 67
mass balancing, 145, 153, 159, 160, 162, 163
mass change, 240, 404
mass distribution, 14, 84, 154, 155, 160, 198, 227, 338, 354, 510
mass matrix, 103, 110, 193, 235, 360, 361, 413
 –, reduced, 407
mass parameter, 14–25, 103, 113, 151, 161, 169, 192, 374, 384, 402
mass production, 157, 158
master degree of freedom, 406
material damping, 44, 315, 324, 445, 476
material fatigue, 222

Index

mating gears, 30, 279
matrix, 400, 425
  –, antisymmetrical, 446
  –, reduced, 417, 425
  –, skew symmetrical, 69, 73, 79
matrix element, 362, 371, 372
matrix notation, 68, 356
maximum moment, 233, 257, 280
maximum value, 252, 275–277, 424, 434, 454
mean value, 224, 275
measure, 162, 505–510
measurement, 3, 5–7, 10, 15, 16, 22–24, 49, 56, 179, 189, 198, 279, 312, 432, 515, 519
measuring matrix, 515
mechanical work, 144
mechanism, 100, 101, 127, 144, 148, 295, 299, 506
  –, constrained, 100
  –, mass balancing, 145
  –, multi-link, 114
  –, planar, 104, 112, 160
  –, reactive force, 106
  –, rigid, 8
  –, spatial, 68
  – with two drives, 105
  –, with varying transmission ratio, 144, 295
mechanism schematic, 299
mechanism synthesis, 507
meshing frequency, 247, 303, 305, 306, 324, 509
method of harmonic balance, 469
milk centrifuge, 329
mill, 93
milling, 498
milling machine, 460
minimal model, 9, 181, 240, 261, 313, 328, 406, 492, 501
mobility, 100, 108
modal analysis, 52, 248, 260, 383, 384
modal coordinates, 378, 379, 442, 461
modal excitation force, 342, 510
modal hammer, 436
modal mass, 236, 379, 380, 391, 414, 416, 427, 437, 508
modal matrix, 247, 377, 379, 380, 382, 390, 393, 402, 421, 445, 446
modal oscillator, 273, 380, 427, 435
modal spring constant, 236, 379, 380, 391, 427, 437
modal transformation, 379
mode shape, 227, 238, 320, 321, 345, 376, 377, 384, 389, 393, 394, 402, 404, 414, 415, 421, 430, 431, 450, 460

  –, antisymmetrical, 393, 419
  –, forced, 430, 439, 440, 450, 454
  –, symmetrical, 393, 419
model
  –, constrained, 229
  –, unconstrained, 229
model generation, 7, 9, 248, 307, 348
  –, 6
model level, 7, 8, 10, 13
  –, model, 8
modeling, 2, 3, 8, 10, 29, 38, 304, 355, 366, 384, 426
moment, 121, 392
  –, elastic, 256
  –, kinetic, 88, 91, 147
  –, kinetostatic, 224, 225, 233, 234, 297, 301, 303
  –, processing, 131
moment curve, 301, 390
moment distribution, 12, 157, 225
moment of inertia, 14, 17, 19, 22, 82–84, 133, 253, 301, 321, 322
  –, reduced, 114, 117, 118, 120, 121, 126, 130, 134, 136–138, 160, 305, 399, 482
momentum theorem, *see* angular momentum theorem, *see* center-of-gravity theorem
motion excitation, 185, 229, 261, 305, 307, 308
motion program, 123, 268, 270, 299–302
motor, 8, 21, 54, 60, 123, 127, 139, 251, 273
motor characteristic, 59, 101, 139, 141
motor shaft, 135, 136, 207
motor torque, 59, 396, 459
motor vehicle, 15, 16
motorbike, 31, 76, 208, 223, 263–266
mounting
  –, gimbal, 75
multi-link mechanism, 114
multibody dynamics, 68, 101, 356
multicylinder machine, 163, 165–167, 241

natural circular frequency, *see* natural frequency, 195, 380
natural frequency, 238, 246, 307, 386, 387, 393, 420, 446
  –, lowest, 8, 67, 198, 215, 238, 240, 307, 346, 351, 376, 399
natural vibration, 375, 378, 469
neck bearing, 315, 328, 329
needle bearings, 34
Newton's second law, 87, 89
node, 237, 240, 402, 404, 432, 439, 440, 450, 509
noise, 142, 248, 486, 491, 507

non-dimensionalized amplitude function, 285, 287, 290
non-harmonic excitation, 52, 452
nondimensionalized amplitude function, 46, 185–188, 201, 451, 463, 472
nonlinear effect, 2, 467, 470, 476, 479, 488
nonlinear equation, 295, 296, 476
nonlinear oscillator, 295, 466–468, 471
nonlinearity, 41, 465, 466, 503
normal mode shape, 510, 521
normalization, 230, 236, 377, 401
number of cycles, 36, 222
numerical integration, 122, 128, 297, 426, 483

observability, 516
offset printing machine, 244
open-loop control, 514
operating cycle, 122, 123, 125, 132, 135
operating speed, 162, 189, 190, 265, 272, 278, 285, 315, 475
opposed-action frame, 478, 479
orthogonality, 378, 379, 397
oscillating conveyor, 357, 387, 439, 466, 473, 476, 477, 479, 481, 486
oscillator, 316
 –, constrained, 229
 –, modal, 273, 380, 427, 435
 –, nonlinear, 295, 466–468, 471
 –, parameter-excited, 298
 –, periodically excited, 454
 –, self-excited, 9
 –, unconstrained, 229, 230
oscillator chain, 227, 234, 235
output link, 150, 301
overall system, 363, 369, 388
overhead crane, 372, 395
overloading, 231, 234

packaging machine, 188
parameter, 6, 7, 9, 12–14, 347, 348
parameter change, 15, 242, 244, 345, 400, 402, 403
parameter influence, 150, 339, 347
parameter value, 6, 42, 51, 60, 171, 242, 248, 269, 301, 353, 478, 479, 483
parametric excitation, 295, 297, 298, 303, 305
parametric resonance, 297–299
partial derivative, 104, 373
partitioning, 425
passing through resonance, 315, 316, 467
patent, 210, 292, 488, 505, 506
pendulum, 203, 295, 470
 – in the centrifugal force field, 283
 –, physical, 17, 293, 485

pendulum absorber, 292, 293, 295, 306, 308
periodic excitation, 8, 190, 259, 427, 452, 454
periodicity condition, 55, 217, 219, 468
phase angle, 260, 428
phase curve, 470, 493
phase frequency response, 449, 451, 464
phase jump, 432, 440
phase plane, 470
phase shift, 184
piston, 54, 137, 138, 163, 168, 263
piston compressor, 216
piston engine, 54, 187, 223
pitch point, 107
pitching vibration, 205, 394
planar mechanism, 104, 112, 160
planetary gear, 106, 108, 109
plunger, 307, 309
polar diagram, 149, 161
position coordinates, 382
position function, 103, 113, 114, 116, 120, 137, 170
 – first-order, 103, 119, 137
 – zeroth-order, 119, 137
position vector, 69, 358
power, 115, 123
 –, kinetic, 145, 148
power balancing, 172
power line frequency, 440
power loom, 142, 143
predetermined breaking point, 509
preload, 260, 384, 387, 466, 470, 475, 490, 507, 509
press, 135, 143, 211, 260, 299, 303
press drive, 122, 129, 130, 132
principal axis, 14, 24, 85, 86, 154, 156, 195, 216, 465
 –, central, 85, 156, 487
Principal coordinate, *see* modal coordinate
principal coordinate, 379, 382, 391, 426
principal elastic axis, 192
principal moment of inertia, 24, 40, 85
principal stiffness, 26, 27, 500
principle of conservation of angular momentum, 87–89, 160, 381
principle of conservation of linear momentum, 87, 160, 213, 219, 381
principle of superposition, 384, 513
principle of virtual work, 104
principle of work and energy, 489
printing machine, 223, 244
printing mechanism, 244, 247
processing machine, 148, 188, 223, 481, 482, 508
processing moment, 131

Index   541

product of inertia, 22, 84, 156, 192
progressive characteristic, 474
proof of operating strength, 122
proof of safety, 252
pseudo resonance, 432, 443

qualitative match, 419

rail, 251, 253
range of the natural frequencies, 9
rattle, 303
Rayleigh damping, 445, 460
Rayleigh quotient, 342, 380, 399, 400
reaction, 89, 97, 106, 309, 426
real system, 6, 347
reciprocating engine, 223, 242, 243, 259
reciprocating saw, 206
rectangular impulse, 274, 433, 435, 436
rectangular matrix, 102, 363, 406, 408
rectifying effect, 467
reduced length, 29
reduced moment of inertia, 114, 117, 118, 120, 121, 126, 130, 134, 136–138, 160, 305, 482
reduced system, 425
reduction, 226, 405, 411, 416
reference system, *see* coordinate system
 –, body-fixed, 68, 75, 99
regenerative effect, 498, 499
regulation, 155, 506
relative damping, 48, 50, 286
relative movement, 43
residual vibration, 275, 277, 279, 437
residuum, 452, 463
resonance, 260, 285, 324, 428, 508
 – higher-order, 260, 265, 453
 – mth order, 453
 –, passing through, 187
 –, subharmonic, 475
resonance amplitude, 42, 247, 262, 285, 456
resonance case, 428, 429, 454
resonance curve, 48, 62, 63, 189, 264, 280, 453, 454, 456, 466, 479, 481
 –, measured, 265
resonance order, 280, 283, 460
resonance peak, 271, 289, 454, 460, 464, 483
resonance point, 205, 220, 222, 264, 271, 278, 281, 298, 315, 316, 430, 450, 454
resonance range, 240, 285
restoring force
 –, elastic, 361
 –, nonlinear, 465, 469, 471
restoring function
 –, nonlinear, 470

revolute joint, 143
rigid body, 69, 70, 152, 255
rigid machine, 67, 100, 141, 198, 224
rigid-body mechanism, 8, 100–102, 104
rigid-body motion, 282, 375, 445
rigid-body system, 8, 11, 68, 105, 152, 224, 229, 257, 294, 376
robot, 1, 68, 102, 188, 356, 515
robust operation, 477
rod, 15, 18–20, 31, 38, 138, 225, 359, 360
rod-shaped link, 151
roller, 77, 90, 92, 308
roller bearing, 34, 35, 349, 350, 466
rolling pendulum, 19
rotary offset press, 178
rotary viscosity damper, 290
rotation
 –, planar, 70
 –, spatial, 72
rotation in opposite direction, 323
 –, synchronous, 325
rotational backlash, 231
rotational energy, 82, 257, 422
rotational frequency, 247
rotational speed
 –, critical, 261
rotational transformation matrix, 71, 73, 75, 79, 84
rotor, 76, 93, 94, 96, 109, 153, 154, 315, 320, 330, 335, 347, 475
 –, elastic-shaft, 157
 –, rotationally symmetrical, 111
 –, unbalanced, 484
rough calculation, 18, 197, 213, 228, 346
round-off error, 128, 389, 394
rubber spring, 36–39, 42, 50, 197, 198, 288, 444
rule, 404, 505, 507, 509
Runge-Kutta method, 383

safety, 5, 38, 177, 197, 252, 311
Salomon pendulum absorber, 306
same direction of rotation, 323, 325, 331, 332, 340, 351, 354
 –, synchronous, 335, 337
scale factor, 236
scattering of parameter values, 387
schematic, 121, 130, 133, 178, 263, 300, 409, 482
screen, 315, 384, 466, 473, 486
screw motion, 74
second approximation, 473
self-centering, 312
self-excited vibration, 9, 349, 490, 497, 500

self-locking, 503
self-synchronization, 467, 484, 486, 488
sensitivity, 384, 400, 412
sensitivity coefficient, 236, 237, 242, 243, 248, 250, 401–403, 414
sensitivity matrix, 402
sensor, 506, 515
separation of variables, 338
sequence of calculations, 260
sequence of motions, 124
series expansion, 126, 136, 436
service life, 29, 34, 154, 281, 505
service period, 231, 292
setup type, 182
sewing machine, 142, 143
shaft section, 27, 28
shape factor, 37, 38
shear center, 338
shear deformation, 340
shear modulus, 27, 36, 38, 251, 253, 360
shock absorber, 466
Shore hardness, 36, 37
shutdown test, 123
side effect, 153
similarity, 7, 189, 476
simplification, 237, 388
simulation, 58, 482, 521
simulation program, 128, 267, 520
SimulationX®, 140, 243, 305, 467, 477, 479, 481, 494, 495
singular, 376, 398, 399
skeleton line, 472, 475
slave degree of freedom, 406
slider-crank mechanism, 56, 114, 134, 136, 137, 163, 165, 168, 172, 174, 476, 482
slippage, 29, 59, 135, 457
small parameter method, 467, 470
software, 5, 24, 29, 30, 68, 85, 101, 121, 150, 157, 163, 238, 246, 260, 263, 346, 351, 356, 383, 384, 433, 510
solution
 –, analytical, 124, 136, 162
 –, particular, 46
 –, steady-state, 262, 429, 432
spatial mechanism, 68
spectrum, 8, 58, 149, 162, 190, 386, 453, 466, 467
speed, 340
 –, critical, 283, 311, 323, 324, 354
spin drier, 76, 324
splined shaft, 28
spring characteristic, 25, 479, 488
 –, degressive, 465
 –, kinked, 488

–, nonlinear, 466, 488
–, progressive, 465, 488
–, static, 39
–, stepped, 476, 488
spring compensator, 303
spring constant, 31, 32, 37
 –, actual, 386
 –, dynamic, 38, 41
 –, mean, 469–471, 488
 –, modal, 236, 379, 380, 391, 427, 437
spring matrix, 193, 358
spring moment, 120, 121
spring parameter, 32, 244, 263, 346, 374, 402, 496
spur gear mechanism, 303–305
stability, 297, 516
stability condition, 497, 503, 504
stability plot, 500, 501
standard, 2, 30, 155, 179, 180, 311
start-up, 8, 122, 251, 276–278, 315, 433, 483, 484
start-up process, 122, 251, 267, 277
starting, 5, 59, 125, 128, 251, 277, 396, 461
starting and braking, 122, 124, 231, 272, 433, 481
starting point, 5, 107, 109, 365, 507
state space, 511
static calculation advocate, 443
static indeterminacy, 405, 507
steady state, 8, 46, 59, 127, 217, 294, 447
steady-state operation, 43, 129, 135, 162, 188
steady-state solution, 184, 188, 201, 219, 262, 427, 429, 432, 448
steady-state vibration, 471, 494
Steiner's theorem, 17, 85, 91, 97
step function, 278
stepping mechanism, 268–270
stepping motion, 271
stepping motor, 273
Stewart platform, 102
stick-slip vibration, 491, 493, 495
stiffened system, 425
stiffness change, 240, 405
stiffness matrix, 193, 199, 235, 358, 361, 366, 367, 371, 376, 518
 –, asymmetrical, 491
 –, partitioned, 417
 –, reduced, 406, 412
stimulation, 298, 446, 491, 498
strength, 38, 122, 128, 158, 281, 315, 509
string, 340, 490
structural change, 510
structure endangerment, 180
strut, 420, 424

sub-assembly, 29, 31, 42, 362, 383
subharmonic, 306, 467, 474, 475, 479
subsoil, 31, 145, 178, 182, 185, 196, 197, 199, 206, 212, 217
substructure, 248, 362, 363, 366, 368–371, 373, 406
subsystem, 346, 352, 363, 394
supercritical, 266, 328, 486
superharmonic, 467, 474
superposition, 44, 383, 392, 427, 436
superposition principle, 453, 467
support, 178, 320, 324, 408, 411
support structure, 368, 369, 393, 416, 418, 506
symmetry, 16, 24, 38, 40, 86, 194–196, 199, 371, 388
synchronization, 2, 467, 484, 487
synthesis, 162, 505, 507
 –, dynamic, 507
system
 –, condensed, 406, 407
 –, conservative, 399
 –, damped, 445
 –, linear, 512
 –, reduced, 420, 425
 –, stiffened, 425
system dynamics, 3, 6, 511
system matrix, 512, 515, 517

Taylor series, 64, 296, 503
tensioning device, 457
tensor, 14, 21–24, 38, 40, 68, 73, 75, 82, 85
textile machine, 188, 385
textile mandril, 328, 412, 414, 455
Timoshenko beam, 339
tire, 31, 52, 466
tolerance range, 269
tool, 460, 499
tooth number, 30, 279, 305
tooth stiffness, 12, 30, 31, 303
torque jump, 272, 273, 275
torsion, 13, 14, 20, 26, 29, 31, 384
torsion spring, 27, 237, 248, 296
torsional damping constant, 483
torsional moment, 225, 275, 276, 282
 –, kinetostatic, 225
 –, residual, 278
torsional oscillator, 13, 20, 229, 235, 237, 241, 273, 395
torsional spring constant, 31, 257, 305, 483
torsional stress, 252, 259
torsional vibration, 205, 223, 297, 487
torsional vibration damper, 444
tower crane, 10, 11, 31
tractor tire, 51, 52

transcendental equation, 65, 167, 282, 308, 499
transfer manipulator, 299
transfer matrix, 311
transient excitation, 8, 56, 272
translational vibration, 487, 488
transmission ratio, 67, 104, 136, 167, 295, 482, 483
transverse shear, 340
trolley, 372
truss structure, 405
tuning, 182
 –, high, 182, 185
 –, low, 181, 182, 185, 191, 209, 486
turbogenerator, 157
two-mass system, 201, 213, 228

unbalance, 154, 156, 157, 171, 175, 205, 443, 486
unbalance excitation, 185, 323, 486, 487
unbalanced rotor, 484

V-belt, 340, 457
VDI Guideline, 51, 106, 123, 142, 149, 162, 164, 174, 179, 180, 203, 260, 268, 283, 299, 303
vehicle, 15, 16, 248, 367
vehicle drive, 224, 266
vehicle engine, 196
velocity, 73, 74, 492
velocity distribution, 80, 82
velocity jump, 217, 437, 460
vibrating compactor, 316, 473, 486
vibrating machine, 316, 486
vibration
 –, amplitude-modulated, 298
 –, bending, 12, 55, 315, 338, 347, 424
 –, damped, 42, 259, 444–464
 –, electromechanical, 59
 –, forced, 46, 50, 149, 200, 259–283, 305, 314, 426–443
 –, free, 381, 392, 445, 446, 462
 –, longitudinal, 307
 –, self-excited, 349, 490, 497, 500
 –, stick-slip, 491, 495
 –, torsional, 20, 51, 54, 205, 252, 305, 444, 487
 –, translational, 4, 457, 487, 488
vibration absorption, 200, 202, 433, 439
vibration exciter, 439, 486, 487
vibration force, 105, 506
vibration intensity scale, 179, 180
vibration isolation, 177, 179, 181, 182, 197
vibration isolator, 204
vibration load, 233

vibration moment, 224, 301
vibrational energy, 257
vicinity of resonance, 259
virtual work, 104, 108, 253, 407, 457, 478, 517
VISCO-damper, 51, 212
viscosity, 35, 54, 123, 290, 292, 349
viscous damping, 42–47, 203, 221, 222, 259, 290, 298, 448, 475
viscous rotary damper, 290–292

viscous torsional vibration damper, 54

wear and tear, 231, 486, 491
weighing, 15
work
 –, mechanical, 46, 47, 144, 509
 –, virtual, 104, 108, 253, 407, 457, 478, 517
work machine, 127
workpiece, 47, 50–53, 299, 434, 460, 498, 499